Demographic Methods and Concepts

Demographic Methods and Concepts

Donald T. Rowland

Reader in Population Studies
The Australian National University

OXFORD
UNIVERSITY PRESS

Oxford University Press, Inc., publishes works that further
Oxford University's objective of excellence
in research, scholarship, and education.

Oxford New York
Auckland Cape Town Dar es Salaam Hong Kong Karachi
Kuala Lumpur Madrid Melbourne Mexico City Nairobi
New Delhi Shanghai Taipei Toronto

With offices in
Argentina Austria Brazil Chile Czech Republic France Greece
Guatemala Hungary Italy Japan Poland Portugal Singapore
South Korea Switzerland Thailand Turkey Ukraine Vietnam

Published in the United States
by Oxford University Press Inc., New York

First published 2003

Reprinted with corrections 2006

A catalogue record of this book is available from the British Library

Library of Congress Cataloguing in Publication Data
Data available

ISBN-13: 978-0-19-875263-9
ISBN-10: 0-19-875263-6

Typeset by SNP Best-set Typesetter Ltd., Hong Kong
Printed in Great Britain
on acid-free paper by
Biddles Ltd., Guildford and King's Lynn

Preface

'. . . the methods of population analysis are less formidable than they often seem to the uninitiated once the underlying logic of the procedures is understood.'

(Dudley Kirk, in Barclay 1966: vi)

This book seeks to enable social science students and researchers to learn demographic methods without training in mathematics or statistics. Making demographic methods more accessible, and helping to disseminate demographic skills beyond the preserve of population specialists, are significant considerations today given the abundance, wide availability and relevance of information about population characteristics and changes. Further aims of this book are:

- To present concise background on conceptual, theoretical and practical considerations which are important in the interpretation of demographic data and indices. For example, the demographic transition, explained in Chapter 1, provides a theoretical context for interpreting statistical information on changes in national populations.
- To present methods for the analysis of population at regional and local levels, as well as at a national scale, since applications to subnational populations are of particular interest to large numbers of people who use demographic methods in geography and planning.
- To provide, through a series of exercises in each chapter, training in the use of spreadsheets for demographic work, employing the widely available program Microsoft Excel. Spreadsheet skills have become essential to proficiency in working with population statistics, as well as with social statistics generally.
- To facilitate understanding of principles, exploration of examples, and the application of demographic techniques – through a set of seventeen Excel computer modules, supplied on the compact disk at the back of the book. The modules are designed to be easy to use, requiring no background in computing, and are described in text boxes at relevant points in the chapters.

The Excel modules

Age Structure Data Base, pp. 80–1
Age Structure Simulations, pp. 102–7
Cohort Component Projections, pp. 449–54
Computer Assisted Learning Exercise on Abridged Life Tables, p. 287
Demographic Transition, pp. 19–20
Epidemiologic Transition, pp. 187–91
Growth Rates, pp. 67–8
Measures of Population Distribution, pp. 357–60
Migration in the United States, pp. 395–6
Population Clocks, pp. 55–7
Population Dynamics, pp. 22–3
Population Momentum, pp. 328–30
Proximate Determinants of Fertility, pp. 226–9
Pyramid Builder, pp. 161–2
Relational Model Life Tables, pp. 320–1
Standardization of Rates, pp. 124–5
West Model Life Tables and Stable Populations, pp. 316–9

The Introduction explains the nature and purpose of the book in more detail. The general aim is to present the range of demographic methods, together with associated concepts and theories, that are most immediately useful to people interested in working with published statistics on population from censuses, official surveys and other sources. The Website for this book provides supplementary materials for readers, including a glossary of the key terms; the address is:

www.oup.com/uk/best.textbooks/geography/rowland/

Acknowledgements

This book has arisen out of teaching demography to social science students at the Australian National University (ANU). I acknowledge with gratitude the assistance of colleagues, and the feedback, through formal surveys as well as informally, from many hundreds of students who have taken my courses on *Population Analysis*, *Population and Society*, *Population Research*, *Population and Australia* and *Contemporary Society*. The first versions of the computer applications in the book were developed in the mid-1990s, with funding from the Commonwealth Government's former Committee for the Advancement of University Teaching. I particularly acknowledge the contribution to this work of fellow demographer Lualhati Bost.

In preparing the manuscript, I benefited substantially from comments on drafts received from: Tom Burch, University of Western Ontario; Dudley Poston, Texas A & M University; Anne Gauthier, University of Calgary; Andrew Hinde, University of Southampton; Shailendra Jain, Australian Bureau of Statistics; Ross Steele, Flinders University of South Australia; John Salt, University College London; Jeromey Temple, Australian National University; and Frans Willekens, University of Groningen. I acknowledge their help most sincerely. I am very grateful as well to Angela Griffin and Jonathan Crowe, commissioning editors at Oxford University Press, for their advice and assistance. My wife, Jenny, and our daughters, Adele and Madeleine, gave their support throughout the manuscript preparation, a process which, in conjunction with other work, spanned several years.

Finally, I am very grateful to the following for permission to use copyright material:

Carol Jenner and Mike Mohrman of the Office of Financial Management, Olympia, Washington for the Map of Population Density in Census Blocks, Washington State, 2000 Census.

Ceri Peach, School of Geography and the Environment, Oxford University, for the Map of the Indian Settlement Pattern in Greater London Wards, 1991.

Christopher Kocmoud, Texas Center for Applied Technology, College Station, Texas, for the United States Population Cartogram, 1996.

Elsevier Science, Oxford, for material from Bongaarts, John and Potter, Robert G. 1983. *Fertility, Biology and Behaviour: An Analysis of Proximate Determinants*. New York: Academic Press, Tables 4.2 and 4.3, pp. 88–91.

Elsevier Science, Oxford, for material from Coale, Ansley, J. and Demeny, Paul 1983. *Regional Model Life Tables and Stable Populations* (2nd edition). New York: Academic Press, pp. 42–54.

Etienne van de Walle, John Knodel and the Editorial Office of *Demography*, for permission to use material from van de Walle, Etienne and Knodel, John 1970. 'Teaching Population Dynamics with a Simulation Exercise'. *Demography*, 7 (4): 433–48.

Graphics Press, Cheshire, Connecticut for permission to use the reproduction of Minard's map from Tufte, Edward R. 1983. *The Visual Display of Quantitative Information*. Cheshire Connecticut: Graphics Press, p. 41.

Hodder and Stoughton Limited, for the graph of population change in Egypt from Hollingsworth, T. H. 1969. *Historical Demography*. London: The Sources of History Limited in association with Hodder and Stoughton Limited, p. 311.

Kluwer Publications, London, for statistics from Mitchell B. R. 1975. *European Historical Statistics 1750–1970*. London: Macmillan, pp. 111–12 (data for Norway and Sweden).

Oxford University Press, Oxford, for statistics from Chesnais, Jean-Claude 1992. *The Demographic Transition: Stages, Patterns and Economic Implications: a Longitudinal Study of Sixty-Seven Countries Covering the Period 1720–1984*. Oxford: Clarendon Press, pp. 518–31 and 556–67 (data for France, Norway, Sweden, England and Wales and the United States).

The Office of Population Research, Princeton University for material from Coale, Ansley and Guo, Guang 1990. 'New Regional Model Life Tables at High Expectation of Life'. *Population Index*, 56, pp. 31–3.

The Population Division of the Department of Economic and Social Affairs of the United Nations Secretariat for statistics from *World Population Prospects: The 2000 Revision*, Vol. I, *Comprehensive Tables*, and Vol. II, *The Sex and Age Distribution of Populations*. New York: United Nations.

The Publications Board, United Nations, New York, for material from United Nations 2000. *Demographic Yearbook: Historical Supplement 1948–1997*, New York: United Nations (CD) (data on age groups and age-specific death rates or Egypt, Japan, Kuwait, Mexico, Philippines, the United Kingdom and the United States).

The Royal Statistical Society, London, for the copy of the Society's emblem.

Thomson Publishing Services, Andover, Hampshire, for permission to reproduce maps from Lawton, Richard 1986. 'Population', in John Langton and R. J. Morris (editors), *Atlas of Industrializing Britain 1780–1914*. London: Methuen, pp. 21 & 29.

U.S. Census Bureau, Population Division, Washington D.C. for material from United States Census Bureau 2001. 'Geographical Mobility March 1999 to

March 2000, Detailed Tables', *Current Population Reports*, P20-538 (data on general mobility by region, sex and age, March 1999 to March 2000).

I have sought to trace all current copyright holders and to give full acknowledgement of sources. If there are any instances where this has been overlooked, the publisher will be very pleased to rectify omissions at the earliest opportunity.

Don Rowland

February, 2003

Contents

Preface v
Acknowledgements vii

Introduction **1**

Aims and scope 2
Key considerations 4
Summary 7
How to use this book 8

Section 1 **Population dynamics**

1 **Population change** **13**
1.1 Origins of demography 14
1.2 The demographic transition 16
1.3 Sources of data 24
1.4 Components of change 29
1.5 Comparative measures 31
1.6 Basic measures of change 33
1.7 Interpreting changes 35
1.8 Conclusion 37
Study resources 39

2 **Population growth and decline** **45**
2.1 Concepts of growth 46
2.2 Analysing growth 57
2.3 Geometric growth 61
2.4 Exponential growth 64
2.5 Growth and replacement 68
2.6 Conclusion 70
Study resources 71

3 **Age–sex composition** **76**
3.1 Statistics on age 79
3.2 Population pyramids 81

3.3 Further methods of data visualization 83
3.4 Summary measures 85
3.5 Concepts in the interpretation of age–sex composition 97
3.6 Conclusion 107
Study resources 109

Section 2 **Analytical approaches**

4 **Comparing populations** **119**
4.1 Comparing rates 121
4.2 Direct standardization 125
4.3 Indirect standardization 130
4.4 Period and cohort analyses 135
4.5 Conclusion 142
Study resources 143

5 **Demographic writing** **150**
5.1 Types of studies 152
5.2 Starting demographic research 157
5.3 Interpreting information 158
5.4 Writing a paper 161
5.5 Conclusion 167
Study resources 169

Section 3 **Vital processes**

6 **Mortality and health** **179**
6.1 Sources of statistics on mortality and health 180
6.2 The epidemiologic transition 185
6.3 Measures of mortality 192
6.4 Health and illness 206
6.5 Conclusion 211
Study resources 213

7 **Fertility and the family** **220**
7.1 Family change 222
7.2 The second demographic transition 223
7.3 Period and cohort approaches 229
7.4 Period measures of fertility 230
7.5 Synthetic cohort measures of fertility and replacement 238
7.6 Real cohort measures of fertility 250
7.7 Measures of marriage and divorce 251

7.8 Conclusion 255
Study resources 256

Section 4 Demographic models

8 Life tables **265**
8.1 The first life tables 266
8.2 Types of life tables 267
8.3 Constructing complete life tables 270
8.4 Constructing abridged life tables 285
8.5 Conclusion 292
Study resources 294

9 Stable and stationary models **300**
9.1 Stable populations 301
9.2 Constructing stable population models 303
9.3 Model life tables 312
9.4 Model stable populations 325
9.5 Population momentum 327
9.6 Population ageing 332
9.7 Further applications of stable and stationary models 334
9.8 Conclusion 337
Study resources 338

Section 5 Spatial patterns and processes

10 Population distribution **347**
10.1 Where people live 348
10.2 Types of spatial units 349
10.3 Measures of population distribution 352
10.4 Housing 363
10.5 Population mapping and GIS 367
10.6 Types of population maps 368
10.7 Conclusion 376
Study resources 377

11 Migration **384**
11.1 Concepts and theories 386
11.2 Migration statistics 391
11.3 Migration rates 394
11.4 Migration effectiveness 400
11.5 Migration expectancy 401
11.6 Estimating migration 403

11.7 Calculating survival ratios 418
11.8 Conclusion 421
Study resources 422

Section 6 Applied demography

12 Population projections and estimates **429**
12.1 Applications and issues 433
12.2 Population estimates 436
12.3 Projection methods 437
12.4 Calculating cohort component projections 442
12.5 Elaborating the basic cohort component model 447
12.6 Further methods for sub-national populations 455
12.7 Conclusion 462
Study resources 464

13 Population composition **469**
13.1 Projecting population composition 470
13.2 Labour force projections 471
13.3 Projections of households and families 473
13.4 Summary measures of population composition 479
13.5 Multivariate measures 490
13.6 Conclusion 495
Study resources 497

Appendices

A Basic maths 505
B Using the Excel modules 507
C Introduction to Excel 509
D Answers to exercises 513

Bibliography 525
Index 540

Introduction

OUTLINE

Aims and scope
Key considerations
Summary
How to use this book

MAIN POINTS

Because of the wide relevance of population statistics in planning and research, this book seeks to make demographic methods generally accessible.

Basic premises are the desirability of a concept-based approach, to foster interpretation skills, and a computer-based approach, to facilitate learning and efficient use of demographic data and methods.

Aims and scope

Demographic Methods and Concepts aims to make accessible the most commonly needed techniques for working with statistical materials on population. The book is intended as a text for social science students and as a reference work for people who use population statistics in their employment.

To understand the methods, little mathematical background is required apart from some basic knowledge and an interest in acquiring skills to undertake quantitative work. Appendix A summarizes the maths needed, including logarithms and powers and roots of numbers. The most common misconceptions about demographic methods are that they are necessarily 'very mathematical' and that they are useful only to people who wish to become specialists in demographic statistics. Yet the methods most widely needed do not call for mathematical expertise, and they have applications across a range of subject areas. Demographic methods are important in the work of geographers, sociologists, economists, administrators, market researchers and planners, as well as demographers and actuaries. The concepts that underlie the methods are also essential to the interpretation of population data in many different contexts.

Recent developments

Several changes have contributed to the wider use of information on population. Particularly important has been the great expansion of statistical collections, together with better ease of access to them, especially from computer files and the Internet. People no longer need the support of computer programmers or special funding to utilize detailed data on population trends and characteristics. The range and scope of information available today offers greater potential for different kinds of investigations, and greater opportunities to undertake 'desktop demography' using a personal computer.

Further developments are the mounting interest in practical applications of demography and the wider use of demographic techniques in other fields, including geography, planning and marketing. These trends are themselves related to the greater availability of population statistics, which has increased the potential for demographic analysis to contribute to the work of governments, organizations and businesses. Demography has had a long-standing concern for improving human welfare, but practical and applied emphases are evident more than ever in research and teaching on population. *Demographic Methods and Concepts* seeks to complement these trends, through explaining the range of demographic techniques commonly needed in studies of populations at national and sub-national levels.

Today, the ways and means are more open than ever before to use population information in advancing understanding of the nature, determinants and consequences of varied changes in societies. The analysis of population data can address questions as diverse as: Who have lifestyles that exclude family formation? Where do social problems arising from immigration have their greatest impact? What will be the size and composition of the labour force in the future? Why does the demand vary from place to place for particular kinds of housing and consumer goods? How are population changes affecting government funding of urban development and social welfare?

Graunt's 'Observations'

This diversity echoes the range of interests of demography's founder, an Englishman named John Graunt (1620–74). He was the first to realize that population statistics reveal much about the nature of society, the changes taking place within it and the issues arising for government and policy-making. In his pioneering 90-page book, *Natural and Political Observations Made Upon the Bills of Mortality*, Graunt (1662) sought to persuade others of this. He quickly gained recognition for his creative insights through election to membership of the Royal Society in London. Quotations from Graunt head each chapter of this book – to illustrate the origins of the demographic imagination and the breadth of the founder's achievements. His work deserves to be better known today for its impressive scope and vision of the potential uses of demographic data:

> Graunt's work covers so wide an area of interest that it may be said that a large part of demography was born all at once. The developments that occurred subsequently were in the nature of consolidation. (Cox 1970: 298)

Contents

Guiding the selection of topics for this book was the goal of providing a substantial grounding in the main areas of inquiry. There has been a strong emphasis in demography on national developments, but applied and interdisciplinary interests in population often call for attention to sub-national patterns and trends. Hence the book deals not only with core topics for national-level studies, but also with matters of prime interest at the local and regional levels, including population distribution, internal migration and small area projections and estimates.

The chapters are grouped in the following sections:

1. **Population dynamics** This section provides an overview of methods and concepts in the study of population change. The chapters explain the main starting points for demographic studies – in terms of concepts, sources of information and basic methods – together with the principal approaches and techniques in the analysis of population growth and age structure changes.
2. **Analytical approaches** The second section introduces key features of demography's approaches to analysing population changes, together with strategies for interpreting population statistics. The subject matter includes the period and cohort perspectives in demography, techniques for comparing rates –

including standardization – and general principles in demographic writing and analysis.

3. **Vital processes** Here the focus is demographic methods for studying mortality and fertility, together with examples of the theories, concepts and issues that the methods address, including the epidemiologic transition and the second demographic transition.
4. **Demographic models** This section discusses the demographic models of widest interest in the analysis of populations. The chapters cover the construction and applications of stationary and stable population models. Specific sections include the calculation of single-year and abridged life tables, the derivation of intrinsic growth rates and stable age distributions, and the nature of population momentum and population ageing.
5. **Spatial patterns and processes** Approaches to the study of spatial aspects of population are of considerable interest to many users of demographic information. The chapters in this section cover the main methods of measuring and mapping population distribution and migration, and techniques for estimating migration from basic demographic data.
6. **Applied demography** This final section is concerned with techniques employed in planning, policy-making and commercial applications, especially methods of projecting and estimating total populations and methods of projecting and analysing population composition.

Excel modules for computer assisted learning are introduced in text boxes at relevant points in the chapters. All of the modules are on the accompanying CD. Appendix B provides additional information on how to use the Excel modules.

At the end of each chapter is a set of study resources that may be used according to the requirements of readers and the availability of computing equipment. The study resources consist of key terms from the chapter, further reading, Internet resources, practical exercises and revision questions, and spreadsheet exercises. Details of these are given in the final section of this Introduction, under 'How to use this book.'

Key considerations

Apart from meeting the needs of a general readership, rather than one with mathematical expertise, there were other considerations influencing the contents of the book and the treatment of the subject material. The first was the desirability of providing a concept-based approach.

Demographic concepts

For many students, the most difficult aspect of working with demographic methods is not understanding the methods themselves, but learning to interpret population statistics. Reading and practice are necessary to develop interpretative skills, yet progress can be facilitated greatly through studying demographic methods in conjunction with the concepts and theories to which they relate. For example, the demographic transition is essential to understanding national trends in population growth rates, age structures, birth rates and death rates (see Chapter 1).

Methods courses, however, are commonly taught separately from those dealing with population issues and trends, and students may need to discover for themselves many of the links between the methods and broader subject matter in which they are embedded. Indeed, some students study demographic techniques only to complement other units on statistics or research methods. As a result, they face considerable difficulties when interpreting demographic data and indices.

To facilitate acquisition of skills in interpreting data, this book discusses methods in the context of concepts and theories that clarify their nature and applications. The intention is to discuss conceptual material succinctly, yet sufficiently to illustrate linkages between techniques and the topics that they address. Chapter 5, and some sections of other chapters, also provide a specific focus on practical approaches to interpreting demographic data.

Desktop demography

The second consideration was the emergence of 'desktop demography', entailing electronic access to information and analysis of demographic data on desktop computers. In many countries, it has become vital for social science students to be computer literate and to be able to use desktop computers for the everyday tasks of accessing, storing, analysing and graphing statistical information. Graphics-based spreadsheet programs, such as Excel, have revolutionized demographic instruction and practice, superseding former mainframe and DOS programs for demographic calculations. They provide the most versatile and easily mastered approaches to graphing and analysing population data, enabling students to carry out analytical work with greater ease and accuracy. They also open up great opportunities for experimentation and exploration of information, all the more since they have become a major platform for using electronic data files.

To facilitate the use of spreadsheets for demographic analysis, the 'Study resources' section at the end each chapter includes a spreadsheet exercise. The exercises provide a cumulative course of instruction in Excel as well as in demographic techniques. Each exercise gives practice in relation to the program features that are most useful for each topic. Thus the book contains many examples of how to use Excel to construct graphs and calculate demographic rates, indices, models and projections.

The first exercise assumes no previous knowledge of Excel, and subsequent exercises explain new functions as they are needed. Each exercise is intended to be completed within about 50 minutes; students familiar with Excel will finish the early exercises more quickly. Working through the spreadsheet exercises reinforces other study using calculators, enhances satisfaction with the learning process and extends skills and versatility for future employment. A summary reference document on Excel, for use with the exercises, is provided in Appendix C.

A closely related consideration in the writing of this book was the desirability of providing access to computer-based teaching and learning materials, taking advantage of opportunities that technology offers in approaches to classroom teaching and individual study. Thus, in addition to the spreadsheet exercises, the book provides a series of Excel modules for the visualization of concepts and data, together with the calculation of graphs, demographic models and projections. Some of the modules, including 'cohort component projections', are also valuable research tools, commercial versions of which can cost many times the price of this book.

Use of most of the modules calls for little computing knowledge, apart from being able to start Excel and load a file (see Appendix B). Each module is an Excel 'workbook' (a set of spreadsheets) featuring on-screen buttons, menus and animated graphics, as well as options to print results and summary materials. They are supplied on the book's compact disk. The author has pursued a minimalist, and easily mastered, approach to the design of the modules, which may be readily extended or adapted to other purposes, for instance through adding data for different countries. The first versions of the modules were developed in the 1990s with funding assistance from the Australian Government's former Committee for the Advancement of University Teaching.

In preparing the modules, the main criteria were that they should:

- reinforce understanding of key concepts and methods, especially where computer-based approaches can enhance conventional means of explanation;
- be reasonably self-contained and self-explanatory;
- not duplicate what can be accomplished on a printed page;
- be suitable for use in different contexts, such as for lectures, tutorial discussions and individual study.

Overall, the modules extend opportunities for students to explore demographic, concepts, methods and data. Working with the modules in pairs or small groups provides opportunities for discussion and enhances understanding. The modules are described separately in text boxes at relevant points in the chapters. Like the spreadsheet exercises, they are a recommended study option, but they may be omitted if suitable computing facilities are unavailable.

Flexible delivery

Finally, the growing importance of flexible learning in higher education is also creating a need for teaching and learning materials appropriate to different styles of course delivery, as well as self-instruction. The traditional emphasis on campus-based teaching through lectures is giving way to a more diversified approach to course delivery, including off-campus and distance education.

In this new environment, formal lectures are less important. Greater emphasis is given to active engagement with the subject material, or 'learning by doing', and 'deep learning', which entails integrating information. The opposite of the latter is 'surface learning', of which a defining characteristic is memorizing unconnected pieces of information (Laurillard 1993; Wade 1994). The requirements of flexible learning, where students' work is independent of fixed times and places, and 'active learning', requiring students to think about information, not just record it, were considerations in the writing of this book, giving rise to the following strategies:

- seeking to provide full explanations and examples of calculations, rather than assuming that lecturers will always be present to provide additional elaboration;
- providing easy-to-use computer modules, and sets of activities arising from them, to encourage active learning;
- presenting step-by-step instructions for the spreadsheet exercises, to enable students to become familiar with the main computing procedures relevant to demographic work, without heavy reliance on a tutor;
- providing self-test exercises and lists of key terms for revision at the end of each chapter.

Summary

Since so much information on population is publicly available, yet under-utilized, it is important that knowledge of the techniques of population investigation be part of the research skills of more social scientists. Possession of a range of complementary research skills increases the potential for searching inquiry. Demographic methods assist in exploring and explaining some of our richest and most reliable sources of information about changes in societies. Also, there are many applied and practical subject areas requiring expertise in using demographic information. These include: planning for housing, schools and urban services; public policy formulation in education, employment and health; and decision-making for businesses and private organizations.

This book covers the methods most commonly needed to analyse demographic information at national, regional and local levels. While the book does not assume any previous knowledge of demography, quantitative methods or computing, experience has shown that students from varied backgrounds readily master the technical content and acquire substantial analytical skills.

The book places some emphasis on learning computing techniques that enhance the breadth and depth of investigations, as well as enjoyment and proficiency in working with population statistics. Spreadsheet approaches are becoming the basis for demographic training wherever computers are available. The spreadsheet-based exercises and modules supplement the text in relation to extending the range of examples available, providing visual demonstrations of principles and performing often-needed calculations, such as population projections and construction of age–sex pyramids.

The book also emphasizes the conceptual background that is essential in the application of demographic methods and the interpretation of population data for varied purposes. Overall, the book aims to enrich instruction in courses dealing with population, through presenting concept-based and computer-based approaches to mastering the most widely used demographic techniques.

How to use this Book

The chapters

The chapters in Sections 1–3 cover major topics of general interest; headings and subheadings within chapters facilitate selection of material appropriate to particular needs. Section 4 introduces core subject matter of a somewhat more technical nature that is also necessary to understanding methods discussed in the later chapters. The contents of Sections 5 and 6 are especially relevant to geographical and applied studies.

1. **Contents summary** As a guide to the content of each chapter, a summary at the start outlines the subject matter and learning objectives, especially the methods and concepts of particular interest, as well as listing the associated computer applications – the Excel modules and spreadsheet exercises.
2. **The text** Initial sections of each chapter are mainly concerned with introducing the significance of the topic, summarizing important conceptual and theoretical considerations and discussing data sources, as appropriate. Later sections focus on the methods employed in analysis and research. A reading of the text would normally precede work on the exercises or use of the computer modules.
3. **Figures and tables** The book contains many figures to illustrate the subject material as well as many tables presenting worked examples and formulas.
4. **Boxes** These provide supplementary information on important topics, together with descriptions of the computer modules. Study of them is intended to extend or enhance understanding of the subject material.
5. **Computer modules** Excel modules are constructed for the visualization of concepts and data, and the calculation of graphs, demographic models and projections. They have applications in individual study and classroom discussion; the more sophisticated modules are also useful tools for analysis and research.

The study resources

The 'study resources,' at the end of each chapter, contain the following materials to assist learning:

1. **Key terms** A list of the theories, concepts and techniques presented in the chapter, that is, the main items to be understood. Definitions of the key terms are provided in the Glossary on the book's website: www.oup.com/uk/best.textbooks/geography/rowland/
2. **Further reading** Key references useful for further reading on the subject matter of the chapter.
3. **Internet resources** Web sites, published by major organizations, containing statistics or reference materials particularly relevant to the chapter. These are intended for reference, or to assist in answering questions in some of the exercises. The Web resources are limited to those of major organizations; hence the information referred to should be locatable even if Web addresses change. Because of frequent updating of information on the Internet, it is not possible to guarantee that information will remain at specific locations on the Web.
4. **Exercises** These provide experience in using the methods explained in each chapter, as well as learning activities enquiring into the nature of data sources and research practices. The calculations may be performed with either a calculator or a computer. Answers to specific questions are given in Appendix D.
5. **Spreadsheet exercises** Spreadsheet exercises are presented at the end of each chapter; together they comprise a cumulative course of instruction in Excel as well as in demographic techniques. They consolidate understanding of particular methods and facilitate the acquisition of computing skills for analysing and graphing population statistics.

Section I

Population dynamics

This section provides an overview of methods and concepts in the study of population change. The chapters explain the main starting points for demographic studies – in terms of concepts, sources of information and basic methods – together with the principal approaches and techniques in the analysis of population growth and age structure changes.

Chapter 1 **Population change**

Chapter 2 **Population growth and decline**

Chapter 3 **Age–sex composition**

1 Population change

a true Accompt of people is necessary for the Government, and Trade of them, and for their peace and plenty.

(Graunt 1662: 16)

OUTLINE

1.1 Origins of demography
1.2 The demographic transition
1.3 Sources of data
1.4 Components of change
1.5 Comparative measures
1.6 Basic measures of change
1.7 Interpreting changes
1.8 Conclusion
Study resources

LEARNING OBJECTIVES

To understand:

- the broad setting of studies of national population change
- the basic methods for analysing population change

To become aware of:

- the main sources of population statistics
- general approaches to interpreting population statistics

COMPUTER APPLICATIONS

Excel modules: Demographic Transition.xls (Box 1.2)
Population Dynamics.xls (Box 1.3)

Spreadsheet exercise 1: Basic demographic rates (See Study resources)

Demography arose from a realization that population statistics could reveal much about the nature of societies and the changes taking place within them. Yet the censuses that the Egyptians, Babylonians, Romans and Chinese conducted in ancient times never spurred such reasoning; no one perceived them as more than head counts, mainly of men eligible for taxation, work or military service. What finally inspired the invention of demography was not censuses of the living, but lists of the dead, with supplementary information on weddings and christenings.

This chapter introduces the study of population change, first through sections on demography's origins in the seventeenth century and its twentieth century attempt at a 'grand theory' of change – the demographic transition. Origins and theory convey much about the purpose of demographic analysis and the ways of approaching the interpretation of population statistics. This sets the scene for the ensuing sections on statistical sources, methods and interpretation. Overall, the chapter is concerned with general approaches to the study of population change in terms of over-arching concepts, basic measures, major sources of information and general principles in interpreting population statistics.

1.1 Origins of demography

After the fourteenth century Black Death killed at least a quarter of the population of Europe, and around 50 million in Europe and Asia overall, there followed centuries with further devastating epidemics of bubonic plague. Early in the sixteenth century, an ordinance required parish priests in London to compile weekly lists of deaths from plague, called the Bills of Mortality. These were intended initially to identify outbreaks and areas for quarantine. Later, other causes of death were included, as well as weddings and christenings and the collection was extended to cover all English parishes (Pollard et al. 1990: 7–8). Disastrous plagues struck London in 1603 and again in 1625. In the latter year an estimated one quarter of the population of London died (Boorstin 1984: 667). Interest in population at the time centred on the effects of epidemics on population numbers, together with the new field of 'political arithmetic', concerned with estimating national wealth.

Yet seventeenth century London produced the founder of demography, John Graunt (1620–74), a 'prosperous and intelligent' cloth merchant who became interested in the Bills of Mortality. He had no scientific training, and he 'knew not by what accident' he was moved to begin the studies that led to his book, the title page of which is shown in Figure 1.1. As mentioned in the Introduction, Graunt's *Natural and Political Observations Made upon the Bills of Mortality*, quickly brought him scientific recognition. Graunt survived London's last major outbreak of bubonic plague in 1665, later the subject of Daniel Defoe's (1722) novel *A Journal of the Plague Year*, but in 1666 the Great Fire of London destroyed his house and ruined him economically (Boorstin 1984: 667). Except among professional demographers, Graunt's work is little known today (Pressat 1973: 4).

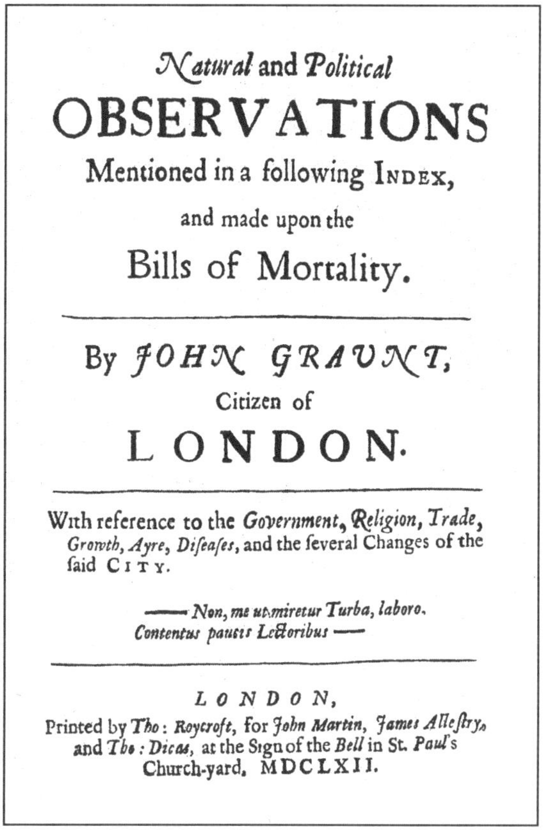

Natural and Political
OBSERVATIONS
Mentioned in a following INDEX,
and made upon the
Bills of Mortality.

By JOHN GRAUNT,
Citizen of
LONDON.

With reference to the *Government*, *Religion*, *Trade*, *Growth*, *Ayre*, *Diseases*, and the several Changes of the said CITY.

—— *Non, me ut miretur Turba, laboro,*
Contentus paucis Lectoribus ——

LONDON,
Printed by *Tho*: *Roycroft*, for *John Martin*, *James Allestry*, and *Tho*: *Dicas*, at the Sign of the *Bell* in St. *Paul's* Church-yard, MDCLXII.

Figure 1.1 Title page of Graunt's 'Observations'.
Source: Graunt (1662).

Graunt's book ranged across many important questions including causes of death, proportions surviving at different ages, health and the environment, the balance of the sexes, family size, age structure, employment, population estimates, population growth and its components and the need for social statistics in public administration. Cambridge demographer Peter Laslett described Graunt as 'ranking among the great natural scientists of the early years of the Royal Society' and his little book as 'universally recognized as a work of genius'. Graunt demonstrated the potential for systematic study of population and he is the acknowledged founder of demography (Laslett 1973: 1).

The appearance of Graunt's book transformed perceptions of population statistics. His ideas stimulated interest in medical statistics on the causes of illness and death, in tables on the statistical likelihood of people's survival – for the life insurance industry – and in the establishment of national censuses and statistical offices. Fruition of some of these ideas, notably the conduct of regular national censuses, took more than a hundred years, but the ideas originated with Graunt.

BOX 1.1 **Defining demography**

The word 'demography' is a combination of two Greek words literally meaning 'description of the people'. It was first used in 1855 by a Belgian statistician, Achille Guillard, in his book *Elements of Human Statistics or Comparative Demography* (Borrie 1973: 75).

There are various definitions of demography. Shorter definitions are easier to remember, easier to communicate and adequate for most purposes. Examples include:

- Demography is the study of human populations (McFalls 1998: 4).
- Demography is the science of population (Weeks, 2002: 4).
- Demography is the study of population processes and characteristics.

The processes include growth, fertility, mortality, migration and population ageing, while the characteristics are as varied as age, sex, birthplace, family structure, health, education, and occupations. The core subject matter of demography is well established, but time brings new research frontiers and shifts in emphasis, as evident in present attention to environmental issues, the status of women, the AIDS epidemic and applied demography.

Longer definitions seek to encompass the breadth of the field of inquiry:

- Demography is a science concerned with the analysis of the size, distribution, structure, characteristics, and processes of a population (Weeks 1994: 25).
- Demography is the study of human populations in relation to the changes brought by the interplay of births, deaths and migration (Pressat 1985: 54).
- Demography, in the most precise sense of the term, is the quantitative study of human populations and the changes in them that result from births, deaths and migrations (Ross 1982: 147).
- Demography is the study of the size, territorial distribution, and composition of population, changes therein, and the components of such changes (Hauser and Duncan 1959: 31).
- 'Demography is generally considered an interdisciplinary subject with strong roots in sociology and weaker, but still important, connections with economics, statistics, geography, human ecology, biology, medicine and human genetics. It is rarely thought of as a completely separate discipline, but rather as an interstitial subject or as a subdivision of one of the major fields.' (Kirk 1972: 348).

1.2 The demographic transition

Alongside its tradition of empirical research, springing from the insights of John Graunt, demography has built a substantial body of conceptual knowledge. Its purpose is to aid description, to identify patterns, to explain developments and, ultimately, to predict them. Demography's predominant theoretical preoccupation in the twentieth century was the theory of the demographic transition. Despite its name, the demographic transition is not so much a theory as a set of generalizations from observed trends. The explanatory and predictive value of transition theory is quite limited, but it is the direct forebear of some of the contemporary discipline's main theoretical interests.

The descriptive and pedagogic value of the demographic transition, moreover, is undisputed; it has long held its place as 'the central preoccupation of modern demography', 'the eternal theme in demography' (Caldwell 1996: 321) and 'one of the best-documented generalizations in the social sciences' (Kirk 1996: 361). It is scarcely possible to describe national trends in births and deaths and growth rates, or changes in national age structures, without reference to the demographic transition. Indeed Geoffrey McNicoll (1992: 404) commented that: 'The greater part of modern demographic research . . . has been directed at understanding the changes that constitute demographic transition and its aftermath.' As a general framework for description and comparison of population changes it is unequalled. History has shown that there are many patterns of transition and varied explanations, not least because policy interventions have had a great impact on the course of events. Amid this diversity, a major point of reference has been the general pattern of change through time in Western Europe.

The development of the original or 'classical' demographic transition theory was due especially to work, in the 1940s and early 1950s, by Frank Notestein, Director of the Office of Population Research at Princeton University (Woods 1982b). The theory originated from observations of long-run changes in birth and death rates in European countries, as illustrated by the idealized pattern in Figure 1.2. The crude rates in the graph denote the number of births or deaths per thousand population.

The demographic transition refers to the movement of death and birth rates in a society, from a situation where both are high – in the pre-transition stage – to one

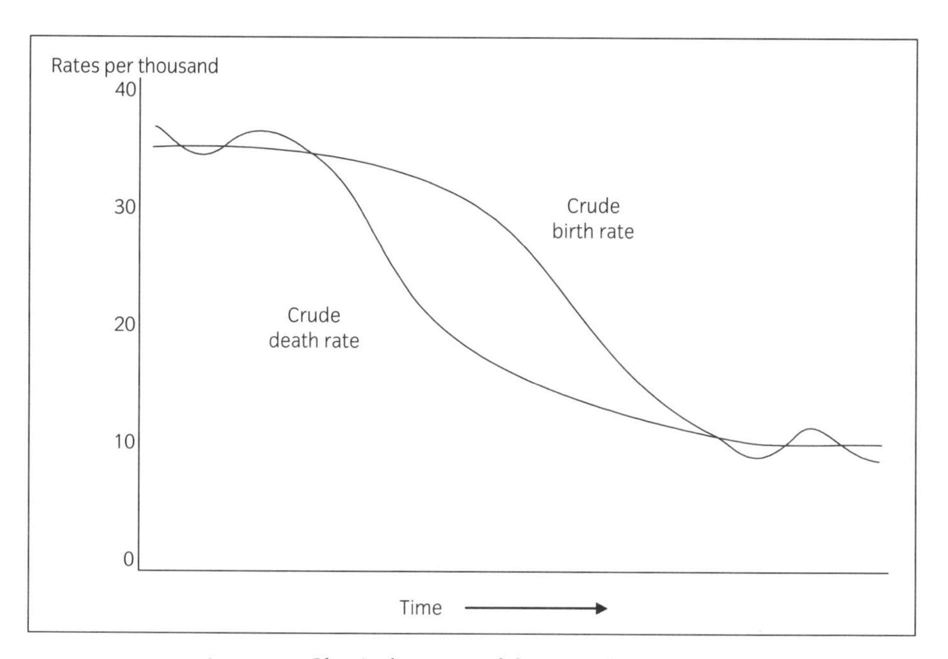

Figure 1.2 Classical pattern of demographic transition.

where both are low – in the post-transition stage. The interval separating the two is the transition itself, during which substantial and rapid population growth often occurs, because births exceed deaths. In Figure 1.2, birth and death rates before and after the transition are approximately in equilibrium, each offsetting the effect of the other.

Stages

In the pre-transition stage, death rates fluctuated from high to extreme while birth rates were high but less changeable. Typically there were 30 or more births and deaths annually per thousand people, and growth was low. Times of sustained population growth did occur in the past, as in Europe in the Middle Ages, but epidemics, famines and wars ultimately depleted gains. Thus the transition 'theory' envisages fluctuating death rates in the pre-transition period producing, in the long term, a near balance between births and deaths. On average, women bore six or more children, but over a quarter of them would have died even before their first birthday. By the mid-twentieth century there were few countries still in the pre-transition stage.

Declining death rates, especially among the young, marked the beginning of the transition. In nineteenth century Europe, these changes arose from improvements in living standards and the environment, which brought better nutrition and fewer deaths from infectious diseases. Population growth rates increased as the numbers of births exceeded the numbers of deaths. A later fall in the birth rate was associated with trends in marriage or birth control, or both of these together. The underlying reasons for changes in marriage and adoption of birth control have been subjects of continuing debate.

Figure 1.3 illustrates the transition stage in Japan. Since birth rates were substantially higher than death rates over most of the period depicted, population numbers in Japan grew from 35 million in 1875 to 127 million in 2000, as shown on the right-hand axis of the graph. The last two peaks in the death rate were associated with war deaths and the influenza pandemic that occurred after the First World War. The death rate rose gradually during the 1980s and 1990s, because of the ageing of Japan's population. The *Demographic Transition* module (Box 1.2) presents further illustrations of the course of the demographic transition. The *Population Dynamics* module (Box 1.3) permits experiments in changing the course of the demographic transition, to reveal consequences for population growth.

Eventual convergence of birth and death rates (Figures 1.2 and 1.3) signals the end of the transition and the coming of the post-transition period. Just as high birth and death rates were in balance before the transition, so too low birth and death rates balance each other after it, resulting in little growth through *natural increase* – the excess of births over deaths. Low birth rates reflect the prevalence of small families, especially the two-child family. Yet without an appreciable proportion of couples having larger families, population decline becomes likely. This is because extra numbers are needed to counterbalance the unmarried, the married but infertile, and couples who have only one child (Coale 1973: 59). Thus, whereas

BOX 1.2 **Demographic transition module (Demographic Transition.xls)**

This module is a database on the past and projected course of the demographic transition, for selected countries and regions. It illustrates contrasts in national experience of the demographic transition since 1950, together with historical trends for Norway, Sweden, England and Wales, and the United States.

Trends since 1950, one country displays United Nations' estimates and projections, of total numbers and crude birth and death rates, for 20 countries and regions.
Data source: Population Division of the Department of Economic and Social Affairs of the United Nations Secretariat 2001. *World Population Prospects: The 2000 Revision*, Vol. I, *Comprehensive Tables*. New York: United Nations.

Trends since 1950, two countries/variables compares crude birth and death rates for two countries using United Nations' estimates and 'medium variant' projections. It can also show the CBRs and CDRs together for one country.
Data source: as above.

Historical trends displays crude birth and death rates, together with rates of natural increase, since 1750 for France, Norway and Sweden, and for England and Wales and the United States since 1800 and 1900 respectively. The list of countries in the menus changes for different periods, according to the availability of data; gaps appear in the menus when there are no data for a particular country.
Data sources: Chesnais, Jean-Claude 1992. *The Demographic Transition: Stages, Patterns, and Economic Implications*. New York: Clarendon Press; Mitchell, B. R. 1975. *European Historical Statistics 1750–1970*. London: Macmillan; Population Reference Bureau (various years), *World Population Wallchart*; US Bureau of the Census 1976. *The Statistical History of the United States: From Colonial Times to the Present*. New York: Basic Books; United Nations (various years), *Demographic Yearbook*; United Nations 2000, *Demographic Yearbook, Historical Supplement 1948–1997 (CD)* (for further details, see spreadsheet 'Data2' in the module).

Instructions

To use the module, insert the CD in the drive, start Excel and open the file *Demographic Transition.xls*. Enable macros when prompted.

- Click one of the buttons to open a display, e.g. Historical Trends.
- Click **Menu** to change the data displayed in the graph.
- Click **Home** to return to this display.
- Click **Exit** to close the module.

See Appendix B for further advice on using the Excel modules.

continued

BOX 1.2 continued

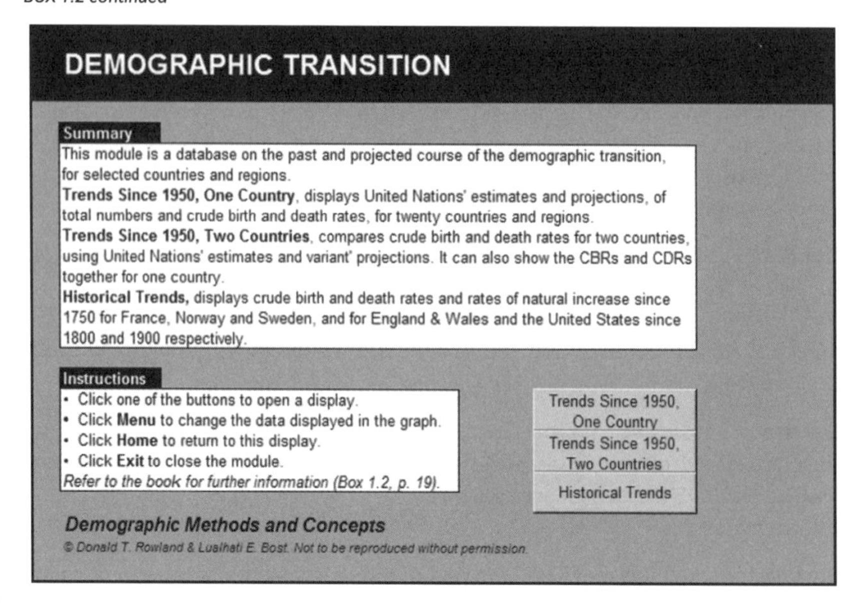

Illustrative applications

Questions to consider in using the module:

1. Referring to the values of the CBR and CDR, at what stage of the transition was the country in 1950? What stage is it at now (pre-transition, transition or post-transition)?
2. To what extent is the country's experience of the demographic transition comparable with the generalized model depicted in Figure 1.2? Is the gap between the birth and death rates greater; are the declines in fertility and mortality faster or slower?
3. Is there any evidence of an upturn in the crude death rate? Why is such an upturn expected in ageing populations?
4. Do the data for the late 1940s and the 1950s show any evidence of a baby boom, indicated by a rise in the crude birth rate?
5. Is there evidence of present or prospective population decline in any of the figures? Is it due to a higher death rate, a lower birth rate, or both?
6. Why were crude death rates so variable in the eighteenth and nineteenth centuries? Why did most of the variations disappear? What were the causes of later peaks?

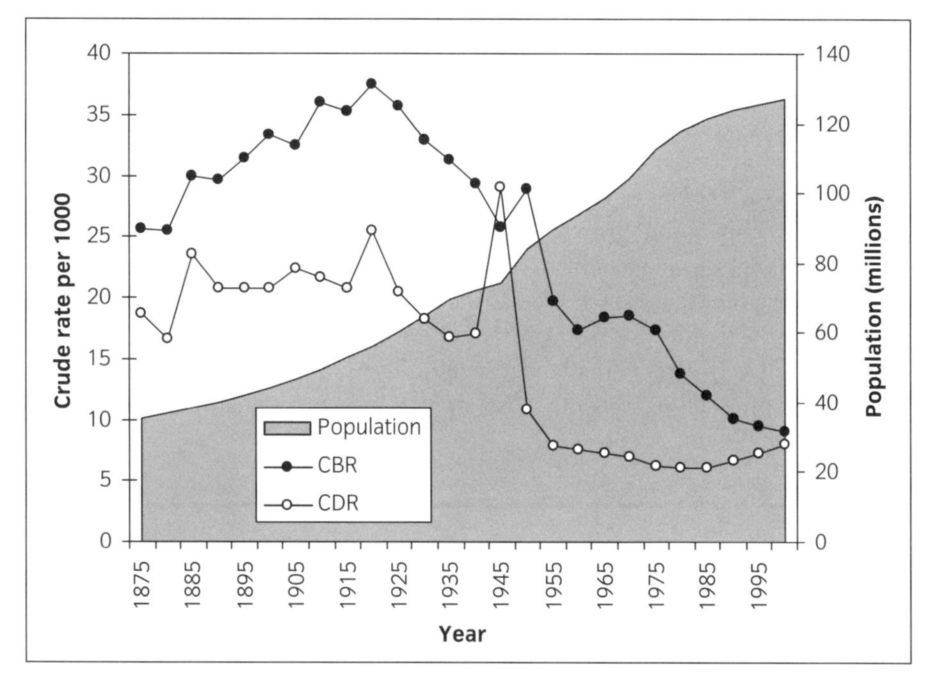

Figure 1.3 Demographic transition in Japan, 1875–2000.

Notes: The data refer to points at five year intervals. CBR = crude birth rate; CDR = crude death rate.

Data sources: Statistics Bureau, Government of Japan 1994. *Japan Statistical Yearbook 1993/94*. Table 2.1; United Nations 2000: *Demographic Yearbook, Historical Supplement 1948–1997* (CD); Population Reference Bureau, *2000 World Population Data Sheet*.

the classical transition envisages a final period of demographic equilibrium, with growth rates fluctuating around zero, long-term population decline is equally possible. Indeed, developments in the last quarter of the twentieth century brought a realization that the demographic tranquillity envisaged in classical transition theory had not eventuated.

Causes

Classical transition theory linked changes in mortality and fertility decline with each other and with urbanization and industrialization. It is now recognized, however, that birth and death rates can decline in a wide range of settings, rather than only in the context of economic development. In recent decades, policies fostering social development – through provision of basic, community-based health services, together with primary education for girls as well as boys – have fostered lower birth and death rates in less developed countries. Moreover, Coale (1973) identified three pre-conditions for fertility decline, none of which require economic development, namely: (1) fertility control must be viewed as ethically and morally acceptable; (2) couples must perceive fertility control as economically and socially advantageous; (3) there must be knowledge and effective use of methods of fertility control. Essentially, these pre-conditions state that couples must be 'ready, willing and able' (Watkins 1987, cited by Lucas 1994: 24).

The missing component of population change in the classical demographic transition is migration, which is an important contributor to national population growth in a number of countries, including the United States, Canada and Australia. In such contexts, trends in birth and death rates do not account for overall national population growth. Historically, migration from rural areas in Europe is thought to have influenced both the amount of population growth occurring during the transition as well as the timing of fertility decline. Where there were few opportunities for outward migration from rural areas, fertility decline occurred relatively early (Friedlander, 1969). The *Population Dynamics* module includes an option illustrating the effects of migration on populations with various levels of natural increase (Box 1.3).

BOX 1.3 **Population dynamics module (Population Dynamics.xls)**

This module demonstrates the effects on natural increase and total population growth of changes in rates of fertility, mortality and migration. It provides straightforward simulations of the course of the demographic transition under different conditions. It can also illustrate the effects of population crises due to epidemics and famines.

The module works by calculating trend lines for the CBR and CDR through three points: for year 0, year 100 and year I, an 'intermediate year' anywhere between them. The normal setting for year I is 50.

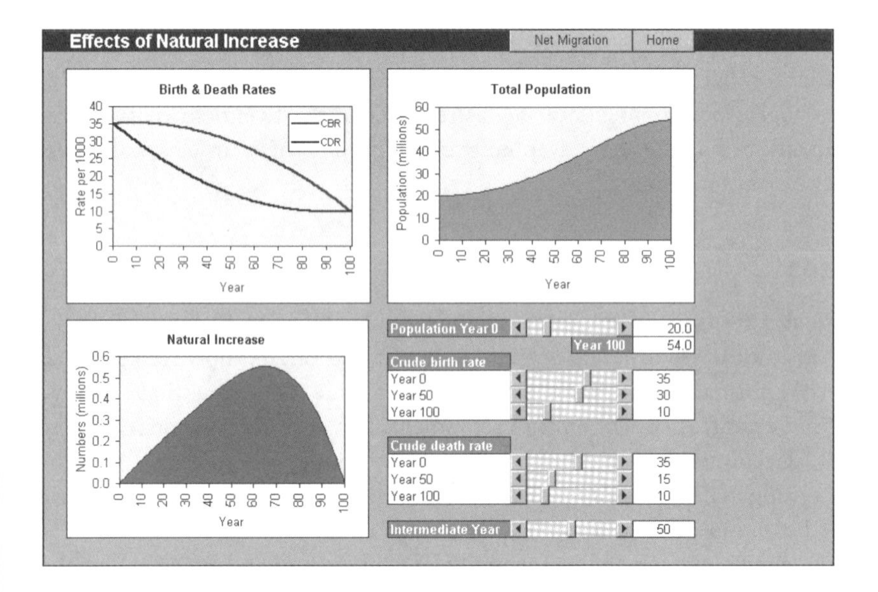

continued

BOX 1.3 continued

Instructions

To use the module, insert the CD in the drive, start Excel and open the file *Population Dynamics.xls*. Enable macros when prompted.

- Click **Natural Increase** to view the effects of changes in crude birth and death rates. On the screen are scroll bars to change the initial population size, the CBRs and CDRs (at the start and end of the period, as well as at the intermediate year), and a final scroll bar to change the intermediate year. The default for the intermediate year is the mid-point of the 100 year interval; setting it earlier or later concentrates changes nearer to the start or end of the interval.
- Click **Net Migration** to examine the added effect of a constant rate of net migration.

See Appendix B for further advice on using the Excel modules.

Illustrative applications

Natural Increase Display

Different scenarios are created by moving the scroll bars to adjust the crude birth and death rates at year 0, year 100 and the intermediate year. As adjustments are made, the graphs update to show the effects of the new settings on trends in the crude birth and death rates, natural increase and population numbers. Note particularly the effects of the changes on population numbers at year 100. Some patterns of transition produce a 'population explosion', others relatively modest growth. The following table presents illustrative settings to display a number of major trends and events through time.

	Example	Population Year 0 (millions)	CBR per 1000	CDR per 1000	Intermediate Year (I)
1	Delayed decline in fertility	100	Year 0 35 Year 50 30 Year 100 10	Year 0 35 Year 50 15 Year 100 10	50
2	Accelerated decline in fertility	Same as 1	Year 0 35 Year 50 20 Year 100 10	Same as 1	Same as 1
3	Fertility decline below replacement	Same as 1	Year 0 15 Year 50 10 Year 100 8	Year 0 12 Year 50 10 Year 100 12	Same as 1
4	War, pestilence and famine	Same as 1	Year 0 35 Year 50 30 Year 100 35	Year 0 35 Year 50 50 Year 100 35	Same as 1

Natural Increase and Net Migration Display

In this option, the scroll bar to change the intermediate year is replaced by a scroll bar to set the net migration rate per 1000 (year I remains at 50). Inward and outward migration can have a considerable impact on population growth. To observe these effects, repeat the first three examples above, using net migration rates of +5, 0 and –5 per 1000.

Outcomes

One of the achievements of transition theory was that it predicted the *population explosion*, the acceleration of population growth following rapid mortality decline. However, in the 1940s nobody expected the fall in mortality in less developed countries to be so rapid and the rate of growth to be so high (Teitlebaum 1975). The original expectation was that mortality decline would be closely associated with overall development, but reductions that took many decades in Europe occurred within a single decade in some less developed countries.

While the transition has brought almost overwhelming growth to some countries, for others it has led ultimately to the complete opposite, owing to very low fertility and deaths exceeding births. For many years now, European governments have been concerned about the prospect of population decline. As early as 1984, member countries of the then European Economic Community adopted a resolution on the need for measures to promote population growth in Europe. The resolution expressed concern about the declining share of Europe's population within the world total, and ensuing effects on 'Europe's standing and influence in the world' (Anon 1984: 569). Ironically, even in the anti-natalist post-transition period, political influence is still equated with population numbers.

1.3 Sources of data

The demographic transition provides a basis for comparing national demographic trends, such as in terms of whether birth and death rates are high or low, diverging or converging, changeable or relatively stable. The source materials for such investigations are readily available in published statistics. Whereas Graunt had no choice but to work with limited and defective data, contemporary sources of population statistics are comprehensive and reliable in comparison.

Awareness of the availability of information is essential in approaching demographic studies. Knowing where to look for information, and being able to make informed choices of data in relation to content and quality, are important demographic skills. For more advanced work, complementary skills in designing and conducting surveys and qualitative studies enhance opportunities to obtain new insights from innovative approaches (see, for example, Moser and Kalton 1975; Minichiello 1990).

Demographic data cover demographic processes and events, such as births, deaths, migration, marriage and divorce. They also include wide-ranging information on population characteristics, such as sex, age, marital status, birthplaces, occupations and education. The principal sources for the study of population are censuses, vital statistics and population surveys. This section summarizes features of the most readily available data sources, together with the main publications on cross-national information for comparative studies. Other chapters discuss data sources in relation to particular topics, such as mortality (Chapter 6), population distribution (Chapter 10) and migration (Chapter 11).

Censuses

A census is a national enumeration of a population at the same time. The taking of regular censuses dates from 1790 in the United States, 1801 in England and Wales, 1871 in Canada and 1881 in Australia (Spiegelman 1973:1). Censuses are an expensive undertaking requiring years of planning, a large staff of field workers and years to compile and publish results. The aim is to establish the size, distribution and composition of the population through an accurate count of the number of persons and households and their characteristics. Ideally, a census achieves a simultaneous count of the population, enumerating everyone at one date.

The late twentieth century brought important innovations in the means of publishing and disseminating census results. Printed volumes were once the mainstay of census output, supplemented, in the 1960s, by computer printouts and computer tapes for specialist research. Subsequent developments, especially the CD ROM, graphics-based mapping and statistical software and the Internet, have greatly extended the versatility of census information and public access to it.

Major purposes of a census are to determine electoral arrangements on the basis of population, and to provide baseline data for planning and administration. The numbers and characteristics of people are necessary considerations in many aspects of planning, including: monitoring of the demand for schools, housing and health services; studying the impact of changes in the labour force and immigration; conducting market analyses of the demand for goods and services; and determining the most appropriate location for public amenities and shopping centres. The census is also the chief source of information on small areas and small groups, because of its complete coverage of the population.

Census counts are of two types. Most common is the *de facto census*, which counts people wherever they happen to be within the country at the time. The alternative is the *de jure census*, which counts people where they usually live. The latter is more complex, for in order to obtain statistics on the de jure, or resident, population of each community, information has to be redistributed for people enumerated away from home. Computer processing facilitates such operations and, in practice, de facto censuses commonly ask for information about usual address. Thus data from one census can be compiled for both the de facto and resident populations. This makes possible the assembly of information on internal migration (see Chapter 11), as well as on resident population numbers that are often needed for planning and population projections. Some places have large visitor populations – as revealed in comparisons of the de facto and de jure counts – and these comprise a further distinctive consideration in planning.

In the 1980s, national censuses covered about 94 per cent of the world's population (McFalls 1998: 31). The most populous countries, China and India, each with more than a billion people (i.e. a thousand million) continue to mobilize

BOX 1.4 **Limits of the census**

A population census aims for complete coverage of a country's population and is the largest and most complex of statistical collections. Nevertheless, the census has a number of major limitations relating to:

- ***Accuracy*** The content of the census is limited by the kinds of information that can be provided quickly and accurately, either from a self-completed form or from a short interview with a census collector.
- ***Privacy*** The content is limited by considerations of privacy. Laws prohibit intrusive questioning – such as on details of personal wealth – and census files cannot be linked with other databases. Census tabulations also have to be designed to prevent the identification of individuals. This makes impossible the publication of detailed cross-tabulations at the small area level, such as tables for census tracts showing age by sex by ethnic group by occupation by income.
- ***Immediacy*** Although national censuses may be conducted every five or ten years, they do not provide information that is sufficiently up-to-date for all purposes. Processing time delays the appearance of the first results for many months after the census, and the full set of census publications may take several years to complete. For current information on topics such as employment and unemployment, monthly or quarterly surveys remain essential.
- ***Coverage*** Census questions are limited to topics that can be readily answered and are needed for studies of small areas and groups. Because of the high cost of conducting a census, it is important to avoid non-response to questions and the wastage of resources entailed in collecting information relevant mainly at the state, provincial or country levels. National sample surveys are a more efficient and economical means of collecting information on more specialized topics, as well as topics for which data are not required for small areas.

resources to take full censuses. Yet developments in the late twentieth century brought a suspension or cessation of census-taking in parts of the European Union. Influential here have been the high cost of traditional censuses, the practicality of assembling 'virtual censuses' – where there are adequate administrative data, surveys and continuous registrations of births, deaths and migrations (Laroche 1993, cited in Lucas and Meyer 1994: 7) – and concerns about respondent burden and public apprehension regarding potential misuse of census information on individuals.

Other sources

Whereas censuses collect information on entire populations at points in time, registration systems collect information on individuals when they experience the vital events of birth, marriage or death. These data are commonly compiled for publication in annual *vital statistics* bulletins. Since the information collected may be quite extensive, vital statistics publications can contain substantial cross-tabulated information. Statistics on births, for example, potentially include details about the newborn (sex, birth weight, place of birth), the mother (age, birthplace, marital status, occupation) and the father (age, birthplace, marital status, occupation). Demographic rates and indices are also presented in vital statistics publications along with the raw numbers and details of the components of population growth. Less than half the world's population live in countries that are able to obtain near-complete registration of vital events (McFalls 1998: 31).

Narrowly defined, vital statistics refer to events that add to, or subtract from, the membership of the population. Births and deaths meet this requirement, but other life-shaping events, such as marriage, divorce and adoptions, are commonly included in the subject matter of vital statistics – all are products of official registration systems. Migration statistics (see Chapter 11) too are sometimes regarded as vital statistics; they are obtained from records of international arrivals and departures, which can include details about the characteristics of movers, such as sex, age, birthplace, occupation and duration of stay.

Supplementing censuses and vital statistics as sources of demographic data are a wide range of population surveys, covering topics such as employment, families, health and housing. These have become an increasingly important source for demographic inquiry; 47 per cent of articles in the leading American journal *Demography* have been based solely on survey data (Teachman et al. 1993: 524–6). Like the technological advances in the publication of census information, as well as vital statistics, the growth in the range and frequency of population surveys is a defining feature of the enriched environment for research on populations that has developed since the 1970s. Information about large-scale surveys, such as the *Current Population Survey* in the United States, is available from Web sites of national statistical organizations (see 'Internet resources' at the end of the chapter).

Statistical agencies in individual countries also publish compendiums of statistical information, such as *Social Trends* (for Australia, Britain, Canada and New Zealand), the *Statistical Abstract of the United States* and national yearbooks. These can be found in the statistical agencies' own catalogues or in their lists of publications on the Internet. Since yearbooks and other compendiums contain wide-ranging summary statistics and descriptive material, they are often valuable sources to consult when interpreting national population changes.

In using the various sources of demographic data, questions concerning data coverage and quality inevitably arise. For example, in less developed countries, birth statistics may omit many children who were born at home or who died in infancy. Similarly, marriage and international migration figures are prone to incompleteness, the former because marriage registration omits consensual or de facto unions, the latter because of illegal or uncontrolled movement across national borders. Survey figures are also subject to sources of error, and findings may not be truly representative of the general population. An evaluation of the nature and quality of the data is a necessary prelude to many studies.

Cross-national data

For comparative studies of population there are several major sources of cross-national data (Table 1.1). Computer modules in this and other chapters present, in graphical form, a selection of information from these sources (see Box 1.2). One of the best known is the United Nations *Demographic Yearbook*, published almost every year since 1948 with statistics for the majority of countries. Standard tables in this series include coverage of population growth, birth and death rates, age composition and urban and rural numbers. Most volumes also contain extra information on a 'special subject' such as fertility, mortality, migration and census statistics – these include more detail than the standard tables. Cumulative lists of special subjects appear at the front of each volume together with the detailed contents. The best starting point for locating information, however, is the index at the back of the *Demographic Yearbook*, since this includes references to the contents of all previous volumes, including the special subjects. The *Demographic Yearbook Historical Supplement 1948–1997* (United Nations 2000), on CD, contains a substantial body of statistics for many countries through time.

Population Trends and Prospects, appearing biennially, is another publication of the United Nations Population Division in New York. In contrast to the *Demographic Yearbook*, which focuses mainly on the most recent information, this publication features data for many countries of the world from 1950 to the present, in five-year intervals, together with projections. The volumes contain statistics on population numbers, growth rates, birth and death rates, life expectancy and other key indicators. The projections have three 'variants' – high, medium and low. The medium series is most widely cited in articles on prospective trends. A companion volume, entitled *The Sex and Age Distribution of the World Populations*, contains age structure data for countries and world regions over the same time period. Other major publications with comparative demographic data are the World Bank's *World Development Report*, the World Health Organization's *World Health Statistics Annual* and Eurostat's *Demographic Statistics*, an annual publication providing information on member countries of the European Union (See

Table 1.1 Summary of sources of cross-national data on demographic changes

Publisher	Title
Eurostat	*Demographic Statistics*
Population Reference Bureau	*World Population Data Sheet*
United Nations	*Demographic Yearbook*
United Nations	*Population Trends and Prospects*
United Nations	*The Sex and Age Distribution of the World Populations*
World Health Organization	*World Health Statistics Annual*
World Bank	*World Development Report*

Table 1.1). Further demographic data sources for Europe are reviewed in Coleman (1999).

The Population Reference Bureau publishes wall charts and booklets annually providing population totals and summary indices for countries of the world (Population Reference Bureau 2001b) and the states of the United States (Population Reference Bureau 2001a). Other organizations, such as ESCAP (Economic and Social Commission for Asia and the Pacific) in Bangkok, and the South Pacific Commission in Noumea, also produce wall charts providing summary demographic data for countries in their respective regions.

Evaluation of the nature and quality of the source statistics is an integral part of demographic analysis; hence users of the sources listed in Table 1.1 need to be prepared to consult accompanying technical commentaries and footnotes to ascertain the meaning and reliability of particular statistics. Occasionally, the caveats themselves are insufficient, and published sources may contain errors. Recognition of such problems may require close acquaintance with the demographic data for the country in question, but figures that appear inconsistent can sometimes be verified readily through comparison with other sources.

1.4 **Components of change**

Vital statistics on births and deaths, together with census data on population totals, are the basic data needed for studying the demographic transition and comparing trends through time. The difference between the numbers of births and deaths comprises the population's *natural increase*. Where migration has little impact on population numbers, natural increase accounts for overall population growth or decline. Thus the two components of population growth or decline are natural increase, the excess of births over deaths, and *net migration*, the excess or arrivals over departures. The *demographic balancing equation* expresses this as follows:

$$\text{population growth} = \text{natural increase plus net migration}$$

or:

$$\text{population growth} = (\text{births} - \text{deaths}) + (\text{arrivals} - \text{departures})$$

Population growth is the difference between population totals at two dates, such as in 1995 and 2000. This is written as $P_n - P_0$, where P_0 = the initial population (at the start of a period of time, e.g. 1995) and P_n = the end-of-period population or terminal population (after n intervals of time, such as years). The 0 and the n are subscripts – numbers or letters written below the baseline to provide additional information. Thus the balancing equation can be written:

$$(P_n - P_0) = (\text{births} - \text{deaths}) + (\text{arrivals} - \text{departures})$$

and P_n can be expressed as:

$$P_n = P_0 + (\text{births} - \text{deaths}) + (\text{arrivals} - \text{departures})$$

When analysing population changes, an important strategy is to discuss the components of growth, since these show the sources of gains and loses, which in turn can suggest possible explanations of trends. Changes in population numbers are always the outcome of natural increase and net migration, interactions between which are illustrated in the Excel module on *Population Dynamics* (Box 1.3). Natural increase and net migration are the processes immediately responsible for change, whereas social, economic, political or environmental factors are the underlying causes of change. Figure 1.4 illustrates past and projected components of growth for the United States. Since the data for net migration are 'stacked' on top of those for natural increase, the two combined represent the total population increase.

The demographic balancing equation refers to changes in total numbers, rather than changes in growth rates. Total numbers are important in themselves because planning for the provision of goods and services depends substantially on the number of customers or service users. However, to compare populations in different times and places, other measures – such as rates and percentages – are needed to show the relative pace, or relative size, of changes in different population totals.

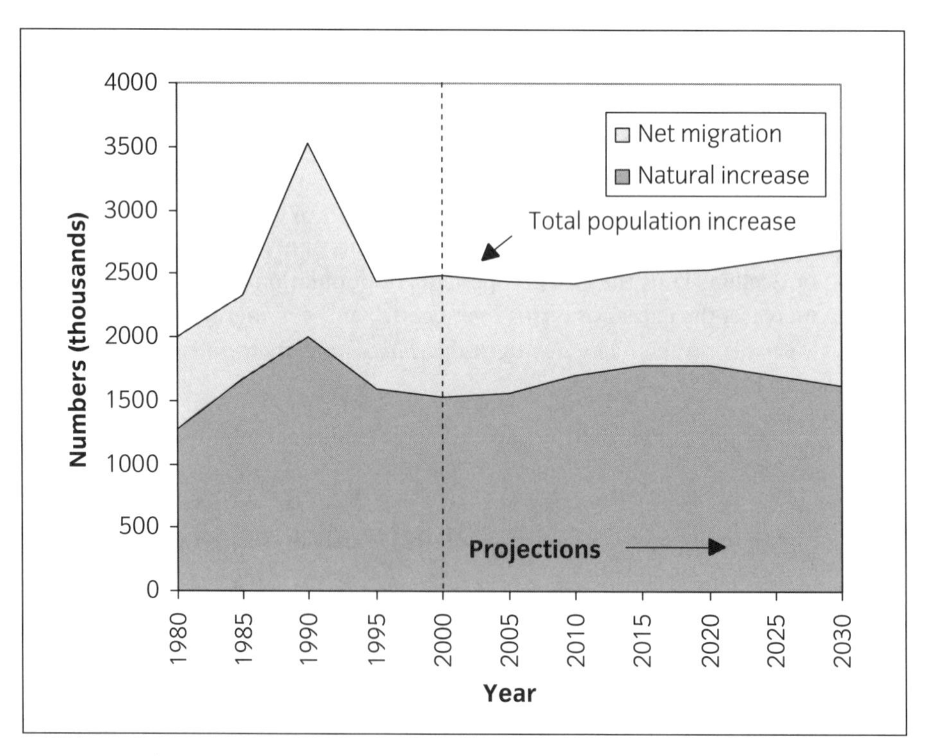

Figure 1.4 Observed and projected components of growth in the United States, 1980–2030.

Note: The data refer to points at five year intervals; figures beyond 2000 are projected.

Data source: *Statistical Abstract of the United States, 2000* (http://www.census.gov/prod/2001pubs/statab/sec01.pdf).

1.5 Comparative measures

Demographic rates

Rates are the most widely used comparative measures of population change. Ideally, demographic rates show the relationship between the number of demographic events (in the numerator) and *the population at risk* of experiencing them (in the denominator). The population at risk is the population that could potentially experience a particular event, such as giving birth or migrating, in a specific period of time. The concept of exposure to risk (Cox 1970: 22–23) is important in demography, since measures based on the population at risk are more accurate and more informative than those based simply on the total population. For instance the population at risk of having a baby in the year 2000 comprised only females of reproductive age at that time. Being 'at risk' does not imply that a particular event is necessarily unwelcome; rather it reflects wider usage of an expression that originated in mortality studies, where there is special interest in the risk of dying (Pressat 1985: 74).

Often, the most convenient approximation to the population at risk is the total population at mid-year. Mid-year is assumed to be the point by which half the changes have occurred. It could be misleading to relate changes to the start-of-year or end-of-year populations if substantial growth or decline took place. If the population on 1st January was 10 000 and on 31st December 40 000, each total would produce a markedly different rate.

When unavailable in published statistics, the mid-year population may be calculated as the mean, or average of the population at the start and end of the year:

$$(P_0 + P_n)/2 \quad \text{e.g. } (10000 + 40000)/2 = 25000$$

Alternatively, the mid-year population can be calculated as the initial population plus half the numerical growth during the period:

$$P_0 + \frac{1}{2}(P_n - P_0) \quad \text{e.g. } 10000 + 0.5(40000 - 10000) = 25000$$

This leads to the more general expression for the population at any point in a year, i.e. the initial population plus the required fraction of the annual increase:

$$P_0 + \frac{1}{n}(P_n - P_0)$$

where n is the number of periods into which the year is divided.

For example, to obtain the population on the 31st January use either:

$$P_0 + \frac{1}{12}(P_n - P_0)$$

which is based on the average monthly change, or:

$$P_0 + \frac{31}{365}(P_n - P_0)$$

which is based on the average daily change.

Rates are commonly multiplied by 100 or 1000 to produce figures greater than 1, since whole numbers are more easily comprehended than decimal fractions. Some rates for rare events, such as deaths from certain diseases, are specified per 100 000 or even per million to reach whole numbers.

Other types of measures

When the population at risk is unavailable, the derived figure is strictly a ratio rather than a rate. A ratio expresses the size of a number relative to another convenient number. One of the best-known in demography is the *sex ratio*, the number of males per hundred females (males/females × 100). Although a simple measure, it is relevant to understanding matters as diverse as marriage opportunities, the representation of women in the labour force and, at later ages, the consequences for family support of sex differences in life expectancy. Denominators for ratios are selected according to the availability of data and ease of understanding.

Rates and ratios are essential bases for many comparative studies, but equally common are proportions and percentages. A proportion is a ratio in which the denominator includes the numerator. For example, the proportion of older people in the population of France at the 1990 Census was 0.15 – obtained by dividing the numbers aged 65 and over in France (8 347 869) by the total population of France (56 634 299). Since proportions are necessarily decimal fractions, such ratios are often multiplied by 100 to produce percentages, which are easier to read; thus 15 per cent of the population of France in 1990 were in the older ages.

A further major type of comparative measure is the probability, which ranges between 0 and 1. There are statistical arguments about the correct definition of a probability, but for practical purposes a probability is the ratio of events in a fixed period of time to the initial population at risk. Probabilities in demographic work are always based on the initial population. The probability of dying at age 100, for example, is based on the number of people who celebrated their 100th birthdays (the initial population). If 2500 women reached 100 and 700 of them died before turning 101, their probability of dying at age 100 would be 700/2500 = 0.28. This is a fairly typical figure for centenarian females in more developed countries.

Table 1.2 summarizes the characteristics of the comparative measures discussed in this section. It is often preferable for absolute numbers to accompany such figures to aid comparisons and assist readers to gauge the magnitude of changes. While the table lists the most common types of comparative measures it is not exhaustive. Others include index numbers, rate ratios and averages.

Table 1.2 Types of comparative measures in demography

Rate	The ratio of the number of demographic events (e.g. births) to the population at risk of experiencing the event. In mathematical demography, the denominator is the number of person years lived.
Ratio	The size of a number relative to another convenient number.
Proportion	A ratio in which the denominator includes the numerator.
Percentage	A proportion multiplied by 100.
Probability	The ratio of the number of demographic events to the initial population at risk of experiencing them. In mathematical demography, the denominator is the number of preceding events.

1.6 Basic measures of change

Crude birth and death rates

The above summary of comparative measures provides the context for recognizing the advantages and disadvantages of the main measures of population change used in cross-national studies. This section discusses the basic measures of population change, as employed for examining trends in population growth and its components.

The crude birth rate (CBR) and the crude death rate (CDR) are based on the total population. They are described as 'crude' because they are unrefined in relation to the population at risk. The crude birth and death rates are key descriptors of the demographic transition (Figure 1.2 and Box 1.2), enabling countries to be compared, irrespective of their total populations:

$$\text{Crude birth rate} = \frac{\text{number of live births in a year}}{\text{mid-year population}} \times 1000$$

$$\text{Crude death rate} = \frac{\text{number of deaths in a year}}{\text{mid-year population}} \times 1000$$

However, there are several characteristics of these rates that affect their calculation and interpretation:

- The rates are 'crude' because they make no reference to smaller groups that might better represent the population likely to experience the event. In more developed countries, for example, the majority of deaths are in the older ages.
- Crude rates are calculated for calendar years. If the birth or death data do not refer to a calendar year the average number of events per year is used instead (Barclay 1966: 35–36). Thus the numerator refers either to the number of events in a calendar year – such as births or deaths – or to the average number per year. Similarly, the denominator refers either to the mid-year population, or to the average population for the period.

Thus if the total number of births for the period 1995–2000 was 25 000, the population in 1995 was 300 000 and 350 000 five years later, the average crude birth rate for the period would be: $(25\,000/5)/((300\,000 + 350\,000)/2) \times 1000 = (5000/325\,000) \times 1000 = 15.4$ per thousand. Averaging the number of events gives satisfactory results, but it is preferable to work with annual data if they are available. Use of the mean population assumes that the population changes by a constant number each year, which is commonly an acceptable assumption for intervals of five years or less.

- Despite their use for cross-national comparisons and comparisons through time, the age structure of populations can have a substantial effect on crude rates. For example, a developing country population with a high proportion of young people might be expected to have a relatively low crude death rate. Accordingly, standardization techniques (see Chapter 4) are employed to refine comparisons by removing the influence of age structure differences, as well as other compositional influences.

Since statistics for calculating crude rates are most widely available, the occurrence of other phenomena is sometimes expressed in the same way as a convenient first approximation. Examples are the crude marriage rate and the crude divorce rate (see Chapter 7). Both are based on the total population, taking no account of the numbers of single people of marriageable age, or of the numbers married and, therefore, potentially at risk of separation or divorce.

Rates of natural increase and net migration

From the crude birth and death rates is derived the *(crude) rate of natural increase* (RNI), which is simply the difference between them:

$$\text{RNI} = \text{CBR} - \text{CDR}$$

Alternatively, the rate of natural increase can be obtained as:

$$\text{Rate of natural increase} = \frac{\text{births} - \text{deaths in a year}}{\text{mid-year population}} \times k$$

where $k = 1000$ or 100.

In the demographic histories of most of the now more developed countries, RNI never exceeded 1.5 per cent – the United States was an exception (McNicoll 1984: 181). Less developed countries have had peaks of double this figure, and rates of natural increase above 3 per cent are still common in Africa.

Table 1.3 presents examples of the calculation of basic rates for Japan in the 1950s and 1990s. Although RNI is equal to CBR – CDR in Table 1.3, the result may be divided by 10 to express RNI as a rate per cent. Japan's rate of natural increase was close to zero per cent in the 1990s.

Table 1.3 Calculation of basic demographic rates for Japan

	1950	1990	1997
Births	2 337 507	1 221 585	1 190 000
Deaths	904 876	820 305	917 000
Population	82 900 000	123 478 000	125 638 000
Rates per thousand:			
Crude birth rate	28.2	9.9	9.5
Crude death rate	10.9	6.6	7.3
Rate of natural increase	17.3	3.2	2.2

Data source: United Nations 2000: *Demographic Yearbook, Historical Supplement* 1948–1997 (CD).

For the global population and the populations of countries where international migration is unimportant, natural increase is the only source of growth – or its converse, natural decrease (an excess of deaths over births), is the only source of decline. Yet it is increasingly common for countries to be affected by migration, whether it be permanent family migration and settlement of individuals, or the movement of refugees and guest workers – who may be denied rights of continuing residence and citizenship at the destination. At the sub-national level, net migration is often the major component of population change, reflecting its dominant role in the growth and decline of cities, suburbs and towns.

Like the other rates discussed so far, the *rate of net migration* (RNM) is based on the mid-year population:

$$\text{RNM} = \frac{\text{net migration}}{\text{mid-year population}} \times k \quad (\text{where } k = 1000 \text{ or } 100)$$

By convention, it is usually referred to as a rate rather than a 'crude' rate. Net migration may be obtained from the balancing equation as a residual left after subtracting natural increase from the total population growth. This is equivalent to the 'vital statistics method' of estimating net migration (see Chapter 11).

1.7 Interpreting changes

These simple demographic indices of population change are easily calculated; the more challenging task is to interpret them in ways relevant to research goals and informative to readers. Articles in *Population Today*, from the Population

Reference Bureau in Washington, serve as good examples of demographic writing for a general audience.

The aims of a piece of research normally provide the focus in analytical work, shaping both the structure and content of the analysis. Nevertheless, there are considerations potentially relevant to many kinds of studies; identifying these is a useful first step towards embarking on analytical writing. Table 1.4 outlines some general considerations in the interpretation of information on demographic change.

Reading about the place, time and subject material in question is an indispensable early step in any study. Knowing what others have written, and the ideas they have used, enriches understanding and increases the likelihood of an informed

Table 1.4 Considerations in the interpretation of population changes

Relevance	What is the problem, theme, issue or hypothesis that the data are intended to address? What are the concerns and interests of the audience for whom the analysis is to be written?
Reliability	Are the source statistics of high quality and are the measures calculated from them reliable? Crude death rates, for instance, may be affected by the age structure of the population.
Conceptual framework	Does the subject material belong within an existing conceptual or theoretical framework, such as the demographic transition, that might assist in clarifying the nature and significance of the statistics?
Comparisons	Are the figures high or low compared with those for other places or other periods of time? Why are they similar or different?
Immediate causes	Is there information on the components of population change to enable an assessment to be made of the relative contribution of natural increase and net migration?
Underlying causes	From background reading and knowledge of the immediate causes of change, is it possible to suggest likely underlying causes of developments, such as social, economic or political changes?
Consequences	Are the observed changes likely to have any appreciable effects on the nature and composition of the community or society in question?
Implications	Are there issues arising that need to be addressed through public policy initiatives, such as in education, health, housing and social welfare? Do the population trends have implications for business activities?
Prospective developments	What is the likely course of change in coming years? Will prospective developments affect the composition of the population and the outlook for society, its institutions and enterprises?

analysis. A second essential is a systematic approach to writing about research findings. Even a structure as simple as 'introduction', 'middle' and 'conclusion' is better than no structure at all, and may be entirely suitable for shorter studies. Beyond these two requirements, there are many issues potentially relevant to the interpretation of population changes, as summarized in Table 1.4.

The table highlights a series of starting points for thinking through the subject material. These entail reflection on both the quality of the information and its meaning as evidence of the phenomenon under investigation. Conceptual frameworks and comparisons are important ways of increasing understanding by placing information in a broader perspective. Going beyond description to consider explanations – immediate and underlying causes – greatly increases the value of analytical writing, provided that there is adequate supporting evidence. Similarly, discussing implications and prospective developments may be essential in studies intended to inform policy-making and planning.

Training in demography should foster the development of a sense of purpose in analytical work, and an appreciation of the potential of demographic data to address questions about the nature of populations and the changes affecting them: the goal is to provide an informed opinion of the quality and meaning of the data. Nevertheless data are sometimes so limited, or of such doubtful quality, that answers to questions will inevitably be incomplete or tentative.

Occasionally, interpretation is avoided or passed over on the grounds that 'the data speak for themselves', or the task degenerates to merely verbalizing statistics already presented in a table. Neither the absence of comment, nor the peppering of the text with statistics, does justice to the task of demographic analysis. Drawing on the content of Table 1.4, questions to ask in such situations include: What have other authors said about the same phenomenon? What patterns or trends are evident? What factors explain the situation shown in the data? How do the figures compare with others for earlier times or other places? What are the implications of the data for future developments? If Graunt, like his contemporaries, had thought that the statistics in the Bills of Mortality merely spoke for themselves, he would never have written his *Observations*.

1.8 Conclusion

Demographic knowledge and skills are needed to access our richest and most reliable sources of statistical information about societies and the changes they are experiencing. Demographic analysis calls for knowledge of the sources of information, the methods of analysing change and the conceptual frameworks that bestow meaning and purpose. It also requires skills in the interpretation and presentation of data. This chapter has introduced each of these matters as a foundation for the book as a whole. The chapter has also shown that demography originated from Graunt's recognition of the potential of statistics on population to inform understanding of the nature of society. Box 1.5 quotes Graunt's own account of the beginnings of his research, revealing his insight into matters which, until then, people had taken for granted.

BOX 1.5 **John Graunt on the beginnings of his 'Observations'**

Having been born, and bred in the City of *London*, and having always observed, that most of them who constantly took in the weekly Bills of *Mortality*, made little other use of them, then to look at the foot, how the *Burials* increased, or decreased; And, among the *Casualties*, what had happened rare, and extraordinary in the week currant: so as they might take the same as a *Text* to talk upon, in the next Company; . . .

2. Now, I thought that the Wisdom of our City had certainly designed the laudable practice of takeing, and distributing these Accompts, for other, and greater uses then those above-mentioned, or at least, that some other uses might be made of them: And thereupon I casting mine Eye upon so many of the . . . General *Bills*, as next came to hand, I found encouragement from them, to look out all the *Bills* I could, and (to be short) to furnish my self with as much matter of that kind, even as the Hall of the *Parish-Clerks* could afford me; the which, when I had reduced into Tables (the Copies whereof are here inserted) so as to have a view of the whole together, in order to the more ready comparing of one *Year*, *Season*, *Parish*, or other *Division* of the City, with another, in respect of all the *Burials*, and *Christnings*, and of all the *Diseases*, and *Casualties* happening in each of them respectively; I did then begin, not onely to examine the Conceits, Opinions, and Conjectures, which upon view of a few scattered *Bills* I had taken up; but did also admit new ones, as I found reason, and occasion from my *Tables*.

3. Moreover, finding some Truths, and not commonly- believed Opinions, to arise from my Meditations upon these neglected *Papers*, I proceeded further, to consider what benefit the knowledge of the same would bring to the World; that I might not engage myself in idle, and useless Speculations, but . . . present the World with some real fruit from those ayrie Blossoms. (Graunt 1662: 17–18)

Study resources

KEY TERMS

Census
Components of growth
Crude birth rate
Crude death rate
De facto census
De jure census
Demographic balancing equation
Demographic transition
Demography
Mid-year population
Natural increase
Net migration
Percentages
Population at risk
Population surveys
Probabilities
Rate of natural increase
Rate of net migration
Rates
Ratios
Vital statistics

FURTHER READING

This section provides a general bibliography of text books, reference works and journals on demography and demographic methods.

General reference works

Pressat, Roland. 1985. *The Dictionary of Demography*, edited by Christopher Wilson. Oxford: Blackwell.

Ross, John A. (editor in chief). 1982. *International Encyclopedia of Population*. New York: Free Press.

Texts on demographic methods

Barclay, George W. 1966. *Techniques of Population Analysis*. New York: John Wiley.

Cox, Peter, R. 1970. *Demography*. Cambridge: Cambridge University Press.

Henry, Louis. 1976. *Population: Analysis and Models*. London: Edward Arnold.

Newell, Colin. 1988. *Methods and Models in Demography*. London: Frances Pinter.

Palmore, James A. and Gardner, Robert W. 1983. *Measuring Mortality, Fertility and Natural Increase: a Self-Teaching Guide to Elementary Measures*. Honolulu: East-West Population Institute, East-West Center.

Pollard, A. H., Yusuf, Farhat, and Pollard, G. N. 1990. *Demographic Techniques* (third edition). Sydney: Pergamon Press.

Pressat, Roland. 1972. *Demographic Analysis*. Chicago: Aldine Atherton.

Shryock, Henry S. and Siegel, Jacob S. 1973. *The Methods and Materials of Demography* (two volumes). Washington: US Bureau of the Census, US Government Printing Office.

Spiegelman, Mortimer. 1973. *Introduction to Demography*. Cambridge, MA: Harvard University Press.

Texts on demographic methods (advanced/mathematical)

Bogue, Donald J. et al. 1993. *Readings in Population Research Methodology* (eight volumes). Chicago: Social Development Center.

Halli, Shiva S. and Rao, K. Vaninadha. 1992. *Advanced Techniques of Population Analysis*. New York: Plenum Press.

Hinde, Andrew. 1998. *Demographic Methods*. London: Arnold.

Namboodiri, Krishnan. 1991. *Demographic Analysis: a Stochastic Approach*. San Diego: Academic Press.

Preston, Samuel H., Heuveline, Patrick, and Guillot, Michel. 2001. *Demography: Measuring and Modeling Population Processes*. Oxford: Blackwell.

Smith, David P. 1992. *Formal Demography*. New York: Plenum Press.

Wunsch, G. J. and Termote, M. G. 1978. *Introduction to Demographic Analysis: Principles and Methods*. New York: Plenum.

Texts on population

Daugherty, Helen G. and Kammeyer, Kenneth C. W. 1995. *An Introduction to Population*. New York: Guilford Press.

Lucas, David and Meyer, Paul. 1994. *Beginning Population Studies* (second edition). Canberra: National Centre for Development Studies, Australian National University.

Nam, Charles B. 1994. *Understanding Population Change*. Itasca, IL: F. E. Peacock.

Stockwell, Edward G. and Groat, H. Theodore. 1984. *World Population: An Introduction to Demography*. New York: Franklin Watts.

Weeks, John. R. 2002. *Population: an Introduction to Concepts and Issues* (eighth edition). Belmont, CA: Wadsworth.

Woods, Robert. 1982. *Theoretical Population Geography*. London: Longman.

Yaukey, David and Anderton, Douglas L. 2001. *Demography: the Study of Human Population* (second edition). New York: Waveland Press.

Weinstein, Jay and Pillai, Vijayan K. 2001. *Demography: The Science of Population*. Boston: Allyn and Bacon.

Zopf, P. 1984. *Population: An Introduction to Social Demography*. Palo Alto, CA: Mayfield Publishing Co.

Periodicals

American Demographics

Population Today

Journals

Demography
Demographic Research (free, available at http://www.demographic-research.org/)
European Journal of Population
Genus
Health Transition Review
International Migration
International Migration Review
Journal of Population Research
Population (in French; 'An English Selection' has been published annually since 1991)
Population and Development Review
Population Bulletin (Population Reference Bureau)
Population Bulletin of the United Nations
Population Index
Population Research and Policy Review
Population Studies

John Graunt

Graunt, John, 1662. *Natural and Political Observations Made Upon the Bills of Mortality*, edited by W. F. Willcox, 1939. Baltimore : Johns Hopkins University Press.

Laslett, Peter (editor). 1973. *The Earliest Classics: John Graunt and Gregory King*, Pioneers of Demography Series. Farnborough, Hants: Gregg.

INTERNET RESOURCES

Subject	Source and Internet Address
'Gateways' to demographic information	**The Australian National University** http://demography.anu.edu.au/VirtualLibrary/ **NIDI – Netherlands Interdisciplinary Demographic Institute** http://www.nidi.nl/ **Population Reference Bureau** http://www.prb.org **POPNET – Population Network, Population Reference Bureau** http://www.popnet.org/ **POPIN – United Nations Population Information Network** http://www.undp.org/popin/
Bibliographic sources	**Population Index** http://popindex.princeton.edu/ **Popline – Johns Hopkins University** http://www.jhuccp.org/popline/popline.stm
Electronic Journals (access mainly by subscription)	**American Demographics** http://www.demographics.com/ **Population and Development Review** http://www.popcouncil.org/pdr/ **Demography** http://muse.jhu.edu/journals/demography/

Subject	Source and Internet Address
	JSTOR (Journal Storage Archive), includes population titles http://www.jstor.org/
Organizations	**Australian Bureau of Statistics** http://www.abs.gov.au/ **Eurostat (Statistical Office of the European Union)** http://europa.eu.int/comm/eurostat/ **INED (Institut national d'études démographiques)** http://www.ined.fr/englishversion/index.html **Statistics Canada** http://www.statcan.ca/start.html **UK National Statistics** http://www.statistics.gov.uk/ **US Census Bureau** http://www.census.gov/

Note: the location of information on the Internet is subject to change; to find material that has moved, search from the home page of the organization concerned.

EXERCISES

1. Why is the demographic transition likely to be a misleading point of reference in explaining contemporary population changes at the regional and local levels?
2. Why is the demographic 'balancing equation' so named?
3. Distinguish between demographic rates and ratios.
4. Why are demographic rates often used in preference to figures on total numbers of births and deaths?
5. Why is the mid-year population commonly used as the denominator for demographic rates?
6. Referring to the table in the following spreadsheet exercise on basic demographic rates, why did the United States have a higher crude death rate in 1999 than Sri Lanka?
7. What are 'crude' rates? Using your knowledge of the nature of crude rates, write down formulas, in words, for the crude marriage rate and the crude divorce rate. Explain any disadvantages of the two rates.
8. When was your country's last population census? Name a publication that describes the nature and content of your country's census.
9. Identify a catalogue, or catalogues, of statistical publications on population for your country. Name the publications, and/or give Internet addresses.
10. Make a list of the types of information available on your country from the most recent issue of the United Nations *Demographic Yearbook*.

SPREADSHEET EXERCISE 1: BASIC DEMOGRAPHIC RATES

Complete the following spreadsheet table for the United States and Sri Lanka. Use formulas in place of the figures in bold.

The notes below provide a step-by-step approach for beginners. The computing notes in Appendix C cover the main Excel procedures and are intended to provide further assistance as needed.

	A	B	C	D	E	F	G
1	**Table 1: Selected Statistics for the United States and Sri Lanka**						
2							
3				United	Sri	United	Sri
4				States	Lanka	States	Lanka
5				1980	1980	1999	1999
6							
7	Mid-year population ('000)			227660	14740	272500	19000
8	Live births ('000)			3598	407	4088	361
9	Deaths ('000)			1985	89	2453	133
10							
11	Crude birth rate			**15.8**	**27.6**	**15.0**	**19.0**
12	Crude death rate			**8.7**	**6.0**	**9.0**	**7.0**
13	Rate of natural increase			**7.1**	**21.6**	**6.0**	**12.0**
14							
15							

Typing table headings

1. Click the mouse on the cell where you want the title to start, then type normally and press Enter when you finish the line.
2. Next, click the mouse on the cell where you want to type the first row heading and proceed in the same way. To begin the second row heading, either click the mouse on the required cell, or use the arrow keys to position the cursor.
3. To type the first column heading, type 'United' in one cell, move down to the next cell and type 'States' (alternatively, to fit two lines in one cell, use Format Cells Alignment Wrap Text).
4. If you make a mistake and wish to start again, position the cursor on the required cell (using the mouse or arrow keys) and simply retype.
5. To make minor alterations to typing see Section 3.2 of Appendix C.

Typing the data

6. Click the mouse on the cell where you want to type the first number. Type the number, then use the mouse or the arrow keys to go to the next cell.

7. It is not necessary to press Enter after typing each cell (except when you have been editing the cell).

Entering the formulas

8. Position the cursor on the cell for the United States CBR.
9. Type an = sign to start the formula.
10. Click on the cell for United States live births (the cell ID will appear in the formula at the top of the screen), then type / for division.
11. Click on the cell for United States mid-year population (the cell ID will appear in the formula at the top of the screen), then type * for multiplication.
12. Type 1000 and press Enter to finish the formula.
13. Type the formulas for the CDR and the RNI in the same way.
14. To obtain the figures for Sri Lanka, copy the formulas from the column for the United States:
15. Highlight the cells containing the formulas for the United States (i.e. click the mouse on the cell with the CBR then, holding the left mouse button down, highlight the other two cells).
16. Click the Copy button (or choose Edit Copy from the Menu) – dashes will appear around the cells.
17. Click on the cell that is to contain the CBR for Sri Lanka and press Enter. The formulas will be copied across and the answers will appear.
18. To change the number of decimal places displayed, highlight the cells containing the numbers you wish to change, choose Format Cells Number and either select the option required, or type 0.0 in the menu box to display numbers rounded to one decimal place. Alternatively use the Increase Decimal Places or Decrease Decimal Places buttons on the Formatting toolbar (see Section 2.5 of Appendix C).
19. Print a copy of the finished table and save the completed file.

Review of Excel procedures

- Entering text (Appendix C, Section 2.1)
- Entering values (Appendix C, Section 2.4)
- Entering formulas (Appendix C, Section 2.6)
- Copying formulas (Appendix C, Section 2.8)
- To become familiar with the location of Excel's help files, click Help. There are numerous help files answering specific questions.

2 Population growth and decline

Adam and Eve in 5610 years might have, by the ordinary proportion of Procreation, begotten more people, then are now probably upon the face of the earth.

(Graunt 1662: 14)

OUTLINE

2.1 Concepts of growth
2.2 Analysing growth
2.3 Geometric growth
2.4 Exponential growth
2.5 Growth and replacement
2.6 Conclusion
Study resources

LEARNING OBJECTIVES

To understand:

- the main concepts used in the study of population growth
- basic approaches to the study of population growth
- applications of the geometric growth formula
- applications of the exponential growth formula
- the nature of processes of population turnover and replacement

COMPUTER APPLICATIONS

Excel modules: Population Clocks.xls (Box 2.2)
Growth.xls (Box 2.3)

Spreadsheet exercise 2: Geometric growth rates (See Study resources)

The methods of analysing changes in population numbers are applicable to all scales of demographic investigation, from local to global. Applying them, and interpreting the results, calls for some acquaintance with the concepts upon which they are founded, such as geometric and exponential change. Understanding the concepts also facilitates informed reading of the literature on population trends, including authors' arguments about the nature of world population growth and its probable future.

In the twentieth century, dire warnings and apocalyptic forecasts for the planet followed from the assumption that a particular pattern of increase, such as exponential growth, would persist in the future. The twentieth century also brought forth national population decline, or at least a fear of it, as a recurring theme in some more developed countries. The continuation of low birth rates seems destined to give decline greater prominence during the twenty-first century. Assumptions about constant rates leading to extinction, however, will be as untenable as those about constant rates creating ever burgeoning numbers.

This chapter begins with a summary of the principal concepts of growth. Their main uses are in the everyday analysis of population change at local, regional and national scales. After introducing the concepts, much of the chapter is concerned with the principal measures of growth and decline, together with their applications.

2.1 Concepts of growth

Since the late eighteenth century, various concepts have had their heyday in the writings of the prophets of population. Each has different implications for the nature of the relationship between population size, resource use and environmental impacts. The concepts are mathematically based and authors have employed them to make predictions through extrapolating past developments. This assumes that the concepts actually have theoretical significance: if past trends in population growth conformed to a mathematical principle there would be solid justification for making predictions on the basis of that principle. Unfortunately, contemporary population growth shows no such conformity. The ensuing discussion briefly introduces the best-known mathematical patterns of growth: arithmetic, geometric, exponential and logistic. Particular authors have seen all but the first of these as representing an underlying principle of population growth. The computer modules for this chapter – *Population Clocks.xls* and *Growth.xls* – illustrate the divergence resulting from different types and rates of growth. *Growth.xls* also calculates population sizes and rates from entered data, while *Population Clocks.xls* – which displays functioning population clocks – provides the option of viewing long-run outcomes in accelerated time.

Arithmetic growth

A population growing arithmetically would increase by a constant number of people in each period. If a population of 5000 grows by 100 annually, its size over successive years will be: 5100, 5200, 5300, . . . Since the annual increases are the same, these numbers form an arithmetic progression or series; the size of the population in the example is also increasing at a constant *arithmetic rate* of 2 per cent (100/5000 = 0.02 or 2 per cent). Arithmetic growth is analogous to 'simple interest', whereby interest is paid only on the initial sum deposited, the principal, rather than on accumulating savings. Five per cent simple interest on $100 merely returns a constant $5 interest every year.

Arithmetic change produces a linear trend in population growth – following a straight line rather than a curve. This is illustrated in Figure 2.1 (panel 1) where the

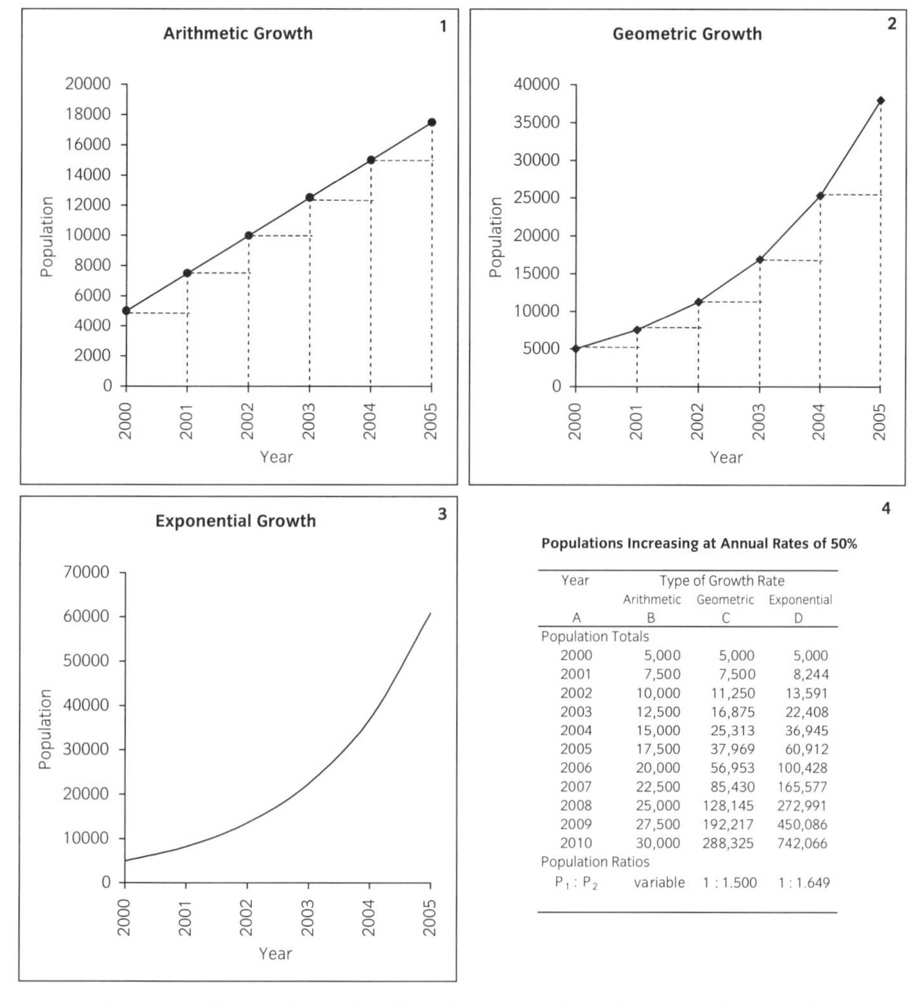

Populations Increasing at Annual Rates of 50%

Year	Type of Growth Rate		
	Arithmetic	Geometric	Exponential
A	B	C	D
Population Totals			
2000	5,000	5,000	5,000
2001	7,500	7,500	8,244
2002	10,000	11,250	13,591
2003	12,500	16,875	22,408
2004	15,000	25,313	36,945
2005	17,500	37,969	60,912
2006	20,000	56,953	100,428
2007	22,500	85,430	165,577
2008	25,000	128,145	272,991
2009	27,500	192,217	450,086
2010	30,000	288,325	742,066
Population Ratios			
$P_1 : P_2$	variable	1 : 1.500	1 : 1.649

Figure 2.1 Comparison of arithmetic, geometric and exponential growth.

dotted lines show the constant increases added at the end of every year: the increments or steps are of equal size. Figure 2.1 depicts unusually high annual growth rates of 50 per cent in order to illustrate contrasting outcomes from arithmetic, geometric and exponential growth in a short time span. Normally, these differences are evident only in the long term, and may be minimal over a five year period.

Real examples of arithmetic population growth are uncommon and necessarily limited to settings where administrators have the option of establishing targets to increase or decrease numbers by a constant amount. For instance during a phase of expansion, budgets could support a linear increase in numbers – such as of employees in an organization, enrolments at a university or soldiers in an army. Despite its limitations for describing actual population changes, arithmetic growth is the basis for a widely used measure in demography, namely *average annual increase*. It was also an important concept in Malthus's (1798) population theory (Figure 2.2).

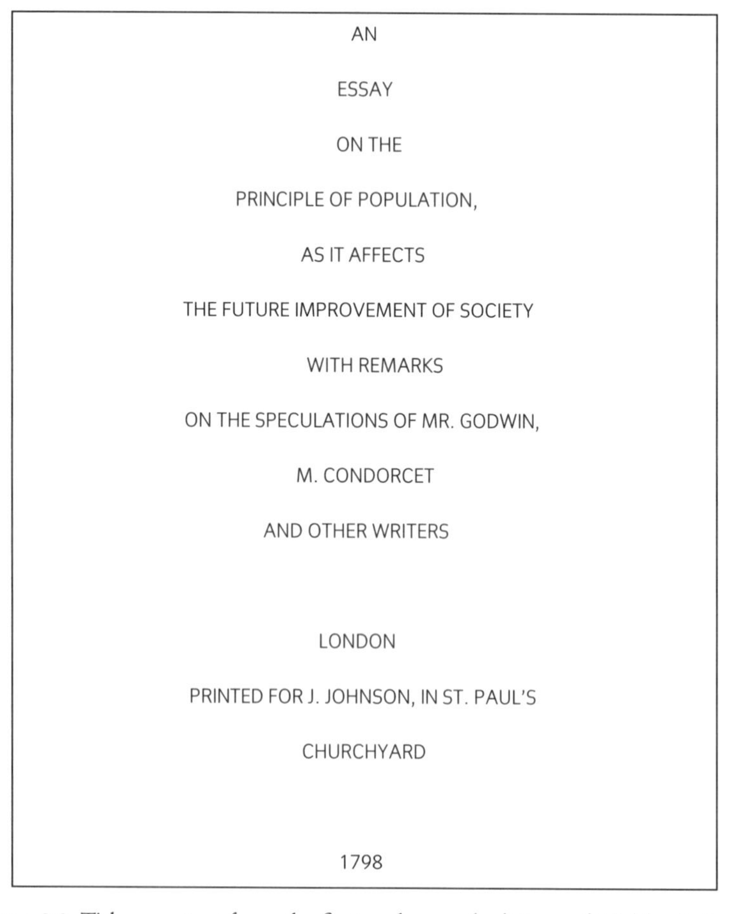
AN

ESSAY

ON THE

PRINCIPLE OF POPULATION,

AS IT AFFECTS

THE FUTURE IMPROVEMENT OF SOCIETY

WITH REMARKS

ON THE SPECULATIONS OF MR. GODWIN,

M. CONDORCET

AND OTHER WRITERS

LONDON

PRINTED FOR J. JOHNSON, IN ST. PAUL'S

CHURCHYARD

1798

Figure 2.2 Title page text from the first and second editions of Malthus's Essay.

First edition (published anonymously 1798).
Second edition 1803.
Source: Castles (1998: 4).

AN

ESSAY

ON THE

PRINCIPLE OF POPULATION;

A VIEW ON ITS PAST AND PRESENT EFFECTS

ON

HUMAN HAPPINESS;

WITH AN INQUIRY INTO OUR PROSPECTS RESPECTING THE FUTURE REMOVAL

OR MITIGATION OF THE EVILS WHICH IT OCCASSIONS.

A NEW EDITION, VERY MUCH ENLARGED

BY T.R. MALTHUS, A.M

FELLOW OF JESUS COLLEGE, CAMBRIDGE

LONDON

PRINTED FOR J. JOHNSON, IN ST. PAUL'S CHURCHYARD,

BY T. BENSLEY, HOLT COURT, FLEET STREET,

1803

Figure 2.2 *continued*

Thomas Robert Malthus (1766–1834), an English clergyman, and professor of history and political economy, is the best known of demography's pioneers. His *Essay on the Principle of Population* (1798) is one of the most important and influential books ever written on the subject of human numbers (Grebenik 1989: 1); it was the first to stimulate interest in the relationship between population and resources (Keyfitz 1972). Malthus wrote to oppose prevailing ideas about the benefits of population growth. His 'gloomy thesis' was that population growth threatened prosperity because it inevitably outran increases in food supplies. He reasoned that, at best, agricultural production grew arithmetically, by the same amount every 25 years, while population grew geometrically, doubling every 25 years (Malthus 1798: 16). Thus the implications of his 'principle of population' arose from the differences between arithmetic and geometric growth.

Geometric growth

Geometric growth quickly leads to greater numbers. Comparison of panels 1 and 2 in Figure 2.1 indicates that, whereas arithmetic growth entails constant increments, geometric growth entails ever larger increments. In geometric growth, population increments become larger because increases are self-reinforcing. This is illustrated in developing countries with high birth rates, where expanding generations of children eventually become expanding new generations of parents who, in turn, beget even larger generations of children.

Geometric population growth is analogous to the growth of a bank balance receiving compound interest. According to this, the interest is calculated each year with reference to the principal plus previous interest payments, thereby yielding a far greater return over time than simple interest. The geometric growth rate in demography is calculated using the 'compound interest formula'.

Under arithmetic growth, successive population totals differ from one another by a constant amount. Under geometric growth they differ by a constant ratio (Table 2.1). In other words, the population totals for successive years form a geometric progression in which the ratio of adjacent totals remains constant. In panel 4 of Figure 2.1, which presents the data for the charts, this ratio is 1:1.50. In the example, successive populations are 50 per cent larger, because the annual growth rate is 50 per cent.

The main problem with describing population growth as a geometric progression is that actual populations seldom increase at constant rates. Populations typically have different (geometric) growth rates from year to year. Nevertheless, the misleading assumption of population growth as tending towards a geometric

Table 2.1 Summary of concepts of population growth

Type of growth	Description of trend	Growth rates	Absolute increments	Ratio of adjacent populations
Arithmetic	growth through constant increments at constant intervals	constant	constant	changing
Geometric	growth compounding at constant intervals	constant	changing	constant
Exponential	growth compounding continuously	constant	changing	constant
Logistic	growth rates changing in relation to population size	changing	changing	changing

progression is important in Malthus's 'principle of population'. The same assumption is present also in some contemporary writing – where recent growth rates are held constant to show extreme outcomes. The imbalance resulting from the assumed arithmetic expansion of food production and geometric expansion of human numbers led Malthus to conclude that populations tend to outgrow their resources. His 'principle of population' was that human numbers ultimately press upon the available means of subsistence.

The notion of diverging trends from arithmetic and geometric growth proved untenable as a description of the course of change in the relationship between population and food supplies. Population growth does not conform to a geometric progression, nor do populations actually grow geometrically, adding increments at constant intervals. In small populations, changes occur at irregular intervals depending on the timing of births, deaths and migrations. In larger populations, changes may occur almost continuously – not just at yearly intervals. Recognition of this led to a focus on exponential growth, which more accurately describes the continuous and cumulative nature of population growth.

Exponential growth

Exponential growth refers to the situation where growth compounds continuously – at every instant of time. Accordingly, it is sometimes called 'instantaneous growth' (Krebs 1978). Geometric growth is a special case of exponential growth (Pressat 1985: 74), because compounding occurs at intervals much longer than an instant. The shorter the interval over which increments occur, the faster the population increases – just as the balance in a bank account with daily interest grows more quickly than one with yearly interest.

Figure 2.1 (panel 3) shows that exponential growth produces a smooth curve with no steps between increments, because change is continuous. By 2010, exponential growth at an annual rate of 50 per cent results in a total of nearly three-quarters of a million, compared with 288 325 from geometric growth – compounding annually – and 30 000 from arithmetic growth. As in geometric growth, the ratio between adjacent populations is constant: 5000 : 8244, and 8244 : 13 591, for example, are equivalent to ratios of 1 : 1.649. In other words, numbers increase by about 65 per cent annually. Continuous compounding explains why a 50 per cent annual growth rate actually creates a 65 per cent increase (see Box 2.1).

Exponential growth was prominent in one of the late twentieth century statements of a mathematical principle of population growth, namely the book *The Limits to Growth* (Meadows et al. 1974, first published 1972) and its sequel *Beyond the Limits* (Meadows et al. 1992). Whereas Malthus believed that populations tend to grow geometrically, the authors of the *Limits to Growth* argued that the world's population is growing exponentially, and consumption of resources is also increasing exponentially (Meadows et al. 1974: 25). They went further in saying that the world's population is really increasing 'super-exponentially',

BOX 2.1 **A paradox of growth**

A seeming paradox is that a 100 per cent per annum exponential growth rate produces more than a 100 per cent increase in a year. The reason is that growth would amount to 100 per cent over a single interval of a year but, with exponential growth, there is an infinite number of intervals over which growth is compounding. For example, if a 100 per cent annual increase on an initial population of 100 was split between two intervals of six months, the rate for each interval would be 50 per cent ($r = 0.5$). Using the geometric growth formula, the population would be 150 (100×1.5) at mid-year and 225 (100×1.5^2) at the end of the year. Similarly, calculating the increments quarterly (i.e. 25 per cent growth per quarter) the total population would be 244 (100×1.25^4) at the end of the year. Thus the greater the number of points at which increments are added, the greater the growth. Exponential growth yields higher outcomes for the same reasons that daily interest payments cause a bank balance to grow faster than monthly or yearly payments (see Krebs 1978).

by which they meant that far from being constant, growth rates have increased through time such that 'the population curve is rising even faster than it would if growth were strictly exponential' (Meadows et al. 1974: 34).

To explain the implications of exponential growth for population numbers the authors quoted the following story (Meadows et al. 1974: 29; 1992: 18):

> There is an old Persian legend about a clever courtier who presented a beautiful chessboard to his king and requested that the king give him in return 1 grain of rice for the first square on the board, 2 grains of rice for the second square, 4 grains for the third, and so forth. The king readily agreed and ordered rice to be brought from his stores. By the fortieth square a million million rice grains had to be brought from the storerooms. The king's entire rice supply was exhausted long before he reached the sixty-fourth square. Exponential increase is deceptive because it generates immense numbers very quickly.

This story, however, is misleading when applied to population growth. It implies that numbers tend to keep doubling at short intervals indefinitely. Population growth does not occur in this way, since the rates seldom remain constant for long. Also, national population doublings normally take decades rather than years. The world's population growth rate has never exceeded 2.1 per cent (doubling time 33 years), and the growth rate has been falling since the mid-1960s (Figure 2.3); in 2001 it was estimated to be 1.3 per cent (Population Reference Bureau 2001b).

Calculating exponential growth rates gives due recognition to the continuous nature of growth in large populations in a period of time, but it is unrealistic to assume that the rates form an exponential series. In the United States and Britain, the growth rates of their populations have varied greatly over time (Figure 2.3): there is no evidence to support the notion that predictions of their future populations can assume a constant exponential growth rate. Vastly different predictions result

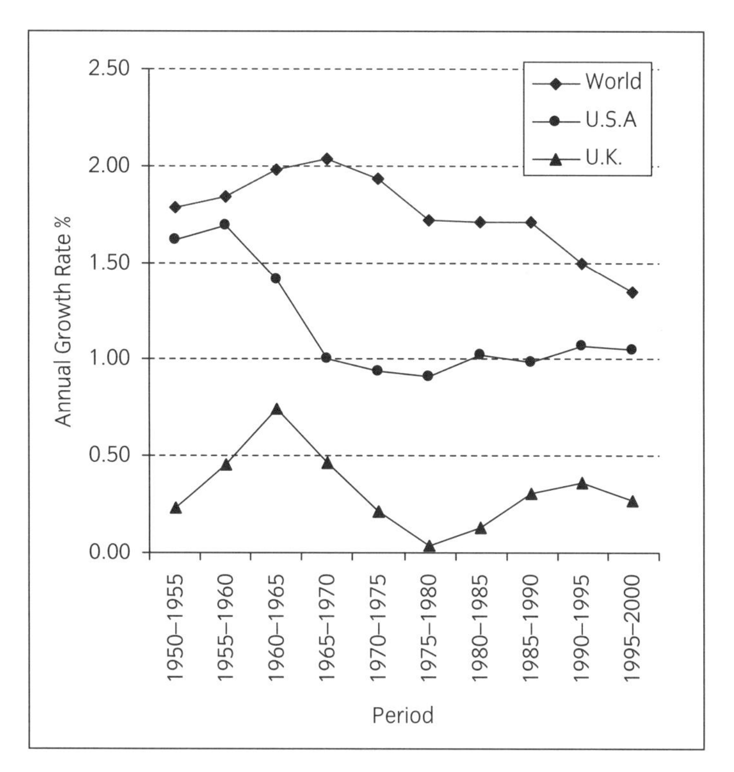

Figure 2.3 Population growth rates, 1950–2000: World, USA and UK.
Source: United Nations 2001. *World Population Prospects: the 2000 Revision* (CD).

according to the date and the rate selected as the starting point. Thus the main problem in describing population growth as an exponential series is the assumption that the rates remain constant through time.

Constant rates of decline raise the same problems as constant rates of growth. A global population extinction scenario, based on contemporary negative growth rates for Germany, shows the human race disappearing by 2400, and the population of the more developed countries vanishing about 150 years earlier. This scenario requires a constant exponential growth rate of –2 per cent to emerge in the twenty-first century (Bourgeois-Pichat 1989: 71ff). Yet declining populations may be expected to experience varying demographic rates as a result of unpredictable shifts in fertility and migration, together with variations in the size of *birth cohorts* – groups born in the same period – reaching different stages of life through time.

Despite these criticisms, constant exponential growth is consistent with constant birth and death rates; hence, the assumption that population growth forms an exponential series has had important applications in demography, especially in the development of stable population models. These have been vital in many areas, such as clarifying the role of fertility and mortality change in the evolution of age structures, projecting long-run outcomes of processes and enabling demographic estimation from incomplete or defective data (see Chapter 9).

The logistic curve

Recognition that populations cannot grow indefinitely has led to interest in other mathematical approaches to representing population growth and defining its upper limit. One of the best known is the logistic curve, first discovered in 1838, but rediscovered and popularized by an American geneticist, Raymond Pearl, and his colleagues in the 1920s (Shryock and Siegel 1973: 382).

The curve is based on Pearl's observations of the growth in the numbers of fruit flies (*Drosophila melanogaster*) under experimental conditions. Pearl later sought to demonstrate that the growth of the human population follows the logistic curve. Pearl proclaimed the logistic curve to be the universal law of population growth (Krebs 1978: 191). Application of the logistic curve to predict the United States population from 1920 to 1940 produced good results, but later events demonstrated that the United States was not simply following a logistic trend (Plane and Rogerson, 1994: 64). Even updating Pearl's work and using data to 1960, as in Figure 2.4, still produces low forecasts of population growth. The population of the United States in 2000 was 275.6 million (Population Reference Bureau 2000), a total far higher than the upper limit of the logistic curve (250.5 million).

Nevertheless the logistic curve corresponds with textbook diagrams depicting the historical course of population growth during the demographic transition

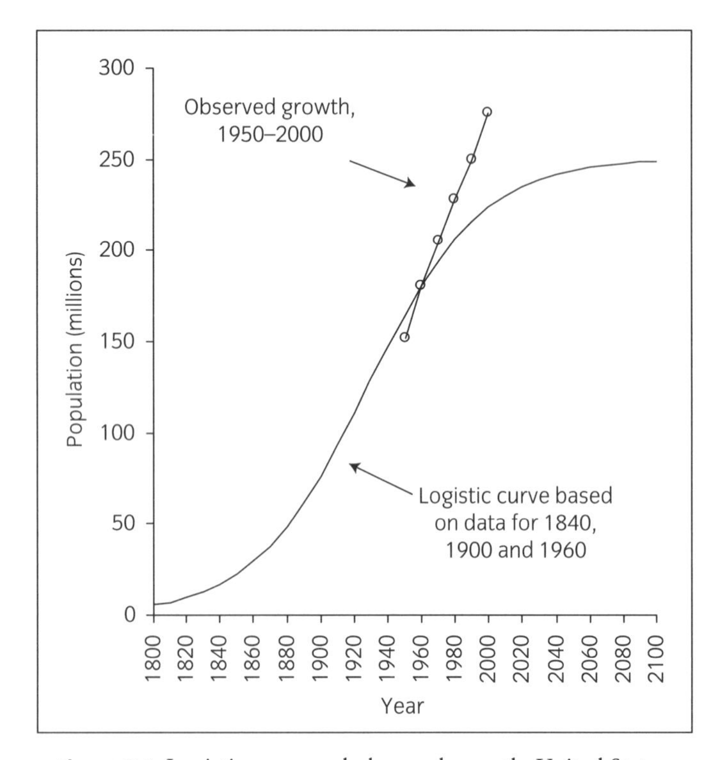

Figure 2.4 Logistic curve and observed growth, United States.

Data source: after Jain (1979: 92–4).

(Plane and Rogerson 1994: 62–3). The curve denotes an initial period of slow growth, followed by a phase of rapid growth – due to falling death rates – and finally a plateau in population numbers arising from low birth rates. The 'S-shaped' curve is also consistent with certain patterns of the diffusion through time in the numbers adopting innovations – with small increases at first, then larger increases and finally a tapering off. In population studies, potential topics for study in comparison with an S curve include changes through time in the proportion of people aware of sexually transmitted diseases, the proportion of newly married couples using contraception, and the proportion of households owning consumer goods, such as microwave ovens or computers.

Overall, an S-shaped trend (Figure 2.4) is a more realistic depiction of long-run national or global population growth than straight lines or exponential curves. It recognizes that numbers cannot increase indefinitely and that social goals and environmental constraints might be expected to bring a slowing of growth. Its main disadvantage is that, like other mathematical concepts, it cannot predict the future, because it is not founded on an explanation of changes. Moreover, despite the greater sophistication of the logistic curve in recognizing changing rates of growth

BOX 2.2 **Population clocks module (*Population Clocks.xls*)**

Population clocks, on office walls and Web pages, usually display changes in population numbers, for countries or the world, occurring in real time. This module can function in the same way, but it also provides opportunities to:

- accelerate time, to reveal long range outcomes;
- set the initial population to any total;
- change growth rates, showing the effects of lower or higher rates of change;
- compare populations growing at different exponential rates.

The module has an opening display, with menu items and instructions, and two clock displays.

Clock 1 A single clock with controls for changing the initial population, the rate of growth and the clock speed per second (in seconds, minutes, hours, days, weeks or years). An animated bar graph compares the initial and current population totals, and tables display current features of population change.

Clock 2 A dual clock for comparing two populations growing at different rates. An animated pie graph shows their changing shares of the combined totals.

Instructions

To use the module, insert the CD in the drive, start Excel and open the file *Population Clocks.xls*. Enable macros when prompted.

continued

BOX 2.2 continued

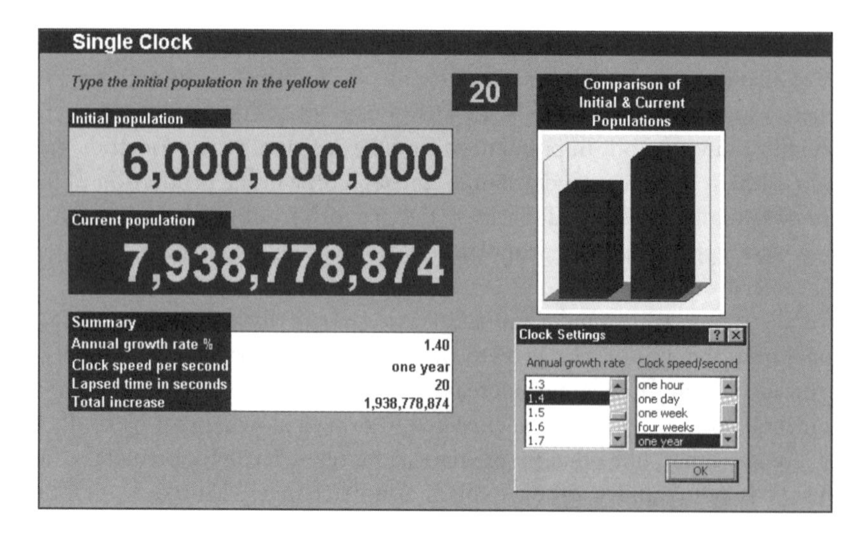

- Click **Single Clock** or **Dual Clock** to open a display.
- Enter the starting population(s) in the yellow cell(s).
- Click **Menu** to set the growth rate(s) and clock speed.
- Click **Reset** to set the starting time to zero.
- Click **Start** to run the clock.
- To stop the clock, press the **Esc** key and click **End** on the menu.
- Click **Home** to return to the opening screen.

See Appendix B for further advice on using the Excel modules.

Illustrative applications

Clock 1 (Single): World population growth

- Set the population and growth rate to the current figures (do not type commas when entering numbers).
- Set the clock speed to real time (1 second per second) and observe how the world's numbers are changing.
- Stop the clock by pressing the Esc key and clicking End on the menu box that appears. Then click reset to set the lapsed time to zero.
- Set the clock speed to 1 year per second and run the clock to show future numbers. Observe how long the population takes to double.
- Repeat the previous two steps to compare the doubling time for other growth rates, such as 1 per cent.

continued

BOX 2.2 continued

Clock 1 (Single): National population growth

- Use the clock to illustrate current national growth trends and alternative scenarios with lower or higher growth rates. Run at 1 year per second to demonstrate long-run outcomes.

Clock 1 (Single): Urban population growth

- Set the total population and growth rate to that of a rapidly growing major city and run the clock to obtain projections of future population totals. Discuss whether it is plausible to assume constant growth rates.

Clock 2 (Dual): Differential growth of ethnic groups

- When ethnic groups grow at different rates, be it tribes in African countries, indigenous and non-indigenous peoples in Fiji, or ethnic minority groups generally, the changing balance of numbers can be a source of concern, tension or conflict. To illustrate outcomes of different growth rates, set the total population to 1 million, the first growth rate to 3 per cent and the second growth rate to 1 per cent. Set the clock speed to 1 year per second, and observe the changes in the two groups' shares of the total population. The clock assumes that the two populations are equally represented at the start.

Clock 2 (Dual): Differential growth of regions

- Use a similar approach to compare outcomes of differential regional growth, such as rural versus urban or growing versus declining regions.

over time, the other simpler concepts serve as the bases for the most commonly used methods of analysing population growth.

2.2 Analysing growth

Population size

Population numbers can evoke mental images of densely packed masses of human beings. One of the most vivid is Paul Ehrlich's description of a taxi-ride in Delhi (Ehrlich 1970):

I have understood the population explosion intellectually for a long time. I came to understand it emotionally one stinking hot night in Delhi a couple of years ago. My wife and daughter and I were returning to our hotel in an ancient taxi. The seats were hopping with fleas. The only functional gear was third. As we crawled through the city, we entered a crowded slum area. The temperature was well over 100, and the air was a haze of dust and smoke. The streets seemed alive with people. People eating, people washing, people sleeping. People visiting, arguing, and screaming. People thrusting their hands through the taxi window, begging. People defecating and urinating. People clinging to buses. People herding animals. People, people, people, people. As we moved through the mob, hand horn squawking, the dust, noise, heat, and cooking fires gave the scene a hellish aspect. Would we ever get to our hotel? All three of us were, frankly, frightened. It seemed that anything could happen – but, of course, nothing did. Old India hands will laugh at our reaction. We were just some over-privileged tourists, unaccustomed to the sights and sounds of India. Perhaps, but since that night I've known the *feel* of overpopulation.

Despite this image of a crowded world, the entire global population of 6 billion in 1999 would 'fit' into a megalopolis, or super-city, the size of the American state of Texas (land area =678 358 km^2, 261 914 square miles), which is larger than France (551 494 km$^{2)}$ but smaller than Pakistan (796 094 km^2). In this scenario, the world's population would be living at a density of about 9000 persons per square kilometre, a density a little below that of New York City as a whole in 1990 (9154 persons per square kilometre) and far less than that of the borough of Manhattan, which has a population density of 26 000 per square kilometre. The scenario obviously takes no account of the nature of the land and standards of living, nor of the vastly greater territorial demands of individuals in terms of their consumption of resources and impact on the environment. Yet envisioning the global population in a 'Texas City' does serve to show, vis-à-vis Ehrlich's anecdote, that population numbers are open to contrasting interpretations, depending partly how phlegmatic or emotive the writer's stance.

Consideration of population size is central to the debate about the future of the world's population, but it is also an essential aspect of the day-to-day analysis of population change. For businesses, the size of the market, or potential market, is vital, while in government administration there is ongoing concern for the numbers requiring physical infrastructure and services. Total population numbers are often the overriding consideration in planning.

Basic measures of growth and decline

Beyond population size, basic measures of change are frequently needed to describe developments. Table 2.2 provides a summary of these measures, together with examples based on World Bank projections for Mexico. The chief advantages of the measures are the ease with which they can be calculated and understood. As discussed in Chapter 1, P_0 represents the population at the start of a period and P_n the population at its end, that is after n intervals of time (e.g. days,

Table 2.2 Basic measures of population growth and decline

Formulas	Examples
Definitions	*Mexico*
P_0 = population at the start, e.g. year zero	P_0 = 98 787 000 (Year 2000)
P_n = population at the end, e.g. after n years	P_n = 162 356 000 (Year 2050) n = 50 years (mid-2000 to mid-2050)
n = number of intervals (e.g. years) between P_0 and P_n	
1 *Absolute change*	
$P_n - P_0$	162 356 000 − 98 787 000 = 63 569 000
2 *Percentage change*	
$\left(\frac{P_n - P_0}{P_0}\right) \times 100$	63 569 000/98 787 000 × 100 = 64.35%
3 *Average annual increase*	
$\frac{P_n - P_0}{n}$	63 569 000/50 = 1 271 380
4 *Arithmetic growth rate*	
$\left(\frac{P_n - P_0}{n}\right) \div P_0 \times 100$	1 271 380/98 787 000 × 100 = 1.29%

Data source: World Bank (1994: 343)

months or, more usually, years). In the set of examples for Mexico, n is set at 50 years.

1. *Absolute change* is simply the difference between the size of the population at the start and end of a period of time. This represents the *net growth* of the population. It is not a measure to be ignored because of its simplicity: sometimes the main implications of population changes arise from the extra numbers to be fed, housed, educated, employed or supported. Table 2.2 presents an example of absolute change in numbers in Mexico, projected for the first half of the twenty-first century. The calculation reveals that, in the 50 years, Mexico could gain the equivalent of the population of the United Kingdom or France in the year 2000.
2. *Percentage change* (absolute change/initial population × 100) provides a means of comparing developments in different populations or different periods of time

by gauging the amount of change relative to initial numbers. Percentage change is always based on numbers at the start of the period. It is a measure frequently used in census atlases and other maps of intercensal population change (see Chapter 10), because percentages are comparable and intelligible to a wide readership.

3. *Average annual increase (or decrease)* is based on the concept of arithmetic change. It assumes constant annual gains or losses – ignoring the notion that growth is self-reinforcing. The measure provides a rough-and-ready indication of the numbers being added to a population, but the longer the period of observation, or the greater the amount of change, the more likely the average is misleading. Average annual change is sometimes used to provide a quick estimate or projection of the size of a population in a particular year, but the same caveats apply. Estimates of average annual increase really have claims to accuracy only when the interval is short or where little change has occurred. The Mexico example shows an average annual increase from 2000 to 2050 about equivalent to adding, every year, the population of Hawaii.
4. The *arithmetic growth rate* is included in Table 2.2 mainly for comparison with other rates: it is seldom used in demographic calculations. The rate compares the annual average increase with the size of the initial population. The result is multiplied by 100, since population growth rates are usually expressed as rates per cent.

In calculating measures of population change there are several potential sources of error arising from the nature of the data. Firstly, boundary changes are common and may call for many adjustments to data for geographical analyses of population change within cities or across regions. For example, incorporation of new territory, such as a neighbouring town, can greatly change the population of a city. The main strategy in such situations is to use constant boundaries, typically the most recent, for establishing the size of the population at the start and the end of the period.

Secondly, changes in the dates of censuses and population estimates affect the length of the interval, which should be expressed as accurately as possible (Pollard et al. 1990: 21). Thus if the first estimate is for 30th June 2000 and the second for 10th August 2001, the length of the interval (n) is 1 year plus 41 days, or 1.11 years.

Thirdly, definition changes can have dramatic effects that may be difficult to resolve. Statistics on the growth rate of the labour force will vary greatly if the first population includes the unemployed, part-time workers and unpaid workers while the second population includes only people in full-time paid employment. Similarly, the apparent growth of ethnic minority groups will vary according to whether the source data are based on self-identification, degree of descent, birthplace, language or religion, or a combination of these. Inclusion or exclusion of children of mixed descent can also affect results considerably. Adjustments may be essential to achieve a consistent definition of the population at the start and end.

2.3 Geometric growth

Beyond the basics, more accurate measures of population growth avoid the limiting assumption of constant annual increments or decrements and facilitate better estimates of intermediate values. Measures of population change based on the geometric growth rate are widely used, as are those based on the exponential growth rate. The former are slightly easier to understand and calculate yet yield similar results where the time interval is short or the rate is within the typical range for national populations (from about –1 to 4 per cent). Also, the collection of official statistics on population size is inevitably periodic – at regular intervals such as census dates – and the intervening pattern of change need not be continuous and cumulative, especially where migration contributes substantially. Measuring growth somewhat crudely, as a geometric process, acknowledges the nature of the source data and the imprecision of estimates between dates.

Rates of change are not the only needed measures from the geometric growth formula. Analyses of population trends often require other information about the past or future size of the population, the time required to reach a given size and the population's doubling time. Table 2.3 therefore presents variations of the geometric growth formula for calculating the initial and end-of-period populations, together with time intervals. As in Table 2.2, definitions are at the top, together with the values employed in the accompanying examples. The footnotes to the table provide brief explanations of the derivation of each formula. Some of the formulas need to be calculated using logarithms; the table shows the logarithmic form of all of the formulas for comparison, as well as for use where a scientific calculator or computer is not available. Results of calculations vary according to the number of decimal places retained in intermediate steps. The examples in Table 2.3 provide sufficient detail to produce results matching the data for the United States at the top of the table.

1. *End of period population* Formula 1 in Table 2.3 has applications in estimating intercensal numbers, or projecting the future population, assuming that the growth rate remains constant. This assumption becomes increasingly untenable as n increases, because growth rates typically vary through time. In the table, example 1 estimates population numbers in 2010 from information on the initial population in 1990 together with the rate of growth and the time period. The growth rate per cent must be divided by 100. Changing the value of n would permit estimates of the size of the United States population in other years, such as 1999 ($n = 9$) and 2009 ($n = 19$).
2. *Initial population* The second formula works backwards in time to calculate what past population numbers would have been if numbers had been growing at a particular rate. The assumption of constant growth rates is too constraining to allow back-projection to proceed very far. Thus, reconstructing the past in this way is best confined to shorter intervals, all the more so in populations

Table 2.3 Measures of population change from the geometric growth rate

Formulas	Examples
Definitions	*United States*
P_0 = population at the start	$P_0 = 250.4$ (millions, mid-1990)
P_n = population at the end	$P_n = 297.2$ (millions, mid 2010)
n = number of intervals between P_0 and P_n	$n = 20$ years
r = annual growth rate	$r = 0.86041\%$
1 *End of period population*	
$P_n = P_0\,(1+r)^n$	$P_n = 250.4 \times (1.008\,604\,1)^{20}$ $= 297.2$
or	or
$\log P_n = \log P_0 + \log(1+r) \times n$	$\log P_n = 2.398\,63 + 0.074\,41$ $\therefore\ P_n = 297.2$
2 *Initial population*	
$P_0 = \dfrac{P_n}{(1+r)^n}$	$P_0 = 297.2/(1.008\,604\,1)^{20}$ $= 250.4$
or	or
$\log P_0 = \log P_n - \log(1+r) \times n$	$\log P_0 = 2.473\,05 - 0.074\,41$ $\therefore\ P_0 = 250.4$
3 *Geometric growth rate*	
$r = \sqrt[n]{\dfrac{P_n}{P_0}} - 1$	$r = (297.2/250.4)^{(1/20)} - 1$ $= 0.008\,604\,1$ or $0.860\,41\%$
or	or
$\log(1+r) = \dfrac{\log\left(\dfrac{P_n}{P_0}\right)}{n}$	$\log(1+r) = 0.003\,720\,724$ $\therefore\ r = 0.860\,41\%$
4 *Interval between two populations*	
$n = \dfrac{\log\left(\dfrac{P_n}{P_0}\right)}{\log(1+r)}$	$n = 0.074\,414\,481/0.003\,720\,724$ $= 20$ years
5 *Doubling time*	
$n = \dfrac{\log 2}{\log(1+r)}$	$n = 0.301\,03/0.003\,720\,724$ $= 80.9$ years

Derivation of formulas:
1. End of period population: 'compound interest' formula.
2. Initial population: divide both sides of formula 1 by $(1+r)^n$.
3. Geometric growth rate: divide both sides of formula 1 by P_0 and make r the subject.
4. Interval: make n the subject of the logarithmic form of formula 3.
5. Doubling time: in formula 4, $(P_n/P_0) = 2$, since the population doubles.

where substantial growth derives from migration – the most variable and unpredictable of all components of population change. The example calculates the 1990 population of the United States from information on the projected population in 2010, the growth rate and the time interval. As before, using other values of n would provide estimates for other years. Also, expressing n as a fraction, such as 5.5 years, can yield estimates for points in time other than the mid-year, to which the figure for 2010 refers.

3. *Growth rate* The growth rate provides an informative basis for comparisons between countries, since it gives an impression of stage of development in terms of progress through the demographic transition. However, the contribution of migration to national population growth is not known accurately for some countries, and cross-national data on rates of natural increase are sometimes published instead of figures on overall growth rates.

   In contemporary national populations, the highest growth rates occur where the demographic transition is yet to be completed, but more developed countries receiving substantial flows of international migrants, such as the United States and Australia, also have rates higher than other 'post-transition' populations. New towns and suburbs have the most spectacular growth rates on account of low initial numbers, but their growth rates typically fall as the population increases. Canberra, the capital of Australia, had an annual growth rate that peaked at 14 per cent during the 1960s, because of substantial government investment in establishing the new city. Subsequently, Canberra's growth rate declined rapidly from this peak, mainly because the larger the population, the greater the increments needed to sustain a particular rate.

   In the United States example in Table 2.3, the growth rate calculation proceeds either by finding the 20th root of P_n/P_0, or by raising P_n/P_0 to the power of 1/20 (the reciprocal of the power). Appendix A, on basic maths, describes how to calculate powers and roots of numbers.

4. *Interval between two populations* This denotes the time taken to change from one specified total to another. Thus the example in the table addresses the question of how long it will take a population to reach a particular size. Applications arise in planning when seeking to anticipate the date that numbers will reach consequential levels. For instance, population size determines the rights of communities to self-government or separate political representation. The passing of size thresholds, such as a million or a billion, are also occasions for reflection and action. The world's population probably reached 6 billion around the end of 1999; formula 4 can provide a first estimate of when it will reach 7 billion. The example shows the derivation of n from the United States data.

   It is important to note that finding the length of the interval, n, is not identical to estimating the year in which a total was reached. To find the year, add n to the date to which P_0 refers – including fractions of a year as necessary. For example, if n equals 15.6 years and P_0 refers to a census held at mid-year 1990, the required date will be 1990.5 + 15.6 = 2006.1, that is the year 2006. Ignoring the census date would lead to the error of specifying 2005 as the year. A decimal fraction can potentially be used to specify a particular day, but such precision is usually spurious.

5. *Doubling time* After the growth rate, the doubling time is the most widely mentioned measure of national and global population growth, because it is readily comprehended. In 2000, the world population doubling time was 51 years, compared with 49 years in 1999 (Population Reference Bureau 1999: 2; 2000: 2). Transitional populations in Africa have most of the fastest contemporary doubling times of 20–23 years.

   Malthus (1798: 16) made effective use of doubling time to convey the implications of population growth:

   > In the United States of America, where the means of subsistence have been more ample, the manners of the people more pure, and consequently the checks to early marriages fewer, than in any of the modern states of Europe, the population has been found to double itself in twenty-five years.

   A well-known shortcut formula for calculating doubling times is to divide 70 by the growth rate per cent. Thus annual growth rates of 2 per cent and 1 per cent give doubling times of 35 years and 70 years respectively. This approximation, which assumes an exponential, rather than geometric, growth rate, is improved slightly by using 69 instead of 70. The approximation is based on the doubling time formula from the exponential growth rate, as discussed below.

2.4 Exponential growth

Exponential growth yields higher totals than geometric growth (given P_0, r and n), especially over longer periods of time or where the rates are high (Figure 2.1). This is because exponential growth is compounding at every moment, rather than at fixed intervals. Conversely, exponential growth rates are always lower than geometric rates from the same P_0, P_n and n.

Examples of exponential change are given in Table 2.4, where most of the formulas are expressed in natural logarithms (base e). Alternatively, the formulas may be expressed using common logarithms, but are less concise. Like π ($\pi = 3.14159$), e is a mathematical constant (e = 2.71828). As in the calculations based on geometric growth, each term (P_0, P_n, n or r) can be calculated from the other three.

Since the natural logarithm of 2 is 0.69315, the formula for the doubling time under exponential growth (Table 2.4, formula 5) is approximately $0.69/r$. Multiplying top and bottom by 100 gives $69/r$ per cent or, rounding to the nearest ten, $70/r$ per cent. Thus is derived the shortcut formula for the doubling time. Similarly,

Table 2.4 Measures of population change from the exponential growth rate

Formulas	Examples
Definitions	*Pakistan*
P_0 = population at the start	$P_0 = 112.4$ (millions, mid-1990)
P_n = population at the end	$P_n = 146.5$ (millions, mid-1999)
n = number of intervals between P_0 and P_n	$n = 9$ years
r = annual growth rate	$r = 2.94401\%$
ln = natural logarithm	
e is a constant (2.71828)	
1 *End of period population*	
$P_n = P_0 e^{rn}$	$P_n = 112.4 \times 1.30338$ $= 146.5$
or	or
$\ln P_n = \ln P_0 + rn$	$\ln P_n = 4.72206 + 0.26496$ $= 4.98702$ $\therefore P_n = 146.5$
2 *Initial population*	
$P_0 = \dfrac{P_n}{e^{rn}}$	$P_0 = 146.5/1.30338$ $= 112.4$
or	or
$\ln P_0 = \ln P_n - rn$	$\ln P_0 = 4.98702 - 0.26496$ $= 4.72206$ $\therefore P_0 = 112.4$
3 *Exponential growth rate*	
$r = \dfrac{\ln\left(\dfrac{P_n}{P_0}\right)}{n}$	$r = \ln(1.30338)/n$ $= 0.02944$ or 2.944%
4 *Interval between two populations*	
$n = \dfrac{\ln\left(\dfrac{P_n}{P_0}\right)}{r}$	$n = \ln(1.30338)/0.0294401$ $= 9$ years
5 *Doubling time*	
$n = \dfrac{\ln 2}{r}$	$n = 0.69315/0.0294401$ $= 23.5$ years

Data sources: World Bank (1994); Population Reference Bureau (1999).

the tripling time of a population is equal to the natural logarithm of 3 (1.098 61) divided by r, or approximately $110/r$ per cent.

The examples show that in the 1990s, Pakistan, one of the world's most populous countries, had a high growth rate of 2.9 per cent, with a doubling time of just 24 years. Changing the values of n or r in formulas 1 and 2 provides a straightforward approach to estimating and projecting the population. For instance, the 1996 population of Pakistan may be estimated by setting n to 6 in formula 1. Similarly, setting P_0 to the 1999 total and r to 2 per cent, would provide a projection at this lower rate of growth for any value of n.

Exponential growth rates are also employed in estimating past populations, but preferably only where there is evidence to support assumptions. An example of the misapplication of this technique is an estimate in 1959 of the number of people who had ever lived in the world (see Bourgeois-Pichat 1989: 81–2). The estimate assumed that the world's population had followed an exponential trend from 600 000 BC to the present, resulting in figures of between 3400 billion humans ever born, assuming one couple at the start, and 5300 billion assuming 500 couples at the start. This ignored the likely rises and falls in population growth rates over time, as illustrated in Hollingsworth's reconstruction of the population change in Egypt (Figure 2.5). An alternative estimate, based on varied sources of evidence, placed the figure at 80 billion humans ever born by mid-1987 – the year in which the world's total population reached 5 billion (Bourgeois-Pichat 1989: 81).

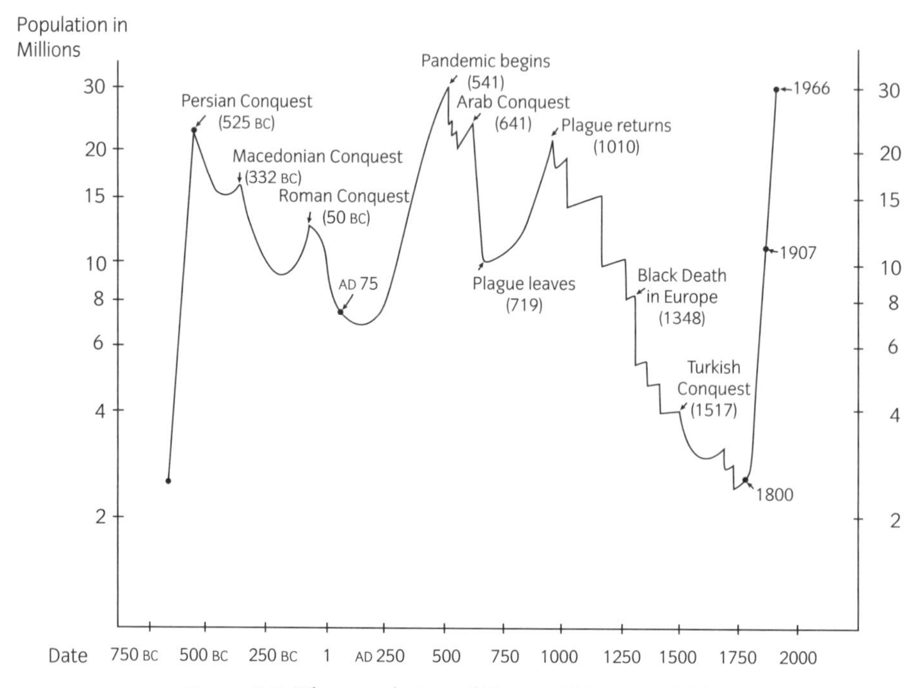

Figure 2.5 The population of Egypt, 664 BC–AD 1966.

Source: Hollingsworth (1969, 311). See also Cox (1970: 311–312).

BOX 2.3 **Growth rates module (*Growth.xls*)**

This module compares outcomes of exponential, geometric and arithmetic growth by plotting growth curves and calculating the end-of-period populations. It uses data either selected from menus or typed in. Its main purpose is to provide visual illustrations of the differences between the three concepts of population growth. The module also calculates rates and end-of-period populations from any entered data.

Instructions

To use the module, insert the CD in the drive, start Excel and open the file *Growth.xls*. Enable macros when prompted.

- Click **Menu Data** for comparisons using in-built data.
- Click **Own Data** for comparisons using your own data.
- Click **Rates** to obtain growth rates, or end-of-period populations, from your own data.
- **Home** returns to the opening display.
- **Menu** brings up the settings menu.

See Appendix B for further advice on using the Excel modules.

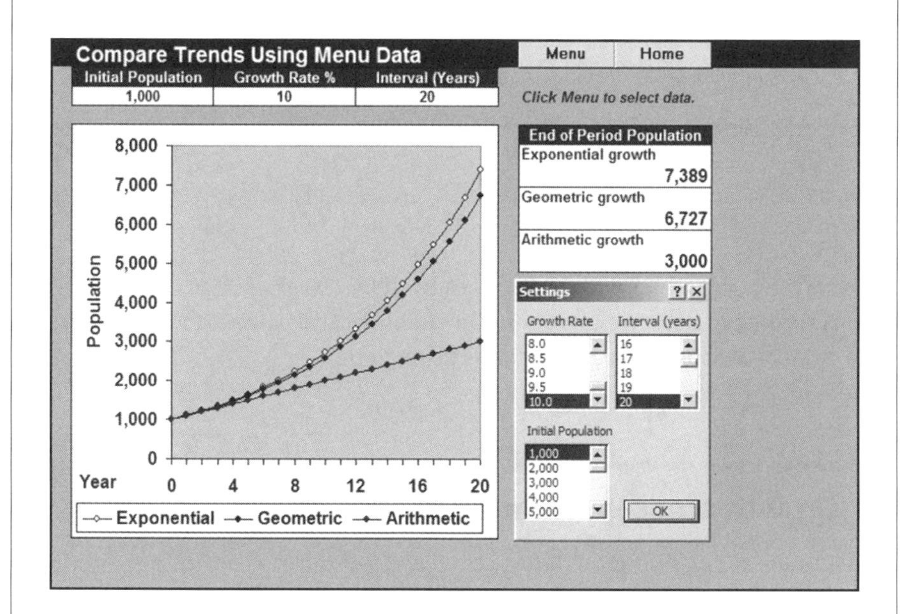

continued

BOX 2.3 continued

Illustrative applications

Questions to explore using the module include:

1. Under what circumstances is annual average growth a satisfactory measure of change? Answer: when the interval is short or the growth rate is low. The module illustrates this by showing that the growth curves coincide in such circumstances.
2. In demographic calculations, will the findings be different according to whether the growth rate is arithmetic, geometric or exponential? If so why? Again, the answer depends upon whether the interval is short and the growth rate is low.
3. What is the range of growth rates within which the geometric rate is similar to the exponential rate. Answer: in the range of 1.0 to 4.0 per cent, that is in the usual range of growth rates for national populations. These growth rates are often exceeded, however, for regional and local populations.
4. If a population of 1 million is subject to a growth rate of 2.5 per cent for 50 years, will the population at the end of the period differ appreciably according to whether the growth rate is geometric or exponential?
5. When using growth rates to derive population estimates between two censuses, such as the 1990 and 2000 censuses, which is the most appropriate growth rate to choose? Answer: unless the growth rate was very high, it would make little difference which rate was chosen. Some prefer the geometric growth rate, because it is easier to calculate, while others prefer the exponential rate because it recognizes that, in large populations, growth is compounding continuously.

Another estimate is 60.5 billion ever born by 2000 (Weeks 2002: 17). Such figures vary according to assumptions about the number of years human beings have existed as well as population growth rates through time.

2.5 Growth and replacement

Statistics on population growth or decline conceal most of the details of population change. Studies often need to go beyond appropriate measures of growth or decline to include reference to the components of population changes (see Chapter 1) or the characteristics of the population. In 1990 the population of California was 29.8 million, while in 1999 it reached 33.1 million. The difference between these numbers is California's population growth of 3.3 million. Yet this figure actually comprises only the *net growth*, or the excess of gains over losses. Before any absolute increase in a population occurs, it is necessary to make up losses due to deaths and outward migration. In California, these included 2.0 million deaths in 1990–1999 and a domestic net migration loss of 2.2 million (See http://www.census.gov/population/estimates/state). Consequently, many additions are absorbed in the process of *population replacement*. Vast numbers of replacements may be needed even to maintain a population at its original size. The total gains, or additions, represent the *gross growth* of the population. Thus:

gross growth = gains (survivors from births and migration arrivals)

or:

gross growth = net growth + replacements

Figure 2.6 depicts Price's (1976: A57ff) concepts of net growth and gross growth, illustrating the point that the replacement of losses over time substantially reduces the gains from births and inward migration. Little or no population growth may occur in a community which is nonetheless experiencing major changes in its population membership. To describe such a community as 'static' would be misleading, because it requires ongoing replacements to maintain numbers. Substantial changes in population composition can arise even where there is minimal growth or decline, because net growth is the 'the tip of the iceberg' of gross growth. Similarly, population decline is not an unqualified outcome of losses but of a deficit in ongoing replacement processes. Thus the concepts of net growth, gross growth and replacement are valuable in appreciating the dynamic nature of populations and obtaining more realistic interpretations of changes.

A related concept is *population turnover*, which denotes the sum of all losses and gains:

population turnover = losses (deaths & migration departures)
+ gains (survivors from births and migration arrivals)

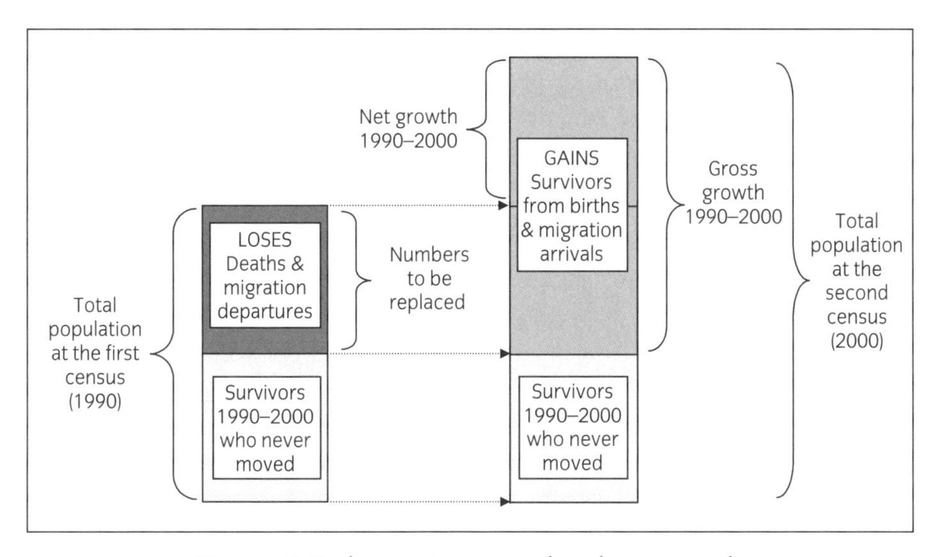

Figure 2.6 Replacement, net growth and gross growth.

Thus, population turnover is equal to losses (the numbers to be replaced) plus gross growth:

population turnover = losses (deaths and migration departures) + gross growth

Population turnover captures the total magnitude of changes in membership that a population experiences – the total of births, deaths, migration arrivals and migration departures – all of which are treated as positive numbers. Shryock and Siegel (1973: 608) describe population turnover as the sum of migration turnover (arrivals plus departures) and natural turnover (births plus deaths). Turnover is even greater in particular population subgroups, such as the aged where it draws attention to:

- the total load that administrative systems bear in maintaining registers of the clientele for pensions and services;
- the difficulties faced in community service provision for the elderly, where the client population is changing continually;
- the frequent changes in the circle of age peers – relatives, friends and acquaintances – with which individual older people must contend.

Overall, an adequate understanding of population dynamics and their implications calls for more than attention to the net growth or net decline in numbers. To obtain a more complete understanding of population change, informative concepts are turnover, replacement, gross growth and net growth.

2.6 **Conclusion**

Growth and decline are essential topics in many national and sub-national population studies; they often serve as a starting point in identifying demographic trends and their implications for planning and policy making. The concepts discussed in this chapter are necessary to an understanding of the literature on demographic change, as well as to making informed use of the range of different measures of population growth. The same concepts and measures have applications also in the important debate on the impact of population changes on the environment.

In developing plans and policies for the future, growth rates cannot be changed at will since they are dependent on a range of other factors, among which the age structure of the population is very influential. *Zero population growth* (ZPG), or a growth rate of zero, is the best-known demographic goal for minimizing environmental damage. Yet zero population growth is well beyond the reach of the less developed countries, because the form of their age structures is such that decades of substantial growth are inevitable. The next chapter discusses the concepts and methods used to describe and summarize features of the age structures of populations.

Study resources

KEY TERMS

- Arithmetic change
- Average annual increase
- Doubling time
- Exponential change
- Geometric change
- Gross growth
- Growth rate
- Logistic curve
- Net growth
- Percentage change
- Population replacement
- Population turnover
- Zero population growth

FURTHER READING

Barclay, George W. 1966. *Techniques of Population Analysis*. Chapter 7, 'Growth of Population', especially pp. 203–211.

Shryock, Henry S., Siegel, Jacob S. et al. 1973. *The Methods and Materials of Demography*, Volume 2, Chapter 13, 'Population Change', pp. 372–388.

Weeks, John R. 2002. *Population: an Introduction to Concepts and Issues*, Chapter 1, 'Introduction to the World's Population', pp. 28–56.

INTERNET RESOURCES

Subject	Source and Internet address
World population growth	**Musee de l'Homme** http://www.popexpo.net/ **United Nations Population Division (The World at Six Billion)** http://www.undp.org/popin/wdtrends/6billion/ **US Census Bureau (World Population Information)** http://www.census.gov/ipc/www/world.html
Population clocks	**US Census Bureau** http://www.census.gov/main/www/popclock.html

Note: the location of information on the Internet is subject to change; to find material that has moved, search from the home page of the organization concerned.

EXERCISES

1 Name one commonly used demographic measure that is based on the assumption of arithmetic change.

2 Distinguish between an arithmetic progression and a geometric progression.

3 Sketch a diagram to illustrate gross growth and net decline where the end-of-period population is smaller than the initial population.

4 Using information from the table below, answer the following questions, using the geometric growth formula for 4(c)–4(e):

Estimated world population 1960–2000 (millions)

Year	Mid-year population
1960	3037
1970	3696
1980	4432
1990	5321
2000	6067

Sources: United Nations 1983: *Demographic Yearbook 1981*; Population Reference Bureau (1990 & 2000). *World Population Data Sheet.*

(a) What was the percentage change in the world's population in each decade, i.e. (i) 1960–70, (ii) 1970–80, (iii) 1980–90 and (iv) 1990–2000?

(b) What was the average annual numerical increase in the population in each decade?

(c) What was the average annual growth rate per cent (geometric) in each decade?

(d) How long would the world's population take to double if the growth rate continued at the average level for each decade?

(e) Using the average growth rates from (c), estimate the year in which the population reached (i) 4 billion and (ii) 5 billion. Assuming a continuation of the 1990–2000 average growth rate, when would the total reach 7 billion?

5 Repeat questions 4(c) using the formula for the exponential growth rate.

6 Construct a table comparing doubling times using the geometric growth formula and the approximation (70/*r* per cent) for rates from 1.0 to 6.0 per cent in steps of 1.0. Does the approximation give satisfactory results?

7 Complete the following table for a population growing at an annual rate of 3 per cent and comment on why the figures differ.

Population growth 2005–2010

Year	Arithmetic growth	Geometric growth	Exponential growth
2005	1 000 000	1 000 000	1 000 000
2006			
2007			
2008			
2009			
2010			

8 For your own country, make a list of the main printed publications, and/or Web sites, that provide the most recent information on national population numbers and components of growth.

9 The larger the population, the greater the increments needed to sustain a constant growth rate. Demonstrate this with an example.

SPREADSHEET EXERCISE 2: POPULATION GROWTH

Calculate the geometric growth rates for the countries in the table; sort the rows of the table into ascending or descending order by growth rate and plot a bar graph of the rates. The notes below provide a step-by-step approach which assumes that Spreadsheet Exercise 1 has been completed. Appendix C covers the main Excel procedures and is intended to provide further assistance as needed.

	A	B	C	D	E	F
1	**Table 1: Population of Selected European Countries, 1990-2000**					
2						
3	Country		1990	2000	Geometric	
4			Population	Population	Growth	
5			(millions)	(millions)	Rate %	
6						
7	Austria		7.7	8.1	**0.5**	
8	Belgium		10.0	10.2	**0.2**	
9	France		56.7	59.4	**0.5**	
10	Germany		79.5	82.1	**0.3**	
11	Greece		10.1	10.6	**0.5**	
12	Italy		57.7	57.8	**0.0**	
13	Netherlands		14.9	15.9	**0.7**	
14	Norway		4.2	4.5	**0.7**	
15	Spain		39.0	39.5	**0.1**	
16	Sweden		8.6	8.9	**0.3**	
17	Switzerland		6.7	7.1	**0.6**	
18	United Kingdom		57.4	59.8	**0.4**	
19						
20						
21	Sources:	Population Reference Bureau 2000. *2000 World*				
22		*Population Data Sheet.*				
23		United Nations 1993. *World Population Prospects,*				
24		*the 1992 Revision.*				
25						

Constructing the table

1. Type Table 1, using formulas to calculate the figures in bold type. Use brackets in the first formula, to ensure the correct order of operations (see Appendix A), then copy the formula down to the other cells (see Appendix C, Section 2.8).

2. Sort the data into ascending or descending order by growth rate:
   - highlight the row numbers to be sorted; choose Data Sort;
   - choose sort by Column E (i.e. the column containing the growth rates);
   - click 'ascending' or 'descending' and click 'OK' or press Enter.
3. Save the spreadsheet, then plot the bar graph of the geometric growth rates.

Plotting the graph

4. Use the mouse to highlight the cells to be plotted, i.e. the country names and the growth rates. If the cells are not in adjacent columns, highlight the first column, then hold down the Ctrl key while highlighting the second column.
5. Click the Chart Wizard button on the standard toolbar. Follow the menu prompts and choose a bar graph (horizontal bars) then click 'Finish', or step through all the formatting options by choosing 'Next'.

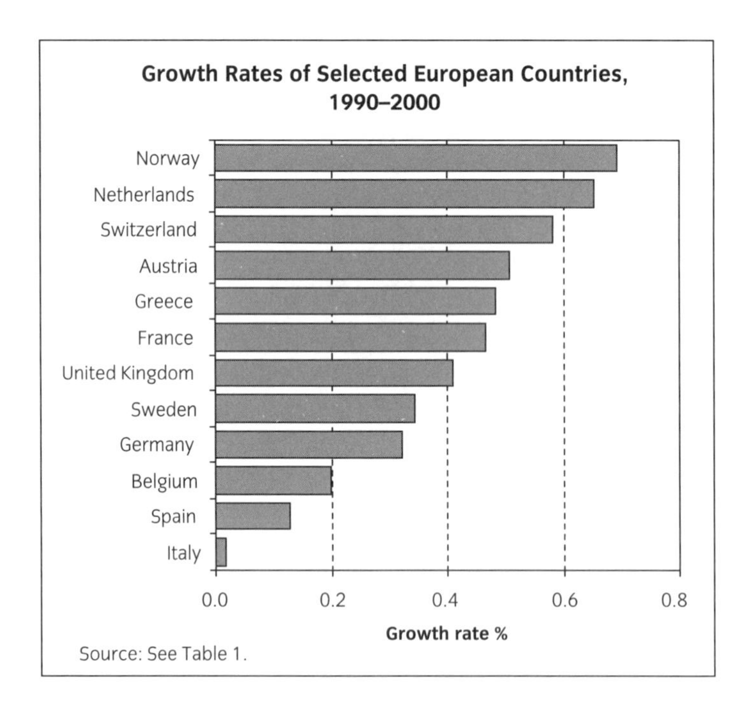

6. Position the graph by dragging it with the mouse. If some of the country names are not displayed, increase the size of the graph by dragging one of the sizing handles (square boxes on the frame around the outside of the graph). If the sizing handles are not displayed, click once on the graph.

7. To position the country names at the low end of the vertical axis:
   - click once on the graph so that a solid box appears around the graph;
   - click the mouse on the vertical axis (a box will appear at each end of the axis);
   - now double click on the vertical axis to bring up the menu for editing the axis; choose Patterns Tick Mark Labels Low (i.e. position the labels at the low end of the scale); also click the Alignment tab and set the Orientation to 0 degrees.
8. To add a title, from the Menu Bar choose Chart Chart Options Titles and type the required title in the box, e.g. Population Growth Rates for Selected Countries, 1990-2000.
9. Label the horizontal and vertical axes in the same way. Note that, in Excel, the normal labelling of the X and Y axes is reversed for horizontal bar charts: the X axis is the vertical axis, while the Y axis is the horizontal axis.
10. To add notes, e.g. 'Source', click near the outer edge of the graph to select the whole graph, then type the information, press Enter and position the text box with the mouse. The text can be moved with the mouse only when there is an edit box around the text. Click on the text if necessary to activate the text box.
11. To change the spacing between bars, click on one of the bars, then double click, or right click, the mouse to bring up the editing menu. Choose Format Data Series Options and change the Gap Width to, for example, 10 or 20.
12. To change the colour or shading:
    - double click one of the bars in the graph to bring up the editing menu;
    - choose Patterns, Area (i.e. choose colour or shading) and click OK;
    - change the plot area shading in the same way (graphs with unshaded backgrounds may print best in black and white).
13. Save and print the spreadsheet.

Review of Excel procedures

- Sorting rows of a table (Data Sort).
- Plotting graphs (highlight data and click the Chart Wizard button).
- Changing the type of chart – click on the chart then choose Chart Chart Type.
- Formatting axes (double click the axis to activate the menu).
- Changing shading (double click a bar to activate the menu).
- Adding titles (Chart Chart Options Titles).

3 Age–sex composition

Every woman may have an Husband, without the allowance of *Polygamy*.

(Graunt 1662: 58)

OUTLINE

3.1 Statistics on age

3.2 Population pyramids

3.3 Further methods of data visualization

3.4 Summary measures

3.5 Concepts in the interprelation of age–sex composition

3.6 Conclusion

Study resources

LEARNING OBJECTIVES

To understand:

- the significance of age–sex composition
- the nature of demographic data on age
- graphical presentation of age–sex data
- the principal statistical methods for summarizing age–sex data
- descriptive and explanatory concepts used in interpreting age–sex composition

COMPUTER APPLICATIONS

Excel modules: Age Structure Database.xls (Box 3.1)
Age Structure Simulations.xls (Box 3.2)

Spreadsheet Exercise 3: Age–sex pyramids (See Study resources)

In seventeenth century England, some contended that a shortage of men, arising from war deaths and emigration to the American colonies, justified allowing polygamy. Graunt (1662) argued against this on the basis of his demographic evidence. He found that, contrary to opinion, the numbers of men and women were similar: more males were born and excess losses of males restored the balance of the sexes. Graunt was also interested in social implications, concluding: 'Where Polygamy is allowed, Wives can be no other then Servants' (ibid. 13). Graunt's pioneering work is the origin of contemporary demographic interest in the balance of the sexes; this has become an important subject of inquiry, including investigations of marriage markets, family formation and equity in access to work and income.

Analysis of the age composition of populations is an essential concomitant of such studies, as well as of demographic investigations generally. At an individual level, when someone marries, or achieves a notable success in their vocation, or dies, the item of greatest curiosity to people is their age. This reflects a pervasive interest in comparing the timing of events with our expectations or 'social timetables' (Neugarten and Hagestad 1976: 35). At an aggregate level, demographers and other social researchers are interested in the age composition of societies, together with the ages at which people engage in certain behaviours. Ages at moving house, marrying, having babies, purchasing products or using services are vital to explaining social trends, targeting markets and planning for the future. Moreover, the age composition of the population is central to understanding the nature and functioning of societies.

In Africa in 2001, for example, around 43 per cent of the population were under 15 years of age, compared with only 18 per cent in Europe (Population Reference Bureau 2001a). The high proportion of children in African countries immediately implies in-built potential for rapid population growth, as well as great and continuing need for investment in education and employment. In contrast, European countries are experiencing rapid growth in the numbers and percentages of older people in their populations, creating concerns about the funding of pensions and health services, as well as allied concerns about diminishing labour supplies and overall population decline.

Age and sex composition are so important to the nature and functioning of societies and communities that population pyramids – graphs depicting the numbers or percentages of males and females in each age group – are instructive starting points for population investigations. Such pyramids not only show the absolute or relative size of each age group, but potentially reveal much about the history of a population and its future prospects.

China's age–sex structure in 1990 preserves the effects of two episodes of sharp decline in the birth rate (Tien et al. 1992: 18ff). The first, between 1959 and 1961, arose from dislocation and famine during Chairman Mao's 'Great Leap Forward' – an intended leap into industrialization that took labour out of food production. The second decline, in the 1970s, predating the One Child Policy (1979), arose from an effective nation-wide 'Later–Longer–Fewer' campaign, promoting

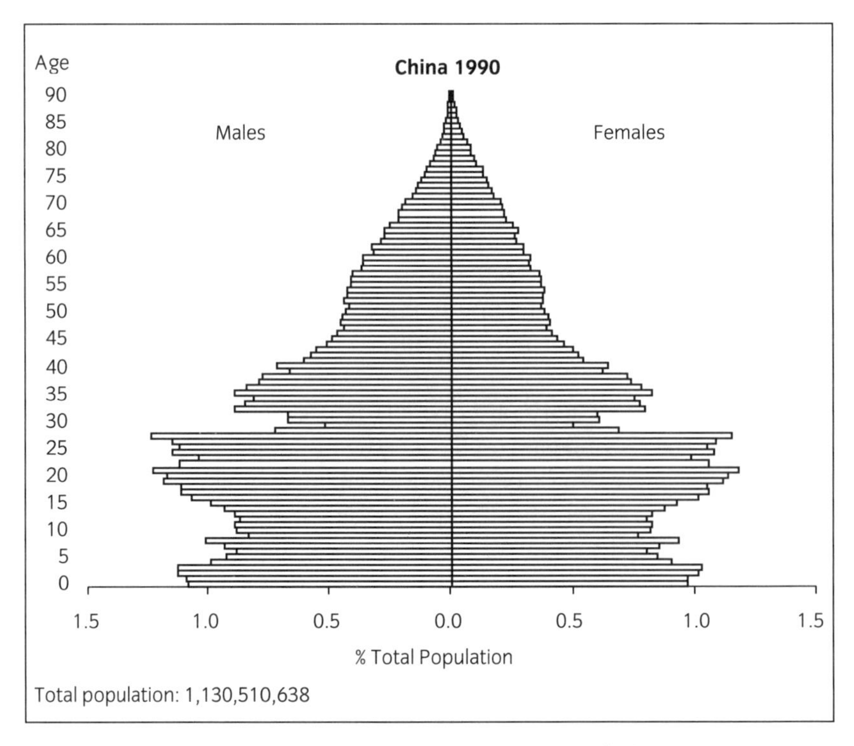

Figure 3.1 Age–sex structure of the population of China, 1990.

Source: Fan Jingjing 1995. *China Population Structure by Age and Sex*. Data User Service Series No. 6. China Publishing House, Beijing.

later marriage and childbearing, longer intervals between births and fewer children. China's birth rate fell from 6 children per woman in 1970 to 3 by 1978, that is even before adoption of the more stringent One Child goal (ibid. 10–11). Age structures further reflect the overall outcome of the processes of fertility, mortality and migration. They therefore provide insight into causes of population change.

This chapter turns now to approaches to graphing and summarizing statistics on age–sex composition. The summary measures include sex ratios, dependency ratios and measures of central age. Later sections introduce approaches to the interpretation of such information with reference to types of age structures and patterns of change during the demographic transition.

The computer modules accompanying this chapter comprise a data base of international age structures (see Box 3.1) and a set of simulations, with animated age pyramids to illustrate processes of age structure change (see Box 3.2). The associated spreadsheet exercise explains the steps in constructing population pyramids using Excel (see 'Study resources'). Age structure is such an important population characteristic that it inevitably enters into the discussion of most other demographic subject areas. Chapter 9, for example, discusses the process of population ageing – in the light of stable and stationary models – and introduces the related concept of population momentum.

3.1 Statistics on age

Population censuses are the main source of statistics on the age–sex composition of populations. In censuses, people are generally required to specify their age at their last birthday. Whereas, in statistical work, figures are rounded to the nearest whole number (see Chapter 2), age is unusual because it is usually rounded down (Newell 1988: 23). Thus a child aged 10 years and 11 months is still described in censuses as a ten year old.

For this reason, census statistics on the numbers at single years of age actually refer to an age range; the ten year olds include everyone from exact age 10, whose birthday was on the census day, to children just one day away from their eleventh birthday. Hence the group aged 10 last birthday includes everyone aged between 10.00 and 10.99 years. Since the single age refers to a twelve month age range, accuracy in certain calculations requires that the mid-point of the range (10.5 years) be used to represent the average age of all members of the group. Examples of this are seen in work on fertility and population models.

For convenience in summarizing and graphing statistics on age, five year age groups are widely employed instead of single year ages. Age group 20–24, for example, is a five year age group referring to everyone 20 years and over who has not had their 25th birthday. Here the class interval is five years and the class limits are 20.00 and 24.99. The mid-point of the group is equal to 20 plus half the class interval (5), i.e. 22.5. Sometimes the mid-point is taken to represent the average age of the group. If the final age group is '65 years and over', or '85 years and over', it is described as an 'open-ended age interval'; its mid-point, if available, is the median age of everyone in the group, the calculation of which is discussed later.

The quality of statistics on age depends substantially upon whether people know when they were born and whether they are inclined to understate or exaggerate their ages. Misstatement of age is evident in preferences for ages ending in zero or 5, which can lead to appreciable *age heaping* or concentration at particular values. Age heaping is conspicuous when data are tabulated and graphed for single years; five year groups conceal most inaccuracies of this kind. Indices to measure the phenomenon are available (Shryock and Siegel 1973: 204–209), but they are needed only infrequently in countries where most of the population is literate and the quality of official statistics is high.

Censuses, estimates for years between censuses, and population projections have created a vast, and regularly updated, volume of statistics on changes in age–sex composition at all geographical levels. The *Age Structure Database* (Box 3.1) presents examples of national and regional age structures, and its animated population pyramids enable changes to be viewed and compared through time. Use of the data base is intended to provide initial familiarization with the subject matter of this chapter through illustrations of contrasting trends through time.

BOX 3.1 Age structure database module (*Age Structure Database.xls*)

This module provides a facility to view national or regional age structures, in five year intervals from 1950 to 2000, together with 'medium variant' projections to 2050. The module plots population pyramids, using percentages or absolute numbers, as well as sex ratios by age. It enables age structures to be compared through time, with the options of setting constant numerical or percentage scales, or using a scroll bar to animate the graphs.

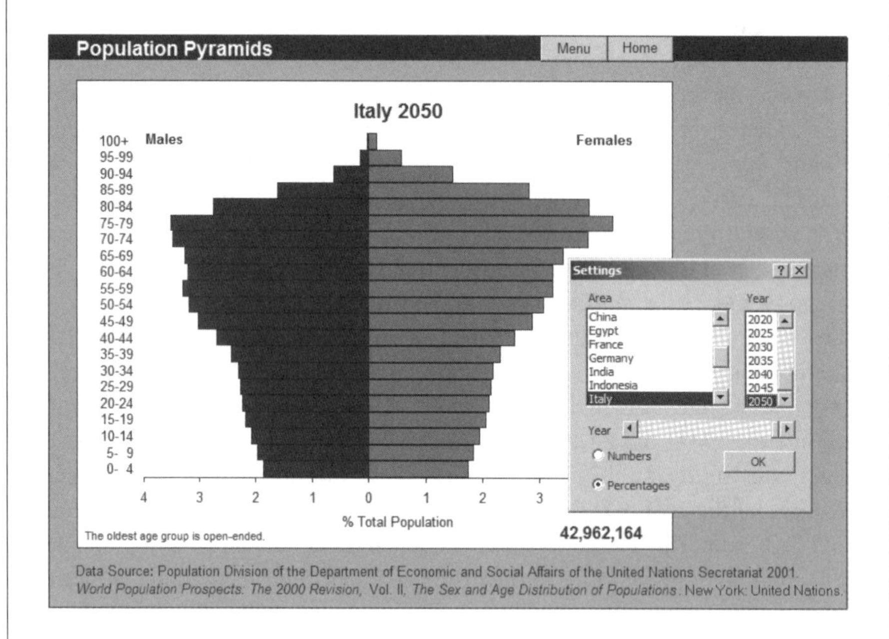

The module is based on selected statistics from a comprehensive data base of age structures for regions and countries, as published in printed form and CD as: Population Division of the Department of Economic and Social Affairs of the United Nations Secretariat 2001: *The Sex and Age Distribution of Populations*, United Nations publications, ST/ESA/Ser.A/199. New York: United Nations.

In the module, the data, reproduced with the permission of the Population Division, refer to the following regions and countries:

World
More developed regions
Less developed regions
Australia
Brazil
Canada
China
Egypt
France
Germany
India
Indonesia
Italy
Japan
Mexico
New Zealand
Pakistan
South Africa
United Kingdom
United States

continued

BOX 3.1 continued

Instructions

To use the module, insert the CD in the drive, start Excel and open the file *Age Structure Database.xls*. Enable macros when prompted.

The total population is shown at the bottom right of each graph. For 1950–1990, the age groups extend to 80 years and over. For later years, the age groups extend to 100 and over.

- Click **Pyramids, Numbers or Percentages** to display any age pyramid.
- Click **Constant Numerical Scale** to display pyramids based on absolute numbers, with the scale optimized for the selected region or country.
- Click **Constant Percentage Scale** to display pyramids with the same percentage scale for all regions and countries.
- Click **Sex Ratios** to display bar charts of sex ratios at each age.
- Click **Menu** to bring up the controls for selecting populations, years and other options.
- Click **Home** to return to the opening display.

See Appendix B for further advice on using the Excel modules.

Illustrative applications

1. Display the age structure of a selected population using (a) numbers and (b) percentages. Discuss the advantages and disadvantages of using each approach.
2. Review the demographic histories and projected futures of a selected population by showing a 'movie' of age structure changes using the scroll bar. Discuss the processes responsible for the observed changes in light of the demographic transition.
3. Examine age-specific sex ratios for a selected population and explain departures from the typical sex ratio at birth of 105 : 100.

3.2 Population pyramids

Census statistics are the main source of the information needed for drawing population pyramids, the most distinctive and best-known of all demographic graphs. This reflects that information on age and sex composition is of wide-ranging importance, and that the visual presentation of these statistics is an indispensable aid to understanding and communication. Population pyramids were formerly time-consuming to draw manually on graph paper, but spreadsheet programs, as well as other more specialized computer graphics packages, have made possible the rapid drawing of them at publication standard. Spreadsheet programs do not include, as an in-built option, population pyramids or back-to-back bar graphs; the steps required to produce them are described in the spreadsheet exercise at the end of this chapter. A later chapter also includes a computer module to create single and superimposed population pyramids from any data (see Box 5.2).

Single pyramids

Known variously as population pyramids, age–sex pyramids, age structures and age profiles, these graphs consist of two bar graphs, one for males the other for females, drawn back-to-back. Several conventions in the construction of age–sex pyramids aid interpretation and facilitate comparisons between graphs from different sources:

- Plot statistics for the youngest ages at the base, those for the oldest at the apex. If the final oldest age group is open ended (e.g. ages 85 and over), it needs to be labelled clearly to indicate this. Plot statistics for males on the left, for females on the right; label each side of the pyramid accordingly.
- Set the ratio of the height of the pyramid to its width between about 1:1 and 1:1.5. Narrow pyramids are less able to reveal variations in numbers or percentages between age groups and minimize apparent variations between populations.
- Use percentages to facilitate comparisons between different populations. The age structure data base permits experimentation in displaying population pyramids based on percentages or absolute numbers (Box 3.1). The percentages should be based on the grand total population, males and females combined, in order to reveal any differences in the relative numbers of each sex in particular age groups. The excess of females in the older ages, for example, is shown adequately only when the percentages are based on the grand total. It is sometimes helpful to state the total population figure somewhere on the graph, so that readers can translate percentages into absolute numbers if required.
- The graph should contain an informative title, a statement of the source of the statistics, and adequate labelling of age-groups and the horizontal axis.

Superimposed and compound pyramids

Comparisons between pyramids can be facilitated by superimposing one upon another. The superimposed pyramids in Figure 3.2 clearly show the similarities and differences between the 2000 and 2100 profiles for the world, most notably the trend towards a more rectangular age structure and the projected increase in the proportion of older people.

Whereas superimposed pyramids facilitate comparisons between two populations, compound pyramids present additional information about the composition of a single population, for example, in terms of marital status, educational attainment or ethnic background. To construct compound pyramids, statistics for the variable in question are needed for each age–sex group. All the percentages – for each age–sex group together with each age–sex and marital status (or other) group – are based on the grand total population. Censuses and surveys are the main sources of cross-classified data, such as age by sex by marital status or age by sex

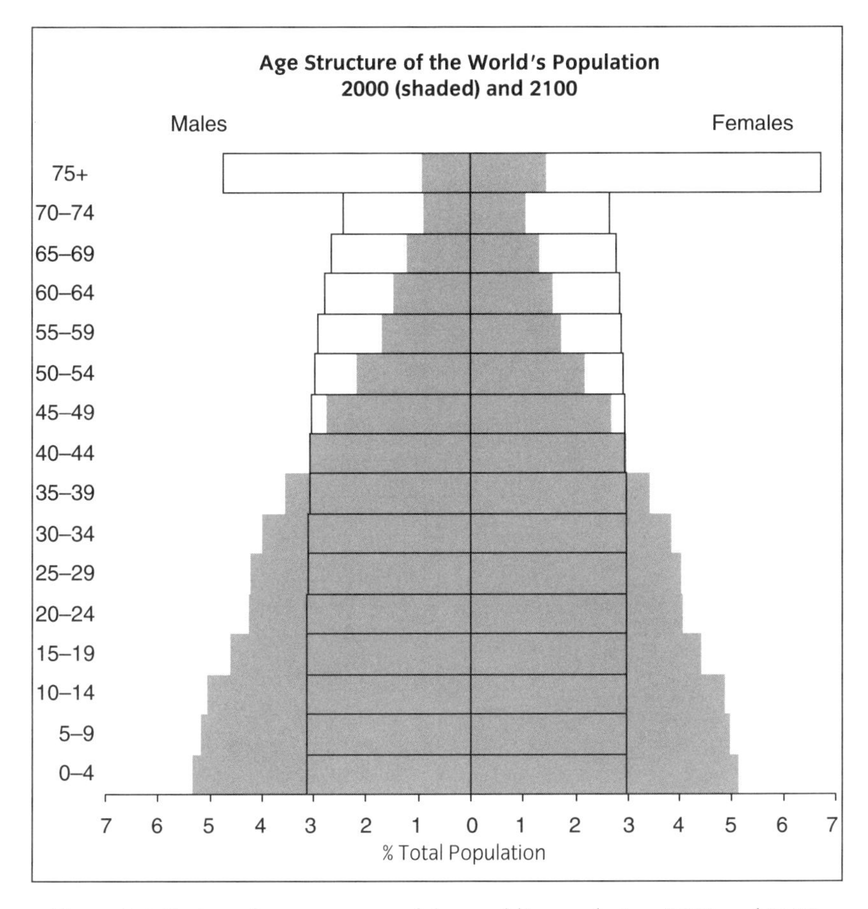

Figure 3.2 Projected age structure of the world's population, 2000 and 2100.
Source: World Bank (1994).

by education. Spreadsheet Exercise 4, in the next chapter, builds on the population pyramids exercise at the end of this chapter, to enable the construction of superimposed and compound pyramids.

3.3 Further methods of data visualization

Despite the dominance of the pyramid for presenting age–sex data, other types of graphs, such as pie graphs (or divided circles) and area graphs, can also assist comparisons or reveal general trends through time. Spreadsheet Exercise 5 covers the preparation of such charts in Excel (see Chapter 5).

Pie graphs of the proportions in child, working and older ages (0–14 years, 15–64 years, 65 and over) provide a quick impression of some major differences

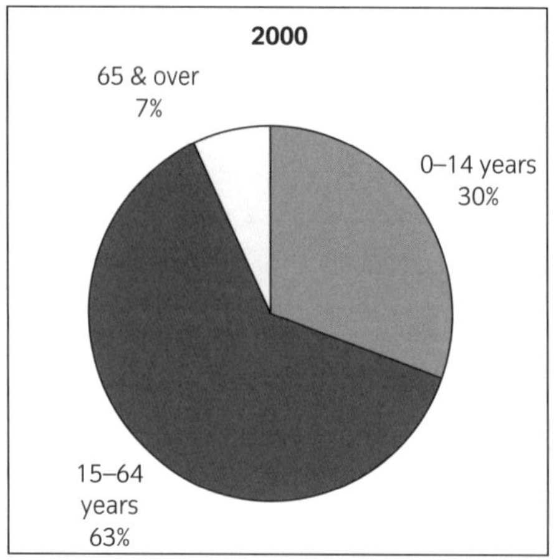

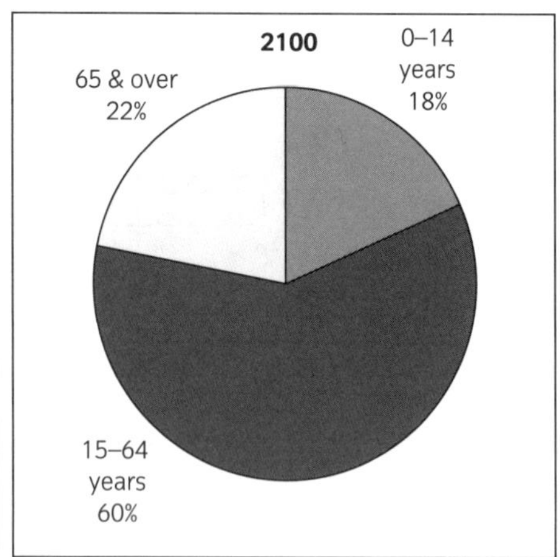

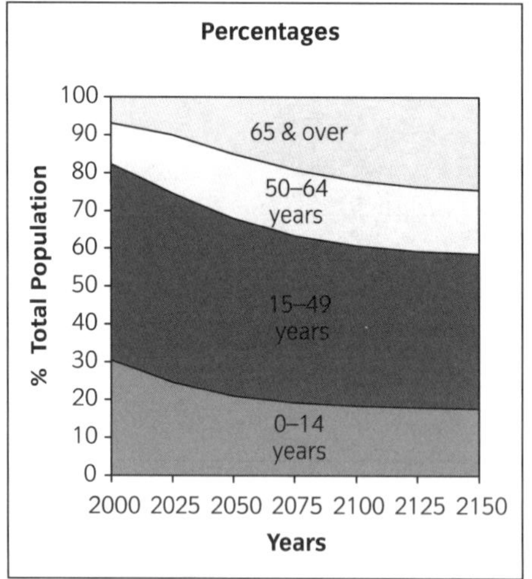

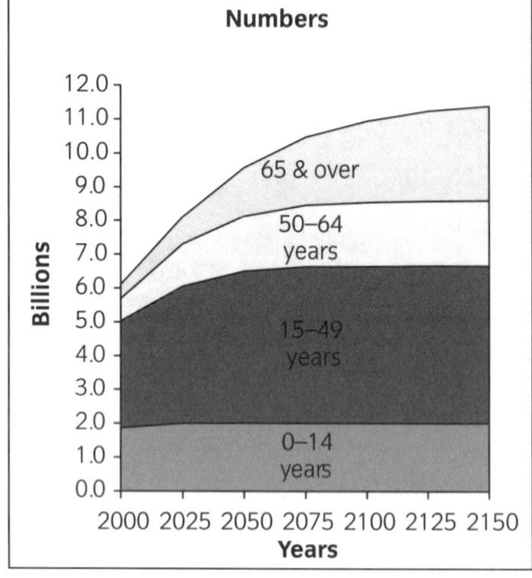

Figure 3.3 Pie and area graphs of projected world age distributions, 2000–2150.
Source: World Bank (1994).

between populations. Pie graphs from the World Bank's (1994) long-range projections of the global population, for example, show the proportions in the older ages growing from 7 to 22 per cent, with corresponding falls mainly in the child age groups (Figure 3.3). Drawing such graphs manually entails starting at the 12 o'clock position and using a protractor to measure 3.6 degrees for every 1 per cent, since there are 360 degrees in a circle. A sector of 72 degrees would represent 20 per cent. Maps incorporating pie charts can depict population composition in different areas. To give an impression of the relative size of the population in each location, such maps may use circles that are proportional to the total numbers they represent (see Figure 10.7).

Area charts of age composition reveal changes through time in the representation of broad age groupings. In area charts, data for different age groups are 'stacked' upon each other, so that the uppermost line of the graph represents either

100 per cent or the size of the total population. Variations in the width of horizontal bands on the graphs show differences in the numbers or percentages in each age group. A potential disadvantage is that area graphs for some data sets give a false impression of smooth changes between the plotted points. Stacked bar graphs (or stacked column charts in Excel), which display values only for the dates to which the statistics refer – without interpolated values between – are more accurate but may lack the visual clarity of an area graph.

Area charts in Figure 3.3 summarize the World Bank's (1994) projections of the global population. Although such long-range projections are merely illustrative, the graphs highlight major outcomes of the assumed trends in birth and death rates. The first area graph, depicting percentages, shows a marked decline in the representation of children, with corresponding increases at older ages: in other words, a marked ageing of the global population. The second graph, depicting absolute numbers, shows substantial population growth overall, but relatively little change through time in the numbers of children: future world population growth is assumed to be concentrated in the working ages and older age groups.

3.4 Summary measures

Beyond the techniques for visualising data, sex ratios and dependency ratios are the simplest and most widely used summary measures of age–sex composition, as listed in Table 3.1. Other summary measures, described later, are the median age of the population and the index of dissimilarity.

Table 3.1 Summary measures of age-sex composition

Measures	Examples
Sex ratio The number of males per hundred females:	United Kingdom, population 75 years and over, 1990 (thousands)
$\frac{\text{number of males}}{\text{number of females}} \times 100$	Males = 1375
or	Females = 2533
$\frac{M}{F} \times 100$	Sex ratio = (1375/2533) × 100 = 54 males per hundred females
Dependency ratio The number children and aged persons per hundred people of working age:	Italy 2050 (thousands) Numbers aged 0–14 = 7454 Numbers aged 15–64 = 25 216 Numbers aged 65+ = 15 136
$\frac{P_{0-14} + P_{65+}}{P_{15-64}} \times 100$	

Table 3.1 *continued*

Measures	Examples
The (total) dependency ratio is also equal to the sum of the child and aged dependency ratios.	Dependency ratio = (7454 + 15 136)/25 216 × 100 = 90
Child dependency ratio The number of children per hundred people of working age:	Italy 2050 (thousands) Numbers aged 0–14 = 7454 Numbers aged 15–64 = 25 216
$\frac{P_{0-14}}{P_{15-64}} \times 100$	Child dependency ratio = 7454/25 216 × 100 = 30
Aged dependency ratio The number of aged people per hundred of working age:	Italy 2050 (thousands) Numbers aged 15–64 = 25 216 Numbers aged 65+ = 15 136
$\frac{P_{65+}}{P_{15-64}} \times 100$	Aged dependency ratio = 15 136/25 216 × 100 = 60
Economic dependency ratio The number of people not in the labour force per hundred in the labour force:	United States resident population 1998 (thousands), excluding armed forces abroad
$\frac{\text{Population not in the labour force}}{\text{Population in the labour force}} \times 100$	Total population not in the labour force (all ages) = 131 405 Total labour force = 138 894 (includes 6210 unemployed) Economic dependency ratio = 131 405/138 894 × 100 = 95
Ageing index The number of the aged per hundred children:	Italy 2050 (thousands) Numbers aged 0–14 = 7454
$\frac{P_{65+}}{P_{0-14}} \times 100$	Numbers aged 65+ = 15 136 Ageing index = 15 136/7 454 × 100 = 203
Caretaker ratio The number of persons aged 80 years and over, per hundred females aged 50–64:	United Kingdom 1990 (thousands) Total population aged 80+ = 2104 Females aged 50–64 = 4568
$\frac{P_{80+}}{P^{f}_{50-64}} \times 100$	Caretaker ratio = 2104/4568 × 100 = 46

Data sources: World Bank (1994); US Census Bureau. *Statistical Abstract of the United States* (1998: 8 & 412); United Nations (1993: 376).

Sex ratios

More boys are born than girls, although nature exacts its 'revenge' in the higher death rates of males generally. The question of why more boys are born is a scientific puzzle, one closely linked with theories, of considerable interest to many intending parents, about the determinants of the sex of offspring. In the fifth century BC, the Greek philosopher Anaxagoras, for example, insisted that boys came from the right testicle and girls from the left. His influence was such that 'centuries later some sonless French aristocrats are said to have endured amputation of the left one' (Ridley 1994: 29).

Contemporary theories, including the coital frequency theory, require more popular interventions, but the question is still unsettled. The coital frequency theory maintains that sperm carrying a Y chromosome are more vigorous, but shorter lived, than those carrying an X chromosome. Hence boys are more likely to be conceived if intercourse occurs soon after ovulation. The greater the frequency of intercourse, the greater the odds of a Y-chromosome sperm reaching the egg. Nevertheless in Britain, although coital frequency declines with age, there is little variation in the sex ratio with birth order or maternal age (ibid: 29). Another theory, which refers to hormone levels, proposes that testosterone and oestrogens promote the conception of boys, while gonadotrophins promote the conception of girls. The hormonal theory is not necessarily inconsistent with the coital frequency theory, since testosterone and oestrogens are higher around ovulation (ibid: 30).

The sex ratio is the ratio of the number of males to the number of females, usually expressed as males per hundred females (Table 3.1). The sex ratio at birth is typically around 105 males per 100 females. Small departures from this level may reflect random variations, while larger departures forewarn of possible data quality issues or social influences.

Low ratios, due to under-reporting of male births, have appeared in censuses of cultures where parents have wished to conceal and protect infant sons from supernatural forces, as in rural Nepal (Rajbanshi and Sharma 1980: 24). In China, high sex ratios at birth have reflected under-reporting of female births, due to preferences for sons, and excess mortality of female infants attributed to neglect. From 1985 to 1987, when the ratios were above 110:100, as many as 500 000 girls were 'missing' from the new-born each year. Around half of these may have been female babies given away for adoption and not reported as a live birth (Tien et al. 1992: 15).

The excess of males in the youngest ages contrasts with the reverse situation in the oldest ages, which is due to female longevity. In the United States in 2001, female life expectancy at birth was 80 years, six years higher than the figure for males (Population Reference Bureau 2001b). The lower death rates of females bring an increasing representation of them in successive ages. Migration also influences sex ratios, especially in the younger working ages where there may be an over-representation of male migrants.

Sex, together with age, is such an indispensable demographic variable that sex ratios have applications in many and varied studies from consumer behaviour to living arrangements, disease prevalence and social inequality. In recent years there has been greater use of the term 'gender ratio' instead of 'sex ratio'. The usual distinction between 'sex' and 'gender' is that the former refers to biological differences, whereas the latter includes social and cultural differences as well. Many differences between males and females are not biological in origin (Giddens 1997: 91). Social and cultural practices shape the roles of men and women in society together with differences, for example, in their access to education and employment (Riley 1997: 4–5). The influence of gender varies between social groups and societies, while the biological characteristic of sex remains the same. Any ratio that simply compares the number of males with the number of females, is strictly a 'sex ratio': the statistic is biologically based. Thus the main demographic journals and periodicals continue to use the term 'sex ratio' in studies of gender issues (ibid: 16). Also, although gender is well recognized as a major influence on population change, difficulties remain in assessing the relative influence of biological and social factors on phenomena such as female longevity.

Dependency ratios

Dependency ratios provide simple summary measures of age composition, with particular reference to relative numbers of supposed 'dependants' and 'supporters', or 'unproductive' and 'productive' groups. The ratios are based on a division of the age range into three broad groupings, namely: children (0–14 years), 'working ages' (15–64 years), and 'old' ages (65 years and over). Varying these ranges, such as through raising or lowering the upper limit of the working ages, can affect the ratios appreciably.

As shown in Table 3.1, the child dependency ratio represents the number of children per hundred 'workers'. Similarly, the aged dependency ratio is the number of people aged 65 years and over, per hundred 'workers', although ages 55 and 60 are sometimes used as the cut-off instead of 65 years. The total dependency ratio compares the sum of child and aged dependants with the numbers in the working ages. In effect, the ratios purport to show how many dependants there are to be supported per 100 people of working age. One use of the ratios is to provide a first indication of comparative dependency burdens in different populations.

As the ratios in Table 3.2 illustrate, total dependency is currently greater in less developed countries – due to their relatively high birth rates and young age profiles. The high proportions of children in these populations impose economic difficulties due to the continuing need to duplicate infrastructure, such as schools, for greater numbers of people. Moreover, the cost to the government of a child in less developed countries is often higher than the cost of an elderly person, because children require education, whereas for older people there are no universal pension schemes and families are expected to bear most of the support for the indigent, the frail and the disabled. By 2050, less developed countries could have dependency

Table 3.2 Illustrative dependency ratios, 1990 and 2050

Region/country	Child	Aged	Total
1990			
More developed regions	32	18	50
Less developed regions	59	8	67
Sweden	27	28	55
India	62	8	70
2050			
More developed regions	30	41	71
Less developed regions	33	21	54
Italy	30	60	90

Data source: World Bank (1994).

ratios similar to those of the more developed countries 60 years earlier. Yet total dependency in more developed countries could continue to rise, especially because of the effects of below-replacement fertility. The figures for Italy in 2050 represent a possible extreme stage in demographic evolution, resulting from very low birth rates and rising aged dependency ratios. Per capita public expenditure on older dependants in more developed countries is thought to be two or three times higher than on children (Easterlin 1996).

Figure 3.4 places these figures in the context of the demographic transition, showing a model of the pattern of change in child, aged and total dependency through time, from pre-transition to post-transition. Child dependency rises during the transition period, then later declines below previous levels. Aged dependency is low over much of the transition period, growing to match the level of child dependency towards the end. The below-replacement fertility scenario, exemplified by the projected figures for Italy in 2050 (Table 3.2), lies beyond the demographic transition and brings a steep rise in aged dependency.

The disadvantages of the dependency ratios are that the age groups are extremely broad, leaving much scope for argument about the meaning of the figures. A key question is: What is dependency? The ratios imply that all children are dependants, whereas some are actually in the paid or unpaid labour force. Also, many people in the working ages are not economically active, whether because they are engaged in domestic duties, or because they are studying, retired, unemployed, or disabled. Similarly, people aged 65 and over are not necessarily dependent in economic or other terms, and many contribute to the economy through voluntary work.

One response to the problem of defining or approximating dependency solely in terms of age has been to propose other ratios focusing more specifically on economic dependency. The economic dependency ratio (Shryock and Siegel 1973: 235 and 358) is the ratio of the economically inactive population to the economically active population of all ages (Table 3.1). However, the new ratio is an economic

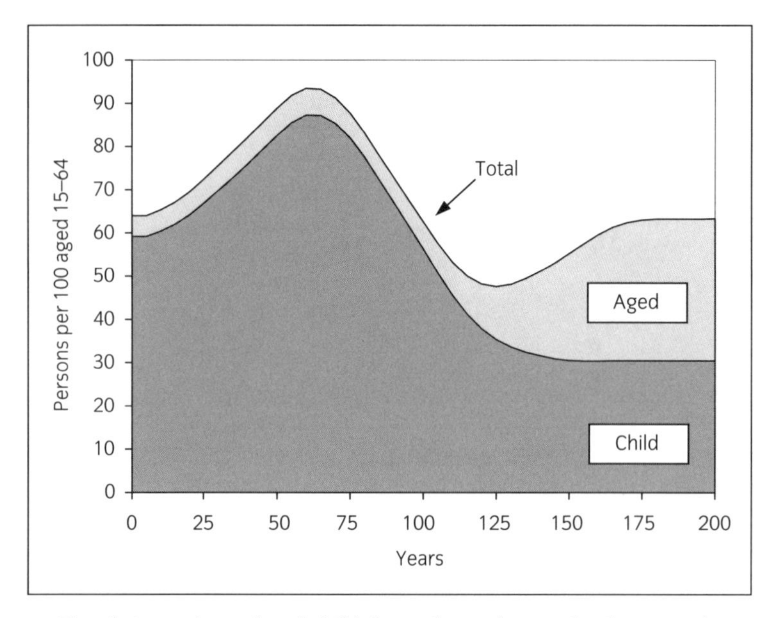

Figure 3.4 Trends in total, aged and child dependency during the demographic transition.

Source: Model data for a 'Western' demographic transition.

measure that is less useful as a general comparative statistic summarizing age structures. Economic dependency ratios are also less comparable over time and from place to place than simpler dependency ratios: they require more refined data, and definitions of economic activity vary through time and from country to country. For instance, the definition of labour force participation may or may not include the unemployed, unpaid helpers and people engaged in varying hours of part-time employment.

Because of the importance of ageing as a demographic phenomenon, researchers have proposed other ratios of one age group to another to describe its implications more fully. One example is the *ageing index* devised to compare age structures of different countries. The ageing index, as illustrated in Table 3.1, is the ratio of the number of persons aged 65 and over, or 60 and over, to the number of children under 15, multiplied by 100 (National Institute on Ageing 1998). The older the population, the higher the index; indices greater than 100 show that older people outnumber children.

Another index of this kind is the *caretaker ratio*. Demographic analyses of dependency mostly compare total numbers of the aged with total numbers in the working ages. This approach is uninformative when addressing the question of dependency in terms of providing instrumental support for frail aged parents. This is because relatives – especially daughters – often assume responsibility for providing expressive (emotional) and instrumental (practical) support, rather than the population at large. In this context, caretaker ratios, the ratio of females 50–64 years to persons 80 years and over (Myers, Torrey and Kinsella 1989), are an improvement on aged dependency ratios, since they approximate the relative numbers of parents and daughters (Table 3.1). However, problems remain in that not all of the aged have daughters, and not all of the carers are in the age range 50–64. Husbands and wives commonly provide the day-to-day care when one partner is frail or disabled.

Measures of central age

Whereas dependency ratios focus on particular features of age composition, other summary measures aim to provide a single number to represent the age structure generally. These can reduce the information contained in cross-tabulations by age and sex to a manageable form, but with considerable loss of detail.

As in statistical analysis generally, measures of central value – the mean, the mode and the median – can be calculated from age statistics to summarize the data. Table 3.3 lists these measures as they are used with ungrouped data, that is data referring to specific individuals. Thus for five people aged exactly 20, 21, 21, 22, and 26, the mean, or average age is 22, the modal age – the most frequently occurring value – is 21, and the median age or middle value is also 21.

Table 3.3 Measures of central age for ungrouped data

Measures	Examples
Mean (average, arithmetic mean)	
$\text{Mean} = \frac{\text{sum of values}}{\text{number of values}}$	The mean age of three children aged 3, 4 and 11 is 6 years.
or	
$\overline{x} = \frac{\sum x}{n}$	
(Σ is the Greek capital sigma, the symbol denoting summation).	
Mode The most frequently occurring value or group.	The modal age of children aged 3, 4, 5, 5, 5, 6, 6 and 7 is 5 years.
Median The middle value in a set of numbers arranged in ascending or descending order.	The median age of three persons aged 10, 25 and 90 years is 25 years. Their mean age is 41.7 years.
If there is an even number of cases, the median is the average of the two middle cases.	The median age of four persons aged 10, 25, 30 and 90 years is 27.5 years.

Median age from grouped data

The examples in Table 3.3 refer to ungrouped data, but censuses and surveys invariably collect information on the numbers in age groups; hence the technique for calculating the median from grouped data is required. The median age from grouped data is a useful summary measure of age structure since, unlike the mean, it is unaffected by extreme values and it can normally be calculated even where the age groups include open ended class intervals, such as 'ages 60 and over'.

The formula for the median age is usually expressed as:

$$\text{Median} = l + \frac{\frac{N}{2} - F}{f} \times i$$

where:

l = lower limit of the class containing the middle case
N = total population
F = cumulative frequency up to the age group containing the middle case
f = frequency of the class containing the middle case
i = the size of the class interval containing the middle case.

Another way of writing the formula is:

$$\text{Median} = l + \left(\frac{N}{2} - F\right) \times (i/f)$$

This alternative groups logical steps and will be referred to in the explanation of the following calculation of the median age from the statistics in Table 3.4.

$$\text{Median} = 25 + \frac{\frac{15629}{2} - 5904}{2219} \times 5$$
$$= 29.3\,\text{years}$$

or

$$\text{Median} = 25 + \left(\frac{15629}{2} - 5904\right) \times \frac{5}{2219}$$
$$= 29.3\,\text{years}$$

Calculation of the median age from grouped data entails three main steps: first, find the middle case; second, find the age group in which the middle case is situated; and third, calculate how far through that age group the middle case is located.

1. List the frequencies – the number of persons in each age group (Table 3.4, Column B).
2. Beside these list the *cumulative frequencies* to each age. Cumulative frequencies are the total of frequencies up to and including the given row. Thus in the

Table 3.4 Age of murder victims, United States 1996

Age group A	Frequency B	Cumulative frequency C
Under 20	3165	3165
20–24	2739	5904
25–29	2219	8123
30–34	1838	9961
35–39	1685	11646
40–44	1212	12858
45–49	877	13735
50–54	539	14274
55–59	362	14636
60 & over	993	15629
Total	15629	

Note: The figures exclude 219 persons whose age was unknown.
Source: *Statistical Abstract of the United States*, 1998, p. 213.

example (Table 3.4) the cumulative frequency to the row for age group 20–24 is $3165 + 2739 = 5904$; similarly the cumulative frequency for age group 25–29 is $3165 + 2739 + 2219 = 8123$. The grand total of the column of figures (N) is the same as the cumulative frequency to the last row (15629).

3. Find the middle case by dividing the total population (N) by 2. The middle case is important because the age of this person – the individual in the middle of the population arranged in ascending order of age – will be the median age of the population. In Table 3.4, the middle case ($N/2$) is 7814.5.
4. Find the age group in which the middle case is located, i.e. with 7814.5 cases above and 7814.5 cases below. Here reference is made to the cumulative frequencies and the values of l, F and f, used in the formula, are obtained. The main step is to find the first cumulative frequency that is greater than $N/2$. In the example, the figure is 8123 for age group 25–29; hence the median is located in that age group. It now follows that $F = 5904$, $f = 2219$ and $l = 25$.
5. Values for all of the terms required are now available and the median age can be calculated by applying the formula, making sure that operations are carried out in the correct order (see 'order of operations' in Appendix A). In effect, the formula measures the middle case's location, which is known to be in the class interval with a lower limit of l and a frequency of f. It is assumed that the population in the group is evenly spread through the class interval. Thus i/f represents the difference in age between adjacent individuals in the group. In the example, $i/f = (5/2219) = 0.00225$ years.

6. $N/2 - F$ measures how far into the age interval the middle case is located. In the example, $N/2 - F = 7814.5 - 5904 = 1910.5$. In other words, starting from age 25, there are 1910.5 people before the middle case. The sum of the age differences between them is $(N/2 - F) \times (i/f) = 1910.5 \times 0.00225 = 4.3$ years.

Therefore the age of the middle case, or the median age, is 25 years plus 4.3 years = 29.3 years. This figure was about 5 years less than the median age of the total United States population: murder victims had a relatively young median age. A question arising from this result is whether similar figures are obtained for male and female and white and black murder victims.

Another application of the median age as a summary statistic is shown in Figure 3.5, which portrays the trend towards population ageing in the United States population over the twentieth century. Clearly, population ageing has been a long-standing trend, but one interrupted for a time as a result of higher birth rates during the baby boom. High birth rates increase the representation of children, whereas low birth rates have the opposite effect, leading to higher percentages in older age groups.

The main disadvantage of the median age as a summary statistic is that it reveals nothing about the details of the age distribution: contrasting age structures sometimes have the same median age. It is also somewhat laborious to calculate. The disadvantages of the median age have sustained interest in other indices, such as the index of dissimilarity, which uses a selected standard population for comparison.

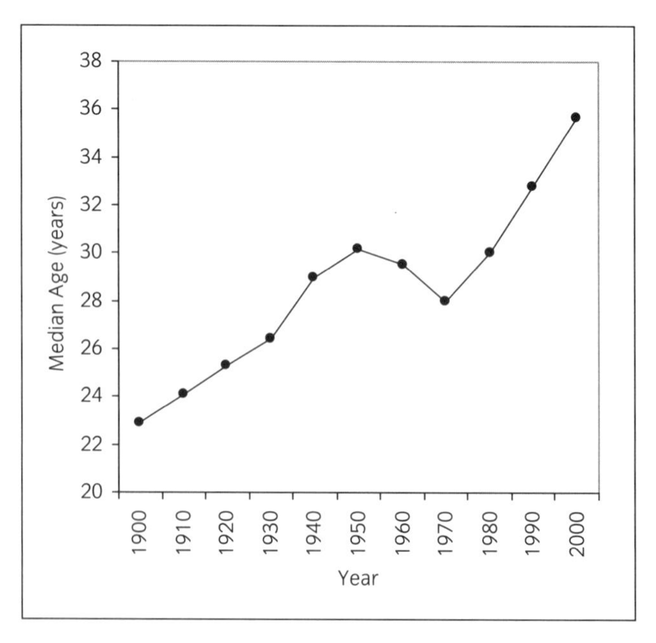

Figure 3.5 Median age of the population of the United States, 1900–2000.

Source: US Census Bureau, *Statistical Abstract of the United States 1998*.

Index of dissimilarity

The index of dissimilarity is a simple yet versatile statistic that may be used to summarize and compare age structures as well as a wide range of other population characteristics. It measures the extent of non-overlap between two percentage distributions, for example, showing the percentage of one population that would need to be redistributed between age groups to match the age distribution of the standard for comparison.

The formula for the index is:

$$I_D = 0.5\sum_{i=1}^{n}|x_i - y_i|$$

where:

the vertical rules denote absolute differences, i.e. treating all differences as positive
x represents the percentages for the standard population – the standard for comparison
y represents the percentages for the population to be compared
i is a data category, such as an age group
n is the number of categories or groups.

The formula states that the index of dissimilarity (I_D) is equal to half the sum (Σ – sigma is the summation sign) of the absolute differences (i.e. treating all differences as positive, as denoted by the vertical rules) between the percentages for the standard population (x) and the percentages for the other population (y) for all age groups (i). Taking half the sum of the absolute differences between the percentages is equivalent to taking the sum of the positive differences or the sum of the negative differences (Shryock and Siegel 1973: 233).

This is illustrated in Table 3.5, where the figures in column B refer to the United States, selected as the standard population, and those in column C to China, the population to be compared. The working is shown in column D of the table.

The percentages for China (column C) are subtracted from the percentages for the United States (column B). For age 0–4 the result is 3.0, i.e. 7.3–10.3, treating the difference as positive; in Excel, =ABS(number) produces this result. The sum of all the differences (28.0) is halved to obtain the index of dissimilarity. The resulting index confirms that the two countries have different age structures: 14 per cent of China's population would need to be in other age groups for its age structure to be the same as that of the United States.

If there are many populations to be considered, the index of dissimilarity provides an efficient way of summarizing differences and highlighting more unusual cases requiring special comment. Using different standard populations can lead to different results, however. Choice of the standard population depends on the purpose of the study, although strong contenders are the total population of which the others are part (e.g. use the total metropolitan population to compare

Table 3.5 Index of dissimilarity: Age structures of the United States and China 1990

Age group A	United States % B	China % C	Absolute (differences) D |B – C|
0–4	7.3	10.3	3.0
5–9	7.3	8.8	1.5
10–14	6.9	8.6	1.7
15–19	7.1	10.6	3.5
20–24	7.5	11.1	3.6
25–29	8.6	9.2	0.7
30–34	8.9	7.4	1.5
35–39	8.0	7.6	0.4
40–44	7.1	5.6	1.5
45–49	5.6	4.3	1.3
50–54	4.6	4.0	0.6
55–59	4.2	3.7	0.5
60–64	4.3	3.0	1.3
65–69	4.1	2.3	1.8
70–74	3.2	1.6	1.6
75–79	2.4	1.0	1.5
80–84	1.6	0.5	1.1
85+	1.2	0.2	1.0
Total	100.0	100.0	28.0
Index of dissimilarity			**14.0**

Sources: United States: http://venus.census.gov/cdrom/lookup/949551316
China: Fan Jingjing (1995).

suburbs), or the population closest to the beginning or end of a development cycle.

When interpreting indices of dissimilarity it is important to note that the greater the number of groups or categories, the higher the probable value of the index. This is because the more detailed the comparison, the more likely it is that differences will occur. Populations compared on the basis of three broad age groupings may be expected to be more similar than when they are compared in terms of 18 age groups, as in Table 3.5.

The theoretical range of the index is from 0 to 100. A zero index would denote that the two percentage distributions were identical, while an index of 100 would

Table 3.6 Illustrations of the range of the index of dissimilarity

Age	Standard population	Population A	Population B	Population C
		% Total population		
0–14	50	50	0	0
15–44	50	50	0	0
45–64	0	0	0	100
65+	0	0	100	0
Total	100	100	100	100
Index of dissimilarity:		**0**	**100**	**100**

indicate that they were totally dissimilar. For example in Table 3.6, the index of dissimilarity for population A is 0, whereas for population B it is 100. It may be observed further that populations with contrasting characteristics can sometimes have the same index of dissimilarity, as instance populations B and C in Table 3.6. Although the theoretical range is up to 100, age structures with all age groups represented will seldom have indices of dissimilarity greater than 40 per cent, which would denote an extreme contrast, such as between the unimodal and triangular age structures in Figure 3.6.

Chapter 10 discusses further applications of the index of dissimilarity in relation to population distribution. As previously mentioned, the index is valuable in an even wider range of contexts including population comparisons in terms of birthplaces, occupations, educational attainment and other characteristics. The principal advantage of the index of dissimilarity is that it provides single summary numbers from a vast amount of detail, thereby drawing attention to similarities and differences between populations. Interpretation of the results still calls for reference to the original statistics, but the index greatly facilitates comparative work.

3.5 Concepts in the interpretation of age–sex composition

The preceding sections have shown that there are various ways of displaying and summarizing information on age–sex composition, including age–sex pyramids, area graphs of age composition, sex ratios, dependency ratios, the median age and the index of dissimilarity. Some of these are demographic techniques while others, such as the median age, involve the application of conventional statistical methods to demographic data. This list is not exhaustive but it represents the best-known approaches. Graphical and statistical techniques need to be selected according to the nature of the investigation. Similarly, the approach to interpreting the information depends on the context.

Sometimes statistics on age–sex composition serve as background to other issues and require little comment in their own right; alternatively, the specific objective may be to analyse the age structure of a community or a country. The process of fulfilling such a task cannot be reduced to ritual performance of a prescribed series of steps – much depends on the aims of the study and the nature of information already available in the literature. Commonly the work will involve: (i) description, (ii) explanation and (iii) consideration of implications, although the emphasis on each will vary.

This section highlights some descriptive terms and concepts needed in the interpretation of age–sex composition, whereas Chapter 5 presents some practical strategies for analysing population pyramids, in the context of general approaches to the interpretation of demographic statistics (see Table 5.4). The particular concern here is with types of age structures – the terminology for describing population pyramids – and with the demographic transition (see Section 1.2), which is central to the explanation of national trends in age structure evolution. These and other concepts such as *population ageing* and *momentum* (see Chapter 9) have applications in studying the implications of age structure change for planning and policy-making. Exploring implications also calls for recognition of interrelationships between features of the age–sex structure and other aspects of the population, such as its growth rate and components of growth, as well as wider linkages between population and social and economic trends. As in all interpretative work, due attention must be given to the published literature on the subject.

Types of age structures

There is no agreed formal terminology for describing age structures, but there are a number of frequently used terms that are valuable in evoking a mental image of the age composition of community, regional or national populations. This partly entails comparing age structures with geometric forms or with the shape of everyday objects. Thus some structures are described as triangular, rectangular or bell-shaped. Complementing these terms are others that are indicative of a population's stage of development ('young', 'mature', or 'old') or which provide a more formal statistical description ('unimodal', 'bimodal', 'symmetrical', 'asymmetrical'). Figure 3.6 illustrates some of the main types of population pyramids. Interpretation of age structures should establish the nature and effects of formative processes – the immediate demographic causes of change. A deeper analysis calls for discussion of the underlying social, economic, political or environmental forces of change.

The demographic transition

The sequence of 'young' (triangular-shaped), 'very young', 'mature' and 'old' (rectangular-shaped) types of pyramids in Figure 3.6 can denote a cycle of change

Description	Examples
Young Broad-based, triangular age profiles with high proportions of children. In the demographic transition, these have high fertility and high mortality	
Very young Very broad-based, triangular profile with very high proportions of children. Associated with large families and declining mortality.	
Mature Transitional between young and old types of profile but still with a relatively high representation of children. Evening up of numbers in younger and middle age groups can denote the persistence of 'replacement level fertility' (see Chapter 7).	
Old Rectangular age profile with similar numbers or percentages in each age group up to those where mortality is high. In national populations this form is indicative of low birth and death rates, while in small communities it denotes demographic heterogeneity – entailing a mixing of groups at different stages of life.	
Undercut Appreciable deficit in the child age groups such as arising from a recent decline in the birth rate.	
Declining Numbers and proportions diminish through younger age groups, denoting long-run persistence of low fertility. The population has 'negative momentum' (see Chapter 9).	
Unimodal Pronounced peak in one age group or a number of adjacent groups. This type of pattern can arise from the inward migration of young adults or recently retired people. Within cities, the presence of institutions, such as hostels and hotels, may produce such age structures in small areas.	
Bimodal Dual peaks, commonly denoting relatively large numbers in parent and child ages in suburbs.	

Figure 3.6 Types of age profiles.

through the demographic transition. Declining mortality makes populations younger, because more children survive. Hence the proportion of children increases early in the transition, transforming young age profiles into younger, wider-based profiles (Table 3.7). Thus, unlike people, populations can experience physical rejuvenation, as illustrated also in the median age graph for the United States (Figure 3.5).

Later in the transition, population ageing occurs as fertility decline reduces the proportion of children. Successive generations then become similar in size, as evident initially in the emergence of 'mature' age structures with similar numbers in parent and child generations. Ultimately 'old' age structures evolve in which the

Table 3.7 Characteristics of populations during and after the demographic transition

	Pre-Transition	Mid-Transition	Post-Transition	Future Declining[1]
Crude birth rate[2]	50.0	45.7	12.9	9.8
Crude death rate[3]	50.0	15.7	12.9	14.8
Annual growth rate %	0.0	3.0	0.0	–0.5
Age structure %				
0–14	36.2	45.4	19.2	15.6
15–64	60.9	52.0	62.3	52.7
65+	2.9	2.6	18.5	31.7
total	100.0	100.0	100.0	100.0
Dependency ratios				
Child[4]	59.0	87.0	31.0	29.6
Aged[5]	5.0	5.0	30.0	60.0
total	64.0	92.0	61.0	89.6
Percentage surviving (females)				
to age 5	46.8	81.7	98.2	99.6
to age 65	7.8	43.3	83.1	94.2
Life expectancy (females)				
at birth	20.0	50.0	75.0	85.0
at age 5	36.6	55.9	71.4	80.3
at age 65	7.5	11.9	15.7	22.2

Notes:
1. Whereas the figures in the other columns derive from demographic models, those in the last column are based on data for Italy, for which the rates refer to 2025–2050, other data to 2050.
2. Crude birth rate: births/population × 1000.
3. Crude death rate: deaths/population × 1000.
4. Child dependency ratio: 0–14/15–64 × 100.
5. Aged dependency ratio: 65+/15–64 × 100.
Sources: Hauser (1976, 66), World Bank (1994, 281), Coale and Demeny (1983), Coale and Guo (1990, 33).

numbers in successive age groups are similar below the advanced ages where mortality is concentrated in post-transitional societies. Overall, fertility decline has the greatest impact on the percentages in older age groups during the demographic transition, especially because it reduces the relative numbers of children. In contrast, mortality decline has a smaller effect on the percentages in older age groups (see Chapter 9), but a dramatic impact on population size, bringing increased numbers through improved survival of infants and children.

One of the disadvantages of the classical transition theory is that it implies that changes are concentrated in the transition phase, whereas age structure transformation actually persists into the so-called post-transition period. It takes many decades before the decline in birth and death rates have their full effects on the numbers and percentages in each age group. Thus growth in the numbers in older ages is typically a delayed result of developments that occurred 60 or more years earlier: the transition in the age structure continues long after completion of the major declines in birth and death rates. These effects are illustrated in the *Age Structure Simulations* module (Box 3.2).

Table 3.7 summarizes age structure changes and related developments over the course of the demographic transition and beyond. Some contemporary populations already have characteristics that are not encompassed within expectations for the post-transition stage. Thus, as well as describing features of the stages of the demographic transition, the table lists potential characteristics of 'future declining' populations, based on a World Bank projection for Italy in 2050. In this scenario population decline is conspicuous, together with very low birth rates, very long life expectancy, and high levels of population ageing. Continuing fertility decline leads to distinctive age structure features, initially including 'undercutting' or deficits at the base of the age pyramid and, ultimately, tapered or 'coffin shaped' age structures with smaller numbers at successively younger ages. Such an eventuality would bring extremely high aged dependency ratios and a return to total dependency ratios similar to those of populations at mid-transition. The *Age Structure Data Base* (Box 3.1) contains examples of the 'future declining' scenario in the population pyramids for Japan and some European countries.

Although reference to the demographic transition and its stages is a useful strategy in interpreting national age structures, it is important to note that immigration and emigration can modify age structures substantially, overriding the impact of changes in birth and death rates. The effects of migration are contributing to divergence in the experiences of societies facing the present reality or the prospect of an excess of deaths over births.

Finally, 'young', 'mature' and 'old' age structures can also emerge in situations where the demographic transition is irrelevant. Suburban age structures, for example, may develop 'mature' or 'old' profiles through a combination of inward migration, outward migration and *ageing in place* (i.e. growing older in the same place of residence), creating a diverse representation of people at different stages of life.

BOX 3.2 **Age structure simulations module (*Age Structure Simulations.xls*)**

This module is an adaptation of a Fortran program and related set of exercises by van de Walle and Knodel (1970). It enables students to study a series of computer simulations which demonstrate the effects on age structures of changes in fertility and mortality. The simulations project populations under different assumptions about birth and death rates. Each model illustrates 100 years of population change by means of population pyramids and accompanying statistics. The models are:

- *Population explosion*: demonstrates the effects of mortality decline.
- *Higher fertility*: demonstrates the effects of an increase in fertility.
- *Desert Island*: consequences ensuing when 1000 young women and 5 young men are marooned on a desert island.
- *Instant transition*: a pre-transition population instantly acquires the fertility and mortality rates of a post-transition population.
- *Population ageing*: transformation of the United States age structure.

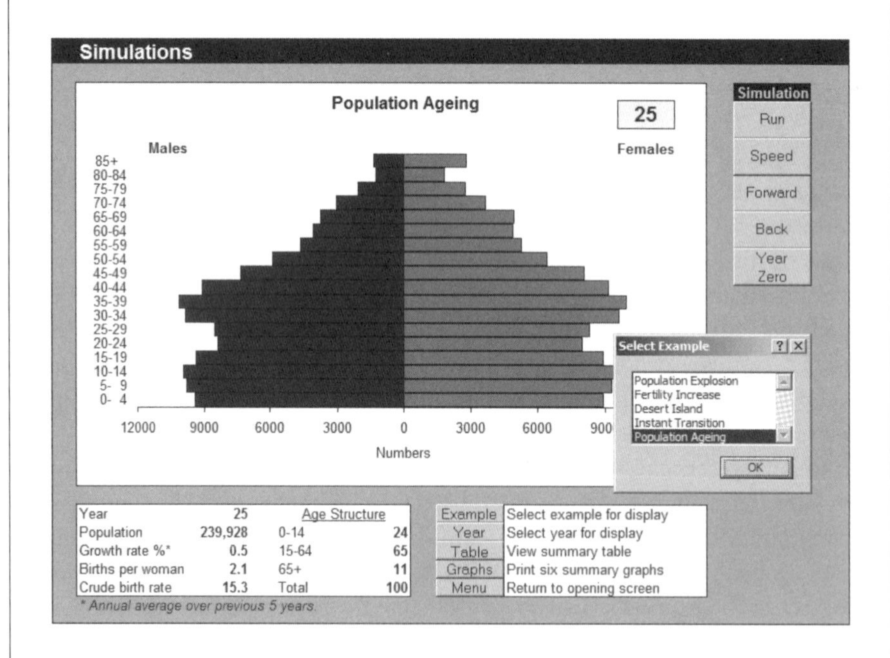

Instructions

To use the module, insert the CD in the drive, start Excel and open the file *Age Structure Simulations.xls*. Enable macros when prompted.

continued

BOX 3.2 continued

- Click **Start** to begin.
- Click **Example** to choose a simulation; the scale of the graph is reset automatically.
- Click **Run** to run a 100 year projection. The **Speed** button changes the speed of the projection. Use the **Forward**, **Back** and **Year Zero** buttons to step through the projections manually.
- The **Tables** button displays summary figures in ten year steps for each simulation.
- Click **Graphs** to print a set of six graphs for 20 year steps in the selected example.
- Click **Home** to return to the opening screen.

See Appendix B for further advice on using the Excel modules.

Characteristics of the simulations

Simulation	Graph scale	Initial age structure	Mortality	Fertility
1 Population explosion	000s	Stationary population with a female life expectancy of 30 years (West level 5)	Age-specific death rates for a population with female life expectancy of 60 (West level 17)	Total births per woman (Total fertility rate) = 4.4
2 Fertility increase	000s	Same as 1	Age-specific death rates for a population with female life expectancy of 30 (West level 5)	Total births per woman = 8.1
3 Desert island	As shown	As shown	Same as 2	Same as 1
4 Instant transition	000s	Same as 1	Same as 1	Total births per woman = 2.4
5 Population ageing	000s	United States population, 1969	Age-specific death rates for a population with female life expectancy of 67	Total births per woman = 2.1 (replacement level)

Notes:
Total fertility rate = number of births per woman, assuming birth rates remain constant.
Age-specific death rate = death rate for a given age group.
'West level 5' and 'West level 17' refer to model life tables in Coale and Demeny (1983).

Illustrative applications

Before viewing each simulation, read the following notes and questions. Write down answers to the questions as you view each simulation. (The answers are at the end.) Technical details about each simulation are given in the above table. Two definitions are needed here:

continued

BOX 3.2 continued

- A *stationary population* is constant in size and age structure – it has zero growth, which classical transition theory envisaged as characterizing pre-transition and post-transition populations.
- A *stable population* has a constant growth rate and a constant age structure; thus a stationary population is a special case of a stable population because it has a constant growth rate of zero.

Populations in which birth and death rates are constant eventually develop a constant growth rate and constant percentages in each age group: in other words they become stable, or stationary if the numbers of births and deaths are equal. In each simulation, the birth and death rates are constant through time. After 100 years, simulations 1 and 2 produce stable populations, while simulations 3, 4 and 5 produce stationary populations.

1 Population explosion

At year zero, this model represents a pre-transition population in which life expectancy at birth is 30 years for females and 28 years for males. The initial population is stationary – it has a zero growth rate and constant percentages in each age group.

In the simulation, the mortality rates immediately fall to those of a population with a life expectancy of 60 years for females and 56 years for males (i.e. a population experiencing the demographic transition). The model illustrates the effects of a dramatic decline in mortality on the age structure and on population size. The population is projected for 100 years under the new mortality conditions in order to show long-run outcomes.

1. What effect does the change in mortality have on the percentages in each age group? Refer, if needed, to the Summary Table by clicking the button labelled 'Table'.
2. How long do the changes take to occur?
3. Does the fall in mortality affect the crude birth rate?
4. Compare the size of the initial population with the numbers after 50 and 100 years.
5. What effect does the decline in mortality have on the numbers aged 65 and over and the percentages in the same ages? Refer to the Summary Table (click the 'Table' button)
6. Did this population grow older as a result of mortality decline? Explain your answer.

2 Fertility increase

This model has the same initial population as the population explosion model. Its mortality rates remain constant at those of a population with life expectancies of 30 years for females and 28 years for males. The birth rate, however, increases immediately to a level roughly comparable in magnitude with the mortality decrease in the previous model. Thus, the rise in fertility produces nearly the same absolute growth in the population after 100 years as did the fall in mortality.

7. What are the immediate and long-run effects of the increase in fertility on the crude birth rate?
8. Do the percentages in different age groups change as a result of the fertility increase?

continued

BOX 3.2 continued

9. How does this effect compare with that of the mortality reduction in the previous simulation?
10. Is the effect of a fertility increase, on the percentages in different age groups, greater during the first or second 50 years?
11. Is the population's age structure (percentages) becoming constant by the end of the 100 years?

3 Desert island

A thousand young women and five young men (all aged 15–19 at the start) are marooned forever on a desert island. What is the future of this population, assuming there is adequate subsistence? Some unusual marriage patterns develop since the simulation assumes that the population has the 'pre-transition' characteristics of high mortality and fertility (female life expectancy at birth = 30 years, while the birth rates imply 4.4 births per woman). Before running the simulation, sketch the age structure of the population after 10 years and after 50 years. In running the simulation, observe the effects of 'waves' in the age structure on the numbers of children born.

12. Is there any resemblance between the initial age structure and that after 100 years?
13. Compare the size of the total population at years 10 and 100.
14. Compare the percentages in the three broad age groups at year 100 with the percentages for year zero in the 'Population Explosion' model and the 'Fertility Increase' model. Why are they similar?
15. Do you agree that when birth and death rates remain constant a population 'forgets' its initial age structure?

4 Instant transition

Here, an instant demographic transition is imagined – the birth and death rates of the population change immediately from those of a pre-transition population to those of a post-transition population. Thus high birth and death rates associated with zero growth change simultaneously to low birth and death rates, which are also destined to yield zero growth in the long term.

16. Does the simultaneous decline in fertility and mortality have much impact on the percentages in different age groups?
17. Does the population become stationary?

5 Population ageing

This simulation starts with the age structure of the United States population at the 1970 census, age-specific mortality for the same period, and a set of age-specific fertility rates which yield replacement level fertility, whereby each generation of mothers produces just sufficient daughters to replace themselves (i.e. a net reproduction rate of 1, see Chapter 7). Note that the oldest age group is 85 years and over.

continued

BOX 3.2 continued

18. If the United States population experienced replacement level fertility from 1970, would the growth rate shift immediately to zero?
19. Would the growth rate tend towards zero in the long run?
20. By how much does the population increase over the 100 years from 1970?
21. Is the population older or younger at the end of the 100 years? Why?
22. Is there a particular type of age structure associated with zero growth?

Supplementary questions

23. One major source of population change is missing from all the simulations. What is it?
24. Each simulation spans a hundred years, which is sufficient time for the operation of constant birth and death rates to produce a stable age structure. Do such age structures have a characteristic shape?
25. The simulations assume constant birth and death rates from year zero. Is this realistic?
26. Contrast the effects of fertility and mortality change on the age structure; are their effects concentrated at particular ages?
27. How long does it take before the operation of constant birth and death rates produces an age structure with percentages that remain fairly constant?

Answers

1. Falling mortality makes the population younger. The percentage aged 0–14 changes from 31 (year 0) to 37 (year 100).
2. The age structure (expressed in percentage terms) is reasonably stable after only 40 years. In absolute terms, each age group grows continuously.
3. The crude birth rate falls slightly because there are more children in the population.
4. Year zero 5.9 million; 50 years 15.6 million; 100 years 43.8 million.
5. The percentage 65 and over remains much the same, but the numbers increase from 276 000 to 2.1 million.
6. No. The population became younger because mortality decline increases the percentage of children.
7. The crude birth rate falls as the percentage of children grows, then levels off as the age structure becomes stable.
8. The percentage of children increases while the percentage in other age groups falls. A fertility increase makes populations younger.
9. Both make the population younger.
10. During the first 50 years. When life expectancy is short, the effects are concentrated in the younger age groups in the age pyramid.
11. Yes. It resembles a stable population growing at a constant rate.

continued

BOX 3.2 continued

12. No, the population forgets its past.
13. 10 = 1922; 100 = 3662.
14. They are stationary populations with constant fertility (4.4 births per woman) and constant mortality (female life expectancy = 30 years). The given birth and death rates ultimately result in zero growth.
15. Yes. The outcome depends on the birth and death rates, not the initial conditions.
16. Yes. Lower fertility makes the population older. The percentage of children falls from 31 to 22. Fertility change overshadows the effects of mortality change (see also Section 9.6).
17. Yes. After 60 years the percentages change little. Replacement level fertility results in a constant age profile and zero growth.
18. No, because of population momentum (see also Section 9.5).
19. Yes, after 60 years the growth rate is approaching zero.
20. The population increases by 32 per cent (see also Section 9.5).
21. Older, because of the lower birth rate, which was previously above replacement.
22. No. The type of profile depends on the particular birth and death rates.
23. International migration.
24. No. See question 22.
25. No. Actual populations do not have constant birth and death rates, although some pre-transition populations may have, if averaged over many years.
26. Fertility and mortality changes mainly affect the youngest age groups.
27. About 70 years.

3.6 Conclusion

Age structures represent the starting point for many studies of population. They are an invaluable guide to the past, present and future of populations. Effects of fertility changes, wars, epidemics and famines, together with policies on immigration and family limitation, are reflected in age structures. Obtaining information on the age structure of a population and plotting the information on a graph is often the first step in seeking to understand the nature of processes affecting populations, and the likely course of development in the future. A nation's age structure can portray a summary of its demographic history, revealing the combined and cumulative impact of all demographic processes. This being so, they are an essential guide to considering prospective developments and policy concerns.

Knowledge of concepts relevant to the description and explanation of age structures are essential points of reference when interpreting the statistics. The task of interpretation can be facilitated though recourse not only to graphs of age–sex pyramids but also to summary indices that assist comparisons and highlight important features. This chapter has discussed the principal methods and introduced concepts in the analysis of age structures. Later chapters extend this material through sections on the mechanics of analysing age–sex pyramids (Chapter 5), the age structures of stationary and stable populations, population ageing (Chapter 9), estimating age-specific net migration (Chapter 11), and projecting populations by age and sex (Chapter 12).

Study resources

KEY TERMS

Age heaping
Aged dependency ratio
Ageing index
Area graphs
Caretaker ratio
Child dependency ratio
Cumulative frequencies
Dependency ratio
Economic dependency ratio
Index of dissimilarity
Mean and modal age
Median age
Open-ended age interval
Pie graphs
Population ageing
Population pyramid
Sex ratio
Symmetrical and asymmetrical age distributions
Unimodal and bimodal age distributions
Young, mature and old populations

FURTHER READING

Bouvier, Leon F. and De Vita, Carol J. 1991. 'The Baby Boom: Entering Midlife'. *Population Bulletin*, 46 (3).

Easterlin, Richard A. 1996. 'Economic and social implications of demographic patterns', in Robert H. Binstock, and Linda K. George (editors), *Handbook of Aging and the Social Sciences* (fourth edition). Academic Press: San Diego, pp. 84–93.

McDonald, Peter and Kippen, Rebecca. 2001. 'Labor Supply Prospects in 16 Developed Countries, 2000–2050'. *Population and Development Review*, 27 (1): 1–32.

Newell, C. 1988. *Methods and Models in Demography*. London, Frances Pinter, Chapter 3, 'Age and Sex Structure', pp. 22–34.

Shryock, Henry S. and Siegel, Jacob S. 1973. *The Methods and Materials of Demography* (two volumes). Washington: US Bureau of the Census, US Government Printing Office, Chapter 7, 'Sex Composition', pp. 188–200 and Chapter 8, 'Age Composition', pp. 201–251.

Weeks, John R. 2002. *Population: an Introduction to Concepts and Issues*. Belmont, CA: Wadsworth, Chapter 8, 'Age and Sex', pp. 293–335.

INTERNET RESOURCES

Subject	Source and Internet address
Document on *World Population Monitoring*	***United Nations Population Division*** http://www.un.org/esa/population/unpop.htm
Population pyramids for any country in a wide range of years	***US Census Bureau*** http://www.census.gov/ipc/www/idbpyr.html
World Population Data Sheet, including summary data on age composition	***Population Reference Bureau*** http://www.prb.org/Content/NavigationMenu/Other_reports/2000-2002/2001_World_Population_Data_Sheet.htm
Animated population pyramids for Canada, provinces and territories	***Statistics Canada*** http://www.statcan.ca/english/kits/animat/pyca.htm
Population Reference Bureau Home Page, includes links to coverage of 'older population and youth' (see Topics list)	***Population Reference Bureau*** http://www.prb.org/
Age–sex data, population pyramids and other demographic information for counties of Minnesota	***State Demographic Center, Minnesota Planning*** http://www.mnplan.state.mn.us/demography/

Note: the location of information on the Internet is subject to change; to find material that has moved, search from the home page of the organization concerned.

EXERCISES

1 Why is it potentially misleading to draw population pyramids in which the percentages are based, not on the total population, but on separate totals for each sex?

2 Explain the advantages and disadvantages of dependency ratios as summary measures of age structure and dependency.

3 Referring to Figure 3.2, identify the demographic processes that would account for the differences between the world's age structure in 2000 and 2100.

4 Suggest reasons why sex ratios and age structures for street blocks and other small areas may be unusual. For example, the sex ratios, or the percentages in particular age groups, may be very high or very low.

5 From the statistics below, on the age structures of Japan and Hong Kong in 1950, calculate the following summary figures for each population:

(a) the sex ratio at ages 30–34, 60–64 and for all ages combined;

(b) the median age of the total populations;

(c) the median age of persons of labour force age, 15–64 years;

(d) child, aged and total dependency ratios;

(e) the ageing index.

Age group	Japan 1950 ('000)		Hong Kong 1950 ('000)	
	Males	Females	Males	Females
0–4	5699	5476	167	149
5–9	4901	4771	67	58
10–14	4452	4344	84	75
15–19	4354	4283	115	99
20–24	3844	3915	127	108
25–29	2807	3379	114	99
30–34	2372	2843	101	85
35–39	2394	2687	83	73
40–44	2205	2287	59	57
45–49	2031	1998	38	44
50–54	1720	1671	25	36
55–59	1380	1375	15	25
60–64	1107	1195	8	16
65+	1737	2398	14	37
Total	41003	42622	1017	961

Source: United Nations 1993. *The Sex and Age Distribution of the World Populations: the 1992 Revision.*

6 From the source statistics for question 5, calculate the index of dissimilarity between the age structures of the two populations, using combined figures for males and females.

7 If you have completed the spreadsheet exercise on drawing population pyramids, use Excel to draw population pyramids (showing percentages) for Japan and Hong Kong in 1950.

8 Write a comparison of the age composition of the populations of Japan and Hong Kong in 1950, making full use of the measures obtained in questions 6 and 7.

9 Draw a population pyramid for your home community, or for a town or city of interest to you. Explain your decision to plot numbers or percentages and write an interpretation of the graph. Discuss potential information sources, and comparisons with other populations, that might improve your understanding of the population pyramid.

SPREADSHEET EXERCISE 3: POPULATION PYRAMIDS

The construction of population pyramids illustrates many of the procedures for creating and formatting graphs in Excel. To begin this exercise, type Table 1 and follow the step-by-step instructions to produce an age–sex pyramid for each population. Write formulas in the cells for all figures shown in bold.

	A	B	C	D	E	F	G	H	I
1	**Table 1: Age Structures of the Populations of Italy and Greece, 1990**								
2									
3	Age	**Italy 1990**				**Greece 1990**			
4	Group	Numbers ('000)		Percentages		Numbers ('000)		Percentages	
5		Males	Females	Males	Females	Males	Females	Males	Females
6				(x-1)				(x-1)	
7	0- 4	1474	1391	**-2.6**	**2.4**	279	269	**-2.8**	**2.7**
8	5- 9	1618	1534	**-2.8**	**2.7**	358	334	**-3.5**	**3.3**
9	10-14	2019	1916	**-3.5**	**3.3**	372	346	**-3.7**	**3.4**
10	15-19	2316	2213	**-4.0**	**3.8**	372	349	**-3.7**	**3.4**
11	20-24	2473	2395	**-4.3**	**4.2**	405	375	**-4.0**	**3.7**
12	25-29	2294	2239	**-4.0**	**3.9**	383	353	**-3.8**	**3.5**
13	30-34	2017	2007	**-3.5**	**3.5**	343	352	**-3.4**	**3.5**
14	35-39	1910	1918	**-3.3**	**3.3**	335	332	**-3.3**	**3.3**
15	40-44	1900	1920	**-3.3**	**3.3**	335	336	**-3.3**	**3.3**
16	45-49	1793	1847	**-3.1**	**3.2**	270	295	**-2.7**	**2.9**
17	50-54	1742	1835	**-3.0**	**3.2**	322	351	**-3.2**	**3.5**
18	55-59	1663	1808	**-2.9**	**3.1**	319	353	**-3.2**	**3.5**
19	60-64	1533	1753	**-2.7**	**3.0**	278	308	**-2.7**	**3.0**
20	65-69	1233	1568	**-2.1**	**2.7**	194	227	**-1.9**	**2.2**
21	70-74	758	1051	**-1.3**	**1.8**	157	199	**-1.6**	**2.0**
22	75-79	717	1106	**-1.2**	**1.9**	136	181	**-1.3**	**1.8**
23	80+	557	1146	**-1.0**	**2.0**	124	181	**-1.2**	**1.8**
24	Total	28017	29647	**-48.6**	**51.4**	4982	5141	**-49.2**	**50.8**
25	Source: United Nations, *Sex and Age 1950-2025*, 1992 Revision.								

Constructing the data table

1. Referring to Table 1, type the age group labels; if Excel converts any of them to dates, type a single inverted comma first to ensure that Excel identifies the label as text. Type the age data and calculate percentages based on the total population. Since the formulas on each row need to refer to cells B24 and C24, use absolute references containing $ signs (see Appendix C, section 2.7). For example, the formula in cell E7 will be: = C7/(B$24 + C$24) * 100. Multiply all the percentages for males by minus one. When drawing the population pyramid, the negative values will place the data for males on the left of the vertical axis, while the positive values for females will appear on the right.

2. To centre headings across two or more columns (e.g. Italy 1990), highlight the cells across which the heading is to be centred and click the 'merge and center' button on the formatting toolbar.

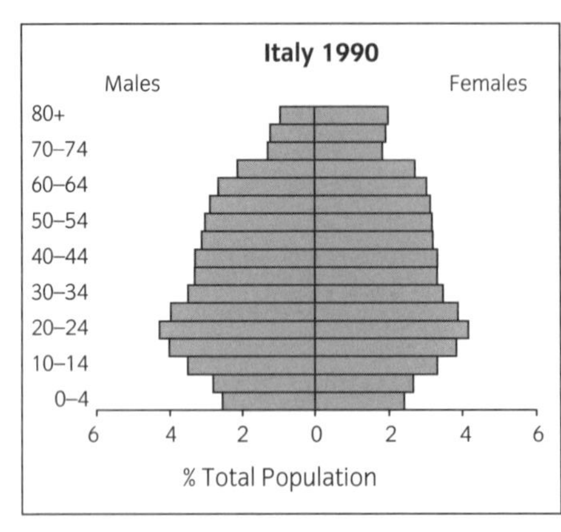

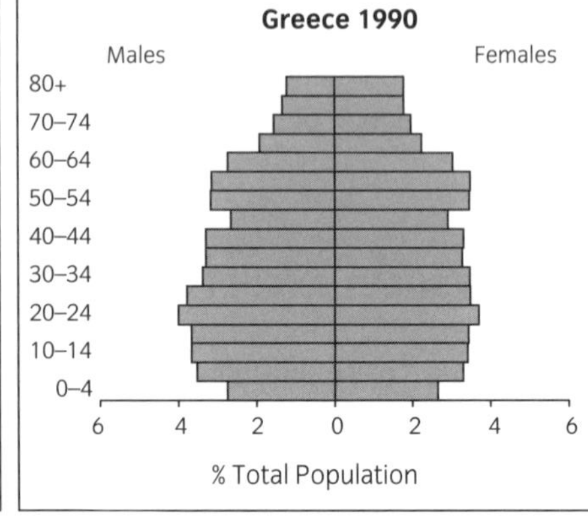

3. Use the borders button on the formatting toolbar to rule lines above, below or beside selected cells. It is advisable to insert lines after the table is finished, however, otherwise unwanted lines may be transferred when copying cells.

Plotting the graph

4. Highlight the age group labels, the negative percentages for males and the percentages for females. (After highlighting the age group labels, hold down the Ctrl key to highlight the other non-adjacent columns.) Do not highlight column headings or column totals. The highlighting should be done without interruption – for example, pressing Ctrl several times can cause unpredictable errors in a graph.
5. Click the Chart Wizard Button and select the first bar chart (other options may produce unsatisfactory results). Step through the Chart Wizard options, adding titles and axis labels (and deleting the legend) as required. When finished, position the chart below the table – move it with the mouse – and resize it with the editing handles.
6. To align the chart with the grid on the spreadsheet, hold down the Alt key while resizing the chart.

Formatting the graph

7. To move the age-group labels from the centre to the left of the graph:
   - click the mouse on the vertical axis (a box will appear at each end of the axis);
   - double click on the vertical axis to bring up the menu of options for editing the axis; choose Patterns Tick Mark Labels Low (i.e. position labels at the low end of the scale). This will place the labels adjacent to the lowest value – the lowest negative percentage for males.
8. To align the bars of the graph, click on one of the bars, then double click the mouse to bring up the editing menu. Choose Format Data Series Options and change the Gap Width to 0, 10 or 20 and set the Overlap to 100.
9. To display the negative numbers for males as positive, click on the horizontal axis, double click to activate the menu, click the Numbers tab, from the Category list choose custom, and in the 'Type' box enter 0.0;0.0 to display positive and negative numbers as positive with one decimal place or 0;0 to display them as whole numbers. Alternatively, choose Category Number and in the Negative Numbers box choose the red number without the minus sign.

Adding titles

10. If you did not add titles and axis labels using the Chart Wizard proceed as follows:
    - chart title: click Chart Chart Options Titles; type a title and press Enter or click OK;
    - label the horizontal axis in the same way (in horizontal bar charts, Excel treats this as the Y axis);
    - to edit the chart title or axis label, click the mouse on the text to be changed, retype and press Enter.

11. To add other text (e.g. 'Males' and 'Females'), click the mouse on any blank section of the chart, type the required text and press Enter. The text will now appear in a box and can be repositioned by clicking the mouse on the frame of the box and dragging it (mouse button down) to the required position; press Enter when finished. Notes and the data source can be added in the same way.

Changing fonts, shading and colours

12. Black and white graphs are usually preferred unless you have a colour printer or you are intending to make coloured slides; coloured charts may print unpredictably on black and white printers.
13. To change fonts: click the required text – edit boxes will appear – highlight the text and click Format Selected Chart Title or Format Selected Axis or Format Selected Axis Title; choose font size and style, then click OK (or press Enter). Selecting the required text and then clicking the right mouse button will also provide access to the relevant menu.
14. To change fonts on the whole chart: click near the perimeter of the chart to activate edit boxes then choose Format Selected Chart Area; choose font size and style; press Enter.
15. To change the colour or pattern of the bars: click on one of the bars, then click Format-Selected Data Series (double clicking on a bar has the same result); from the Patterns tab choose the area colour and pattern you want; click OK (or press Enter).
16. To change the colour of the chart background or plot area: click on the chart background or plot area to activate edit boxes, then choose Format Selected Chart Area or Format-Selected Plot Area. (Double clicking on the background or plot area brings up the same menus). Choose a colour and/or pattern; put on sunglasses and click OK (or press Enter).
17. To change the colour of fonts: click the required text; click the Font Colour button on the Formatting Toolbar and click a colour; click OK (or press Enter).

Creating further charts with the same layout

18. When a chart is completed, further charts, with the same formatting, can be made without having to repeat the previous steps. To do this, click on the chart you wish to use; then on the menu bar click Chart Chart Type Custom Types. In the 'Select from' box (bottom left) choose User Defined, click the Add Button and give the custom chart a name – this will be saved in the list of user defined formats for future use.

    These steps save the formatting of the current chart and enable it to be applied to further charts. For example, when you use the Chart Wizard, choose the Custom Types tab to find the chart formats you have saved. This procedure assumes that you will be working at the same computer next time you want to draw the same kind of chart.
19. An alternative way of creating further charts is to save the original spreadsheet under a new name and type in a new data set. The percentages and the chart will automatically update to show the new data, although the chart title will have to be changed.

20. If you find a mistake in the original spreadsheet data, make the correction and the chart will update automatically.

Printing the chart

21. Click on the chart to select it, and then click the Print Preview button on the Standard Toolbar to see the chart as it will appear when printed. Change the page formatting if necessary by clicking the Setup button. Click Print when finished.

Section II

Analytical approaches

This section introduces key features of demography's approaches to analysing population changes, together with strategies for interpreting population statistics. The subject matter includes the period and cohort perspectives in demography, techniques for comparing rates – especially standardization – and general principles in demographic writing and analysis.

Chapter 4 **Comparing populations**

Chapter 5 **Demographic writing**

Comparing populations

. . . we come next to compare the sickliness, healthfulness and fruitfulness of the several Years, and Seasons, one with another.

(Graunt 1662: 50)

OUTLINE

4.1 Comparing rates
4.2 Direct standardization
4.3 Indirect standardization
4.4 Period and cohort analyses
4.5 Conclusion
Study resources

LEARNING OBJECTIVES

To understand:

- major techniques and concepts for comparing populations
- methods for direct and indirect standardization
- the nature of period analysis and cohort analysis
- the Lexis diagram
- the concept of a synthetic cohort
- the concept of disordered cohort flow

COMPUTER APPLICATIONS

Excel Module: Standardization.xls (Box 4.1)

Spreadsheet Exercise 4: Superimposed and compound population pyramids (See Study resources)

Understanding is deepened through comparing populations, and limited through studying them in isolation. No rate or percentage can be deemed 'high' or 'low', and no set of characteristics can be considered 'more developed' or 'less developed', or 'traditional' or 'modern', for instance, without comparisons with other populations. Concepts and theories, such as the demographic transition, provide a general comparative setting for research, but empirical comparisons are needed to substantiate conclusions and improve explanations. Comparing data for different populations, at national or regional or local levels, is essential in gauging whether populations are distinctive, how much they have changed through time and whether their characteristics are adequately understood. Knowing that 20 per cent of sex workers in Ho Chi Minh City were HIV positive in the year 2000 is an important statistic for policy-makers, but knowing that almost none were HIV positive five years before creates a greater sense of urgency (ESCAP 2001: 3).

This chapter introduces approaches to drawing comparisons in demography, particularly important among which are *period analysis* and *cohort analysis*. Both have broad relevance in research, including applications in studies of population change, social change, and trends in fertility, mortality and migration.

Period analysis employs data for periods of time. Examples include studies of similarities and differences between countries in a given year or of changes in a population over a series of census years, such as 1980, 1990 and 2000. Period analyses are sometimes based just on a single year, but comparisons clarify the significance of findings.

Cohort analysis is less well known by its technical name outside the social sciences, but it features in popular literature on 'the baby boomers', 'lost generations' and 'generation X'. Cohort analysis is 'macro-biography' (Ryder 1965: 859) – the biography of a group rather than an individual – tracing the life experiences of people who were born in the same years (birth cohorts), or who, in the same period of time, married (marriage cohorts) or migrated (migration cohorts). Cohort analysis employs data on the histories of groups, examining their characteristics and behaviour over part or all of their lives. It provides an insightful approach to studying change. Demographers have contributed much to its development, but it is well known in other disciplines such as sociology and social history. Studies of birth cohorts are most prevalent and important because:

- 'People born at roughly the same time tend to experience life course events or social rites of passage at the same time, including puberty, marriage, childbearing, graduation, entrance into the workforce, and death.' (Newman 2000: 476).
- 'People born at the same time also share a common history. A cohort's place in time tells us a lot about the opportunities and constraints placed on its members.' (ibid.). Children born in the first decade of the twenty-first century in the countries of sub-Saharan Africa most afflicted by the AIDS epidemic face sharply contrasting futures compared with cohorts born in the 1960s, when death rates were declining and orphanhood was less prevalent.

To implement period or cohort comparisons, it is preferable to use measures that reveal genuine differences in the occurrence of events, rather than reflecting differences in age structure or other aspects of population composition. For instance, an older population may be expected to have relatively high overall death rates from heart disease, cancer and strokes. Because of the effects of population composition, the least refined demographic measures, including the crude death rate, may not be immediately applicable in comparative studies of populations. Some other, more sophisticated, demographic measures – such as the total fertility rate (see Chapter 7) and life expectancy (see Chapter 8) – do permit valid comparisons, but there are many instances where there is a need to allow for the effects of differences in population composition. Important examples are cross-national, or inter-regional, comparisons of mortality from particular causes of death, because age composition has a great influence on mortality patterns.

Standardization is a versatile technique that addresses this problem of lack of comparability. Although mathematically straightforward, standardization involves the unfamiliar step of giving one population some of the attributes of another, such as giving Kuwait the age structure of the United Kingdom. In 2001, the United Kingdom's crude death rate was 11 per thousand, compared with 2 per thousand for Kuwait, which had one of the lowest crude death rates in the world (Population Reference Bureau 2001a). This did not necessarily mean that people in Kuwait were healthier. The contrast arose mainly from differences in the age composition of the two populations. Other things being equal, the older a population, the greater the annual number of deaths, and the higher the crude death rate (Figure 4.1).

Accordingly, a 'standardized' crude death rate for Kuwait may be calculated by assuming that it has the same age structure as the United Kingdom. This example is discussed later in the chapter, using some detailed data available for 1996. The main purpose of standardization is to facilitate better comparisons of demographic rates, through gauging the effect on them of age structure differences or other differences in population composition. The next section outlines basic methods and issues in comparing rates, as background to the discussion of methods of standardization. Period and cohort analyses are considered later in the chapter.

4.1 Comparing rates

Rate ratios

Comparisons of rates are often undertaken without any further analysis, for instance through a visual inspection of the population growth rates of countries to gain an initial impression of their progress through the demographic transition. More precise comparisons may be obtained through calculating absolute differences between rates, or through employing *rate ratios*. As the name indicates, the latter represent the ratio of two rates. For example, in France the 1950 death rate for males aged 60–64 (25.8 per 1000) was 73 per cent higher than the correspon-

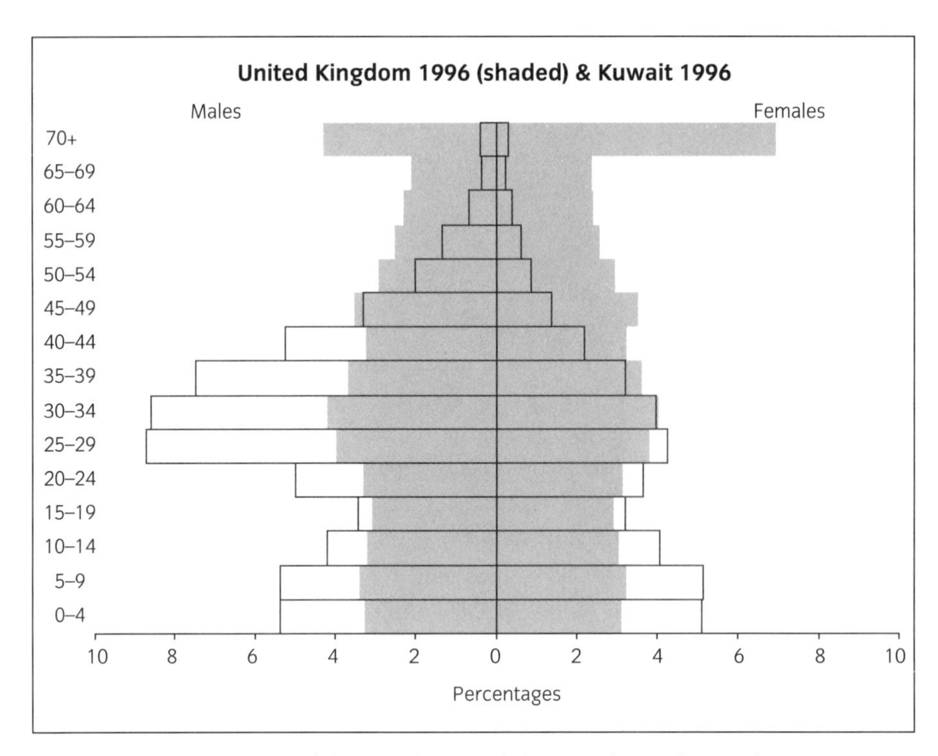

Figure 4.1 Age structures of the populations of the United Kingdom and Kuwait, 1996.

Data source: United Nations 2000. *Demographic Yearbook: Historical Supplement 1948–1997*, New York: United Nations (CD).

ding death rate of females (rate ratio = 25.8/14.9 × 100 = 173), while in 1990 the figure was 273 (rate ratio = 17.5/6.4 × 100 = 273). The rate ratios reveal that although the death rates for both sexes declined, the relative difference in the rates for males and females widened appreciably. This particular rate ratio is known as the *sex ratio of age-specific death rates*. It is discussed further in Chapter 6, together with other examples of rate ratios, namely measures of *relative risk*.

Rate ratios can serve simply to measure the relative magnitude of two rates. More refined approaches seek to calculate rate ratios in which the probable cause of the differences between a pair of rates is identified at the same time, such as by comparing the suicide rate of 30 year old males who are clinically depressed with the suicide rate of those who are not. Such comparisons seek to reveal the role of specific factors in raising or reducing rates. Epidemiologists commonly employ rate ratios to compare the likelihood of getting a disease or dying from one, according to whether people were exposed to a particular health hazard (Rockett 1999: 26). This is a means of determining the *relative risk* of disease or death arising from exposure, for instance, to air or water pollution, radioactive materials, toxic chemicals, alcohol consumption, smoking, a sedentary lifestyle or a high fat diet.

Rate ratios from death rates of nuclear power plant workers and workers in other locations can assist in investigating the relative risk to health of employment in a nuclear power plant. Such studies are very complex, however, as they require detailed information on other matters such as age, duration of employment and the nature of the work, in order to allow for their effects. Other things being equal, younger or newly recruited employees will have had lower exposure to risk. Also, the statistical significance of the differences must also be measured, and further corroborative evidence sought to establish whether a particular health hazard is the real cause of differences (see Rockett 1999; Harper et al. 1994; Kleinbaum et al. 1982).

Effects of population composition

Contrasting rates often arise because two populations differ in their demographic composition. Such differences are more readily investigated than those supposedly due to variations in exposure to hazards or risks. As noted earlier, other things being equal, a population with more older people will have a higher overall death rate from degenerative diseases than a younger population (i.e. deaths from degenerative diseases/total mid-year population × 1000). Comparisons commonly rely, therefore, on standardization techniques to demonstrate the influence of population composition on demographic rates.

Standardization produces new rates that reveal the effect of eliminating the influence of age, or other selected characteristics. Applications include comparisons between crude death rates, or death rates from specific diseases – for countries, regions or social groups. Standardization is also used in studies of fertility and migration. Apart from age, compositional influences examined include sex, education, occupation, income and religion. In the United States, Seventh Day Adventists and Mormons have lower mortality than the general population, probably because of healthier diets and abstaining from tobacco and alcohol (Daugherty and Kammeyer 1995: 150–151). Nevertheless, the most prevalent application of standardization concerns the effect of age composition on death rates.

Why rates differ

When comparing the crude death rates of countries, observed differences will be due to three possible causes, namely:

- *Defective data* for one or both countries, due to under-registration of deaths or other inaccuracies. A seemingly low death rate might reflect that the counting of deaths was incomplete.
- *Differences in age-specific rates of mortality*. Places closer to the start of the demographic transition, or experiencing calamitous events, will have relatively high death rates at each age and elevated crude death rates. High mortality in infancy and early childhood often underlies crude death rates that exceed 20 per thousand. Moreover, wars, terrorism, genocide, famine, floods, earthquakes and epidemics have been continuing factors in excess mortality. Differences in the age-specific rates are responsible for many of the contrasts in crude death rates.

- *Differences in the age composition of the populations*. Generally, the greater the proportion of older people in a population the higher the crude death rate. This leads to the apparent paradox of crude death rates in more developed countries exceeding those in less developed countries. In 2001, crude death rates of less than 10 were prevalent in South and Central America and Asia, whereas rates above this were common in Northern and Western Europe.

Standardization assists in distinguishing the relative importance of (1) differences in the age-specific occurrence of demographic events (deaths, births or migrations), and (2) differences in the composition of populations. It does not address the data quality issue, which has to be examined in advance of standardizing. The Excel module *Standardization.xls* illustrates the technique of standardization and is intended for use in conjunction with the ensuing sections (Box 4.1).

BOX 4.1 **Module on the standardization of rates (*Standardization.xls*)**

This module illustrates direct and indirect standardization, using data on age structures and death rates. The module includes a range of examples from which different standard and comparison populations can be selected, enabling the effects of standardization to be seen in varied settings. Results can be displayed as a simple bar chart (see below), a statistical table showing the source statistics and working, or as a set of charts depicting overall rates, age structures and age-specific death rates for the two populations selected.

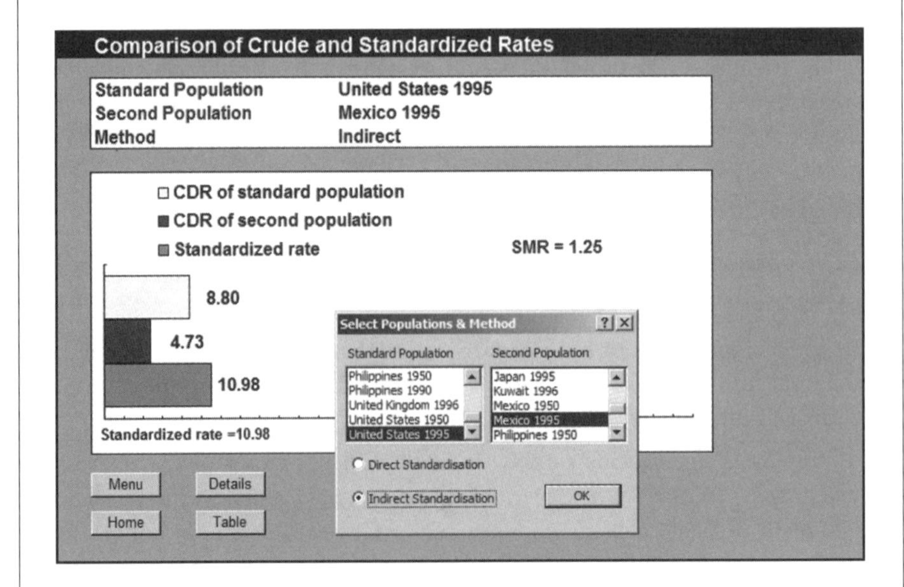

continued

BOX 4.1 continued

Instructions

To use the module, insert the CD in the drive, start Excel and open the file *Standardization.xls*. Enable macros when prompted.

- Click **Rates** to apply direct and indirect standardization to selected data.
- Click **Details** to view graphs of deaths and age structures.
- Click **Table** to see the results in table form.
- Click **Home** to return to the opening display.

Illustrative applications

To view the previous example on the United Kingdom and Kuwait: select the **Rates** screen, click the **Menu** button and set the standard population to the United Kingdom 1996, the second population to Kuwait 1996, and the method to direct standardization. Now consider the following:

1. Why is the UK's crude death rate so much higher than Kuwait's?
2. Why is Kuwait's standardized rate so much higher than its crude death rate?
3. Is the directly standardized rate for Kuwait a more realistic measure of that country's mortality?
4. Refer to the **Details** screen to see whether the charts showing age structures and age-specific death rates clarify the conclusions reached.
5. What will happen to the standardized rates if Kuwait becomes the standard and the UK the second population? Try to answer this question before changing the **Menu** selections.

With reference to other standard and comparative populations listed in the **Menu**, answer the following:

6. (Before making each selection) What are the expected CDRs for each population (high, medium or low?) and what is the expected outcome of direct standardization – will the standardized rate be higher or lower than the second population's CDR? Explain your reasoning. After making each selection, refer to the **Details** screen to verify answers.
7. To what extent do direct and indirect standardization produce different standardized death rates? Select examples from the menus and record answers in table form, together with the standardized mortality ratios.

4.2 Direct standardization

There are two approaches to standardization: 'direct' and 'indirect'. While there are no clear reasons for the use of such adjectives (Barclay 1966: 161), they are convenient labels. Direct standardization employs a 'standard' age structure (or other aspect of population composition), while indirect standardization employs a standard set of *age-specific rates* – calculated for each age group:

$$\text{Age-specific rate} = \frac{\text{number of events (e.g. deaths) in the age group}}{\text{mid-year population in the age group}} \times 1000$$

The derivation of age-specific death rates (ASDRs) for the United Kingdom and Kuwait is illustrated in Table 4.1, where the results were multiplied by 1000 to avoid rates consisting largely of zeros.

Drawing on these data, Table 4.2 presents an example of direct standardization, to produce rates that are adjusted for differences in age structure. In applying standardization to the data for the United Kingdom and Kuwait, the question for resolution is: What would the crude death rate of Kuwait be if that country had the same age structure as the United Kingdom? We expect the standardized rate to be higher, in light of Kuwait's younger age structure (Figure 4.1).

Table 4.1 Source statistics for direct and indirect standardization: Age distributions and age-specific mortality for the United Kingdom and Kuwait, 1996

Ages	Total population		Total deaths		Age-specific death rates per 1000	
	UK	Kuwait	UK	Kuwait	UK	Kuwait
A	B	C	D	E	F D/B × 1000	G E/C × 1000
0–4	3763438	183169	6018	620	1.59907	3.38485
5–9	3905281	184198	552	55	0.14135	0.29859
10–14	3689635	144812	655	51	0.17752	0.35218
15–19	3522276	116271	1745	74	0.49542	0.63644
20–24	3802792	151313	2519	107	0.66241	0.70714
25–29	4577590	227957	3307	136	0.72243	0.59660
30–34	4842576	220695	4321	132	0.89229	0.59811
35–39	4289272	187339	5221	168	1.21722	0.89677
40–44	3803542	129984	7129	191	1.87431	1.46941
45–49	4129737	81840	12187	220	2.95104	2.68817
50–54	3465915	50395	17027	209	4.91270	4.14724
55–59	2986370	34108	24784	249	8.29904	7.30034
60–64	2772244	18889	38472	289	13.87757	15.29991
65–69	2646245	10855	62424	298	23.58965	27.45279
70+	6604552	12156	452536	1016	68.51880	83.58013
					Crude death rates	
Total	58801465	1753981	638897	3815	10.86532	2.17505

Data source: United Nations 2000. *Demographic Yearbook: Historical Supplement 1948–1997*, New York: United Nations (CD).

Table 4.2 Example of direct standardization: Mortality in the United Kingdom and Kuwait 1996

Age Group A	Standard Age Structure UK 1996 B	ASDRs per 1000 Kuwait 1996 C	Expected Deaths D B × (C/1000)
0–4	3763438	3.38485	12739
5–9	3905281	0.29859	1166
10–14	3689635	0.35218	1299
15–19	3522276	0.63644	2242
20–24	3802792	0.70714	2689
25–29	4577590	0.59660	2731
30–34	4842576	0.59811	2896
35–39	4289272	0.89677	3846
40–44	3803542	1.46941	5589
45–49	4129737	2.68817	11101
50–54	3465915	4.14724	14374
55–59	2986370	7.30034	21802
60–64	2772244	15.29991	42415
65–69	2646245	27.45279	72647
70+	6604552	83.58013	552009
Total	58801465		749546

Expected deaths in standard population (UK) = 749546
Total standard population (UK) = 58801465
Age standardized death rate (Kuwait):
(expected deaths/total standard pop. × 1000) = 12.75

Note: direct standardization uses a standard age structure.
Data source: United Nations 2000. *Demographic Yearbook: Historical Supplement 1948–1997*, New York: United Nations (CD).

To calculate a standardized crude rate, the steps are:

1. Select a standard population, whose age distribution will be the standard for comparison. The standard population is normally the population of foremost interest in an investigation. The United Kingdom's age structure will be the most informative standard for a study of mortality in the United Kingdom. Choice of a standard is more open if there is equal interest in members of a group of countries, such as OECD countries, or EU countries, although the age

distribution of the aggregate population is one alternative. Different standard populations lead to different results; hence the standard should be chosen to provide the most useful comparisons.

2. List the numbers in each age group of the standard population (Table 4.2, column B) and the age-specific death rates for the other population (column C). For greater accuracy, work with rates per thousand expressed to several decimal places. For ages 0–4 in Kuwait (Table 4.2), the death rate in 1996 was 0.003 per individual (620/183 169 = 0.003) compared with 3.385 per 1000 (620/183 169 × 1000 = 3.385): the latter figure will yield more accurate results in subsequent calculations. It is difficult to justify appearances of mathematical precision, however, when the quality of the source data is deficient.
3. Calculate the expected number of deaths that would occur in a year if the standard population (UK) experienced the age-specific death rates of the second population (Kuwait). This entails multiplying each age group in the standard population by the corresponding ASDR for the second population – dividing the rates by 1000 if they are expressed as rates per 1000 (Table 4.2, column D). Add the figures to obtain the total expected deaths in the standard population, i.e. 749 546. Using the standard age structure in this way is equivalent to giving Kuwait the age structure of the United Kingdom.
4. Calculate the age-standardized crude death rate. To do this, divide the total expected deaths by the total standard population, then multiply by 1000 (i.e. 749 546/58 801 465 × 1000 = 12.75). The result indicates that Kuwait's crude death rate would be 12.75 per thousand if it had the same age structure as the United Kingdom – far higher than Kuwait's observed crude death rate in 1996 of 2.2 (Tables 4.1 and 4.2).
5. Finally, compare the original crude death rates and the age-standardized rate. In the example, the aim is to compare the United Kingdom's CDR with that of another country to learn whether the former's mortality is relatively high or low. Cross-national comparisons are valuable indicators to governments about their country's standing in relation to measures of survival and population health.

   The calculations reveal that if allowance is made for the United Kingdom's older age structure, Kuwait would have a higher crude death rate of 12.7 compared with 10.9 for the United Kingdom in 1996 (Table 4.1). Further comparisons could be made with other countries besides Kuwait, to establish broadly whether the United Kingdom's death rate compares favourably or unfavourably. The standardization module enables a wide range of other comparisons (Box 4.1).

In summary, direct standardization by age involves applying different age-specific rates to a standard age distribution according to the formula (Shryock and Siegel 1973: 419):

$$\text{standardized rate (direct method)} = \frac{\text{total expected deaths}}{\text{total standard population}} \times 1000$$

$$= \frac{\sum m_a P_a}{P} \times 1000$$

where

P or ΣP_a = the total standard population
P_a = the standard population at each age
m_a = age-specific rates for the second population.

Capitals in the formula denote figures for the standard population. The direct standardized rate is the rate that a population would have if its age structure was the same as that of the standard population. Thus, the age-standardized death rate for the standard population would be the same as its own crude death rate (Armitage and Berry 1987: 403).

Interpreting standardized rates

The main benefit of direct standardization is that it facilitates comparisons by removing the confounding effects on rates of age structure variations, or other differences in population composition. It produces manageable summary statistics that are readily compared with the original figures. An essential check on the accuracy of the calculations is to consider whether the results are consistent with prior expectations.

In interpreting and presenting findings, another question arising concerns whether the standardized rates should be used instead of the original crude rates. Standardized rates provide comparisons, but do not substitute for the original rates. They are constructs that indicate the effects of compositional influences. They vary according to the standard adopted. Kuwait, for instance, does not have the age structure of the United Kingdom. As artificial indices to aid comparisons, standardized rates are often presented in conjunction with the unstandardized rates. Also, standardization by age does not remove all confounding influences, although age is usually the most important compositional influence on mortality (Barclay 1966: 166).

An enlightening alternative, or supplement, to standardizing is to compare the age structures (Figure 4.1) and sets of age-specific rates (Table 4.1) for the populations concerned. Kuwait has a relatively young age structure, but one greatly affected by international migration of men, mainly from other Arab countries for employment in the oil industry and reconstruction following the Gulf War. Kuwait has a small population, but around 10 per cent of the world's oil resources. The young age structure, in conjunction with progress in mortality control resulted in their being nearly 45 000 births in Kuwait in 1996, but fewer than 4000 deaths. Compared with the United Kingdom, Kuwait's age-specific death rates are appreciably higher for children and older people, but lower in the working ages, probably because of the impact of international migration, which is selective of people in good health (Table 4.1).

As Armitage and Berry (1987: 405) note: 'Any method of standardization carries the risk of over-simplification, and the investigator should always compare age-specific rates to see whether the contrasts between populations vary greatly with age.' Standardizing nevertheless offers brevity and a clear basis for comparisons.

Standardizing by more than one variable

Sometimes it is desirable to standardize by more than one variable, such as age and sex. The aim might be to allow for the effect on the CDR of differences in both age composition and the representation of the sexes. Rates standardized by more than one variable are calculated by determining expected deaths separately for each variable and adding them together to obtain total expected deaths. The rates must be specific to each of the variables considered. Just as standardizing by age means using a standard age distribution, standardizing by age and sex means using a standard age–sex distribution.

The example in Table 4.3 illustrates the procedure, which is much the same as that for one variable. There is a massive imbalance in the representation of the sexes in Kuwait, producing a sex ratio of 159 : 100 (Figure 4.1). However, because there are excess numbers of men in the working ages, the female population of Kuwait is actually younger than the male population. Standardizing by age and sex might be expected to reflect this feature more strongly, resulting in a higher standardized rate.

To standardize by age and sex, calculate the expected deaths separately for each sex using the age–sex distribution of the standard population – the United Kingdom (columns B and C) – and age and sex specific death rates (columns D and E) for Kuwait. Add the expected deaths to obtain the total expected deaths (columns F and G), then, as before, divide by the total standard population ($761\,307/58\,801\,465 \times 1000 = 12.95$). The new rate, standardized by age and sex, is slightly higher than the age-standardized rate, because the influence of the younger age structure of the female Kuwaiti population is better recognized. This is confirmed through comparing the crude death rates and standardized death rates calculated separately for the male and female populations of Kuwait: namely 2.2 and 10.3 respectively for males, and 2.1 and 15.5 for females.

4.3 Indirect standardization

An important application of standardization is in comparisons of particular causes of death, such as cardio-vascular disease, to determine whether some populations are at greater risk than others. Typical questions include: Are managers more at

Table 4.3 Example of direct standardization by age and sex: Mortality in the United Kingdom and Kuwait 1996

Age group	Standard age structure United Kingdom		ASDRs per 1000 Kuwait		Expected deaths	
	Males	Females	Males	Females	Males	Females
A	B	C	D	E	F B × (D/1000)	G C × (E/1000)
0–4	1929315	1834123	3.50989	3.25380	6772	5968
5–9	2002186	1903095	0.39345	0.19965	788	380
10–14	1894505	1795130	0.47623	0.22435	902	403
15–19	1809664	1712612	0.89814	0.35621	1625	610
20–24	1950076	1852716	0.87024	0.48452	1697	898
25–29	2339288	2238302	0.65189	0.48285	1525	1081
30–34	2465502	2377074	0.72196	0.32991	1780	784
35–39	2166124	2123148	0.97676	0.71056	2116	1509
40–44	1906308	1897234	1.54056	1.30009	2937	2467
45–49	2064945	2064792	2.73508	2.57561	5648	5318
50–54	1726587	1739328	4.49144	3.35151	7755	5829
55–59	1478284	1508086	7.17866	7.57146	10612	11418
60–64	1354592	1417652	16.49640	13.24313	22346	18774
65–69	1242151	1404094	26.23409	29.23182	32587	41044
70+	2526114	4078438	78.29812	90.21719	197790	367945
Total	28855641	29945824			296879	464428

Expected male deaths = 296879
Expected female deaths = 464428
Total expected deaths = 761307
Total standard population = 58801465
Death rate standardized by age and sex (Kuwait):
(sum of male and female deaths/total standard population × 1000) = 12.95

Data source: United Nations 2000. *Demographic Yearbook: Historical Supplement 1948–1997*, New York: United Nations (CD).

risk of cardio-vascular disease than other workers? Do immigrants from Italy, many of whom follow a 'Mediterranean diet', have lower mortality from this disease than members of other ethnic groups? Does Japan's relatively long-lived population have lower mortality from cardio-vascular disease than that of the United States? Allowing for differences in age composition is an obvious first step in answering such questions, but the application of direct standardization here would call for age and cause (and sex) specific data for all of the populations to be compared. Sometimes such detailed data are unavailable for the required countries, or for the particular regional, ethnic, occupational or other groups in question. Moreover, when available, the data may be difficult to use, requiring extensive editing and adjustment to allow for differences in coverage, scope and definitions.

An alternative strategy is to use indirect standardization. This consists of applying a standard set of rates to the age structures (or other characteristics) of different populations. The procedure requires just one standard set of age-specific rates – for the standard population – together with total deaths (e.g. from a particular cause of death) and statistics on numbers in age groups for the other populations. The statistic most often sought from indirect standardization, however, is not a standardized rate but a standardized ratio.

Using statistics for the United Kingdom and Kuwait, Table 4.4 provides a worked example of indirect standardization to obtain a *standardized mortality ratio* together with an *indirectly standardized death rate*. While these data again address the question of whether Kuwait has higher overall mortality than the United Kingdom, age-specific death rates for a particular disease could be used to examine whether cause-specific mortality was higher in Kuwait than in the United Kingdom. The steps in the calculation are:

1. Choose a standard population and list its age-specific death rates (Table 4.4, column B). As in direct standardization, the standard figures are usually those for the population of greatest interest, or aggregate or average figures for a number of populations to be compared.
2. List the age distribution of the second population (Kuwait) – the one to be compared with the standard population (column C).
3. Calculate expected deaths by multiplying each age group by the corresponding age-specific death rate for the standard population (column D); sum the figures to obtain the total expected deaths. The total shows the number of deaths that would occur in a year if the second population experienced the age-specific rates of the standard population.
4. Calculate the ratio of observed deaths in Kuwait to expected death (i.e. 3815/3459 = 1.10). The result is the *standardized mortality ratio* (SMR); standardized ratios for other phenomena are possible, but that for mortality is most common. The ratio is exactly 1 if the observed and expected deaths are the same. SMRs measure similarities between populations, assuming they all experienced a standard set of age-specific death rates.

In the example, the ratio shows that the observed deaths in Kuwait were 10 per cent higher than they would have been if Kuwait's age-specific death rates were the same as those of the United Kingdom, thereby confirming Kuwait's higher overall mortality. As the example illustrates, the standardized mortality ratio is a valuable

Table 4.4 Example of indirect standardization: Mortality in the United Kingdom and Kuwait 1996

Age group A	Standard ASDRs per 1000 UK 1996 B	Age structure Kuwait 1996 C	Expected deaths D C × (B/1000)
0–4	1.599 07	183 169	293
5–9	0.141 35	184 198	26
10–14	0.177 52	144 812	26
15–19	0.495 42	116 271	58
20–24	0.662 41	151 313	100
25–29	0.722 43	227 957	165
30–34	0.892 29	220 695	197
35–39	1.217 22	187 339	228
40–44	1.874 31	129 984	244
45–49	2.951 04	81 840	242
50–54	4.912 70	50 395	248
55–59	8.299 04	34 108	283
60–64	13.877 57	18 889	262
65–69	23.589 65	10 855	256
70+	68.518 80	12 156	833
Total		1 753 981	3 459

Observed deaths in Kuwait = 3 815
Expected deaths in Kuwait = 3 459
Standardized mortality ratio:
(observed/expected deaths) = 1.10
Indirectly standardized death rate (Kuwait):
(SMR × CDR for UK) = 11.98

Note: indirect standardization uses a standard set of rates.
Data source: United Nations 2000. *Demographic Yearbook: Historical Supplement 1948–1997*, New York: United Nations (CD).

statistic for comparing populations; it is capable of summarizing detailed figures for many countries simply through the application of the same standard set of age-specific rates to the age distribution of their populations. The formula below summarizes the steps (Shryock and Siegel 1973: 409):

$$\text{Standardized mortality ratio} = \frac{\text{observed deaths}}{\text{expected deaths}} \times 100 = \frac{\sum m_a p_a}{\sum M_a p_a} \times 100$$

where

M_a = age-specific death rates for the standard population
m_a = age-specific death rates for the second population
p_a = the second population at each age.

Obtaining standardized ratios is often the most useful outcome of indirect standardization. The interpretation of the ratios is reasonably straightforward, although there are complications when they derive from small numbers (see Shryock and Siegel 1973: 409). It is also possible to multiply the CDR of the standard population by the SMR in order to obtain an indirectly standardized death rate, as in the formula below (Shryock and Siegel 1973: 421). In effect, the crude death rate of the standard population is 'adjusted' upwards or downwards by the SMR. This is shown at the end of Table 4.4, where a figure of 12.0 per thousand was obtained, compared with the directly standardized rate of 12.8 (Table 4.2). The two rates differ because they are based on different age structures.

$$\text{Standardized rate (indirect method)} = \text{SMR} \times \text{CDR} = \left(\frac{d}{\sum M_a p_a} \right) \times M$$

where

d = observed deaths in the second population
M_a = ASDRs of the standard population
p_a = the second population at each age
M = CDR of the standard population.

Thus the main disadvantage of an indirectly standardized rate is that it does not keep age structure constant: the SMR, and the CDR of the standard population, derive from different age distributions. Hence a study comparing many populations would produce indirectly standardized rates where each derived from a different mixture of age structure influences. Nevertheless, since populations are most variable over the entire age range, this problem is greatly reduced if the indirectly standardized rates are needed only for certain age groups, such as the reproductive ages. For calculations involving all ages, direct standardization offers the key advantage over indirect standardization of keeping the age distribution constant in all comparisons.

Applications of standardization in mortality studies are best known, but the technique is important also in comparisons of other rates (see Cox 1970). Standardization is one of the major techniques for comparing rates, and hence for assisting in investigating changes through time and differences between populations, especially using period data. The rest of this chapter discusses demography's two main concepts in the comparative study of populations, namely period analysis and cohort analysis.

4.4 Period and cohort analyses

Time in demographic studies

Contrasts between period and cohort analyses arise from their use of different concepts of time (Table 4.5). Periods are identified as *intervals of time*, commonly a year, although census dates, which refer to *points in time*, can also serve as bases for period studies. A study of the characteristics of the United States population at the 2000 census would constitute a period analysis. Period analysis entails the study of populations in particular years, or longer intervals. It is also described as *cross-sectional analysis*.

In contrast, the synonym *longitudinal analysis* is used for cohort analysis, denoting an interest in tracing the experiences of groups of people over successive years. Cohorts are identified according to the *duration of time* since a defining life event, such as birth or marriage. Thus the 1900s *birth cohort* refers to people born in the first decade of the twentieth century, while the 1900s *marriage cohort* refers to people who married in that decade. Cohort studies often compare the circumstances and behaviour of different cohorts. A cohort analysis comparing the experiences of women born in Singapore in 1925, 1950 and 1975 – in relation to

Table 4.5 Concepts of time in demography

Concept	Description
A *point* in time	A particular date, such the date of a census or the date marking the beginning, middle or end of a calendar year.
A fixed *interval*, or fixed period	A span of time, commonly a single year, a quinquennium or a decade.
Duration	The time elapsed since the identifying event or 'event origin'. For example, age denotes the duration of life since the event of birth. For duration of marriage and duration of stay in a place, time is measured from the day or year of marriage and arrival respectively.

Table 4.6 Main types of cohorts

	Birth cohort	Marriage cohort	Migration cohort
Defining characteristic	Born in the same year(s)	Married in the same year(s)	Arrived in the same year(s)
How identified in statistics	Age (duration of life)	Duration of marriage	Duration of stay
Event origin	Year(s) of birth	Year(s) of marriage	Year(s) of arrival

education, marriage, family size, employment and living arrangements – would reveal considerable differences in the lifestyles and life chances of each birth cohort.

The general definition of a cohort is a group of people who experienced the same demographic event, such as birth, marriage, or migration, during the same period of time, usually a single year or a five year interval (Shryock and Siegal 1973). A concise definition of a cohort is: a group with the same 'event origin'. The event origin identifying each type of cohort is the date or year of birth, marriage or migration.

Table 4.6 defines the main types of cohorts in social science research. Some writers use the term 'age cohort', although usually as an alternative to 'age group'. Strictly speaking, there is no such thing as an 'age cohort' since, unlike birth, age measures duration: it is not a demographic event defining the origin of the cohort.

The Lexis diagram

As Table 4.5 showed, demography's three main concepts of time are points, intervals and durations. These may be compared in the Lexis diagram in Figure 4.2. The Lexis diagram is a square grid depicting the location of demographic events in time. Although its invention is customarily attributed to the German statistician Wilhelm Lexis, there is dispute over his claim to authorship (Vandeschrick 2001). The diagram has had applications particularly in analysing mortality and the relationships between period and cohort data (see Wunsch and Termote 1978).

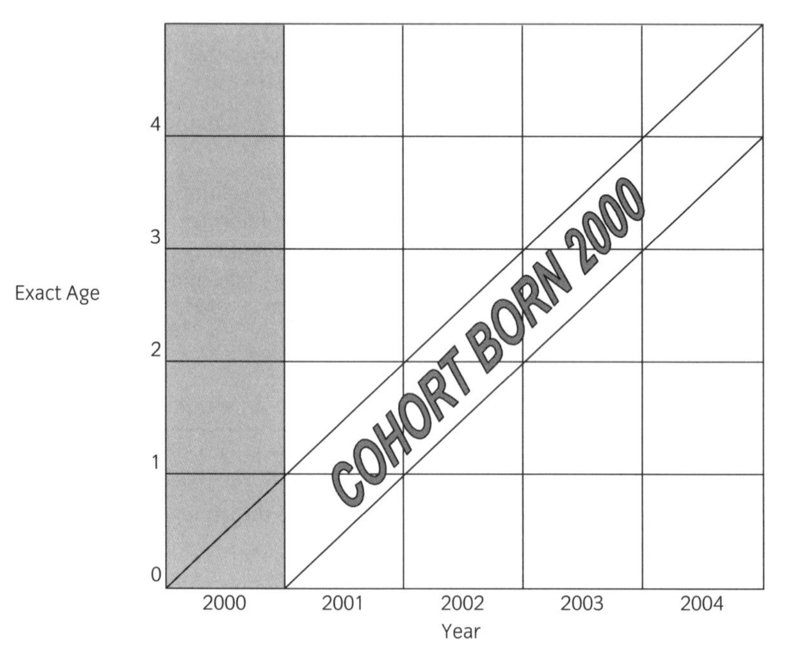

Figure 4.2 A Lexis diagram.

On the Lexis diagram, the horizontal axis depicts calendar time through a series of points and fixed intervals. The points represent dates, such as 1st January, while the intervals are calendar years. The vertical axis represents duration, commonly age or duration of life. Points on the vertical axis are anniversaries of the identifying event, such as birthdays. The scales on the horizontal and vertical axes are the same, producing the square grid.

Diagonal lines across the grid represent the lives of individuals. For example, a child born 1 January 2000 will be 1 year old 1/1/2001, two years old 1/1/2002 and so on. The line connecting these points describes the location of the individual in time and is called a 'life line' or 'line of life'. Individuals born on other days in 2000 have parallel life lines. Thus the diagonal on the Lexis diagram represents the life lines of all individuals belonging to the year 2000 birth cohort.

The Lexis diagram assists in visualizing the differences between period analysis and cohort analysis. Period analysis entails working with data for a vertical cross-section on the diagram of Lexis. The cross-section consists of information for all age groups, or at least a selection of age groups, at the same date or year. The vertical lines on the diagram represent cross-sections at dates, while the shaded vertical band is a cross-section for the calendar year 2000. In contrast, cohort analysis uses data for one group – such as the year 2000 birth cohort, as depicted by the diagonal band. Clearly cohort analysis requires data on the same people over successive years and ages. The difficulty in assembling such data is the main obstacle to cohort analysis.

Period analysis

Comparing the characteristics of populations at one or more points in time is the best-known means to studying differences and changes. There is often considerable interest in obtaining information about a population at a particular time. Thus *period analysis* is the most prevalent approach to demographic inquiry, utilizing census and survey data, vital registration data for calendar years and population projections.

Although period analysis relies on information relating to a 'cross-section' of a population – including all or many age groups in the same year – it can benefit considerably from an awareness of the origins of different groups in the population. The cross-section resembles a 'marble cake' with distinctive layers, rather than a uniform 'blancmange'. Recognizing that a population at any point in time consists of cohorts with varying histories can be a basic step in interpreting cross-sectional data. In this way, period and cohort perspectives become complimentary.

Cohort analysis

Cohorts have distinctive histories, such that groups at particular stages of life differ in numbers and characteristics – as well as needs, wants and resources – from those who went before them and those who will come after. Some differences reflect gradual changes through time, such as improvements in women's access to education and employment, while others constitute demographic shock waves arising from events that had a rapid impact on particular cohorts. Baby booms, immigration booms, divorce booms, wars and economic recessions are examples of events that can have concentrated effects in particular cohorts. Their consequences may call for accelerated adaptations in society as the cohorts progress through the age structure.

The term *generation* is sometimes employed as an alternative to cohort – especially in popular literature whose readers mainly associate cohorts with the army of ancient Rome. This also coincides with the French word for birth cohort, which is *generation*. Yet, 'generation' itself has a range of different meanings in historical, demographic and sociological literature, as summarized in Box 4.2. For demography, Norman Ryder (1972: 546) recommended use of 'generation' to refer to a lineage, rather than a cohort, but this has not become the general usage of the term.

As noted earlier, the birth cohort, a group born in the same year or years, is the most frequently employed unit of cohort analysis. Comparing birth cohorts at the same age or stage of life helps to determine whether they have had experiences in common or whether new developments have occurred. Such comparisons may reveal shifts towards later ages of marriage and childbearing, earlier ages of retirement, and improvements in survival. Studies of the extent of changes in family

BOX 4.2 **What is a generation?**

Meaning	Interpretation
Contemporaries	Generation as people living in a particular historical period.
Lineage	Generation as the line of descent, for example referring to child, parent and grandparent generations in the same family. Ryder (1972: 546) considers this the appropriate use of the term.
Time	Generation as an interval of time, measured as the average difference in age between mothers and daughters. The time separating them, or 'the mean length of a generation', is typically around 29 years (Pressat 1985: 87). In demography, it is calculated as the average age of mothers at the birth of their daughters. The average does not vary greatly between populations. This is because childbearing begins at an early age and finishes late where families are large, but starts later and finishes earlier where families are small.
Age	Generation as an age range or life stage. Found in studies comparing 'older' and 'younger' generations, the focus is on differences between age groups. For example, to describe the aged in Japan as the fastest growing 'generation' is to equate generation with an age range.
Cohort	Generation meaning birth cohort. Ryder (1972: 546) recommended that cohort and generation should not be used interchangeably, but it has become common in contemporary usage.

Source: adapted from Kertzer (1983).

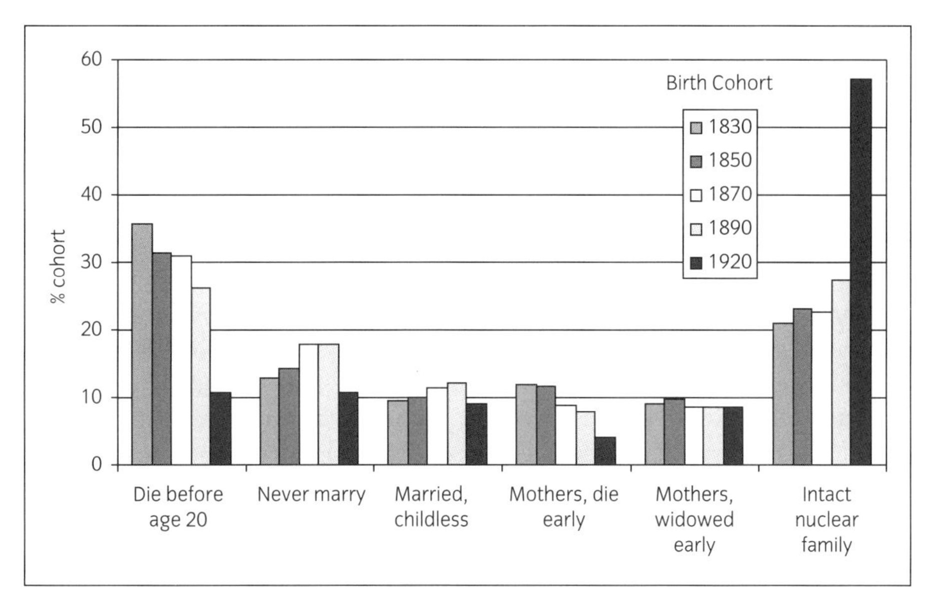

Figure 4.3 Life cycles of cohorts of Massachusetts females.

Data source: Uhlenberg 1979).

building, labour force participation and other aspects of life course experience rely heavily on cohort data wherever they are available.

Uhlenberg's (1979) comparison of cohorts of Massachusetts females illustrates the dramatic changes in life experience that cohort analyses may reveal. Figure 4.3 summarizes the life course experience of females born between 1830 and 1920. Of the cohorts born in the nineteenth century, between 25 and 35 per cent did not survive to age 20. Of those who did survive to this age, many never married, or married but had no children, or died or became widowed before all their children left home. Thus the great majority did not experience an intact nuclear family life cycle entailing marriage, raising children and seeing them all leave home before widowhood occurred. Only for the 1920 cohort was the nuclear family prominent in the cohort's life experience. Uhlenberg's study showed that high death rates, and factors such as economic adversity, made the intact nuclear family an uncommon characteristic of individual life experience for Massachusetts women born in the nineteenth century. The impact of premature mortality was much lower in the 1920 cohort, but even among them the intact nuclear family characterized the lives of less than 60 per cent. The study contradicted notions of the past prevalence of nuclear family living.

Thus cohort analysis traces the changing numbers and characteristics of cohorts over part or all of their lifetimes. This is an informative way of studying change since it permits detailed comparisons of different groups of people according to their age and experience of life stages and events. It also enables present circumstances to be related to antecedent events, such as the effects of wars on cohorts.

Cohorts at the ages for military service suffer the greatest combatant losses in wars. J.M Winter's study of Britain's 'lost generation' of the First World War showed not only the decimation of the cohorts that were in the armed services, but also the particularly heavy losses among the educated elite within these cohorts: they enlisted early and, as officers, suffered the highest casualties. After the First World War there was concern in Britain about the decline of the traditional ruling class, because many of their sons had been killed and others who survived were burnt out as a result of their experiences. Winter's research (1977, 1986) is an insightful example of macro-biography, showing that particular events selectively affect cohorts, and sub-groups within them, potentially with enduring consequences for society.

Thus one of the advantages of cohort analysis is its facility in demonstrating how formative events and social changes affect people differently according to their age, and other characteristics (Kertzer 1983: 45). Cohort analysis also assists in discriminating between lasting and transitory changes. It can answer the question of whether events, such as the Great Depression, had an enduring impact on those most affected at the time, or whether the characteristics of cohorts 'recovered' subsequently (Ryder 1972: 548). Overall, the cohort approach is a valuable means of studying the process of social change through comparisons of cohort careers (Ryder 1965: 844). As noted earlier, the need for suitable data spanning many years is the main problem confronting cohort analysis. Unless survey data are available on respondents' life histories, researchers may have to draw upon a range of different sources.

Synthetic cohorts

The greater availability of cross-sectional information has fostered widespread use of certain demographic indices that combine data for many cohorts at a single date, to provide a convenient summary of the current situation. Examples include the *total fertility rate* (see Chapter 7) and *life expectancy at birth* (see Chapter 8), both of which utilize statistics by age in a calendar year. To obtain the total fertility rate, for instance, *age-specific fertility rates* (the number of births to women aged x per 1000 women aged x) for a calendar year are added, producing a total likened to completed family size or total children ever born per woman. This procedure combines data for different birth cohorts in the same year.

Thus, on the Lexis diagram, a vertical cross-section in a year is used as if it represents the experience of a cohort through time (Figure 4.2). Cohorts based on period data are called *synthetic*, or *fictitious cohorts*, as distinct from *real cohorts* for which there is actual observed information over some years. Only if behaviour remained constant through time would the age-specific figures match the experience of real cohorts. A related problem is that the cross-sectional figures may refer to an atypical year, such as when epidemics, or famines, or fuel shortages in winter, resulted in excessive mortality. This can be ameliorated by deriving synthetic cohort figures from average data for a number of years, as is conventional practice in life table construction (Wunsch and Termote 1978: 60–64).

Interest in the relationships between cohort and period measures of demographic change has led to the development of models and formulas to permit *demographic translation* (Ryder 1964; Wunch and Termote 1978: 53). This entails 'translating' cross-sectional information into cohort information to indicate the long-term situation. Translation thus entails 'making cohort-type inferences from cross-sectional data' (Ryder 1964: 82), rather than simply using synthetic cohort data to summarize the situation at a point in time. The most prevalent difficulty confronting demographic translation is the influence of long-term change.

Disordered cohort flow

An important concept that highlights implications from cohort studies is *disordered cohort flow* (Waring 1975). Changes in society occur as birth cohorts, differing in size or composition, progress from one age group to another. Sometimes the changes are sudden, as unusually distinctive cohorts enter and leave particular stages of life. This creates 'disorder' in the flow of cohorts through the age structure, as opposed to a smooth sequence of changes. Thus disordered cohort flow refers to the movement of distinctive cohorts through the age structure, occasioning rapid changes or 'demographic shock waves' rather than maintenance of a continuous trend.

Some of the best-known demographic shock waves arise from the maturing and ageing of the baby boom cohorts, born in a number of countries after the Second World War. Their relatively large numbers alone have caused, and will continue to cause, surges in demand for age-specific services and support, ranging from maternity wards in hospitals, to places in schools and universities, employment opportunities, housing and consumer goods, and finally pensions and aged care. Ultimately, the demise of the baby boom cohorts will create unprecedented demand in the funeral industry.

To the extent that such shock waves are built into the age structure of a population, changes from cohort flow are more predictable than most developments in society. Similarly, the starting point for determining their implications for public administration and private enterprise is to analyse, and project through time, data cross-classified by age. The *cohort component method* of population projection is well suited to this purpose (see Chapter 12). Cohort characteristics, therefore, provide a notable instance of a basis for forecasting not dependent upon extrapolating long-term developments. Discontinuities or abrupt changes are just as predictable from cohort characteristics as consistent trends.

Other examples of discontinuities in the cohort record for particular countries include the high representation of war veterans, immigrants, childless couples and the well educated in certain cohorts. Their presence foreshadows responsibilities for governments and communities in terms of provision for people with special entitlements or needs, as well as opportunities for businesses in anticipating and adapting to new opportunities in consumer markets. Differences between adjacent cohorts indicate the magnitude of changes. Identifying discontinuities is a way of anticipating upheaval and gaining preparation time.

4.5 Conclusion

'Have you standardized for age?' is a question asked of students and researchers who seem to have overlooked the influence of age structure variations on their findings.

Standardization is a simple, yet powerful, method for removing confounding influences. It has potential applications in the work of social scientists generally through revealing the effects of varied compositional influences and facilitating many kinds of comparisons. This chapter has discussed the usefulness of 'directly' standardized rates, where age-specific rates are available for all the populations under investigation, and 'indirectly' standardized ratios, where they are not.

The chapter has also introduced other techniques and concepts that are relevant to comparative studies generally, namely rate ratios, concepts of time, the Lexis diagram, period analysis, cohort analysis and disordered cohort flow. They provide a foundation for understanding some of the major demographic methods and models discussed later in the book, including cohort fertility, life tables and the cohort-component method of population projection. The concept of a synthetic cohort is especially important in understanding the advantages and disadvantages of cross-sectional measures such as the total fertility rate (Chapter 7) and life expectancy (Chapter 8). Overall, the chapter has introduced concepts and techniques that are important in the comparative study of populations, and hence in the interpretation of demographic data. The next chapter discusses practical considerations arising when undertaking demographic writing.

Study resources

KEY TERMS

Age standardization
Cohort analysis
Cross-sectional analysis
Demographic translation
Direct standardization
Disordered cohort flow
Fictitious cohort
Generation
Indirect standardization
Lexis diagram
Longitudinal analysis
Period analysis
Rate ratio
Real cohort
Standardized mortality ratio
Synthetic cohort
Time: point, interval, duration.

FURTHER READING

Standardization

Armitage, P. and Berry, G. 1987. *Statistical Methods in Medical Research*. Oxford: Blackwell Scientific Publications.

Hinde, Andrew. 1998. *Demographic Methods*. London: Arnold.

Smith, David P. 1992. *Formal Demography*. New York: Plenum Press.

Lexis diagram

Wunsch, G. J. and Termote, M. G. 1978. *Introduction to Demographic Analysis: Principles and Methods*. New York: Plenum.

Period and cohort analysis

Glenn, Norval D. 1977. *Cohort Analysis, Quantitative Applications in the Social Sciences*, Sage University Papers. Beverly Hills: Sage, pp. 5–17.

Hobcraft, J., Menken, Jane, and Preston, Samuel. 1982. 'Age, Period and Cohort Effects in Demography: A Review'. *Population Index*, 48: 4–43.

Ryder, N. B. 1964. 'The Process of Demographic Translation'. *Demography*, 1: 74–82.

Ryder, N. B. 1965. 'The Cohort as a Concept in the Study of Social Change'. *American Sociological Review*, 30: 842–861.

Waring, Joan M. 1975. 'Social Replenishment and Social Change: The Problem of Disordered Cohort Flow'. *American Behavioural Scientist*, 19: 237–256.

INTERNET RESOURCES

Subject	Source and Internet address
Standardized rates of unemployment for OECD countries	**OECD** http://www1.oecd.org/std/lfs.htm
Applications of standardization in comparing disease rates	**British Medical Journal** http://bmj.com/epidem/epid.3.html
Age-standardized suicide rates for New Zealand	**New Zealand Health Information Service** http://www.nzhis.govt.nz/stats/suicidestats-p.html
Standardized rates of coronary heart disease mortality, for the UK and other countries	**British Heart Foundation** http://www.dphpc.ox.ac.uk/bhfhprg/stats/2000/1999/contents.html
Continuing, multidisciplinary studies of birth cohorts in Britain	**Centre for Longitudinal Studies, University of London** http://www.cls.ioe.ac.uk/

Note: the location of information on the Internet is subject to change; to find material that has moved, search from the home page of the organization concerned.

EXERCISES

Age distributions and age-specific mortality, United States and Japan, 1995

Age	Population		Deaths	
	United States 1995	Japan 1995	United States 1995	Japan 1995
0–4	19591148	5995254	35976	7040
5–9	19219956	6540671	3780	1235
10–14	18914532	7477805	4816	1184
15–19	18064517	8557958	15089	3362
20–24	17882118	9895001	19155	5087
25–29	19005343	8788141	22681	4596
30–34	21867796	8126455	35064	5129
35–39	22248914	7822221	46487	6839
40–44	20218805	9006072	55783	12814
45–49	17448898	10618366	65623	24136
50–54	13629862	8921918	77377	32946
55–59	11084606	7953480	96641	44732
60–64	10046478	7475109	138871	68310
65–69	9927958	6396078	204347	89089
70+	23550897	11995717	1489979	615003
Total	262701828	125570246	2311669	921502

Data source: United Nations 2000. *Demographic Yearbook: Historical Supplement 1948–1997*. New York: United Nations (CD).

1. Using the statistics in the above table, and taking the population of the United States as the standard population, calculate the following for Japan:
   (a) The directly standardized death rate.
   (b) The standardized mortality ratio.
   (c) The indirectly standardized death rate.
2. Repeat the above, using Japan as the standard population.
3. Discuss the answers obtained for questions 1 and 2. In doing so, draw comparisons with Tables 4.2 and 4.4, and comment on the observation that use of different standard populations leads to different results.
4. Compile a list of national and international sources of statistics on age-specific mortality.
5. Discuss how to set about using indirect standardization to study the mortality of different occupational, ethnic and religious groups in your own country. Note any potential problems arising in such work.
6. Period and cohort approaches. From three censuses, ten years apart, compile statistics for a selected country on the total numbers of males *or* females in ten year age groups from 0 to 9 to the oldest ten year grouping available. Referring to these figures, comment on:
   (a) Differences in the age structure of the population at the three censuses.
   (b) Differences in the initial sizes of the youngest birth cohorts (at ages 0–9).
   (c) The extent to which cohort sizes change through time (from one ten year age group to another).
   (d) Whether cohort comparisons enhance understanding of the period data.

SPREADSHEET EXERCISE 4: SUPERIMPOSED AND COMPOUND POPULATION PYRAMIDS

Superimposed population pyramids consist of one pyramid overlaid upon another. Compound population pyramids show, within a total population, the age structure of a sub-population, such as the tertiary educated or the foreign-born. Both types of graphs are used for drawing comparisons. This exercise assumes knowledge of the previous spreadsheet exercise on *Population Pyramids*, and is expedited by building on that spreadsheet. The aim is to construct one graph of each type, which can be used in future as a template for plotting other populations. As before, write formulas in the cells for all figures shown in bold.

	A	B	C	D	E	F	G	H	I
1	**Table 1: Age Structures of the Populations of the United States and Mexico, 1990**								
2									
3	Age	**United States 1990**				**Mexico 1990**			
4	Group	Numbers		Percentages		Numbers		Percentages	
5		Males	Females	Males	Females	Males	Females	Males	Females
6				(×-1)				(×-1)	
7	0- 4	9,392,409	8,962,034	**-3.8**	**3.6**	5,569,260	5,349,143	**-6.7**	**6.5**
8	5- 9	9,262,527	8,836,652	**-3.7**	**3.6**	5,317,369	5,136,202	**-6.4**	**6.2**
9	10-14	8,767,167	8,347,082	**-3.5**	**3.4**	5,327,952	5,175,670	**-6.5**	**6.3**
10	15-19	9,102,698	8,651,317	**-3.7**	**3.5**	5,075,483	5,013,815	**-6.1**	**6.1**
11	20-24	9,675,596	9,344,716	**-3.9**	**3.8**	4,028,480	4,144,597	**-4.9**	**5.0**
12	25-29	10,695,936	10,617,109	**-4.3**	**4.3**	3,231,005	3,364,712	**-3.9**	**4.1**
13	30-34	10,876,933	10,985,954	**-4.4**	**4.4**	2,702,641	2,828,178	**-3.3**	**3.4**
14	35-39	9,902,243	10,060,874	**-4.0**	**4.0**	2,209,530	2,320,045	**-2.7**	**2.8**
15	40-44	8,691,984	8,923,802	**-3.5**	**3.6**	1,773,037	1,868,673	**-2.1**	**2.3**
16	45-49	6,810,597	7,061,976	**-2.7**	**2.8**	1,462,551	1,536,229	**-1.8**	**1.9**
17	50-54	5,514,738	5,835,775	**-2.2**	**2.3**	1,169,857	1,240,378	**-1.4**	**1.5**
18	55-59	5,034,370	5,497,386	**-2.0**	**2.2**	953,910	1,031,636	**-1.2**	**1.2**
19	60-64	4,947,047	5,669,120	**-2.0**	**2.3**	748,227	830,118	**-0.9**	**1.0**
20	65-69	4,532,307	5,579,428	**-1.8**	**2.2**	571,540	656,882	**-0.7**	**0.8**
21	70-74	3,409,306	4,585,517	**-1.4**	**1.8**	367,068	435,969	**-0.4**	**0.5**
22	75-79	2,399,768	3,721,601	**-1.0**	**1.5**	256,445	316,400	**-0.3**	**0.4**
23	80+	2,223,792	4,790,112	**-0.9**	**1.9**	247,740	328,340	**-0.3**	**0.4**
24	Total	121,239,418	127,470,455	**-48.7**	**51.3**	41,012,095	41,576,987	**-49.7**	**50.3**
25	Source: United Nations 2000. *Demographic Yearbook, Historical Supplement 1948-1997* (CD).								

Creating a 'User Defined' chart format

1. Open the spreadsheet used for Spreadsheet Exercise 3, and save it under a new name (e.g. Exercise 4) so that you will have copies of both the old and the new files.
2. On the renamed spreadsheet, click on the chart for Italy to activate the chart menus; a black and white chart is best for this exercise.
3. Now click Chart (on the menu bar at the top of the screen) and choose Chart Type Custom Types User Defined Add. Call this 'User Defined' format **Pyramid.** This saves the formatting of the chart under the name 'Pyramid' and enables it to be applied to further charts. This procedure assumes that you will be working at the same computer next time you want to draw the same kind of chart (see also Exercise 3, Section 18).
4. Click on the spreadsheet to exit graph editing. Delete the chart for Greece by clicking on it once and pressing the Delete key.

Data entry

5. In place of the original data from Exercise 3, type the data for the United States and Mexico, omitting the commas. Alternatively, use the previous data for Italy and Greece if preferred.
6. To insert commas in the figures, highlight the data and click Format Cells Number and click the box labelled 'Use 1000 separator (,)'. Alternatively, click the Comma Style button on the formatting toolbar.
7. The percentages for the new populations should appear automatically. If starting Table 1 from scratch, type formulas where the figures are shown in bold, making those for males negative (see Exercise 3, Section 1).

Plotting the graph

8. The chart should now display the statistics for the United States 1990, but the chart title will have to be amended (see Exercise 3, Section 10).
9. To add the data for Mexico to the chart, highlight the percentages for Mexico in Table 1 (omitting the totals), click the copy button on the standard toolbar, click the chart to activate it, then click the paste button on the toolbar. The pyramid for Mexico will now be superimposed on that for the United States.
10. Alternatively, if drawing a new chart for both populations, instead of building on an existing chart, highlight the age-group labels, the percentages for the United States and the percentages for Mexico. (NB: to highlight non-adjacent columns, hold down the Cntrl key and do the highlighting smoothly, without pressing Cntrl more than once). Then:
    - Click the chart Wizard and choose Chart Type Custom Types User Defined Pyramid.
    - If the chart now looks like a rectangle, or some of the bars are truncated, the horizontal scale should be reset. To do this, click on the horizontal axis (an edit box should appear at each end); then click Format Selected Axis Scale (or double click on the axis to bring up the menu) and type new maximum and minimum values – numbers slightly above the maximum and minimum values in the spreadsheet.

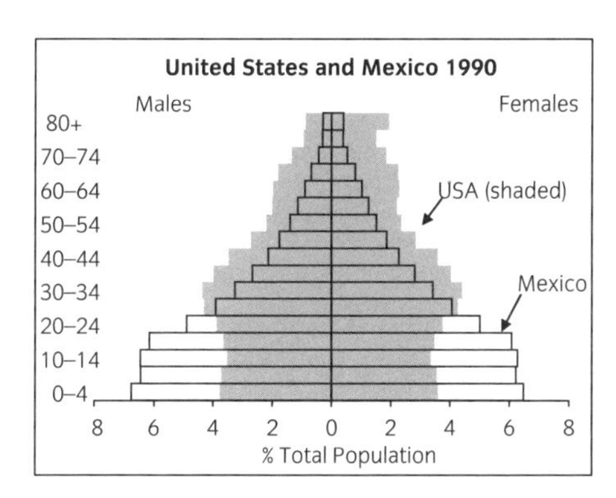

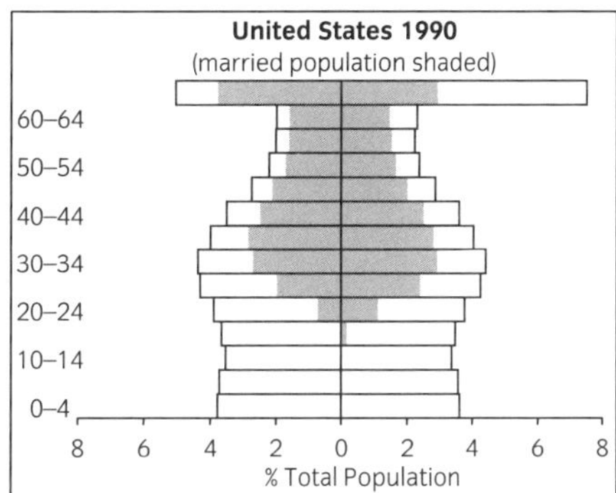

Reformatting the graph

11. Superimposed pyramids can be clearly differentiated if one is shaded while the other is left unshaded. In order to see the outline of both pyramids, it is necessary to make the upper pyramid 'transparent'.
12. Click on the upper pyramid at a point where it does not coincide with the lower one; click Format Selected Data Series Patterns and, in the 'Area' section of the menu, click None then OK – this will enable the shape of both pyramids to be seen. Repeat for the other side of the upper pyramid: click on the other side and click Edit (on the menu bar) then click Repeat Format Data Series.

13. Click on one side of the second pyramid; click Format Selected Data Series Patterns, set Border to None and choose an Area colour to shade the pyramid. Repeat for the other side. Add or reformat titles as described in Exercise 3, Sections 10 and 11.

Adding text and arrows

14. To add arrows, activate the drawing toolbar (View Toolbars Drawing) and click on the arrow button; draw an arrow. An arrow will appear on the screen with edit boxes at each end; click on the edit box at the arrow head and move it to the required position; repeat for the tail of the arrow. Add text to the chart as needed (see Exercise 3, Section 11).

Saving the new chart format

15. When the chart is finished, save its formatting under a name such as **Pyramid2** for future use (see Section 3).

	A	B	C	D	E	F	G	H	I
27	**Table 2: Age Structures of the Total and Married Populations of the United States 1990**								
28									
29	Age	**United States 1990**				**Married Population 1990**			
30	Group	Numbers		Percentages		Numbers		Percentages	
31		Males	Females	Males	Females	Males	Females	Males	Females
32				(x-1)				(x-1)	
33	0- 4	9,392,409	8,962,034	**-3.8**	**3.6**				
34	5- 9	9,262,527	8,836,652	**-3.7**	**3.6**				
35	10-14	8,767,167	8,347,082	**-3.5**	**3.4**				
36	15-19	9,102,698	8,651,317	**-3.7**	**3.5**	177,707	421,080	**-0.1**	**0.2**
37	20-24	9,675,596	9,344,716	**-3.9**	**3.8**	1,794,993	2,792,767	**-0.7**	**1.1**
38	25-29	10,695,936	10,617,109	**-4.3**	**4.3**	4,923,920	5,936,405	**-2.0**	**2.4**
39	30-34	10,876,933	10,985,954	**-4.4**	**4.4**	6,719,234	7,206,307	**-2.7**	**2.9**
40	35-39	9,902,243	10,060,874	**-4.0**	**4.0**	7,044,787	6,947,915	**-2.8**	**2.8**
41	40-44	8,691,984	8,923,802	**-3.5**	**3.6**	6,183,767	6,162,667	**-2.5**	**2.5**
42	45-49	6,810,597	7,061,976	**-2.7**	**2.8**	5,268,012	4,961,062	**-2.1**	**2.0**
43	50-54	5,514,738	5,835,775	**-2.2**	**2.3**	4,265,662	4,099,650	**-1.7**	**1.6**
44	55-59	5,034,370	5,497,386	**-2.0**	**2.2**	4,037,127	3,774,291	**-1.6**	**1.5**
45	60-64	4,947,047	5,669,120	**-2.0**	**2.3**	3,967,364	3,623,934	**-1.6**	**1.5**
46	65+	12,565,173	18,676,658	**-5.1**	**7.5**	9,398,672	7,218,018	**-3.8**	**2.9**
47	Total	**121,239,418**	**127,470,455**	**-48.7**	**51.3**	**53,781,245**	**53,144,096**	**-21.6**	**21.4**
48	Note: Figures on marital status by age available only to ages 65+.								
49	Source: United Nations 2000. *Demographic Yearbook, Historical Supplement 1948-1997* (CD).								

The compound pyramid

16. A compound population pyramid showing the total and married populations of the United States may be constructed from the data in Table 2, which is prepared by copying and editing Table 1. This time, however, the percentages for the second, married, population have to be based on the total United States population.

17. To draw the chart, select the age group labels and the two sets of data, and apply the superimposed pyramid format (Pyramid2) saved earlier. In highlighting the data, the blank cells for children must be included. Amend the titles and formatting as necessary.

18. If it is not possible to select for editing the portion of the graph referring to the married population, right click on the graph itself, choose Format Data Series, click the Series Order tab and use the Move Up and Move Down buttons to change the series order – move the first two data sets to the bottom.
19. To print the charts, see Exercise 3, Section 21.

5 Demographic writing

It may be now asked, to what purpose tends all this laborious buzzling, and groping?
(Graunt 1662: 77)

OUTLINE

5.1 Types of studies
5.2 Starting demographic research
5.3 Interpreting information
5.4 Writing a paper
5.5 Conclusion
Study resources

LEARNING OBJECTIVES

To understand:

- the nature of the principal types of demographic studies
- essential considerations in starting demographic work
- general principles in the interpretation of population statistics
- basic considerations in writing a research paper

COMPUTER APPLICATIONS

Excel module: Pyramid Builder.xls (Box 5.2)

Spreadsheet Exercise 5: Line graphs and pie charts (See Study resources)

In the early 1830s, the British Association for the Advancement of Science established a committee, under the chairmanship of Thomas Robert Malthus, to examine a proposal for the establishment of a statistics section of the Association. This led, in 1834, to the founding of the Statistical Society of London, which later became the Royal Statistical Society (Castles 1986: 6–7). Malthus was a Founding Fellow, but died the same year. As its emblem the Society chose 'a fat neatly bound sheaf of healthy wheat – presumably representing the abundant data collected, and well-tabulated (Figure 5.1). On the binding ribbon was the Society's motto – the Latin words, *Aliis exterendum*, which literally mean "Let others thrash it out."' (Cochrane, cited by Castles 1986: 7).

Figure 5.1 The Emblem of the Royal Statistical Society.

The new Statistical Society was concerned with the collection of statistics for the governance of society. Yet its motto implied a detachment from the ultimate goal of exploring the meaning of the data. This contrasts with the nature of Malthus's own controversial work and is far removed too from the goal that John Graunt set himself, namely to demonstrate the practical applications of his investigations.

This chapter introduces practical considerations in demographic writing and interpretation – steps taken in thrashing out population statistics. Since initial assignments in demographic research typically involve working with census statistics or other published information, the chapter focuses particularly on approaches to writing about *secondary data* (see Box 5.1). The chapter first reviews the main types of demographic studies, including inductive and deductive approaches, to illustrate the varied nature and goals of demographic investigations. Then follows a discussion of strategies in starting demographic research, interpreting information and writing a paper. More detailed references on social research methods and scholarly writing, as well as on data collection and analysis, are listed in the section on 'Further reading'.

The computer module in this chapter, *Pyramid Builder.xls*, is an aid to interpretation, through facilitating the production of single and superimposed age–sex pyramids from any data (see Box 5.2). Population pyramids are a common starting point for studying the nature of a population and drawing comparisons with others. The spreadsheet exercise on 'Line graphs and pie charts' introduces further techniques for visualizing demographic data.

5.1 Types of studies

Statistics on population are relevant to many aspects of social science inquiry. This is especially true today because of the greater range and scope of information collected through censuses, myriad surveys and continuous registration data on births, marriages and deaths. Demographic studies therefore vary greatly in content. They also vary from the largely descriptive to the highly analytic, reflecting the diverse objectives of people working with population statistics. Writing about population statistics requires an awareness of the possible types of demographic studies, in order to focus on the approach best suited to the particular task. This section discusses the main types of demographic studies, as summarized in Table 5.1. Statistical reporting commonly relies on *secondary data*, originally collected for other purposes or for general use – such as censuses. In contrast, deductive research commonly requires *primary data*, that is, information collected specifically to address a particular research problem (Box 5.1).

Statistical reporting

Best known, from its frequent appearance in newspapers and magazines, is statistical reporting (Table 5.1). Here the aim is to produce an article or report that sum-

BOX 5.1 **Primary and secondary data**

Primary data are collected for specific purposes, by or for those who wish to use the information. Later, when others use the same data it becomes secondary data, because the users are no longer those who initiated the collection (Hannagan 1982: 6–7). Thus, secondary data consists of information originally intended for other purposes, or collected by another researcher or organization. It includes data intended for general administrative and research use, including population censuses, vital statistics and surveys undertaken by government statistical agencies. Obtaining primary data is likely to be time-consuming and expensive; it also calls for advance knowledge both of the subject matter and of methods for conducting surveys or other statistical collections.

Secondary data offer great economies in terms of obtaining statistical information. Secondary sources potentially provide vast amounts of information – on different topics, time periods and places – far beyond that obtainable in a special-purpose survey or other primary data source (see Hakim 1987: 20ff). The main disadvantage is that exploiting secondary data may require a compromise between what is wanted and what is available.

Appropriate use of secondary data depends on adequate knowledge of how the statistics were obtained, their reliability and comparability, the definitions of terms, and the extent to which the data truly represent the population being studied (Hannagan 1982: 7). Secondary data for another period, for instance, are unlikely to be satisfactory when the aim is to document present-day population characteristics.

Table 5.1 Types of demographic studies

Type of Study	Description
Statistical reporting	Summarizing and interpreting a *given set of statistics*.
Enhanced statistical reporting	A focused analysis based substantially on *a given set of statistics*.
Inductive research	Seeking, from a body of data, generalizations and explanations concerning *a particular phenomenon*.
Deductive research	Investigating a theory about *a particular phenomenon* through a focused collection and analysis of data.

marizes and interprets a given set of statistics. The data, such as in a newly published statistical bulletin, define the scope and content of the report. Various organizations require such reports from the latest census or survey figures, and a common objective of demographic analysis is to produce a manageable and informative summary from a substantial volume of information. As well, journalists often seek to summarize the most newsworthy developments evident from the latest statistical release. To add depth, or give a human face, to such material, journalists may quote from interviews with authorities in the field or discuss their own 'case study' of an individual whose recent experience illustrates one of the trends or characteristics apparent from the statistics – such as a case study of a homeless person to introduce statistical findings on homelessness.

In statistical reporting, the investigator does not have the opportunity to pursue a more 'scientific' approach through identifying a research question, then collecting information to address the question. Instead, in statistical reporting the reverse process occurs – the study originates from the data. Under these circumstances, there is less reliance on curiosity or a question as the start and *raison d'être* of a piece of work, and there is a risk of reducing the writing to a ritual or routine that contributes little to understanding. Verbalizing data, for example, results in descriptive text full of numbers and conveying little more than the original figures:

There were 4259 migrants from Blefescu. Their ratio of males to females was 54:46. Almost 91 per cent were concentrated in three coastal towns. The largest proportions were counted in the second largest (45 per cent) and largest towns (32 per cent), representing 0.1 per cent and 0.3 per cent of the total population of each centre. While just over 5 per cent were less than 20 years of age, 16 per cent were over 55. The majority of residents (78 per cent) had been there for 15 years or more. Nearly 4 per cent had been there for three years or less. Of persons who had been there for more than three years, 43 per cent had acquired Lilliputian citizenship. In the 5–14 age group, 83 per cent spoke a language other than Lilliputian at home. For the older age groups the figure was 69 per cent.

The above commentary throws little light on the age structure of the population, and does not specifically address major questions about immigrant groups such as whether the migrants had characteristics different from those of the host population and whether assimilation was occurring. Nevertheless, preparing a report on a given set of statistics can be an important exercise – demographic inquiry often involves making sense of many statistics about which a government department or another organization needs to be informed. Statistical reporting becomes more searching when it places the analysis in the context of relevant concepts and previous findings from the literature.

Enhanced statistical reporting

Thus, an enhancement of statistical reporting is developed through analysing a *given* set of statistics with a firm sense of purpose from the outset, such as through investigating a question or problem or pursuing a theme, such as the social integration of immigrants, inequality in access to education, or discrimination in relation to age, sex or ethnicity. This can entail substantial preliminary research, to determine the most profitable pathway towards understanding the information. It may be necessary to omit portions of the original data, if irrelevant to the chosen aims, and seek out additional information to reinforce the line of investigation.

Many demographic studies based on official statistics or other secondary data proceed in this purposeful way and produce valuable results. In statistical reporting the ultimate aim should be to explain, rather than merely describe. The most important exception to this is in publications of statistical agencies whose foremost role is to collect and publish raw data for others to use. Such agencies may follow a policy of *aliis exterendum* – mentioned in the introduction to this chapter – to avoid potential controversies arising from differences of opinion about what the data mean.

Inductive research

Enhanced statistical reporting moves demographic inquiry towards a more purposive and scientific procedure. The first scientific method invented was the method of induction – formulating general principles from detailed information. The method of induction entails investigating a body of information, on a topic that the researcher has chosen, to discover what it might reveal. This has been a common approach in demography, which Dudley Kirk (1972: 348) described as often engaged in 'digesting vast reservoirs of social data'. Similarly, Samuel Preston (1993: 594) noted:

> The close connection between producers, evaluators, and users of demographic data is one of the most attractive features of demography. The interchange helps to produce better data and to avert misinterpretation of existing data. It also affords demographers un-

usual opportunities to affect the content of large-scale data instruments, especially national surveys. In part because of their closeness to data production, demographers are the most inductive of social scientists, focused to a greater extent than other social scientists on careful measurement and cautious interpretation.

Many demographic studies have had as their starting point the aim of investigating a particular phenomenon from statistics collected originally for other purposes, such as vital statistics and historical records. Graunt's study of the Bills of Mortality is an outstanding example, but varied secondary sources, such as population censuses and cross-national statistical compilations, have inspired creative research. The demographic transition, the epidemiologic transition and the second demographic transition are examples of influential theoretical perspectives in demography that evidently arose from scrutiny of empirical data. Inductive research is considered important in the modification of existing theories and the development of new ones, such as through unanticipated or serendipitous findings (Baker 1988: 56).

Inductive research differs from statistical reporting in that the starting point is a topic for inquiry, rather than a given set of data. Also, there is a quest for generalizations and explanations, rather than merely a summary. The range of information studied is not restricted to a particular source, but draws upon as many as seem relevant or are available. In principle, induction implies no pre-conceived assumptions or expectations about the phenomenon under investigation, but this is unlikely. Individuals bring to research their own theoretical, conceptual and empirical knowledge, as well as their own viewpoints.

Deductive research

Demography's predilection for evidence from empirical data arises not only from its use of inductive approaches, but also from a determination to collect information that can be employed in testing theories:

> The strong empirical element in modern demography is to some extent an attempt to test . . . theories and provide them with adequate grounding, but it is also a reaction to, and a suspicion of, grand theory. (Caldwell, 1996: 309)

The deductive approach of 'classical' scientific inquiry envisages that knowledge is advanced through research that begins with a research question and relevant theory, then proceeds systematically through data collection, data analysis and appraisal of the findings. Thus, whereas induction proceeds from observations to the construction of a theory, deduction proceeds from a theory to its testing with observations (Lin 1976: 57ff). A theory is 'a set of logical and empirical statements that provides an explanation of some phenomenon' (Sullivan 1992: 31–32). Theories offer answers to research questions, or explanations of research problems. Hypotheses derive from theories and are 'testable statements describing the relationship between two or more variables'; in other words they declare what the researcher expects to find (Hessler 1992: 9–10).

A theory influencing contemporary demographic thought concerns *convergence*, which envisages that the economic rationality and competitiveness of the industrial mode of production, in a globalized economy, cause all industrial populations to experience a similar pattern of development. This is characterized by strong conformist pressures on individuals, families, and national economic and social policies. If demographic characteristics follow the convergence of economic characteristics and social institutions, uniform demographic patterns should develop, with any contrasts attributable simply to different stages of development (Coleman 1998: 7). Clearly, many hypotheses arise from convergence theory including expectations about similarities between cross-national trends in fertility, marriage, living arrangements and labour-force participation in countries at the same stage of development.

In undertaking deductive research, researchers choose or collect statistical data most relevant to the research question and the hypothesis considered central to answering it. Statistical reporting, in contrast, may take the form of a compendium of 'facts and figures' in which it may be difficult to integrate all of the information and achieve an overall sense of its meaning and importance. Classical scientific inquiry begins with a question or problem, whereas statistical reporting follows from an assignment to 'write up' a particular data set.

Papers developed using the deductive method may be written with a particular logic, first providing background to the research question and relevant theory ('Introduction'), then describing and analysing the data assembled to address the question ('Methods', 'Results'), and finally appraising the findings and their implications ('Discussion'). This structuring assists readers to follow and evaluate the research, as well as to draw comparisons with other studies. Such orderly presentation, however, is often an idealization of the research process, contrasting with the actual experience of conducting the research (McNeill 1985: 121). Modifying ideas, returning to the literature, investigating alternative explanations and examining different ways of analysing the data are commonly a necessary part of research. This does not mean that the conduct of research is haphazard, but ideas may undergo substantial changes through revisiting the supposedly sequential stages of the research process.

Not all demographic studies fit within one of the above categories, but the four types illustrate the point that there are different approaches to research on population, no one of which is suited to all purposes. The deductive method is a model for demographic research where authors have the resources to collect information on a particular phenomenon, but there is no inevitability that the progress of the research will accord with the orderly series of steps envisaged for this type of inquiry. The method of induction is also common in demography, underlying a tradition of empirical research that has been important since Graunt. Statistical reporting similarly has a role to play, and is especially valuable when it focuses on specific issues.

5.2 Starting demographic research

The process of research and writing varies between the different types of studies. Yet all call for systematic work, in which the logic pursued in deductive research is a valuable reference point. The initial steps in demographic research and writing often include formulating the research question, defining the scope of the investigation and assembling data.

A major impetus for research is recognition that knowledge of a particular phenomenon is insufficient as a basis for understanding or action. A research question can arise from many different sources – including previous studies or personal observation, from a problem that an employer wants investigated, or from issues arising in public debate. Formulating a question to address provides the essential focus, creating an ongoing sense of purpose and relevance in the ensuing research and writing. The question, however, must be testable as well as manageable within the time and resources available.

One example of a research question is: Under what circumstances are higher levels of immigration an appropriate response to population ageing and population decline caused by the effects of low birth rates? To address this question, a *research design*, 'a system of test environments or conditions' (Bouma 2000: 8), needs to be chosen according to the further aims of the study. If one such aim is to provide a general, statistical account, an experimental research design might be appropriate, using hypothetical data to examine the effects on age structures and population numbers of various combinations of rates of fertility, mortality and migration. Alternatively, the research design might be a case study of projected futures for a single country, or a comparative research design examining potential outcomes for a number of countries. Major types of research design are: (1) simple case study, (2) longitudinal study, (3) comparison, (4) longitudinal comparison and (5) experiment (Bouma 2000: 88–115).

In deductive research, the answer to the research question should entail explanation rather than description alone. Addressing the question of immigration and low fertility requires not only an analysis of the volume of migration that could compensate numerically for the effects of low fertility (see United Nations 2000a), but also an exploration of whether heightened immigration will be acceptable socially, economically and politically. Thus the question calls for an explanation of the circumstances in which immigration is an appropriate response. Theories concerning the capacity of societies to absorb new settlers represent a basis for developing the explanation.

Identifying a promising research question, relevant theory and an appropriate research design, however, will not comprise an adequate starting point for the investigation until it is confirmed that no other study has covered the same subject material. A literature survey is indispensable at an early stage, not only to establish the state of knowledge about the subject, but also to obtain further insight into the proposed topic. Undertaking a literature survey, and compiling an initial bibliography, is greatly facilitated through electronic data bases such as *Popline* and *Population Index*, as listed in the Internet resources at the end of this chapter. Book and journal collections on the Internet, such as *JSTOR*, also facilitate access to relevant materials. Thus, this type of work calls for *information literacy* – knowing how to find, evaluate and use needed information. Information literacy comprises skills in library research as well as in information technology – such as for accessing electronic data bases and the Internet.

Another early step in the research process is to define the scope of the investigation. This includes defining the population, the time span and geographical scale. Choices depend on the aims of the study and the availability of information. Study populations may vary considerably according to whether particular groups are included or excluded. For instance, inclusion of persons in consensual unions can affect greatly the size of the 'married' population, while inclusion of persons aged less than 65 years have dramatic consequences for the size of the 'aged' population. Substantially different results can also arise from studying the de jure (usual resident) population of a city or region, rather than its de facto (total present) population, such as in places with a large, temporary migrant labour force or many tourists and other visitors.

Taking stock of available statistical materials is also an important step when starting a research project. The nature of secondary data sources can be ascertained from catalogues of official statistics published by national and international statistical agencies, as well as from listings of the holdings of data archives. Even if the intention is to collect new data for a project, it is essential to know the nature of existing statistical resources to avoid duplication and to profit from previous work. Basic concerns arising here include the coverage and quality of the data, together with their comparability through time and from place to place. The types of work undertaken in starting a research project are interdependent: they are not necessarily completed separately. Research is a learning process, rather than a series of predictable stages, and modification of initial ideas is to be expected. If a new survey or other data collection is not required, the preceding steps lead to embarking on interpreting the data and preparing an outline of the paper.

5.3 Interpreting information

Interpreting information on populations depends partly on demographic skills and partly on a range of generic skills such as: working systematically, consulting the literature, addressing conflicting evidence, documenting sources and methods to enable readers to verify results, and keeping within constraints of time and resources. Whether a study relies on the demographic data and indices alone, or on additional statistical analyses, there is a set of common considerations arising in interpreting most types of demographic information, as summarized in Table 5.2

Table 5.2 Considerations in interpreting demographic data

1 Ranges	Examine the range of the data – from highest to lowest values. Is the population varied or fairly uniform in its characteristics?
2 Extremes	Note any extreme values; these may deserve special attention as exceptions or errors.
3 Clusters	Identify clusters of values. Are there natural groupings in the data? For example, are people with low incomes concentrated in particular suburbs, or occupational categories, or in groups with poor health or other social disadvantages?
4 Patterns	Identify patterns of variation, such as whether the data vary by age, ethnicity or education in some regular way. Rates of fertility, mortality and migration, for example, vary by age and other characteristics.
5 Processes	Consider processes of change: • Demographic processes – such as natural increase, net migration, cohort flow and ageing. These are the immediate causes of population changes, whereas social, economic and political forces comprise the underlying causes. • Non-demographic processes – such as globalization, social mobility, social discrimination, changes in the status of women, technological innovation, amendments to laws and policies. Marshalling evidence on the influence of demographic and non-demographic processes is important in developing explanations, or at least partial explanations, of changes.
6 Comparisons	Draw comparisons: • between sub-populations and the total population; • with other times and places.
7 Concepts	Place the discussion in the context of concepts and theories. These extend the comparative context, at the same time providing the most informative basis for generalizations and explanations.
8 Visualization	Draw graphs and maps as appropriate to provide a visual impression of variations in the data. These can greatly assist data interpretation, although normally only a small selection of them need to be included in the final paper.
9 Implications	Discuss implications – for the future, for plans and policies, for further research.
10 Summary points	Describe the main findings, compared with conclusions from previous research. Identify anything new or unexpected. Highlight any insights into the question or problem investigated. Explain data needs for further inquiry.

(see also Table 1.4). Attention to these provides a route to capturing the content and meaning of demographic data, although the emphasis given to each depends on the aims of the research. Beyond these general considerations, the interpretation of demographic information involves detailed analyses of statistical tables and graphs.

Table 5.3 lists practical considerations in designing statistical tables. These are important not only to people reading a paper, but also to users of secondary data, who require a clear description of the tables they want. The main considerations in designing tables are providing concise and accurate labelling of the data through the title and row and column headings, presenting the statistics in their most useful form, and documenting sources and definitions.

The tables in a research paper are usually only a small selection of those produced, chosen because they provide concise evidence of the author's main observations and conclusions. Editing is needed both to limit the number of tables and to limit the rows and columns within tables, so that they are readable and of manageable size. Graphs and maps may serve as a valuable alternative to particular tables. Published demographic research papers commonly contain a maximum of

Table 5.3 Considerations in designing statistical tables

1 Title	The title of the table indicates which population is under consideration, when the data were collected and which characteristics of the population are included.
2 Row and column headings	The headings define the categories into which the data are divided. The number of categories depends on whether the table is intended as a summary or as a means of publishing detailed findings. Broad categories may disguise significant internal variations, while many narrow categories may defeat the purpose of producing a summary. Headings ought to be mutually exclusive, as instance income categories \$10 000–19 999 and \$20 000–29 999 rather than \$10 000–20 000 and \$20 000–30 000.
3 Data	Summary figures are usually presented in an immediately useable form, ready to be read in conjunction with the text. Percentages, rates and ratios facilitate comparisons. Absolute numbers may be important in their own right, such as for planning and marketing applications. Absolute numbers also serve to show sample sizes or the totals from which other figures derive.
4 Source	The source of the information is usually specified at the bottom of the table; it is an essential aid to readers in judging the reliability of the information and in following up further details for themselves.
5 Footnotes	Footnotes may be needed to define terms in the title or row and column headings, or to specify limitations in the scope or accuracy of the information.

two tables or graphs per thousand words. In statistical reporting, more numerous and larger tables may be required if the task includes providing a more concise data set for others to consult.

When interpreting tables, a potential problem is merely expressing data in words, rather than addressing the types of considerations listed in Table 5.2. Causes include lack of familiarity with the population in question, insufficient conceptual knowledge to place the information in a comparative context, and difficulties in making generalizations. The problem can also arise from neglecting the reliability of the data, possible competing explanations of patterns, and relevant matters that the data do not reveal.

Demographic graphs require the same attention to labelling and documentation as tables, and some of the general principles in interpreting demographic data (Table 5.2) are transferable to them as well. Table 5.4 describes key considerations in analysing a population pyramid, one of the most common demographic graphs. Different objectives inevitably call for different approaches. The *Pyramid Builder* module, described in Box 5.2, is an aid to analysing age–sex composition, through facilitating the drawing of graphs.

5.4 Writing a paper

Preparation for the writing of a demographic research paper entails making an outline of its potential contents, through a set of headings and sub-headings denoting the subject matter of the paper or report. An outline is an important aid to

BOX 5.2 **Pyramid builder module (*Pyramid Builder.xls*)**

The *Pyramid Builder* plots single or overlaid population pyramids from any data in five year age groups. Statistics for up to five populations can be entered, or pasted in, at a time. The headings for each population are automatically displayed in drop-down menus, enabling viewing and printing of single pyramids or superimposed pyramids for two populations.

Options include plotting of numbers or percentages, and automatic adjustment of the graph for any final open or closed age interval between 60–64 (or 60+) and 110–114 (or 110+).

The purpose of the module is to facilitate the drawing of age–sex pyramids, which are important aids for interpreting population trends and characteristics. The module does not require knowledge of Excel; the spreadsheet exercises on *Population Pyramids* (Exercise 3) and *Superimposed and Compound Population Pyramids* (Exercise 4), however, explain the underlying principles for producing such graphs.

continued

BOX 5.2 continued

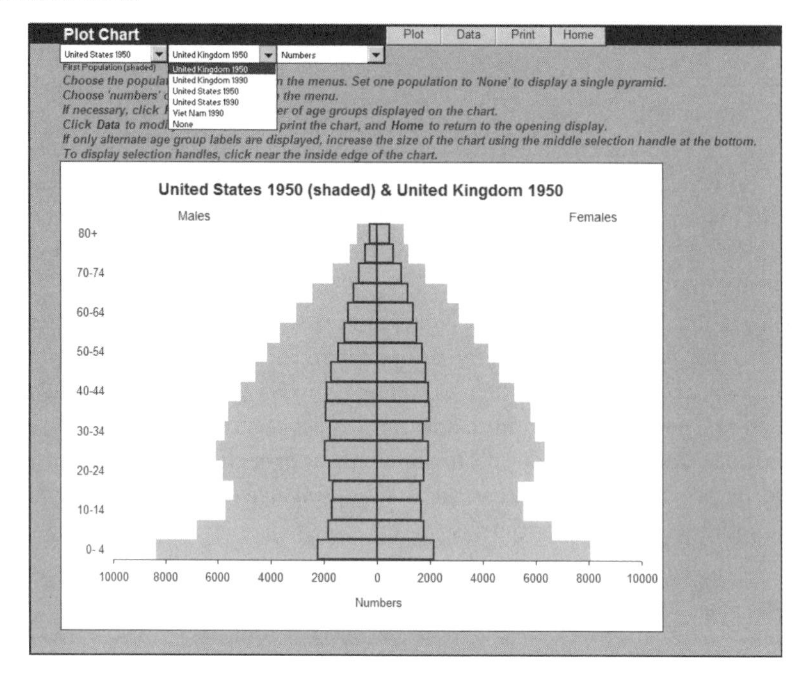

Instructions

To use the module, insert the CD in the drive, start Excel and open the file *Pyramid Builder.xls*. Enable macros when prompted.

- On the opening screen, click **Start** to begin.
- Choose the final age group from the drop down menu (e.g. 65+ or 85+). This will display the required age labels beside the columns in which the age data are to be entered.
- Type the headings and numbers (not percentages) for up to five populations in the yellow cells. Alternatively, to save typing, data from any spreadsheet can readily be copied and pasted into the yellow cells. Do not alter the white cells.
- The headings identifying each population will appear in the menu boxes on the Chart screen.
- Click **Chart** to view the population pyramid(s).
- Click **Plot** to adjust the chart to the required number of age groups.
- Use the other controls to obtain single or overlaid charts showing numbers or percentages.
- Click **Print** to print a copy of the chart. Alternatively, select the chart with the mouse, and choose File Print from the menu at the top of the screen.
- Click **Home** to return to the opening screen.

See Appendix B for further advice on using the Excel modules.

Table 5.4 Considerations in analysing population pyramids

1 Introduction	The time and place to which the data refer; the historical and/or geographical setting; social forces affecting the population; main features of the age–sex structure; unusual features; inherent potential for change; the reliability of the statistics (e.g. undercounting or misreporting of age groups).
2 Middle	
Description	Overall form (e.g. young, mature, old, bimodal, unimodal); location of peaks and troughs; relative numbers of males and females at different ages; overall population size. Small populations are more likely to have irregular age structures or reflect institutional influences, such as the presence in a locality of a hotel, hospital or boarding school.
Explanation	Immediate causes may be discussed with reference to the effects of demographic processes – fertility, mortality, migration and ageing. Fertility decline produces undercutting at the base of the pyramid, migration may cause peaks in the younger working ages, while demographic ageing leads to more rectangular age profiles. Underlying causes – which affect demographic processes – are often more diverse and more difficult to document. They include political interventions (e.g. policies on economic growth, urban development, migration and fertility control), social forces (e.g. trends in marriage, family formation, labour-force participation and retirement) and events (e.g. wars, famines, refugee movements and natural disasters). National trends might be explained also in the context of the demographic transition or other theories. At the regional and local level, migration is such as pervasive influence on changes that it is hazardous to explain features of age profiles with reference to influences that are important nationally. The effects of baby booms and the stage of demographic transition are often irrelevant at a smaller scale, because of the overriding influence of inward and outward migration. Housing type, life cycle stage and ageing-in-place may be especially important at the local level in explaining migration patterns and age structures.
3 Conclusion	Like the introduction, the conclusion is an opportunity to reveal insights into the population or topic under investigation. Three potentially important aspects of a conclusion are summary points, linkages with other subject material, and implications. The summary points may highlight overall features, the most important findings or the most distinctive characteristics of the population. Linkages with other subject material emphasize the relevance of the age-structure analysis to any other aspects of the current study, for instance population growth and family change. This can provide a bridge to other sections or chapters of a report. Implications include prospective developments inherent in the age structure, such as the movement of larger or smaller cohorts into the working ages and retired ages. They also include considerations for government policies and programmes, whether relating to education, employment, health, welfare, migration or reproductive health. Finally, the implications can include ideas for further research, improvement of data collections, or enhancement of concepts and methods.

systematic work and efficient writing – enabling authors to produce a first draft that is as close as possible to the final product, instead of anticipating several time-consuming cycles of cutting, pasting and re-drafting. Authors commonly modify the structure of the paper as the work progresses, but refining the outline assists in maintaining a clear focus and a logical progression of ideas. It is also standard practice to include at least part of the outline – the major section headings – in books, journals and other publications on population. The headings assist readers to ascertain quickly whether an article is relevant to their interests.

The three-fold structure of 'introduction, middle and conclusion' provides a useful start in preparing an outline. Having a more elaborated initial structure, especially for the 'middle', opens up greater possibilities for exploring the data and identifying the main points. A reasonably detailed set of section headings and sub-section headings is likely to be helpful in organizing the subject material, even if there is no intention of including many headings in the finished paper. Some scientific journals, such as in medicine and epidemiology, require authors to adhere to a uniform structure – for example comprising: introduction, methods, results, discussion, bibliography. This assists readers to find information and compare different studies. Other journals allow authors greater flexibility in the structure and content of papers.

Word-processing packages can facilitate the use of outlines through enabling authors to define headings and sub-headings – moving any of which automatically moves the accompanying text. In Microsoft *Word*, for example, this is accomplished using 'styles' to define a hierarchy of headings and 'outline view' to re-arrange them. Information on these topics is found in *Word* by clicking Help and searching for 'outlining'.

Table 5.5 presents an overview of the range of potential contents of demographic reports and papers, although actual contents vary depending on the study's purpose and audience. The table shows, firstly, that summary information may be needed, in the form of an abstract, key words, an executive summary or a list of main findings. Some of these are used as items in bibliographic data bases to facilitate electronic searches for reference material. Even if the summary items listed in Table 5.5 are unnecessary, it can be helpful to authors to reflect on possible key words and main points as an aid to ensuring their work remains focused.

Secondly, the text, the main body of writing, requires an introduction, saying what the paper is about, and a conclusion, summarizing findings or stating implications. The 'middle' part of the text may contain information about previous research on the topic, the sources and methods employed, and the analysis of the data. Sometimes the middle also includes a separate 'discussion' section highlighting the meaning and importance of the findings.

Thirdly, reaching a conclusion or synthesis can be the most difficult part of a study. Like preparing an abstract, writing a conclusion is important, for authors as well as readers, in clarifying aims and outcomes. Sometimes there is no over-arching conclusion, more a set of conclusions – especially in statistical reporting. Yet influential research is often that which discovers one or two key findings –

Table 5.5 Potential contents of demographic reports and papers

Summary information	
Abstract	A paragraph, often 100–200 words, stating the study's purpose, methods and main findings. Abstracts included in electronic data bases greatly enhance the likelihood of finding specific information.
Key words	Several words that identify the subject matter of the paper. These also assist in preparing indexes and bibliographic data bases.
Main points	A list of main findings and conclusions.
Executive summary	A condensed description of the study and its principal findings, usually in prose – for readability – rather than point form. It may further include a summary of recommendations arising from the investigation.
Text	
Introduction	A statement of the aims and scope of the study, together with reference to key concepts or theories. The introduction may outline the general topic to be explored or the specific issue, problem or question to be addressed. People writing for a general audience sometimes begin with an example or case study to introduce the subject material and evoke interest. Mentioning major findings can also draw attention to the paper's importance and encourage detailed scrutiny. Including a brief outline of the contents of the paper helps readers to follow the argument and understand the structure of the study.
Literature review	The literature review discusses previous findings, hypotheses or concepts that are helpful in describing and explaining the setting of the study in terms of existing knowledge.
Sources and methods	This section describes the nature and quality of the source materials and the methods employed in analysing them.
Findings	The findings are the *raison d'être* of the paper and could comprise the greater part of the text. Statistical evidence supporting the findings is provided in tables and graphs – usually just a selection of those compiled and consulted during the analysis.
Discussion	A separate 'discussion' section may follow the 'findings' to place them in a broader perspective and show their importance. Coverage here can include interrelationships between trends, explanations of developments and implications for action or further inquiry. Discussion of 'policy implications' requires particular caution (see Lin 1976: 366ff). Conclusions must be based on evidence, but speculation beyond what the data actually show is sometimes permissible if helpful for further inquiry.
Conclusion	A summary of the main findings in the light of the initial research question and guiding concepts. The conclusion may include (i) an evaluation of the extent to which the present findings differ from those of other studies, (ii) proposals for the refinement of concepts and data collections and (iii) recommendations for policies and strategies. A separate 'conclusion' may be redundant, however, depending on the scope of the 'discussion' section.
Sources of information and assistance	
Bibliography	A complete list of references cited in the text.
Acknowledgements	A sentence or paragraph giving due recognition to individuals and organizations who assisted with the study, including any sources of research funding.

an observation of a pattern of change or variation in a population, an explanation of demographic behaviour, or a concept with applications in later studies. Finally, good workmanship and ethical practice require that all reference materials and sources of assistance be acknowledged at the end of the paper.

Preparation of conference posters (Box 5.3) is an exercise that tests all of the skills entailed in report writing, and is a means by which many new social scientists make their first public presentations. Preparing a conference poster, based on a

BOX 5.3 **Conference posters**

What are conference posters?

Posters are widely used at scientific and professional gatherings as an alternative to the formal presentation of papers. They are intended to convey a summary of the research problem, hypotheses, ideas, methods, findings and issues discussed in a research paper. The posters need to accomplish this task succinctly and in a way that is self-explanatory, as well as interesting.

Posters, like 10–20 minute oral presentations, are a way of conveying research findings to a wide audience. A poster is not an alternative to a full paper: rather, it is a means of initial presentation.

Although posters may be displayed for the duration of a conference, time is usually set aside for authors to discuss their work. A well-organized poster, therefore, functions also as a visual aid for speakers, assisting authors to lead a small audience through the content of the paper and inform them about the findings.

Designing a poster is a challenge to be concise and clear about the aims and substance of a study. Thus preparing a poster is partly an exercise in writing an effective summary. An extra consideration is that imaginative use of graphs and other illustrative material can be much more important in a poster than in a research paper, in order to attract the attention of a passing audience.

Designing a poster

Unless the assistance of a graphic designer is available, poster presenters normally aim for a simple display, neatly and logically laid out, conveying a clear summary of the paper. The name and affiliation of the author and title of the full paper should be included, but the title of the poster may differ – for conciseness or to evoke interest in the main findings.

The following are typical specifications for conference posters:

- *Display area* 1 metre by 1 metre. Posters must normally fit within this space, usually on a single sheet of cardboard mounted with velcro strips or pins.
- *Print size* All text should be easily legible from a distance of one metre. Use at least 18 pt type for the text, and 24 pt or a larger font for section headings. The title at the top of the poster should be larger still, such as 2.5 cm (one inch).

continued

BOX 5.3 continued

- *Number of words* Often no more than 500 words. Some conference organizers recommend 500 words as the limit and encourage keeping the text to 200 words. A small handout with a facsimile of the poster or other notes will help people to keep a record of the work, while a small 'post-box' for business cards can identify people interested in receiving a copy of the full paper.
- *Layout* This should follow a logical sequence, leading the reader through the argument with headings, text and diagrams as appropriate. It is essential that there is no confusion about the order in which the material is to be read. Visual impact can be enhanced through breaking up the display into discrete components: e.g. introduction, method, results, conclusions – each with its own heading. Bullet points are helpful if not used to excess.
- *Illustrations* These include simple, legible graphs, pictures, diagrams and tables if relevant, together with concise captions and legends. Having illustrative material as well as text to refer to can enhance interest and make oral poster presentations easier to deliver.
- *Colour* To improve appearances or add emphasis. Generally, posters on coloured backgrounds, or with coloured headings and illustrations, are visually more interesting than plain black and white.
- *Mounting* Posters are more durable and have a more finished appearance if laminated in plastic.

Further information

An Internet search for 'Conference Posters' (e.g. from http://www.google.com) will produce examples for inspection. Also a visit to the conference sites of organizations such as the Population Association of America will give access to guidelines on the preparation of posters (see 'Internet resources' at the end of the chapter).

published paper by another author, can also provide valuable early experience in demographic writing: such a task is suggested in the set of exercises for this chapter (see 'Study resources').

5.5 Conclusion

This chapter has introduced strategies in the writing of a demographic report or research paper, especially from secondary data sources. Recognizing the type of study required, compared with the alternatives, is a basis for deciding the most suitable approach (Section 5.1). For many studies, formulating the research question is the most important stage. The nature and scope of the investigation also depend substantially on the feasibility of obtaining relevant data (Section 5.2). At the analysis stage, being aware of the range of possible considerations in interpreting information can clarify whether essential information is receiving due attention (Section 5.3). Also, when writing about the findings, an outline is an aid to maintaining a focus on the research question and working systematically (Section 5.4). Beyond these basic considerations, studying research methodology (see 'Further reading'), together with learning by example – through reading books and articles on population – will extend familiarity with the handling of the subject material. Ultimately the quality of demographic analysis and writing depends on enthusiasm for Graunt's 'buzzling [puzzling] and groping'.

Study resources

KEY TERMS

Conference posters
Deductive research
Hypothesis
Inductive research
Information literacy
Outlining
Primary data
Research design
Secondary data
Statistical reporting
Theory

FURTHER READING

Becker, Howard S. 1986. *Writing for Social Scientists: How to Start and Finish Your Thesis, Book or Article*. Chicago: University of Chicago Press.

Betts, K. and Seitz, A. 1994. *Writing Essays and Research Reports*. Melbourne: Nelson.

Blaxter, L., Hughes, C., and Tight, M. 1996. *How to Research*. Buckingham: Open University Press.

Bouma, Gary D. 2000. *The Research Process*. Melbourne: Oxford University Press.

Bryman, Alan. 2001. *Social Research Methods*. Oxford: Oxford University Press.

Gibaldi, Joseph. 1995. *MLA Handbook for Writers of Research Papers*. New York: Modern Language Association of America.

Gibaldi, Joseph. 1998. *MLA Style Manual and Guide to Scholarly Publishing*. New York: Modern Language Association of America.

Hakim, Catherine. 1987. *Research Design: Strategies and Choices in the Design of Social Research*. London: Allen and Unwin.

Hartley, Shirley F. 1982. *Comparing Populations* Belmont, CA: Wadsworth.

Hessler, Richard M. 1992. *Social Research Methods*. St Paul: West Publishing Company.

Lethbridge, Roger. 1991. *Successful Seminars and Poster Presentations*. Melbourne: Longman-Cheshire.

Lewins, F. 1987. *Writing a Thesis*. Canberra: ANU Press.

Lucas, David, McMurray, Christine, and Streatfield, Kim. 1989. *Looking at the Population Literature*. Demography Teaching Notes 6. Canberra: National Centre for Development Studies, The Australian National University.

Van Krieken, Robert et al. 2000. 'Social Research', Chapter 14 in van Krieken et al. *Sociology: Themes and Perspectives*. Sydney: Longman, pp. 585–619.

Windschuttle, Keith and Windschuttle, Elizabeth. 1988. *Writing, Researching, Communicating: Communication Skills for the Information Age*. Sydney: McGraw-Hill.

INTERNET RESOURCES

Subject	Source and Internet address
Archive of surveys and other data for social science research	**Social Science Data Archives, The Australian National University** http://ssda.anu.edu.au/ *Links to other data archives:* http://ssda.anu.edu.au/other/other_archives.html
JSTOR: archive of scholarly journals, including demography journals	**JSTOR** http://www.jstor.org/
POPLINE: abstracts of works covering family planning, health, population law and policy	**Johns Hopkins Centre for Communications Programs** http://db.jhuccp.org/popinform/index.stm
Population Index: bibliography of the world's population literature	**Office of Population Research, Princeton University** http://popindex.princeton.edu/
The conference site for the Population Association of America, including a link to instructions on preparing conference posters	**Population Association of America** http://www.popassoc.org/meetings.html
PORTAL: resources for the humanities and social sciences The section on 'Sociology, Demography and Social Statistics' includes materials on Charles Booth and T.R. Malthus and links to major data bases	**The British Academy** http://www.britac.ac.uk/portal/
SOSIG. A source of selected, high quality Internet information for the social sciences	**Social Science Information Gateway** http://sosig.ac.uk/welcome.html
Research resources for the social sciences	**McGraw-Hill** http://www.socsciresearch.com
Social sciences virtual library	**University of Florida** http://www.clas.ufl.edu/users/gthursby/socsci/index.htm

Note: the location of information on the Internet is subject to change; to find material that has moved, search from the home page of the organization concerned.

EXERCISES

For written work and/or group discussion:

1 Imagine a primary school, in a predominantly English-speaking community, where only half of the children speak English fluently. Briefly describe ways in which each of the following characteristics of the local community might contribute to an explanation of this situation:

- age composition of ethnic groups
- migration
- birth rates
- family size
- spatial distribution of ethnic groups
- housing
- occupations, income, education
- any other factors.

2 Using the *Pyramid Builder.xls* (see Box 5.2), and statistics from sources such as your country's census or the United Nations *Demographic Yearbook*, draw superimposed population pyramids for a country or region at two different dates. Consider whether absolute numbers or percentages are more useful. After completing the pyramids, prepare an interpretation of them.

3 Choose a study of particular interest from the *Population Bulletin* (Population Reference Bureau) or *Population and Development Review* (available from JSTOR, see 'Internet Resources'). Critically examine the section headings used in relation to their relevance to the aims and scope of the study and the extent to which they assist reading and comprehension.

4 Using an article from *Population and Development Review*, prepare a 'conference poster' which conveys the main points (see Box 5.3). Use the poster as the basis for a five minute tutorial presentation about the article.

5 Write a 200 word abstract of a published paper, summarizing its scope, purpose and key findings. (Choose a paper without a published abstract.)

6 Write a critique of the published abstract of a paper, examining whether it conveys sufficient information about the content of the paper.

7 Identify, in point form, essential considerations in a possible assignment about interpreting census statistics on either the age–sex structure, or the occupational characteristics, of the population of a city, town or region with which you are familiar. Arrange the list of points under headings such as:

- formulating a research question;
- making effective use of the literature;
- statistical considerations;
- selecting methods of analysis;
- main sections of the paper.

The objective of this particular exercise is to develop an appreciation of issues that could arise in research, rather than engage in an actual analysis of demographic data.

8 Prepare a report, of up to 2000 words, describing and explaining contemporary population characteristics of a village, suburb, district or small town. Choose either the place in which you live or one that you can visit easily.

Potential contents include: population growth, age structure, socio-economic characteristics (occupation, education and income), ethnic composition, type of housing and whether rented, owned or being purchased. Selection of the subject material, however, will depend on the purpose of the study.

Aim to focus your research on a question or concern relevant to the area, such as:

- The extent of demographic change in the last ten years, and potential implications arising in terms of the need for services and facilities for particular age groups.
- The nature of the area as a place of residence – in terms of having a predominance of families or single people, or a settled or mobile population, or a rich or poor population – and implications arising for local businesses, services and community facilities.
- The numbers and characteristics of disadvantaged groups compared with the rest of the population, together with explanations of where the disadvantaged live and possible consequences of their presence for the wider community.

Employ the following sources of information:

- Maps showing the location of the study area in relation to other places.
- Census reports from libraries or the Internet.
- Field work. Spend some time walking or driving around your study area to gain an impression of the patterns of land use and the age and nature of the housing.
- Articles, books, theses and newspapers.

SPREADSHEET EXERCISE 5: LINE GRAPHS AND PIE CHARTS

This exercise introduces two frequently needed types of graphs, understanding the plotting of which, in conjunction with the two previous exercises on population pyramids, provides a general foundation for preparing graphs in Excel.

Line graphs are commonly employed for plotting continuous data, such as trends through time, where interpolation between points may be possible. Time is usually plotted on the horizontal axis. For non-continuous data, including occupation and income groups, bar or column graphs are more often used, to avoid the impression that there is continuous change between the categories.

Pie charts or divided circles are useful for showing components of a total, although they become hard to read if there are many categories. Categories are represented as sectors, the areas of which are proportional to the values they represent. If percentages are plotted, 1 per cent is represented by a sector with an angle of 3.6 degrees.

Table 1: Projected World and Regional Population Growth (millions)

Year	World	Developing Regions	More Developed Regions
2000	6,116	4,842	1,273
2025	8,126	6,762	1,364
2050	9,586	8,219	1,367
2075	10,491	9,125	1,365
2100	10,968	9,589	1,379
2125	11,265	9,873	1,392
2150	11,411	10,013	1,399

Source: World Bank (1994).

Table 2: Age Composition of Regional Populations, 2000 and 2100 (millions)

	Developing Regions		More Developed Regions	
	2000	2100	2000	2100
Numbers				
0-14	1,614	1,770	252	244
15-64	2,985	5,745	846	800
65+	244	2,074	176	335
Total	4,842	9,589	1,273	1,379
Percentages				
0-14	**33.3**	**18.5**	**19.8**	**17.7**
15-64	**61.6**	**59.9**	**66.5**	**58.0**
65+	**5.0**	**21.6**	**13.8**	**24.3**
Total	**100.0**	**100.0**	**100.0**	**100.0**

Source: World Bank (1994).

Figure 5A

Plotting a line graph

1. Type Table 1, ensuring that the column headings are in a single row – this allows them to be selected along with the statistics when plotting the graph. To display the complete column headings, highlight cells A3 to D3 and click Format Cells Alignment Wrap Text; at the same time set Horizontal and Vertical Text Alignment to Centre and Top respectively.
2. Highlight the row and column labels and the three columns of statistics. Highlighting the column headings enables them to appear in a legend. There should be no blank rows between the column headings and the data. The area highlighted must be a square or a rectangle, with the same number of rows in each column.
3. Click the Chart Wizard and choose a Line Chart with markers. Click Next. If the row headings (Years) are plotted as a line instead of as labels for the horizontal axis: (1) click the Series tab, highlight the series named Year and click the Remove button; (2) click the mouse in the 'Category (X) axis labels' box, highlight cells A4 to A10 on the spreadsheet, then click Next. Finally, add titles and gridlines, or modify other options as required, by working through the tabs on the screen, then click Next and Finish. (For further information on constructing charts in Excel, click Help Contents Working with Charts.)

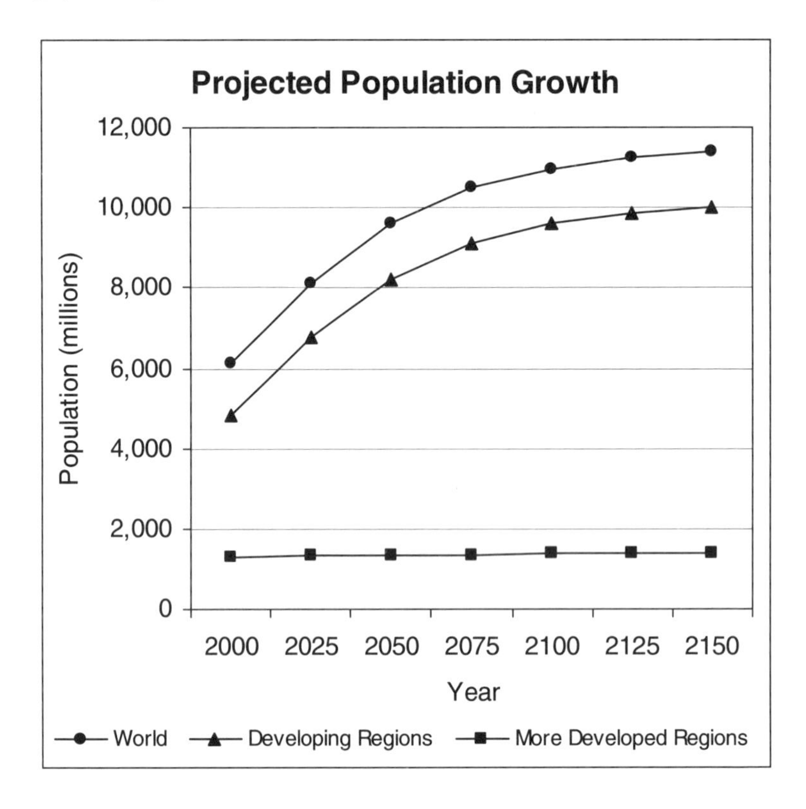

Reformatting a line graph

4. Changing the appearance of the lines: click on one of the plotted lines and choose Format Selected Data Series Patterns (clicking the line with the right mouse button also brings up the menu).
5. Changing the border: click the mouse in a blank space near the edge of the chart – edit boxes will appear around the whole chart. Then choose Format Selected Chart Area Patterns; under Border select the Style Color and Weight of line and click Shadow if wanted.
6. Modify the title and axis labels using Chart Chart Options. To make the X axis labels run horizontally, click on the axis and choose Format Selected Axis Alignment, moving the alignment pointer with the mouse.
7. Modifying a legend: choose Chart Chart Options Legend. Its position can also be changed by clicking on the legend in the chart, and moving it with the left mouse button down. To change the font size within the legend, right click on it to obtain the Format Legend menu, or double click with the left mouse button.
8. Note that different options appear in the menus according to the item selected. Instead of choosing Format from the main menu, double clicking the left mouse button on the item to be formatted will immediately bring up the required menu.

Plotting a pie chart

9. Type the numbers and headings in Table 2 and use formulas to calculate the percentages shown in bold. To centre headings across two or more columns, merge the cells into one by highlighting them and clicking the Merge and Centre button on the formatting toolbar (the button with two arrows and the letter 'a').
10. Highlight the row labels (one column only) then press Cntrl while highlighting the percentages for Developing Countries 2100. Do not highlight the column headings or any of the totals. The row labels will appear in the legend for the pie chart. Click the Chart Wizard, choose the first pie graph and step through the Chart Options as before.
11. Plot a second pie graph for More Developed Regions 2100. Save the spreadsheet.

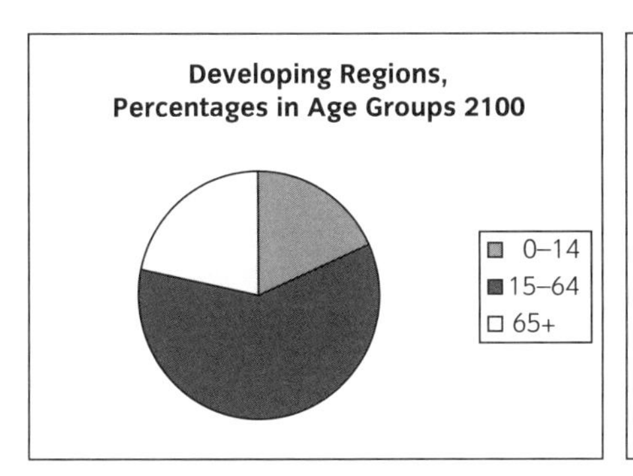

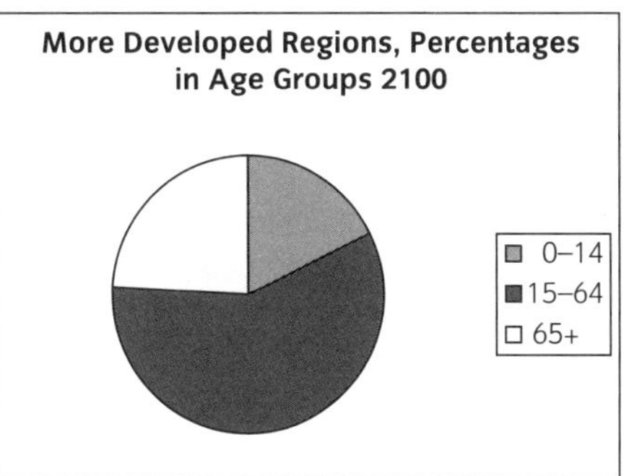

Reformatting a pie chart

12. Modify the title, chart border and legend as described above.
13. To change the shading in sectors: click on a sector and choose Format Selected Data Point Patterns or double click on the sector to bring up the required menu.
14. To 'explode' a sector of the chart (i.e. move it out for emphasis): click on the sector and move the mouse with left button down. To return a sector to its original position, hold down the Alt key while moving the sector.
15. When a chart is completed, further charts, with the same formatting, can be made without having to repeat the previous steps, as described in Exercise 3, Section 18.

Section III

Vital processes

Here, the focus is demographic methods for studying mortality and fertility, together with examples of the theories, concepts and issues that the methods address, including the epidemiologic transition and the second demographic transition.

Chapter 6 **Mortality and health**

Chapter 7 **Fertility and the family**

Mortality and health

As for unhealthiness it may well be supposed, that although seasoned Bodies may, and do live near as long in *London*, as elsewhere, yet new-comers, and Children do not, for the *Smoaks*, *Stinks*, and close *Air* are less healthful then that of the Country; otherwise why do sickly Persons remove into the Country Air? And why are there more old men in Countries then in *London*, *per rata*?

(Graunt 1662: 55–56)

OUTLINE

6.1 Sources of statistics on mortality and health

6.2 The epidemiologic transition

6.3 Measures of mortality

6.4 Health and illness

6.5 Conclusion

Study resources

LEARNING OBJECTIVES

To understand:

- sources and applications of statistics on mortality and health
- the main conceptual framework for the analysis of mortality and health
- recent mortality changes and why particular measures are used
- the principal measures of mortality and mortality differentials
- the concepts of incidence and prevalence and their applications in the analysis of health statistics

COMPUTER APPLICATIONS

Excel module: Epidemiologic Transition.xls (Box 6.1)

Spreadsheet Exercise 6: A graphical data base (See Study resources)

Matters of life and death have been a central interest of demography from its beginnings. Improvements in survival through time have owed much to the collection of mortality statistics, the analysis of trends and the development of explanations of why risks of disease and death vary between regions and social groups. Investigations of the reasons for inequalities in health have been essential in developing preventative strategies, an early illustration of which is John Snow's account of cholera in nineteenth century England:

> Mr Baker, of Staines, who attended two hundred and sixty cases of cholera and diarrhoea in 1849, chiefly among the poor, informed me in a letter with which he favoured me in December of that year, that 'when the patients passed their stools involuntarily the disease evidently spread.' It is amongst the poor, where a whole family live, sleep, cook, eat and wash in a single room, that cholera has been found to spread when once introduced, and still more in those places termed common lodging-houses, in which several families were crowded into a single room. It was amongst the vagrant class, who lived in this crowded state, that cholera was most fatal in 1832; but the Act of Parliament for the regulation of common lodging-houses, has caused the disease to be much less fatal amongst these people in the late epidemics. When, on the other hand, cholera is introduced into the better kind of houses, as it often is . . . it hardly ever spreads from one member of the family to another. The constant use of the hand-basin and towel, and the fact of the apartments for cooking and eating being distinct from the sick room, are the cause of this. (Snow, cited by Harper et al., 1994: 49)

Improvements in survival have brought no waning of such inquiries, partly because of the appearance of new concerns – including the emergence of further strains of hepatitis, the HIV/AIDS pandemic and a resurgence of tuberculosis and malaria – and partly because there remains considerable latitude to address many causes of premature death and disability in all countries. Long life, moreover, is coveted almost above every other personal and social objective, so much so that the wealthier societies devote enormous investment and scientific endeavour 'to squeezing yet another year or two out of life's orange' (Borrie 1977: 19–20).

Working with mortality statistics calls for a knowledge of data sources, mortality trends, and methods of analysis. The first sections of the chapter introduce the sources of statistical information and outline the epidemiologic transition – the principal general theory for interpreting mortality trends and differences. The chapter then discusses the main methods for analysing patterns and trends in mortality and health.

6.1 Sources of statistics on mortality and health

Accurate information on deaths is a prerequisite for improvements in survival. Mortality statistics enable the monitoring of progress in combating diseases and other causes of death, such as road accidents and dangerous working conditions.

At the same time, they facilitate the identification of areas with potential for effective interventions – such as in child health, control of infectious diseases and prevention of lifestyle related illnesses.

Demography has a longer history of interest in mortality than in health, an interest allied with the effects of mortality rates on the growth and structure of national populations. Also, until relatively recently, there was a lack of detailed information on the health of populations, which further explains the pre-occupation with mortality. Although the demographic consequences of non-fatal illnesses and conditions are less obvious at the national level, they are important, such as their impacts on labour force participation and dependency. Mortality statistics often serve as a substitute for health data when appraising the effects of disease prevention programs, since only wealthier countries can afford extensive data collections on health.

National mortality data derive from death certificates, the information on which is sometimes unreliable. This occurs if the physician completing the certificate is unfamiliar with the medical history of the deceased or if there is uncertainty about the cause of death (Lilienfeld and Stolley 1994: 61). Uncertainty can arise if the deceased suffered from interrelated illnesses, such as diabetes with coronary disease, either of which could have proved fatal. Similarly, someone with kidney disease may develop high blood pressure leading to a cerebral haemorrhage – so that there are three contributing causes. Establishing the cause of death is sometimes impossible without recourse to an autopsy and sophisticated medical tests. Other

Table 6.1 Sources of statistics on mortality

Source	Characteristics
Death certificates	The death certificate, completed by a physician, specifies a number of demographic and social characteristics of the deceased and details about the cause of death. Vital statistics data on mortality come from this source. The official death certificate, used for death registration and made available to relatives of the deceased, usually contains only a summary of the information on the original form (sometimes called the medical death certificate), including a statement of the underlying cause of death.
Vital statistics	These include mortality data on the numbers and causes of deaths, together with the age and sex of the deceased. Other characteristics, such as birthplace, marital status, place of residence, education and occupation may also be extracted from the death certificate and made accessible in printed or electronic form.
Cross-national data	Comparative data on mortality are published in the United Nations' *Demographic Yearbook* and the World Health Organization's *World Health Statistics Annual*.

concerns about the specification of causes arise when alcohol, tobacco or other drugs contribute to deaths. Although cirrhosis of the liver, cancer and heart failure may be the identified causes of death in these cases, it is desirable to have additional information on the damage to health due to drugs.

To help in addressing such issues, international practice, as recommended by the World Health Organization, is to base cause of death statistics on the *underlying cause of death*. This is the disease or injury that initiated the train of events leading directly to death. It is the cause of death specified in vital statistics publications. The concept's main limitation is that it does not provide information on the involvement of social and behavioural factors in mortality. Underlying causes of death are classified according the *International Classification of Diseases, Injuries and Causes of Death* (ICD). The World Health Organization revises the classification periodically to take account of refinements in the identification of diseases. Sudden infant death syndrome (SIDS), for example, became a separate category in the ICD, emerging from 'unspecified and ill defined causes of infant mortality'.

Other contributing causes, and the *immediate cause of death*, are also recorded on the medical death certificate and may be made available for analysis, such as through the National Death Index, a data base of death certificates for the United States (Rockett 1999: 13). The *immediate cause of death* is the final disease or condition resulting in death. For example, a coronary occlusion may be the immediate cause of a death in which the underlying cause was diabetes.

Deaths due to 'external causes', including motor vehicle accidents, accidental drowning, falls, suicide and violence, are classified according to the external cause, such as a car accident, rather than the nature of the injury or the antecedent condition, which may have been intoxication. However, modern practice is to record on the medical death certificate details of other circumstances contributing to the death. In this way the contribution of alcohol and drugs to accidental and violent deaths may be assessed. These hidden and extra causes of death formerly passed unrecorded. Today, the issue for mortality research in the more developed countries is less often how much information is collected, but how much is coded and made accessible for research.

Although vital statistics generally identify just a single underlying cause, some countries publish statistics on multiple causes of death – presenting more of the detail contained on the medical death certificate. This recognizes the involvement of combinations of diseases and conditions in many deaths:

> A healthy person goes into hospital to have an operation to set a fractured finger. The patient has an adverse reaction to the anaesthetic and has further complications and dies. The underlying cause of this death according to the rules is fracture. In multiple cause coding, you would be able to see the chain of events that led to the person's death. This particular example could equally apply to any elective surgery for minor ailments and often any admission to hospital where drugs will be prescribed. Deaths due to complications of surgery or hospitalization have been the subject of numerous requests for information. . . . (Australian Bureau of Statistics 1997: 8)

Cancer is likely to be reported as the underlying cause whenever it is present, but this is less likely for heart diseases and other causes (Australian Bureau of Statistics 2000: 37). In old age, a significant proportion of people suffer from multiple illnesses, sometimes including more than one that is life threatening. Ischaemic heart diseases, for example, may be associated with hypertension or diabetes. Also, influenza and pneumonia can be fatal in people already weakened by other illnesses; they are commonly seen, therefore, in lists of multiple causes of death – including those where the underlying cause was cancer, heart disease, respiratory disease or renal failure. For external causes, information on multiple causes offers advantages in identifying causes of injuries, types of injuries and the involvement of drug abuse and self-inflicted harm.

Table 6.2 summarizes sources of national health statistics. Health statistics are potentially very varied in content and, compared with mortality data, are more often dispersed through a range of collections with different aims, content and coverage. National health surveys are the most accessible sources, where they exist, and are most readily used in conjunction with census figures and other statistics for the country as a whole. Health surveys either provide cross-sectional data for a particular time or, if they are repeated with the same respondents at a later time, longitudinal data on cohorts through time.

Disease registries and health system records have become major resources for epidemiologic research and detailed investigations relating to health planning and policy-making. Both sources provide information on specific sub-populations and hence differ greatly in their coverage. Disease registries and health system records have particular strengths as sources of health statistics when they provide cohort data revealing the outcomes for individuals of illnesses and receiving particular treatments.

The great scope and diversity of health-related data calls for detailed knowledge of the nature and quality of particular sources and data items. Response rates and accuracy may vary according to the type of question. General obstacles to adequate reporting of ill health in surveys and hospital and medical records include:

- lack of awareness of having an illness;
- misdiagnosis of symptoms;
- lack of access to, or avoidance of, health care when ill;
- potential shame and social stigma associated with being known to have certain illnesses, such as a sexually transmitted infection or a mental disorder;
- attributing certain symptoms to supernatural forces, rather than illness. In some traditional cultures, common childhood diseases are considered to be due to the displeasure of the gods and, therefore, treatable only through religious means.

Table 6.2 Sources of statistics on health

Source	Characteristics
Health surveys	Sample surveys of the population supply detailed information on many aspects of health and health-related behaviours. Examples include symptoms experienced and actions taken; visits to doctors, dentists and other health professionals; diseases, injuries and disabilities experienced; episodes of hospitalization; habits in relation to diet, exercise, smoking and alcohol consumption; and knowledge of the causes of ill health and the transmission of disease. Both government and private organizations conduct such surveys; those from the former source usually have the widest coverage and the greatest potential for describing the national population.
Disease registries or surveillance systems	These contain information, supplied by physicians, on all new cases of diseases or injuries of special concern for health policy and research. They may include communicable diseases such as HIV/AIDS and tuberculosis, and other diseases – such as cancer and diabetes – that are of particular interest to investigations into causes, prevention strategies, and the effectiveness of treatments.
Health system records	These include records maintained by hospitals, doctors and other health care providers, health insurance organizations, pathology laboratories and screening programs, such as for breast cancer. They contain extensive details on the occurrence of morbidity and the nature and duration of treatments that patients receive. While supplementing other sources, they can also offer insight into the functioning of the health care system, for instance through comparisons between hospitals in terms of conditions treated and duration of stay, or between doctors in terms of the frequency of prescribing medications or making referrals to specialists.
Censuses	Censuses provide the denominators for population-based rates, in the absence of more specific information on the numbers at risk.
Cross-national data	World Health Organization *Health Statistics Annual*

The interpretation of statistics on health and mortality clearly calls, initially, for familiarity with the nature of the data and any inherent statistical problems. Beyond that, interpretation entails comparisons with other populations, with other sub-groups or with other times. Comparisons are most meaningful if made with reference to concepts and theories that facilitate explanations of differences. A broad context for making such comparisons, and developing explanations in studies of mortality and *morbidity* (ill health), is provided by the epidemiologic transition.

6.2 The epidemiologic transition

The first three stages

Abdul Omran's (1971, 1981) epidemiologic transition describes and explains variations in countries' experience of mortality changes through time. It elaborates on the account, in demographic transition theory, of the course of change in the occurrence of diseases and death. Omran recognized four main patterns of mortality change through time, each with three stages.

The names of the stages evoke the nature of conditions affecting mortality: (1) *the age of pestilence and famine*, (2) *the age of receding pandemics* and (3) *the age of degenerative and man-made diseases* (Table 6.3). Humans have lived most of their history in the age of pestilence and famine, which explains the late emergence of sustained population growth. Tables 6.3 and 6.4 summarize the stages and patterns of change. Omran differentiated the three stages with reference to rates and causes of death.

The four patterns, or models of change through time, differ according to the timing, speed and extent of progress through the three stages (Table 6.4). Because of social development and fertility decline in some countries, as well as social and economic dislocation in others, demographic trends in low income countries continue to change. Thus some formerly in the 'Delayed' pattern, such as Thailand and Indonesia, are now better described as experiencing the 'Transitional Variant of the Delayed Model'. Conversely, the recent rise in mortality in many countries of sub-Saharan Africa, because of wars, famines, poverty and the HIV/AIDS epidemic, justify a new 'regressive' model.

Omran envisaged that, over the course of the transition, all social classes benefited, although female survival improved more than male survival. In the context of high mortality, females may have higher death rates than males. Reasons include

Table 6.3 Stages of the epidemiologic transition

1. *The age of pestilence and famine*
Mortality is high and fluctuating, precluding sustained population growth. Life expectancy at birth varies between 20 and 40 years.

2. *The age of receding pandemics*
Mortality declines progressively as epidemics decrease in frequency and magnitude. Average life expectancy at birth rises to 55 years. As the gap between birth and death rates widens, rapid population growth ensues.

3. *The age of degenerative and man-made diseases*
Mortality continues to decline and eventually approaches stability. Life expectancy at birth increases to more than 70 years. Continued maintenance of population growth depends on the birth rate.

Source: Omran (1981).

Table 6.4 Models describing patterns of change through time in the epidemiologic transition

---

1. Classical or Western model
Gradual decline in mortality in response to social economic and environmental improvements. This model denotes the experience of Western societies through the nineteenth and twentieth centuries. Fertility decline is also gradual and protracted.

2. Accelerated variant of the classical model
Slow mortality decline in the nineteenth century, giving way to rapid decline during the twentieth century, as in Japan and Eastern Europe. The acceleration is associated with general social improvements together with sanitary and medical advances. The birth rate also declines rapidly after the Second World War.

3. Delayed model
The delayed model is characteristic of many less developed countries, where there is little mortality decline until after the Second World War. Subsequently, death rates fall rapidly, with modern medical technology and international aid having a substantial influence. Overall gains remain limited, and socio-economic differences in health and survival persist, without social development and improvements in community-based health care. There is a continuation of high infant and child mortality, high birth rates and high rates of population growth.

4. Transitional variant of the delayed model
In this model, mortality decline is also delayed until after the Second World War, then it occurs rapidly and to lower levels than in the delayed model itself. Social development and family planning efforts facilitate a rapid decline in the birth rate. Gains in the control of infant and child mortality do not slacken to the same extent as in the delayed model. As examples, Omran (1981) cited Taiwan, South Korea, Singapore, Hong Kong, Sri Lanka, Mauritius, Jamaica and China.

Source: Omran (1981).

maternal mortality together with lowered resistance to infections owing to frequent child-bearing and prolonged lactation on an inadequate diet. Important also is an under-investment in females in terms of food, financial expenditure, medical care and education. The poor status of females has been an early obstacle to overcome in lowering death rates (Ware 1981: 58ff). When mortality is low, however, biological and lifestyle differences enable females to achieve greater longevity than males.

Another feature of the epidemiologic transition is the trend through time towards an increasing concentration of deaths in older ages. This is often described as the *rectangularization of the survival curve*. The survival curve for a population is depicted on a line graph plotting the proportion of a birth cohort still living at each age. Rectangularization occurs as infant and child mortality decline and the proportions surviving in the younger and middle years are maintained near 100 per cent; hence the curve falls away only in the older ages (see Wilmoth and Horiuchi 1999). Consequently, the shape of the curve becomes more rectangular as survival at young and middle ages improves. This is illustrated in the Excel module on the epidemiologic transition (Box 6.1). Comparing countries with reference

BOX 6.1 The epidemiologic transition module (*Epidemiologic Transition.xls*)

This module illustrates the course of the epidemiologic transition using model data on mortality. It is intended to enable comparisons between mortality patterns at different stages of the transition and to simulate the main patterns of change through time. New statistics on age-specific death rates can also be entered and plotted to draw comparisons.

The model data are based on the 'West' series of model life tables which illustrate the history of mortality change in Western societies (see Chapter 9). The model data show mortality in hypothetical populations numbering 100 000 at age zero. Each birth cohort is assumed to experience constant death rates over its lifetime.

Data sources: Coale and Demeny (1983); Coale and Guo (1990).

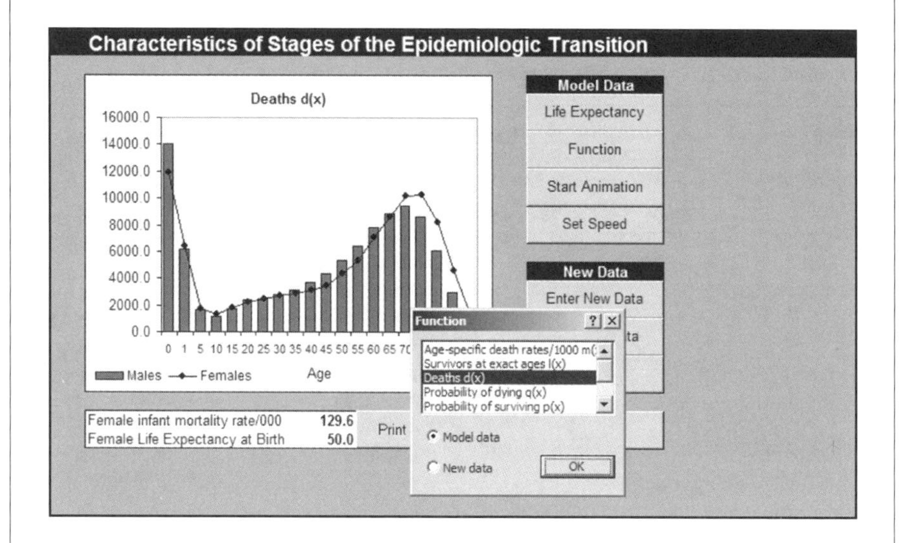

Instructions

To use the module, insert the CD in the drive, start Excel and open the file *Epidemiologic Transition.xls*. Enable macros when prompted.

- Click **Start** on the opening display.
- Click **Life Expectancy** to set female life expectancy at birth for the stage of the epidemiologic transition you wish to view.
- Click **Function** to choose the required measure of mortality.
- Click **Start Animation** to view a 'movie' of progress in mortality from the selected stage. The computer increases female life expectancy at birth in steps of 2.5 years until 85 years is reached. This produces an animated display of whatever measure of mortality has been set using the **Function** button.

continued

BOX 6.1 continued

- Click **Set Speed** to change the speed of the animation.
- Under the heading 'New Data', click **Enter New Data** if you wish to enter a new set of age-specific death rates, from which the mortality measures (life table functions) will be calculated. Click **Return** to go back to the main display. Now, under 'New Data', click **Function** to choose the mortality measure to be calculated and plotted. For example, if you have information for an indigenous population, or a population affected by the HIV/AIDs epidemic, this is a convenient way of visualizing the information and comparing it with the model data.
- Click **Print Summary** to view or print six graphs showing the characteristics of the selected population.
- Click **Home** to return to the opening display.

See Appendix B for further advice on using the Excel modules.

The epidemiologic transition: an expanded model

Stage	Indicative mortality levels			Main causes of death	Major initiatives to reduce mortality
	IMR	e_0	e_{50}		
I High and fluctuating mortality	200	40	10	Infectious disease associated with epidemics, famine and pestilence	Famine relief, improved diet, sanitation, water quality
II Progressive declines in mortality, predominantly in younger ages	100	50	24	Infectious disease, but epidemics less frequent and catastrophic	Public health and sanitation, improved nutrition, hospital development.
III Stable mortality at a low level	12	70	25	Degenerative disease – such as cancer, heart disease – and accidents	Therapy developments, medical advances.
IV Further declines in mortality, especially at older ages	9	75	30	Degenerative disease and accidents	Lifestyle changes (e.g. diet, exercise), medical advances

IMR = infant mortality rate (infant deaths per 1000 births)
e_0 = life expectancy at birth
e_{50} = life expectancy at age 50
Source: adapted from Hugo (1986: 23).

Activities and questions

Work through the steps below, referring to the above table and printing summary materials as needed. The module is intended for lecture room discussions, or for use by small groups or pairs of students. Answers to the questions are at the end.

continued

BOX 6.1 continued

Stage one ('The Age of Pestilence and Famine')

- Click the button labelled **Compare Stages** on the opening screen.
- Display information for a population at Stage 1 of the epidemiologic transition by clicking the **Life Expectancy** button and setting female life expectancy at birth [e(0)] to 30 years.
- Display the life expectancy graph by clicking the **Function** button and choosing 'Life expectancy'. The ages refer to age 0, **1**, 5, 10. . . .

Q1 Study the graph of life expectancy and explain the trend in the figures.

- Display the graph of deaths at each age by clicking the **Function** button and choosing 'Deaths'. The graph of deaths shows the number of males and females (from original cohorts of 100 000 males births and 100 000 female births) who die between different ages (e.g. between 0 and 1, 1 and 5, 5 and 10, 10 and 14 years, etc).

Q2 In the younger ages, at which age does the lowest number of deaths occur? Why are there two peaks in the numbers of deaths?

- Print a set of summary graphs for Stage 1 of the transition by clicking the **Print Summary** button.
- The numbers in five year age groups in the age structure graph, on the printed sheet, represent the survivors from 500 000 female births (100 000 births per annum for 5 years) and 525 000 male births. The figure for males is higher because the number of males at each age was multiplied by 1.05 to allow for the fact that there are about 105 boys born per 100 girls.

Q3 Referring to the age structure graph, explain its triangular form.

- The printed age structure graph shows five year age groups 0–4, 5–9, . . . to 95+, whereas the 'age structure' displayed using the **Function** button shows ages 0–1, 1–4, 5–9, . . . and contains no correction for the sex ratio at birth.

Q4 Why should you expect there to be more females than males in the total population?
Q5 Comment on the 'dependency burden' for families in Stage 1 populations.

Stage two ('The Age of Receding Pandemics')

- Display information for a population at Stage 2 of the epidemiologic transition by clicking the **Life Expectancy** button and setting female life expectancy at birth to 50 years.
- Using the **Function** button, examine the graphs of different measures of mortality (especially e_x, d_x, m_x, and l_x) and the main changes in the pattern of mortality between Stages 1 and 2.

Q6 What changes occur in the causes of death between stages 1 and 2 (refer to the above table 'the epidemiologic transition: an expanded model')?
Q7 What changes occur in the pattern of mortality by age between stages 1 and 2 (switch between life expectancies of 30 and 50 to compare patterns)?

continued

BOX 6.1 continued

Stage three ('The Age of Degenerative Diseases')

- Display mortality trends in a Stage 3 population by setting female life expectancy to 70 years.

Q8 What changes occur in the causes of death between Stages 2 and 3?
Q9 Describe the main changes in the pattern of mortality between Stage 2 and Stage 3.

Stage four ('The Age of Delayed Degenerative Disease')

- Display mortality trends in a future Stage 4 population by setting female life expectancy to 85 years.

Q10 At which ages is there most improvement in mortality between stages 3 and 4?
Q11 At which ages do the death rates of males still differ from those of females?
Q12 For your own country, consider how likely are the changes required to raise female life expectancy to 85 years?

National mortality trends

- Open the module *Demographic Transition.xls* (see Box 1.2), on the opening screen click **Trends Since 1950, Two Countries**, click **Menu** and set both the 'First country' and the 'Second country' to China. Also, set the 'First variable' to CBR and the 'Second variable' to CDR.

Q13 Consider whether the trends in the crude rates for China correspond with the 'classical', 'accelerated', 'delayed' or 'transitional' models of the epidemiologic transition. Compare the figures for China with those for other countries.

Answers

1. Life expectancy declines with age except at ages 0 and 1. The increase in life expectancy in early childhood reflects that survival prospects are much better beyond infancy, where there is greatest vulnerability to infectious diseases.
2. The number of deaths is lowest at age 10 (the immune system matures after age 5). The peaks in infancy and older ages are associated especially with infectious and degenerative diseases respectively, although infectious diseases are also major causes of deaths of adults.
3. High birth and death rates produce a narrow-based triangular age structure with a growth rate of zero. Mortality progressively reduces numbers in each older age group.
4. Females have lower death rates than males at every age. A possible exception is excess female mortality due to maternal deaths.
5. Families have high child dependency and low aged dependency.
6. Most important is the decline in infectious and epidemic diseases.
7. Better survival of infants results in a higher proportion of deaths at older ages. Life expectancy increases at all ages, but at younger ones in particular.

continued

BOX 6.1 continued

8. Infectious diseases are replaced by degenerative diseases as the main causes of death.
9. Childhood mortality falls to a relatively low level and most of the cohort survives to middle age or later life.
10. Infant mortality is about halved; small reductions in mortality occur at every other age.
11. At every age. Females have higher life expectancy than males at all ages.
12. The change requires very low infant mortality, negligible mortality at every other age to age 50, and death rates at older ages reduced by a third to a half compared with a population with a (female) life expectancy of 75.
13. The pattern for China corresponds with Omran's 'transitional variant of the delayed model', since death rates and birth rates were still fairly high in the early 1950s, but subsequently fell rapidly. For diagrams of the trends in the crude rates for each model, see Omran (1981: 173).

to the overall age pattern of survival is a means of determining the extent of mortality control and progress through the transition. The level of infant mortality is also a major point of comparison.

The fourth stage

As in the demographic transition, Omran considered the third stage of the epidemiologic transition to represent a new equilibrium where societies reached the limits to mortality decline. Higher mortality from cancer, for instance, was expected to offset falling death rates from heart diseases, thereby producing little net gain in survival (Omran 1981: 174).

During the 1980s, however, research increasingly questioned whether completion of the demographic transition necessarily heralded the creation of a plateau either in fertility rates (van de Kaa 1987; Davis et al. 1986) or mortality rates. Continuing changes in mortality in post-transition societies prompted the suggestion that 'a new era in epidemiologic history' had begun' (Olshansky and Ault 1986). Whereas stages one to three in Omran's formulation were differentiated by the substitution of one set of diseases for another, this 'new era', or fourth stage, was characterized by a change in the ages at which members of a population succumbed to disease.

Termed 'the age of delayed degenerative diseases', the fourth stage of the epidemiologic transition was distinguished from the third stage mainly by a shift towards older ages at death from degenerative diseases. This trend, in conjunction with continuing improvements in survival at young ages, created a further major decline in death rates (Olshansky and Ault 1986: 360–361).

Other proponents of a fourth stage have drawn attention to the emergence of individual behaviours and lifestyles as significant determinants of morbidity and mortality (Trlin 1994). Although there has been growth in awareness of health risks and the role of diet and exercise in maintaining good health, Rogers and Hackenberg (1987) characterized the distinctive behaviours of the fourth stage not in terms of avoiding risks, but in terms of taking risks. Hence they described the fourth stage as the 'hubristic stage' – in which people, perhaps with a sense of invulnerability or insufficient motivation to do otherwise, engage in reckless or irresponsible behaviour, bringing ruin upon themselves.

The risk-taking includes using addictive drugs, sharing injecting needles and engaging in unsafe sex. Driving at high speeds, drinking excessive amounts of alcohol, avoiding exercise and consuming a diet high in saturated fat, salt and sugar are similarly behaviours entailing health risks that many know well, but often choose to ignore. Thus some major health problems of the fourth stage are thought to be due to personal choice, rather than only to broader social, economic or political influences that are beyond individual control. Nevertheless, advertising and peer-group behaviour remain powerful social forces encouraging unhealthy dietary habits and lifestyles.

A further feature of health issues in long-lived populations is the growing numbers with severely disabling neurodegenerative diseases, including Alzheimer's disease and Parkinson's disease (Broe and Creasy 1995). While unrelated to hubristic behaviour, the rise in cases of senile dementia and other diseases associated with brain ageing is undoubtedly a major feature of the current stage of the epidemiologic transition in developed countries.

Overall, it seems reasonable to recognize a fourth stage of the epidemiologic transition in which occurs: (1) a continuing improvement in survival, notably at older ages; (2) a decline in mortality from certain degenerative diseases, because of delayed onset or better diagnosis and treatment; (3) a greater role of individual irresponsibility, the reasons for which are known only partially; (4) a rising prevalence of degenerative diseases of the brain.

The epidemiologic transition is an example of an important theory relevant to demographic inquiry, but one open to change and development. Conceptual enhancements and advances are ultimately more important than empirical inquiry per se, but usually the two are interdependent. The epidemiologic transition remains a valuable point of departure for research on mortality and health, one that is all the more interesting because of its potential for elaboration, and modification.

6.3 Measures of mortality

The implications of the theory of the epidemiologic transition for mortality analysis are that measures are needed to address various aspects of change:

- Trends in the overall level of mortality, revealing developments through time and differences between countries.
- Trends in the causes of death, since these are linked with the shift from 'the age of receding pandemics' to the 'age of degenerative and man-made diseases' and the 'fourth stage'.
- Differences between age groups, which are key indicators of progress through the transition, such as in terms of the decline in infant mortality and the rectangularization of the survival curve.
- Differences between the sexes, to gauge the extent of convergence in the relative vulnerability of males and females to the main causes of death.
- Differences between socio-economic groups, again to show the extent of convergence or continuing inequality in the face of death.
- Trends in the prevalence of health and illness in the population.

The above features are the subject matter of the principal measures of mortality and morbidity. The most commonly used of these are easily understood and calculated. This section and the next introduce them, together with examples of their applications.

Crude death rate

The best-known general measure of mortality in a population is the crude death rate, previously discussed in Chapter 1. This is simply the number of deaths in a year divided by the mid-year population, then multiplied by 1000. Diagrams depicting the course of the demographic transition and the epidemiologic transition (see Figure 1.2) invariably include a line showing the crude death rate.

A basic knowledge of trends through time in the crude death rate enables contemporary figures to be placed in perspective, and judgements made about whether overall mortality is high or low. High rates, of 30 per thousand or more, occur in the early part of the epidemiologic transition while low rates, of 10 per thousand or less, occur late in the transition (Omran 1981). However, the crude death rate actually increases very late in the transition, because of the rising proportion of the population in the older ages.

Little is known about crude death rates before about 1800, because of incomplete recording of deaths until the establishment of national registration systems. It is argued that crude death rates above about 45 per 1000 are unsustainable for long periods, because the maximum crude birth rate is 45 to 50 per thousand (Pressat 1972: 71). Statistical recording of deaths does go back sufficiently far to demonstrate that considerable short-term fluctuations in death rates were characteristic in past times (see Box 1.2). Crude death rates for Sweden indicate normal mortality levels in past times of 30 or more per thousand – fluctuating to 40 or 50 per 1000 in years of epidemics. This situation typifies 'the age of pestilence and famine'.

Until the recent HIV/AIDS pandemic, the highest known mortality from disease was from the fourteenth century Black Death in Europe and Asia. Many of the early 'statistics' on deaths were grossly exaggerated, since some chroniclers used numbers more as adjectives, to represent 'many', rather than as actual estimates. Nevertheless, in the pandemic plague of 1348–1350, at least a quarter of the population of Europe died. Famine has also produced episodes of extreme mortality.

The decline and stabilization of crude death rates followed control of famines and epidemic diseases such as plague, cholera and smallpox. There is no country now with an average crude death rate in excess of 30 per thousand. Some countries in sub-Saharan Africa have the highest mortality in the range 20–25 per thousand. Crude death rates in the United States, Britain, France, Germany and Italy are all in the low range of 10 or less (Population Reference Bureau 2001a). The times and places with the worst mortality usually have the most deficient statistical collections.

Other overall measures of mortality and survival, such as life expectancy, come from life tables, which are discussed in Chapter 8. Standardization of rates (see Chapter 4) is often used to make crude death rates more comparable, by eliminating the effects of differences in age structure or other aspects of population composition. Rather than relying solely on measures for total populations, however, cross-national comparisons often use key indicators for particular sections of the population. This entails the study of *mortality differentials*, or mortality differences, between population groups defined in terms of age, sex, marital status, socio-economic characteristics or place of residence.

Mortality differentials appear in the statistics in the form of differences in death rates between groups. Socio-economically disadvantaged groups tend to have the highest mortality from communicable diseases as well as from the so-called 'diseases of affluence', such as cardio-vascular disease. A sign of a country's development is the removal of inequality in the face of death which Pressat (1973: 37) described as 'the worst form of injustice suffered by the human race'. The following sections discuss the main measures of differential mortality.

Age-specific death rate

The distribution of deaths by age and sex in developed countries tends to resemble the mushroom cloud of an atomic bomb blast (Figure 6.1). The 'stem' of the cloud is comprised of the relatively few deaths before middle age, while the large 'cap' reflects the predominance of deaths in the older ages. The cap is asymmetrical, however, because males typically die at younger ages than females.

In most populations, the death rate begins at a relatively high level for infants under one year, declines to its lowest levels for children of school age and slowly increases thereafter.

$$\text{Age-specific death rate} = \frac{\text{deaths in calendar year at age } x}{\text{mid-year population aged } x} \times 1000$$

Table 6.5 Basic measures of mortality

Measures	Examples
Crude death rate (CDR)	Egypt 1990 (estimates)
$\text{Crude death rate} = \dfrac{\text{number of deaths in a year}}{\text{mid-year population}} \times 1000$	Deaths: 534 000 Mid-year pop.: 52 426 000 CDR: 10.2
or	
$\text{CDR} = \dfrac{D}{P} \times 1000$	
Age-specific death rate (ASDR)	Russian Federation 1995
$\text{Age-specific death rate} = \dfrac{\text{deaths in calendar year at age } x}{\text{mid-year population aged } x} \times 1000$	
or	
$\text{ASDR} = \dfrac{D_x}{P_x} \times 1000$	

Russian Federation 1995

	Mid-Year Pop. Males	Deaths	ASDRs
45–49	4 499 834	86 647	19.3
50–54	2 667 863	72 457	27.2
55–59	4 296 680	145 374	33.8
60–64	2 909 878	133 360	45.8
65–69	2 859 230	170 747	59.7
70–74	1 209 717	91 776	75.9
75–79	609 267	64 710	106.2

Table 6.5 *continued*

Measures	Examples		
Infant mortality rate (IMR)	Egypt 1990 (estimates)		
Infant mortaliy rate = $\frac{\text{deaths under 1 year of age}}{\text{total live births in a calendar year}} \times 1000$		Infant deaths	63 813
		Births	1 737 000
		IMR =	36.7
or			
$IMR = \frac{D_0}{B} \times 1000$			
Neonatal mortality rate	Australia 1996		
Neonatal mortality rate = $\frac{\text{deaths in the first 28 days after birth}}{\text{total live births in a calendar year}} \times 1000$		Neonatal deaths	759
		Live births	253 567
or		Neonatal mortality rate =	3.0
Neonatal mortality rate = $\frac{D_{0-28\ \text{days}}}{B} \times 1000$			
Post-neonatal mortality rate	Australia 1996		
Post-Neonatal mortality rate		Infant deaths	1 460
$= \frac{\text{deaths between 29 days and one year after birth}}{\text{total live births in the calendar year}} \times 1000$		Infant deaths minus neonatal deaths	701
		Live births	253 567
		Post-neonatal mortality rate =	2.8

or

$$\text{Post-neonatal mortality rate} = \frac{D_{29-365\,\text{days}}}{B} \times 1000$$

*Perinatal mortality rate**

Perinatal mortality rate

$$= \frac{\text{annual deaths between 28 weeks gestation and 28 days after birth}}{\text{total live births and stillbirths in the year}} \times 1000$$

or

$$\text{Perinatal mortality rate} = \frac{D^f + D_{0-28\text{days}}}{B + D^f} \times 1000$$

where D^f = late fetal deaths

Australia 1996	
Stillbirths (late fetal deaths)	1 411
Neonatal deaths	759
Live births	253 567
Perinatal mortality rate =	8.5

Stillbirth rate

$$\text{Stillbirth rate} = \frac{\text{annual deaths from 28 weeks gestation}}{\text{total live births and stillbirths in the year}} \times 1000$$

or

$$\text{Stillbirth rate} = \frac{D^f}{B + D^f} \times 1000$$

where D^f = late fetal deaths

Australia 1996	
Stillbirths (late fetal deaths)	1 411
Live births	253 567
Stillbirth rate =	5.5

* Australia uses the definition of perinatal mortality as the sum of stillbirths and neonatal deaths, rather than stillbirths and deaths in the first seven days after birth.
Sources: Shryock and and Siegel (1973); United Nations (1993, 2000b); Australian Bureau of Statistics 1997. *Causes of Death, Australia 1996*.

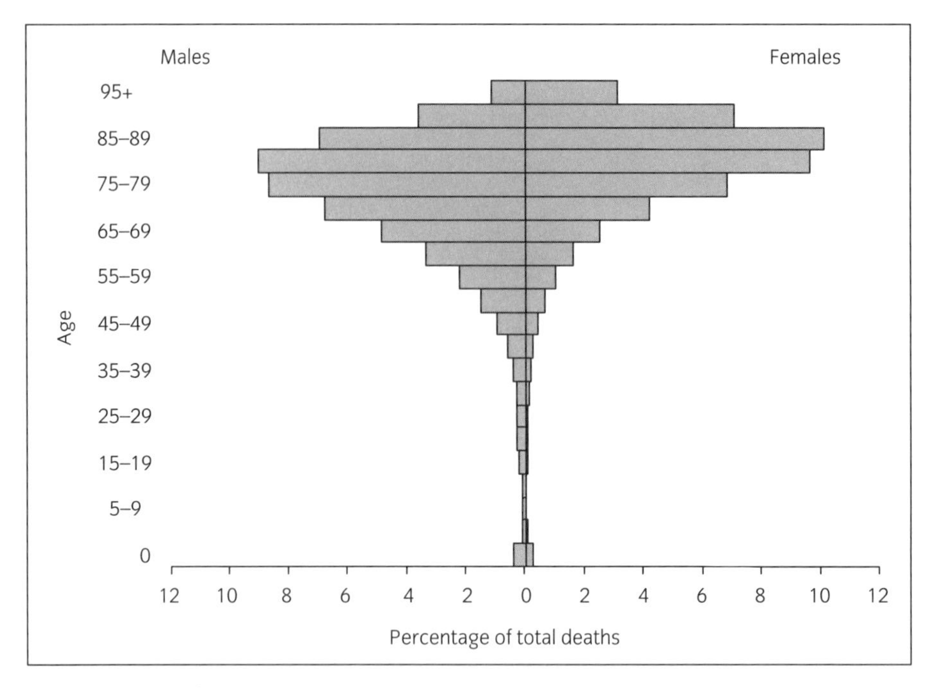

Figure 6.1 Distribution of deaths by age and sex in a long-lived population (Stationary model data: female life expectancy at birth = 80 years).

Note: the first two age groups are <1 and 1–4 years.

The *age-specific death rate* (ASDR) measures the incidence of death at each age (Table 6.5). 'Specific' rates in demography refer to a group in the population with a particular characteristic, such as belonging to a certain age group. The result is normally multiplied by 1000 to avoid having to interpret very small numbers or decimal fractions. Figures in the numerator usually come from vital statistics publications, while those in the denominator are from censuses, population estimates or surveys. The mid-year population is chosen as the denominator, since it approximates the average number at risk of mortality during the year. The ASDRs may refer to single year ages or to grouped ages, such as 60–64 and 65–69.

Infant mortality rate

Historically, so many deaths have occurred early in life that saving of infants under 1 year of age has been a major public health issue, calling for accurate statistics and rigorous measures to identify the extent of the problem and monitor progress. Today, even in long-lived populations, there persists an abiding interest in obtaining detailed information on the numbers and causes of mortality during infancy. In developed country populations the death rate in the first year of life is exceeded only among people of middle age or older.

The best-known and most widely available measure of mortality in early life is the *infant mortality rate* (IMR). Because of its availability and focus on a vital area of change, the infant mortality rate has become a key indicator of demographic development and health conditions in different countries (Figure 6.2).

$$\text{Infant mortality rate} = \frac{\text{deaths under 1 year of age}}{\text{total live births in a calendar year}} \times 1000$$

The infant mortality rate is calculated by dividing the total deaths under 1 year of age by the total live births in a calendar year. It is expressed as a rate per 1000.

Clearly, the infant mortality rate differs from an age-specific death rate because the denominator is live births, rather than the mid-year population. While it would be possible to use a denominator consisting of the mid-year population under age one, the denominator would be accurate only for census years. Live births are an appropriate denominator for any year. A more important reason for avoiding the mid-year population is that the majority of infant deaths usually occur in the first days and weeks of life. Since the deaths are not evenly distributed over the first 12 months, the mid-year population is not a valid indicator of the average size of the population at risk of infant mortality.

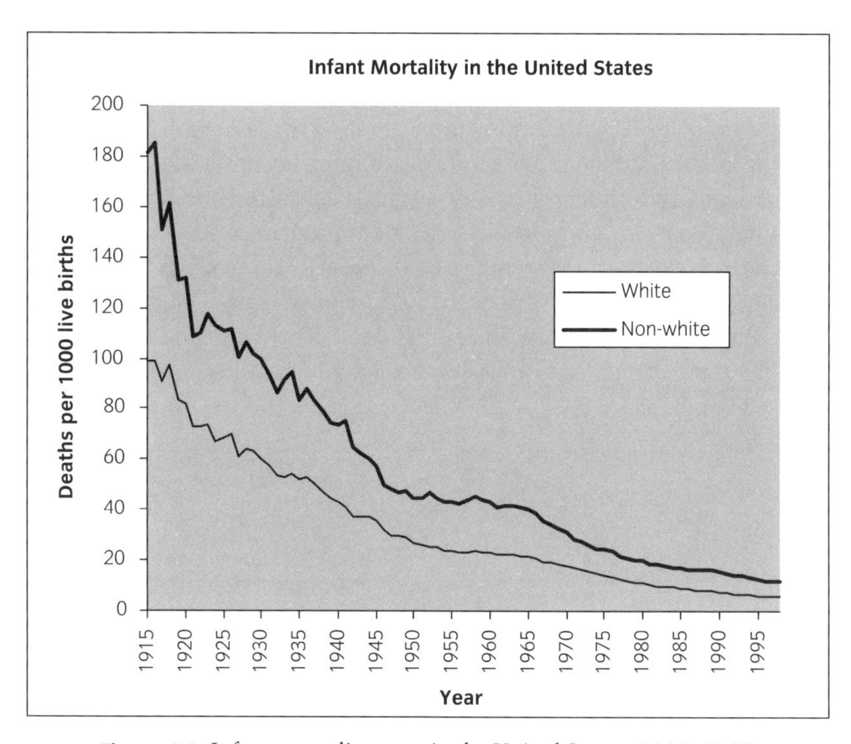

Figure 6.2 Infant mortality rates in the United States, 1915–1998.

Data sources:
1915 to 1969: US Census Bureau 1975. *Historical Statistics of the United States, Colonial Times to 1970*, Bicentennial Edition, Part 1;
1970 and 1975 to 1998: National Center for Health Statistics 2000. 'Deaths: Final Data for 1998,' by S. L. Murphy. *National Vital Statistics Report*, 48, (11);
1971 to 1974: US Department of Health and Human Services, Public Health Service, National Center for Health Statistics 1986. *Vital Statistics of the United States 1981*, Volume 2, 'Mortality'.

Although the IMR is referred to as a 'rate', it is more a probability measure based on the initial population. The live births in a calendar year better represent the population at risk. Nevertheless, there are two reasons why the infant mortality rate does not precisely relate the number of events to the population at risk:

- First, the migration of persons under 1 year of age may affect the numerator (the number of deaths) but not the denominator (live births in the area). Thus if a group of refugees cross a border into another country, their children will not have been included in any count of the destination's live births, but any infant deaths among the refugees will augment the receiving country's infant mortality rate. At the national level, this is normally only a minor influence.
- Second, while the denominator refers to live births in a calendar year, the numerator will include some deaths of children born in the previous year, since they will still be under one year of age (Figure 6.3). Again, it is not normally a major concern, unless there have been sharp variations in annual births or the number of infant deaths. Calculating the infant mortality rate from the average of births and deaths over several years is a response to such problems (Palmore and Gardner 1983: 126).

In pre-industrial times infant mortality was very high – at a mean level of 250 to 300 per thousand (Pressat 1985: 88), meaning that a quarter or more of each birth cohort died in their first year. Another quarter died between ages 1 and 20. In the West, the historical decline in the crude death rate was not initially accompanied by very marked improvements in infant survival. This mainly occurred in the twentieth century. Today, infant mortality rates of more than 50/1000 are regarded as high, but rates in excess of 100 per thousand persist in parts of sub-Saharan Africa. In contrast, the infant mortality rates for all developed regions was 8 per thousand in 2001 (Population Reference Bureau 2001a). The Epidemiologic Transition module illustrates changes in death rates through time at all ages, including infancy (see Box 6.1).

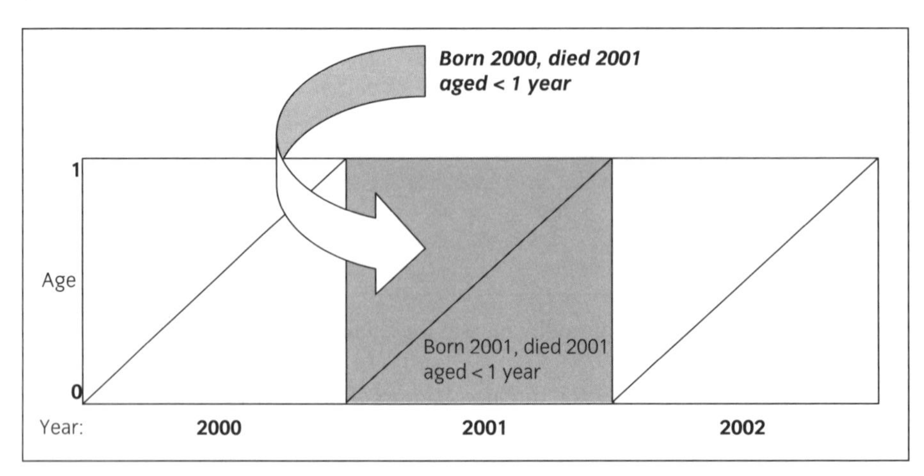

Figure 6.3 Lexis diagram showing cohorts contributing to the 2001 infant mortality rate.

The historical decline in infant mortality has been especially dependent on control of *exogenous causes* of death. Exogenous causes are sometimes described as the 'social component of mortality' in that they are potentially preventable. Exogenous causes are factors external to the body, especially infections and accidents. Contrasted with exogenous causes are *endogenous causes* of death. These are internal agents operating within the body, leading to biological defects in the new-born, as well as degenerative diseases of later life.

There are difficulties in maintaining the exogenous–endogenous dichotomy, however, as the health of babies is related partly to their mothers' nutrition and general health during pregnancy, together with specifically biological factors. Also, conditions such as sudden infant death syndrome have both endogenous and exogenous risk factors. At the older ages, many external influences, such as smoking and diet, are implicated in the occurrence of so-called degenerative diseases, including cancers, strokes and heart disease. Some now consider that 'there is probably no cause of death or disease that is not influenced by environmental or behavioural factors' (Preston et al. 2001: 194).

Further measures of mortality in early life

The exogenous–endogenous dichotomy has, nonetheless, served as a basis for refining measures of mortality in early life, especially through seeking to specify the period in which endogenous factors have their greatest impact. One distinction is between *neonatal* and *post-neonatal* deaths (Figure 6.4). Neonatal mortality refers to deaths in the first 28 days of life, and is customarily assumed to be the period in which deaths from endogenous causes are concentrated.

Post-neonatal deaths occur between 28 days and the first birthday. In the past, post-neonatal deaths were mainly due to exogenous causes, such as gastro-intestinal and respiratory infections. Because of the elimination of these causes in more developed countries, together with improvements in the care of low birth-weight babies, post-neonatal mortality no longer serves as a proxy for infant mortality from exogenous causes (Poston and Rogers 1985).

Nevertheless, neonatal and post-neonatal mortality rates still have general applications for demographic measurement in varied settings. The two rates are defined as follows:

$$\text{Neonatal mortality rate} = \frac{\text{deaths in the first 28 days after birth}}{\text{total live births in a calendar year}} \times 1000$$

$$\text{Post-neonatal mortality rate} = \frac{\text{deaths between 29 days and one year after birth}}{\text{total live births in the calendar year}} \times 1000$$

The *neonatal mortality rate* refers to deaths in the first 28 days after birth. Here the population at risk is the number of live births. The risk of death is especially high

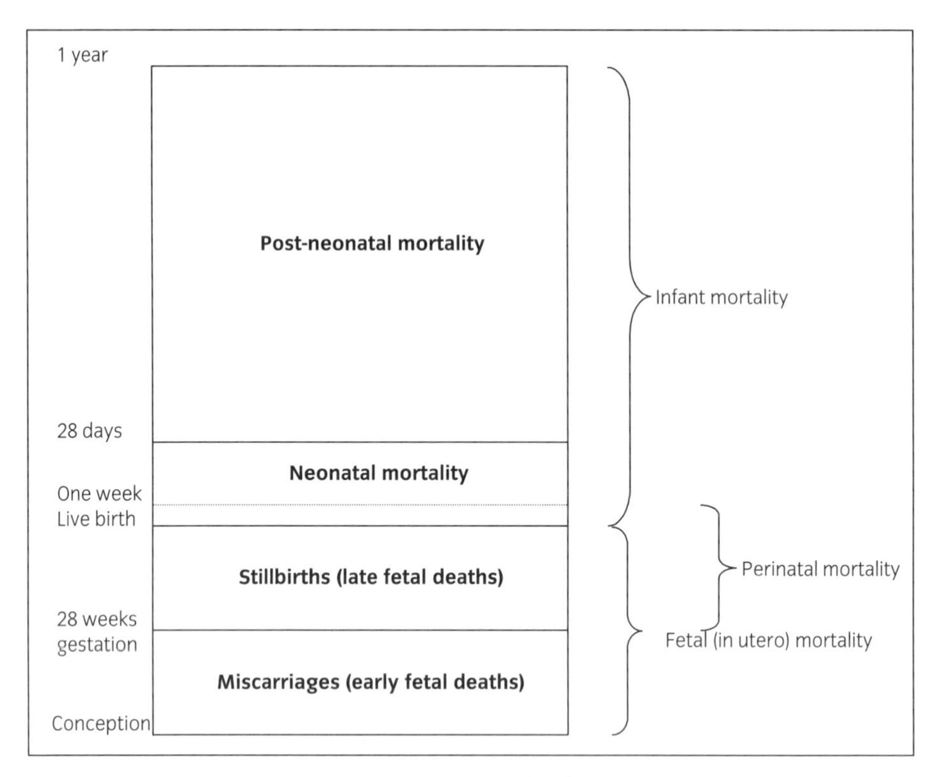

Figure 6.4 Concepts in early life mortality.

in the first day and first week of life, but remains high beyond this. Accordingly, the neonatal mortality rate captures the period of greatest risk.

The endogenous causes of mortality in the week after birth are similar to the causes of *stillbirths* or *late fetal deaths*. Hence the two may be combined into the *perinatal mortality rate*, which refers to deaths in the period immediately before or after birth. Perinatal mortality tends to be highest among mothers aged less than 20 or over 35 years, who are outside the optimal age range for childbearing.

$$\text{Perinatal mortality rate} = \frac{\text{annual deaths between 28 weeks gestation and 7 days after birth}}{\text{total live births and stillbirths in the year}} \times 1000$$

In national data collections, the definition of the perinatal period varies, sometimes including the first month after birth rather than just the first week (Preston et al. 2001: 37). Although 40 weeks is the full term of pregnancy, the definition of the perinatal period assumes that babies are capable of surviving outside the womb after 28 weeks. Hence the use of 28 weeks gestation in the formula. Some statistical collections, however, now use other intervals such as 20 or 22 weeks. The availability of intensive care for premature babies enables a shorter gestation age to define a potentially viable fetus. When the gestation time is unknown, a birth weight of 400 or 500 grams, or a body length of 25 cm crown to heel, may be used to define whether the fetus reached the perinatal period (Australian Bureau of Statistics 2000: 57 and 68).

The stillbirth rate is based on the sum of live births and stillbirths, that is, a full count of the population at risk of this event:

$$\text{Stillbirth rate} = \frac{\text{annual deaths from 28 weeks gestation}}{\text{total live births and stillbirths in the year}} \times 1000$$

Deaths at an earlier stage of pregnancy are described as *miscarriages* (or *early fetal deaths*): these are not registered as deaths, but are sometimes identified in hospital statistics on obstetric procedures. Similarly, induced abortions do not figure in the mortality statistics, but appear in separate statistical collections.

In summary, given the high rates of mortality in early life, there has been particular attention to improving measurement as a basis for evaluating the reasons for the losses and monitoring progress. The development of methods of analysing mortality among the very young has entailed a process of refinement, identifying different periods associated with different risk factors. Formerly, a useful distinction was the notion that deaths due to endogenous factors were concentrated in the perinatal period, but in the late stages of the epidemiologic transition this is less apparent. Nevertheless, the concepts employed in the study of early life mortality continue to provide an informative basis for comparing mortality trends.

Cause-specific death rates

Whereas changes in age-specific and infant mortality rates provide a broad indication of progress made in controlling diseases, cause-specific death rates supply the detail needed to identify and address specific health problems. Since the numbers of deaths from individual causes are often low, cause-specific death rates are conventionally expressed as rates per 100 000:

$$\text{Cause-specific death rate} = \frac{\text{deaths in calendar year from a particular cause}}{\text{total mid-year population}} \times 100000$$

The denominator of the cause-specific death rate is not strictly the population at risk, because the risk of dying from many causes varies by age. Bowel cancer, cirrhosis of the liver, strokes and ischaemic heart disease, occur mainly in people of middle age or older. Diseases of the reproductive organs are specific to each sex, while the so-called degenerative diseases of later life tend to kill males at younger ages than females. Cause-specific death rates, therefore, are mostly calculated by sex and age.

In countries with low mortality, further progress in improving survival at all ages still depends on monitoring the causes of death, identifying social and medical interventions, and undertaking prevention programs. One focus of concern is the gap between the survival of males and females.

Sex differentials

In contemporary more developed countries, the average life expectancy of females is greater than that of males, partly because of biological factors, and partly because of behavioural differences. Men generally smoke more tobacco, drink more alcohol, have more motor vehicle accidents, engage in more dangerous occupations and are more prone to suicide. There is also scientific support for the biological superiority of females: in most species of animals females live longer than males. Among humans, males have the greatest susceptibility to life-threatening diseases (Lopez 1983).

Whether excess male mortality is preventable depends on the involvement of lifestyle factors. In more developed countries, over half the excess mortality of males is due to two causes: ischaemic heart disease and cancer of the lung, both of which are closely related to lifestyles (ibid.). Female hormones also provide protection from coronary artery/ischaemic heart disease until menopause, resulting in a ten year delay, compared with males, in the onset of heightened risks of death from this cause.

Simply comparing the number of male deaths with the number of female deaths can be misleading, because the sex ratios affect these figures. Since more male babies are born, other things being equal we would expect more male deaths. To avoid the effect of sex ratio imbalances in such comparisons, demographers use a rate ratio, the *sex ratio of the age-specific death rates*, which measures the phenomenon of *male excess mortality* (Pressat 1972: 78–82). This is obtained by dividing the male ASDR by the female ASDR:

$$\text{Sex ratio of age-specific death rates} = \frac{\text{male death rate at age } x}{\text{female death rate at age } x} \times 100$$

Thus a sex ratio of 150 would denote that the male death rate was 50 per cent higher than the corresponding rate for females. In low mortality populations, excess male mortality is virtually universal, although exceptions can occur at extreme ages. Figure 6.5 shows the extent of excess male mortality in the United States, highlighting the peak figures among young adults. Although mortality is quite low at these ages, car accidents, other accidents and violence, together with greater susceptibility to disease, result in death rates for young men that are two to three times higher than those for young women.

In the more developed countries, maternal mortality was formerly a significant cause of female deaths. The rates declined rapidly after the Second World War, through improvements in antenatal care and advances in the control and prevention of infections in childbirth. In some less developed countries, maternal mortality is still one of the main causes of death for women in the reproductive ages. The conditions responsible are the same as in industrialized countries in the first half of the twentieth century – including haemorrhage, infection, toxaemia, and eclampsia (Rosenfield and Maine 1985). The maternal mortality rate refers to deaths from causes connected with pregnancy, labour or the puerperium – the period of ap-

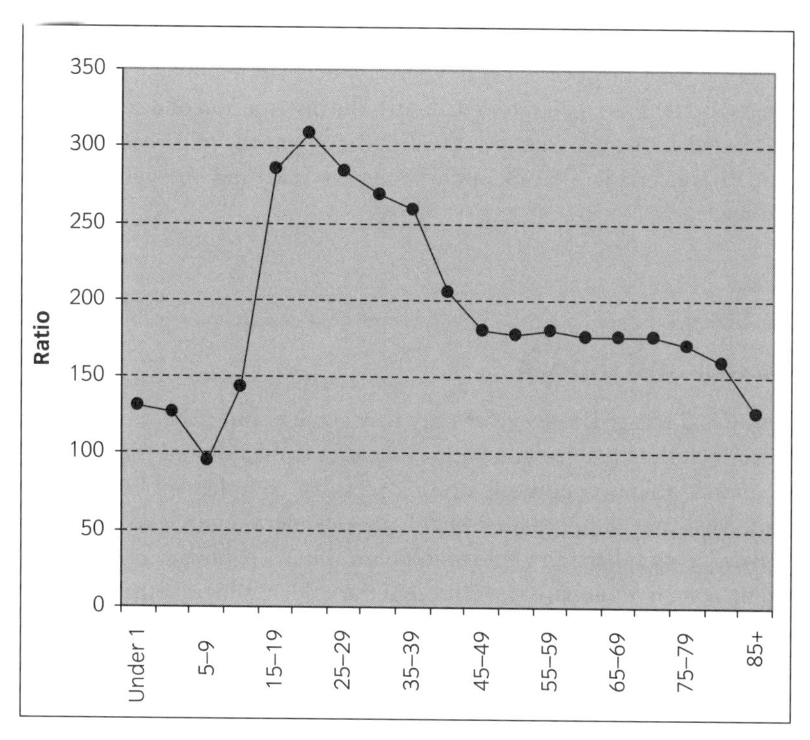

Figure 6.5 Sex ratios of age-specific death rates, United States, 1990.

Source: United Nations 2000. *Demographic Yearbook: Historical Supplement 1948–1997*, New York: United Nations (CD).

proximately six weeks after childbirth, during which the uterus returns to normal (Shryock and Siegel 1973: 418):

$$\text{Maternal mortality rate} = \frac{D}{B} \times k$$

where

D = maternal deaths
B = live births
k = 10 000 or 100 000.

Socio-economic differentials

Variations by age, sex and cause are the first essentials in documenting the nature of mortality in a population. Deeper investigation of causes entails the analysis of death rates by social and economic characteristics of the deceased, such as their ethnic group, marital status, education, occupation, and income. The reasoning here is that the risk of death varies according to people's lifestyles and life chances. For example, death rates specific for age, sex and marital status tend to show that married people have lower death rates than people of other marital statuses at all ages and for both sexes (Shryock and Siegel 1973: 407). Marriage is selective of the healthy and results in better care.

Table 6.6 shows death rates by cause varying substantially according to marital status: the married of both sexes have lower death rates for all causes. The requirements to calculate such rates are, firstly, the distribution of deaths by the characteristic in question and, secondly, the distribution of the mid-year population by the same characteristic. The denominators for the rates are most reliable for census years.

6.4 Health and illness

The measures discussed so far refer only to mortality, but the epidemiologic transition also includes changes in health. Measures of health and the occurrence of disease and disabling conditions are a necessary complement to the mortality measures. Measures of population health are of general interest to demographers, sociologists, geographers and epidemiologists. Interdisciplinary concerns here include comparing national progress through the epidemiologic transition, and identifying social and spatial variations within countries in patterns of disease and mortality.

Demographic research on health in less developed countries has contributed substantially to an interdisciplinary effort to understand questions such as: How can the health of children and mothers be improved? How can basic health services be made acceptable and accessible to the population as a whole? What are the social circumstances facilitating the spread of communicable diseases, such as sexually transmitted infections (STIs) and HIV/AIDS? Measures of health and illness are an important component of this work. Health research in poorer countries generally has to rely on the results of small surveys.

In the more developed countries, the coming of the fourth stage of the epidemiologic transition has seen disabling illnesses and conditions of older people begin to rival causes of death as priority concerns for health research and health policy. In the fourth stage, there is particular interest in restraining cost increases for health care systems, improving the quality of life of older people, and promoting avoidance of health risks among young and old. Environmental improvements, such as reducing air pollution, continue to have a role in health promotion even late in the epidemiologic transition, but the fourth stage witnesses increasing attention to unhealthy behaviours and lifestyles. Targeted risks include smoking, lack of exercise, obesity, high blood pressure and high levels of cholesterol in the blood. Each of these may have adverse long-term consequences for individuals' survival and quality life, as well as for the financing of health services.

Incidence and prevalence

Incidence and prevalence are basic concepts in establishing how frequently new cases of diseases and disabilities occur in a population, and whether long-term

Table 6.6 Age-specific death rates for selected ages, causes of death and marital statuses, Australia 1999

Sex:	Males				Females			
Age group:	45–54 years		55–64 years		45–54 years		55–64 years	
Marital status: Cause of death	Never married	Married	Never married	Married	Never married	Married	Never married	Married
Malignant neoplasms	135	98	609	329	177	101	492	247
Ischaemic heart diseases	134	46	498	147	28	8	158	44
Cerebrovascular diseases	19	7	81	27	14	6	69	22
Chronic lower respiratory diseases	15	2	101	22	10	4	50	16
Transport accidents	15	7	19	9	14	3	6	6
Intentional self-harm	61	16	47	14	19	5	9	3
Drug related deaths	62	9	43	9	22	4	13	4
Total	703	228	1884	681	402	161	1103	422

Rates per 100 000 male or female estimated resident population by marital status.
Source: Australian Bureau of Statistics 2000. *Deaths 1999*, p. 39.

impairment is typical. Incidence measures the number of new cases, prevalence the total number of cases. Thus incidence denotes the risk of developing a disease, while prevalence denotes the risk of having a disease (Lilienfeld and Stolley 1994: 126). At certain times, chickenpox has a high incidence among children, but during an outbreak the prevalence of chickenpox may not rise greatly, because children generally recover quickly. In contrast, incurable but non-communicable and non-life-threatening diseases, such as arthritis, will have a prevalence considerably greater than the incidence. Data on the incidence and prevalence of illnesses and injuries can make an important contribution to planning service provision and establishing whether public health policies are effective in addressing the causes of health problems.

Whereas mortality data supply information only on fatal conditions, incidence and prevalence rates have the added advantage of encompassing conditions that are not life-threatening but potentially costly to individuals and the community. The costs may be physical and emotional in terms of suffering, disabilities and dependency, or financial in terms of days of employment lost, insurance payments, and usage of hospitals and other health services. The bone disease osteoporosis and the eye diseases trachoma, cataract and glaucoma, can incapacitate large numbers of people, without appreciable effects on how long they live. Even minor illnesses of short duration, notably the common cold, have a major economic impact through absences from work.

The incidence rate is calculated by dividing the number of new cases occurring in a specified period by the mid-period population:

$$\text{Incidence rate} = \frac{\text{new cases}}{\text{mid-period population}} \times k$$

where $k = 1000$ or $100\,000$.

The incidence rate measures the likelihood, or risk, of developing a disease during a specified period of time. It can be made specific for age, sex, or any other characteristic (Lilienfeld and Stolley 1994: 109–111). If the period is less than a year, converting the data to an annual estimate may aid comparisons. This is achieved by multiplying the number of new cases by the ratio of 365 to the number of days in the period of observation (Spiegelman 1973: 189). In the absence of precise information on the population at risk, the mid-period population is often used as the denominator (Rockett 1999: 19).

Cases included in the measurement of incidence also comprise part of the total cases used to measure prevalence. The prevalence rate refers to the total number of cases of a health condition occurring in a period of time:

$$\text{Prevalence rate} = \frac{\text{total cases}}{\text{mid-period population}} \times k$$

where $k = 1000$ or $100\,000$.

A prevalence rate reveals the frequency of an illness in a population; it therefore has applications in planning health services. The total cases include people who had the illness during all or part of the period of observation – including those who recovered or died from the illness during that time. The period of observation may be a point in time or an interval. A basic distinction, therefore, is between *point prevalence* and *period prevalence*. The former uses data for a point in time, such as a single day, whereas the latter is based on data for an extended period of observation, such as a month. Point prevalence is useful for measuring the prevalence of *chronic illnesses* of long duration – that is lasting months or years. Extending the period of observation beyond a day would have little effect on the total cases of a chronic illness. Period prevalence is more relevant for measuring the prevalence of *acute illnesses*, arising suddenly and lasting only days or weeks (Christie et al. 1997: 5).

Figure 6.6 illustrates the differences between prevalence and incidence for an acute illness, such as influenza; it gives an example of period prevalence over two weeks. Calculating point prevalence, such as at the mid-point of the period of observation, would have understated the number of cases of the acute illness. A

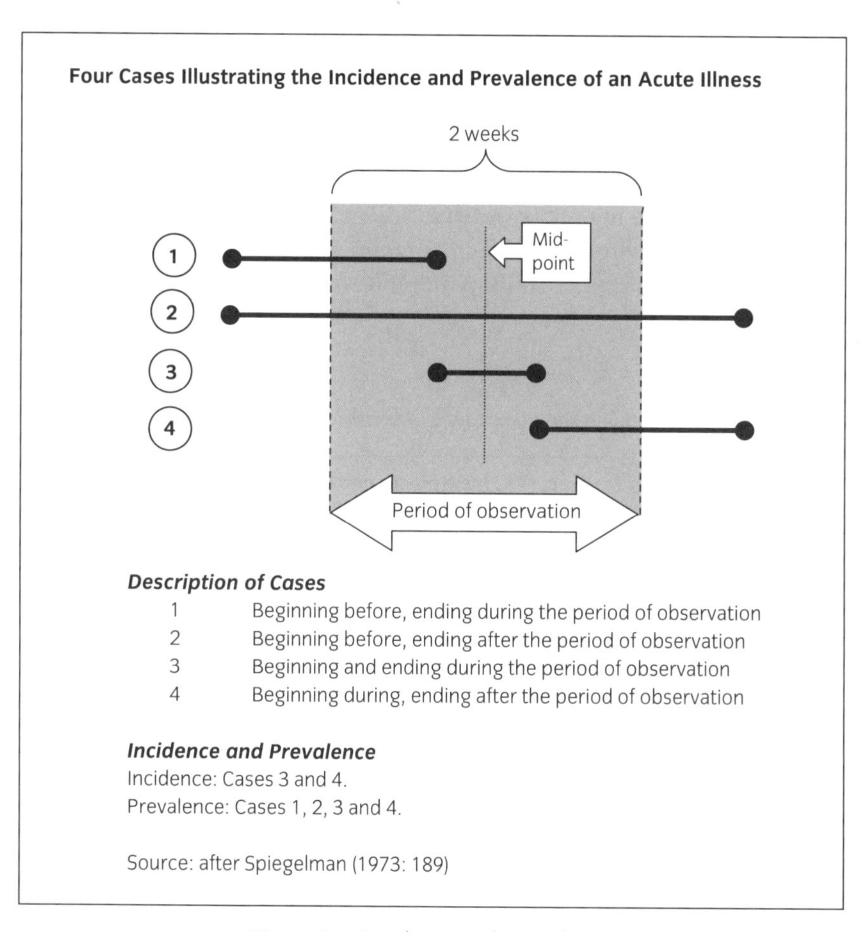

Figure 6.6 Incidence and prevalence.

suitable period of observation for an acute illness may be the average duration of the illness. An individual experiencing two episodes of the same illness, such as a recurrence of influenza, may be counted as two cases if the primary interest is the number of episodes of illness, rather than the number of persons affected.

Risk factors

The fourth stage of the epidemiologic transition has brought heightened concern for the prevalence of risk factors and risk-taking behaviours in populations. The collection and analysis of information on these topics is an area of interdisciplinary activity.

Table 6.7 summarizes risk factors for mortality and morbidity in terms of some broad categories, highlighting the wide range of influences under scrutiny. Demographic research has had a particular concern with the first two categories, extending to the others more recently. The ascribed characteristics arise from biology, ageing and other factors over which individuals have no influence. Advancing age is the most prevalent risk factor for disease and death in the late stages of the epidemiologic transition. The acquired characteristics, which include education and income, affect not only people's awareness of health risks and health maintenance strategies, but also their ability to afford appropriate health care. Risks related to life history are very varied, such as living at some time near an asbestos mine, past exposure to dust and toxic chemicals in employment, or exposure to health hazards while engaged in military service.

The lifestyle and environmental factors represent areas with the greatest potential for improvement through public policy interventions. Like ascribed character-

Table 6.7 Factors influencing the occurrence of morbidity and mortality

Ascribed characteristics	Age, sex, birthplace, parentage.
Acquired characteristics	Socio-economic status, marital status, family status.
Life history	Places of residence, occupational history, military service.
Lifestyle	Diet, exercise, smoking, use of motor vehicles, consumption of alcohol and drugs, exposure to sexually transmitted infections, exposure to other sources of infection.
Environment	Exposure to air, water and soil pollution, exposure to sources of radiation and chemical or radioactive contamination.
Medical history	Presence of other disease; past exposure to disease, trauma and injury.
Medical care	Access to doctors, pharmaceuticals, screening programmes, pathology tests, health services, hospital care, medical technology.

istics, an individual's medical history is unchangeable, although adverse effects may be ameliorated through lifestyle modification or medical care. For many, economic considerations remain a significant constraint on access to medical care even in the fourth stage of the transition.

Analyses of morbidity and mortality now extend to investigations of many and varied characteristics, reaching far beyond demography's traditional concern with mortality differentials in terms of demographic and social characteristics. Demography has joined forces with other disciplines, including epidemiology, in addressing areas of mutual interest.

Relative risk

To examine the effects of presumed causes of disease, ratios of incidence rates or death rates (i.e. rate ratios) may be used to measure *relative risk* (Lilienfeld and Stolley 1994: 200; Christie et al. 1997: 37):

$$\text{Relative risk (incidence)} = \frac{\text{incidence rate from disease X of persons exposed to the risk factor}}{\text{incidence rate from disease X of persons not exposed to the risk factor}}$$

$$\text{Relative risk (mortality)} = \frac{\text{mortality rate from disease X of persons exposed to the risk factor}}{\text{mortality rate from disease X of persons not exposed to the risk factor}}$$

An American study, for instance, obtained a relative risk of 10 comparing the incidence of lung cancer among smokers (numerator) and non-smokers (denominator); in other words smokers were ten times more likely to develop the disease (Lilienfeld and Stolley 1994: 200). Types of risk factors analysed through such means may include being a war veteran, taking a contraceptive pill, drug addiction, or any of the many other items summarized in Table 6.7. The same kinds of ratios can also be employed to study the health benefits of certain behaviours or treatments, such as a Mediterranean diet or prescribed medications. Relative risks of less than one suggest a protective effect.

6.5 Conclusion

There are other important and accessible measures of health that do not require medical assessment or pathology tests – such as self-rated health, the body mass index, measures of proficiency in activities of daily living (ADL measures) and measures of disability. Such indices are included in many large scale surveys and have become part of the data resources that demographers and other social scientists access for studies of the health of populations. Mortality and health, like fertility and the family, comprise a subject area where interdisciplinary research effort is important, taking advantage of different analytical skills and disciplinary perspectives. Studies of mortality and morbidity inform planning and policy making, thereby contributing to progress in averting premature death, preventing disease and injury and designing appropriate health services.

As long life becomes a prospect for the majority, the maintenance of health at older ages emerges as a major concern. Death rates from heart diseases have declined spectacularly in more developed countries and a greater percentage of people are living longer with arthritis, vision and hearing loss, mental impairments and other non-fatal, but disabling, diseases of later life. In this context, research on issues relating to health and morbidity assume even greater importance.

Study resources

KEY TERMS

- Acute illness
- Age-specific death rate
- Cause-specific death rate
- Chronic illness
- Crude death rate
- Endogenous causes of death
- Epidemiologic transition
- Exogenous causes of death
- Fetal deaths
- Immediate cause of death
- Incidence of a disease
- Incidence rate
- Index of excess male mortality
- Stillbirth rate
- Male excess mortality
- Maternal mortality rate
- Morbidity
- Mortality
- Mortality differentials
- Neonatal mortality rate
- Perinatal mortality rate
- Post-neonatal mortality rate
- Prevalence of a disease
- Prevalence rate
- Rectangularization of the survival curve
- Relative risk
- Sex ratio of age-specific death rates
- Underlying cause of death

FURTHER READING

Harper, Andrew C., Holman, C.D.J., and Dawes, Vivienne P. 1994. *The Health of Populations*. Melbourne: Churchill Livingstone.

Hinde, Andrew. 1998. *Demographic Methods*. London: Arnold, Chapter 2, pp. 8–18.

Murray, Christopher J.L. and Chen, Lincoln C. 1992. 'Understanding Morbidity Change'. *Population and Development Review*, 18: 481–503.

Palmore, James A. and Gardner, Robert W. 1983. *Measuring Mortality, Fertility and Natural Increase: a Self-Teaching Guide to Elementary Measures*. Honolulu: East-West Population Institute., East-West Center, Chapter 2, pp. 9–58.

Pol, Louis G. and Thomas, Richard K. 2000. *The Demography of Health and Health Care*. New York: Plenum Press.

Rockett, Ian R.H. 1999. 'Population and Health: An Introduction to Epidemiology' (second edition). *Population Bulletin*, 54 (4). Washington: Population Reference Bureau.

Weeks, John R. 2002. *Population: an Introduction to Concepts and Issues*. Belmont, CA: Wadsworth, Chapter 4, 'Mortality', pp. 117–165.

INTERNET RESOURCES

Subject	Source and Internet address
WHO website	**World Health Organization (WHO)** http://www.who.ch
International data base on the health and living conditions of children	**UNICEF** http://www.unicef.org/statis/
Statistical Abstract of the United States	**US Census Bureau** http://www.census.gov/prod/2001pubs/statab/sec01.pdf
National Institutes of Health	**US Department of Health and Human Services, National Institutes of Health (NIH)** http://www.nih.gov/
Gateway to extensive information on mortality, health and vital statistics, especially for the United States	**US Department of Health and Human Services, National Centre for Health Statistics** http://www.cdc.gov/nchs/
The gateway to statistics from over 100 US Federal agencies, including statistics on health.	**US Government** http://www.fedstats.gov/
United States vital statistics	**US Department of Health and Human Services, National Centre for Health Statistics** http://www.cdc.gov/nchs/nvss.htm
Australian health statistics and research	**Australian Institute of Health and Welfare (AIHW)** http://www.aihw.gov.au/ *Links to related organizations:* http://www.aihw.gov.au/links.html

Note: the location of information on the Internet is subject to change; to find material that has moved, search from the home page of the organization concerned.

EXERCISES

Population numbers and deaths by age and sex, United States 1990 (thousands)

Age	Population		Deaths	
	Males	Females	Males	Females
Under 1	1645	1573	22	16
1–4	7748	7390	4	3
5–9	9263	8837	2	2
10–14	8767	8347	3	2
15–19	9103	8651	12	4
20–24	9676	9345	16	5
25–29	10696	10617	20	7
30–34	10877	10986	24	9
35–39	9902	10061	28	11
40–44	8692	8924	30	15
45–49	6811	7062	33	19

Population numbers and deaths by age and sex, United States 1990 (thousands) *continued*

Age	Population		Deaths	
	Males	Females	Males	Females
50–54	5515	5836	42	25
55–59	5034	5497	61	37
60–64	4947	5669	94	61
65–69	4532	5579	128	89
70–74	3409	4586	148	113
75–79	2400	3722	158	143
80–84	1366	2568	138	163
85+	858	2222	152	311
Total	121241	127472	1115	1035

Source: United Nations 2000. *Demographic Yearbook: Historical Supplement 1948–1997*, New York: United Nations (CD).

1 From the above table, calculate:

(a) the crude death rate for males, females and the total population;

(b) the age-specific death rates for males, females and the total population;

(c) the index of excess male mortality for each age.

2 Write a brief commentary on the results obtained in question 1; include consideration of the potential difference between the infant mortality rate and the age-specific death rate for those less than 1 year old.

3 For your country of residence, compile a list of sources of mortality data, including a two or three line summary of the principal contents of each source. Discuss the advantages and disadvantages of each source. The sources may include printed and electronic data on total deaths, causes of death, and characteristics of the deceased. The discussion should include reference to issues relating to underlying causes and multiple causes of death.

4 Using data for your country of residence, compile a set of death rates – by age, sex and major causes – for people in the five year age groups 20–24 and 25–29. Write a commentary on the differences you observe, using demographic indices and graphs as appropriate. Answers should include the sex ratio of the age-specific death rates, discussion of the main reasons for excess male mortality, and comments on whether the excess is apparent for all causes.

5 If a depressed person jumps off a bridge and drowns, what will be the single cause of death recorded in vital statistics?

6 Using demographic indices as appropriate, prepare an analysis of Table 6.6 on age-specific death rates for selected ages, causes of death and marital statuses. Answers could employ rate ratios and graphs to compare the figures. Explanations of the differences should refer to the literature on marital status differentials in mortality.

7 Propose definitions of (a) death and (b) health and explain your reasoning. Provide a list of key references.

SPREADSHEET EXERCISE 6: A GRAPHICAL DATA BASE

This exercise introduces a simple means of displaying graphically the contents of a data base, including some of the techniques used in the Excel modules on the CD. Although the exercise utilizes only a small data base, the approach is equally applicable to much larger undertakings. The exercise introduces the following techniques:

- linked worksheets
- the index function
- drop-down menu boxes

The exercise requires three worksheets. While it is necessary for tables and graphs to be on separate worksheets only when a data base is very large, this exercise illustrates the potential advantages of utilizing linked worksheets.

	A	B	C	D	E	F	G
1	**Table 1: Selected Demographic Statistics for the United States, 1950-1996**						
2							
3	Year	Population	Births	Deaths	Infant Deaths		
4	1950	152,271,000	3,571,928	1,456,626	104,355		
5	1960	180,671,000	4,257,850	1,711,982	110,873		
6	1970	205,052,000	3,731,386	1,921,031	74,667		
7	1980	227,726,000	3,612,258	1,985,540	45,526		
8	1990	249,907,000	4,158,212	2,148,463	38,100		
9	1996	265,283,783	3,914,953	2,322,256	28,237		
10							
11	Source: United Nations 2000. *Demographic Yearbook, Historical Supplement 1948-1997* (CD).						

Sheet 1: Source data

1. On Sheet1, type Table 1, omitting the commas within numbers. Insert the commas later using the Comma Style button on the formatting toolbar.

Sheet 2: Demographic rates

2. Create Table 2 as follows:
   - highlight the entire table on Sheet1 and click the copy button;
   - click the tab for Sheet2, choose cell A1 and click the paste button;
   - edit the table title and column headings; fit the column headings into a single cell using text wrapping (see Spreadsheet Exercise 5, Section 1);
   - replace each number in the table with a formula to calculate rates from the statistics in Sheet1. For example, to calculate the crude birth rate in 1950: click cell B4 in Table 2, type =, click the tab for Sheet1 and click cell C4, type / (division sign), click cell B4, type *1000, press Enter. This formula can now be copied down to calculate all the CBRs in Table 2.

	A	B	C	D	E	F	G	H
1	Table 2: Vital Rates and Infant Mortality Rates for the United States, 1950-1996							
2								
3	Year	Crude birth rate	Crude death rate	Rate of natural increase	Infant mortality rate			
4	1950	23.5	9.6	13.9	29.2			
5	1960	23.6	9.5	14.1	26.0			
6	1970	18.2	9.4	8.8	20.0			
7	1980	15.9	8.7	7.1	12.6			
8	1990	16.6	8.6	8.0	9.2			
9	1996	14.8	8.8	6.0	7.2			
10								
11	Source: see Table 1.							
12								
13	Selected Row							
14	6							
15								
16	Table 3: Data Selected for Display							
17	Year	Crude birth rate	Crude death rate	Rate of natural increase	Infant mortality rate			
18	1996	14.8	8.8	6.0	7.2			
19								
20								
21	=INDEX(A4:A9,$A$14)				=INDEX(E4:E9,$A$14)			
22								

3. On Sheet2, below Table 2, type ‘Selected Row’ (in cell A13). In cell A14, type any number between 1 and 6. The data base has six rows of statistics to be displayed, for the years 1950–1996. Cell A14 indicates the current row chosen for display.

Table 3: Data selected for display

4. Copy the column headings in Table 2 to Table 3
5. Table 3 also contains a row of formulas in which the index function selects data from Table 2. The data selected for Table 3 depend on the value of the number specified under the heading ‘Selected Row’ (1 means select row one, 2 means select row 2, etc).
6. To type the first formula in cell A18 of Table 3:
   - =INDEX(
   - highlight cells A4 to A9 in Table 2 (i.e. the range of cells from which to choose);
   - type a comma;
   - click on the number under the heading ‘Selected Row’ and press F4 to insert two dollar signs in the cell reference (making this reference absolute or unchangeable);
   - close the bracket and press Enter.
7. Copy this formula to the other columns in Table 3. If you change the value of the ‘Selected Row’, the values in Table 3 will change. The syntax for the INDEX function is: =INDEX(list from which to choose; location of the selected item in the list).

Sheet 3: Graph of the selected data set

8. The next step is to create a chart displaying the selected data:
   - highlight the column headings and data in Table 3, omitting column A;
   - click the Chart Wizard Button and follow the instructions on the screen to create the type of graph you want;
   - paste the graph into Sheet 3;
   - align the graph borders with the grid on the worksheet by holding down the Alt key when resizing the graph;
   - make the vertical scale of the graph constant to aid comparisons – set the maximum value to 30 to accommodate the highest value in Table 2 (and Table 3). To do this, click on the vertical axis, double click the mouse, choose the Scale tab in the Format Axis menu, and set the maximum value to 30.

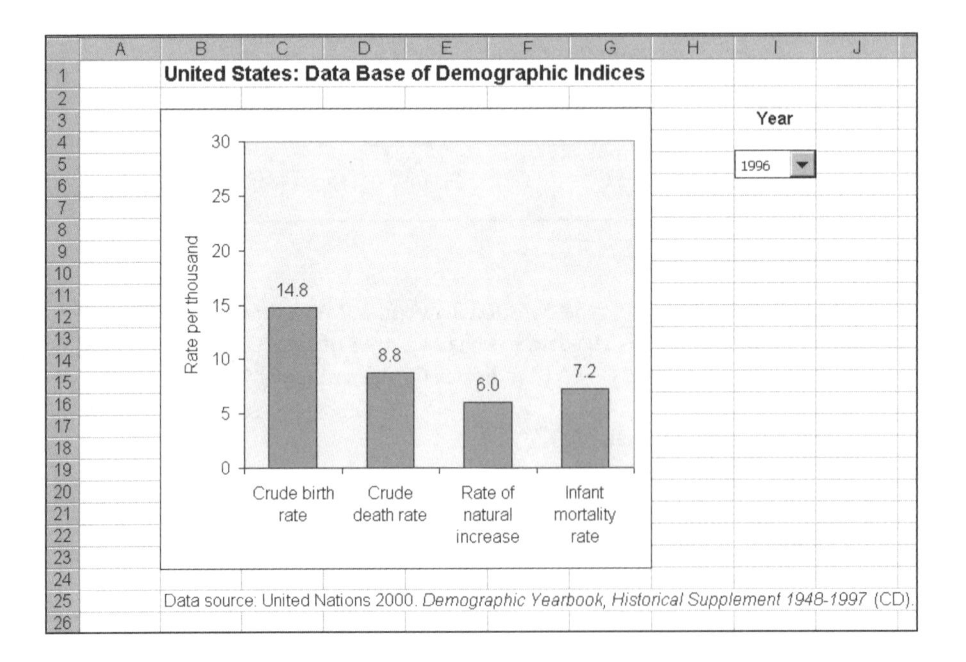

Menu box

9. The final step involves creating a means of changing the information displayed in the graph:
   - click View Toolbars and choose the Forms toolbar;
   - on the Forms toolbar click the 'combo box' button (it looks like a drop-down menu);
   - draw a rectangle on Sheet3 the size of the required menu box. Avoid drawing the menu box on the graph itself.

Editing the menu box

10. Click the menu box with the right mouse button so that edit boxes appear around it (these can be used to resize the menu box); from the list that appears upon right clicking the menu box, choose Format Control.

    Alternatively, on the Forms Toolbar click the 'Control Properties' button (the button has a hand pointing to a document); the Format Control dialog box will appear. (Note that the menu box you drew must be highlighted (i.e. edit boxes visible) otherwise the 'control properties' button will not be accessible).
11. Now click the mouse in the 'Input range' box and, on Sheet2, Table 2, highlight the row headings (i.e. the years 1950–1996). If the Format Object box is in the way, move it by clicking the mouse on the title bar and moving the box while holding down the left mouse button. The input range box specifies the information to appear in the drop-down menu box; in this case it will be a list of years.
12. Click in the Cell Link box and, on Sheet2, click the cell under the heading 'Selected Row'(i.e. A14), set the Drop Down Lines to 6 (the number of years), then click OK in the Format Control dialog box. The value of the cell link will change according to the year selected from the menu.
13. Click in a blank cell of Sheet3 to deselect the drop-down menu. If, later, you need to alter the settings for the menu box, simply repeat steps 10–13.
14. The menu box is now ready to use. Clicking the menu box arrow and choosing a year will change the data displayed in the graph, as well as in Table 3.

For discussion

Consider how the workbook just constructed could be extended to enhance its usefulness for analysing populations and presenting information visually.

7 Fertility and the family

As to the causes of Barrenness in *London*, I say, that although there should be none extraordinary in the Native *Air* of the place, yet the intemperance in feeding, and especially the Adulteries and Fornications, supposed more frequent in *London* then elsewhere, do certainly hinder breeding. For a Woman, admitting 10 Men, is so far from having ten times as many Children, that she hath none at all.

(Graunt 1662: 56)

OUTLINE

7.1 Family change

7.2 The second demographic transition

7.3 Period and cohort approaches

7.4 Period measures of fertility

7.5 Synthetic cohort measures of fertility and replacement

7.6 Real cohort measures of fertility

7.7 Measures of marriage and divorce

7.8 Conclusion

Study resources

LEARNING OBJECTIVES

To become aware of:

- the broad context of family change within which fertility indices are interpreted
- applications of period and cohort approaches

To understand:

- the principal techniques of fertility analysis, including period, synthetic cohort and real cohort measures
- the most commonly needed measures of trends in marriage and divorce

COMPUTER APPLICATIONS

Excel module: Proximate Determinants.xls (Box 7.1)

Spreadsheet Exercise 7: Measures of fertility (see Study resources)

Fertility and the family are of great importance in contemporary demographic research, so much so that over 40 per cent of the articles published in the journal *Demography* have dealt with these subjects (Teachman et al. 1993: 526). The extent of past changes, together with contemporary diversity in family building behaviour, have called for a continual renewal of understanding of developments that profoundly affect the lives of individuals and the nature of societies.

Fertility and family formation also comprise one of the areas of greatest discontinuity between national policies and individual goals. In less developed countries, the preferences of many for large families can run counter to national goals to limit population growth. This contrasts with the situation in the more developed countries where low birth rates have raised concerns about population decline, contraction in the size of the labour force, excessive levels of population ageing and erosion of overall population size and national influence in world affairs.

Demographic data and methods provide a society-wide perspective on the outcome of innumerable family changes, clarifying trends, the extent of differences in behaviour between social groups, and consequences for society. All demographic processes affect the family – migration and mortality, as well as fertility. The family is a nexus, where life-shaping decisions of individuals and couples have their effect, and where much of the meaning and importance of demographic processes become manifest.

Contemporary demography is greatly interested in family decision processes, to assist in explaining why different people marry, cohabit or stay single, and why many couples have no children or only one, compared with the norm of two. Despite the extent of changes, the well-being of the family is still perceived as essential to the well-being of society. The family remains 'the basic social institution', 'the natural and fundamental group unit of society', 'the cornerstone of society'. For example, rises and falls in the birth rate have always been topics for intense scrutiny and debate in the social sciences as they are symptoms of changes in society, often of far-reaching proportions. A particular concern is whether they denote new lifestyles – affecting the balance between family, work and leisure – or shifting life chances due to events – including economic restructuring, booms and recessions – or the sometimes subtle workings of social inequality and discrimination. Understanding of the family is, therefore, a basis for perceiving the nature of forces of change in societies.

This chapter first introduces the context within which demographic data on fertility and the family need to be interpreted. Essentially, this context consists of trends in marriage, living arrangements (household composition), fertility and life cycle experience. The initial sections, on 'Family Change' and 'The Second Demographic Transition', aim to encapsulate major features of this setting. The list of 'Further reading', at the end of the chapter specifies other sources of background on trends and explanations. Later sections introduce the principal methods used in the analysis of these phenomena.

7.1 Family change

Popular interpretations of the contemporary family tend to compare a present state of disarray with a supposed nineteenth century 'golden age', when most people married early, had many children, stayed together all their lives, and supported grandparents under the one roof. Nevertheless, the supportive extended family household describes 'the family of Western nostalgia' (Goode 1963) rather than past reality. Historical research on household composition and marriage has shown the stereotype to be false.

Before the twentieth century, the survival of a grandparent generation within Western families was unusual:

> In nineteenth-century France, marriages averaged 20 years. Parents died approximately when children reached maturity, and the successive generations hardly had the opportunity to become acquainted. Only recently have the generations overlapped. . . .
> (Tabah 1980: 359)

Nevertheless, regardless of mortality, it was typical in the past for different generations of Western families to live separately. The majority of pre-industrial family households in England and north-western Europe were not extended: the dominant household form was the nuclear family, consisting of parents and their children (McDonald 1995: 9).

Even the notion that early marriage and childbearing were universal in Western countries during the nineteenth century has been disproved. Hajnal's (1965) famous study of 'the European marriage pattern' found that women, as well as men, had a high average age at marriage (25 to 30 years for women), and a high proportion (10 to 20 per cent) of men and women never married. The pattern existed from at least the middle of the 18th century, that is well before Malthus's 1798 essay portraying humankind as reproducing to the full extent that subsistence allowed. While this marriage pattern was unique to European populations, it was repeated among immigrant European populations in other countries. The ethic of the economically 'proper time to marry', that is when income and savings were sufficient to support a family, was a key determinant (McDonald 1995: 2).

Whereas the extended family has been a mythical nineteenth century ideal, the nuclear family has been widely considered to be a twentieth century norm, typifying the experience of family life in Western societies. However, early death or widowhood, and never marrying or childlessness (Figure 7.1), meant that the intact nuclear family was far from characterizing the life cycle experience of all in the early decades of the twentieth century. In a number of countries the baby boom after the Second World War, due especially to early and more universal marriage, created a short-lived heyday for the nuclear family. In later decades, cohabitation, separation, divorce and remarriage added to the complexities of life cycle experience, countering the chances of having a 'typical', lifelong nuclear family existence.

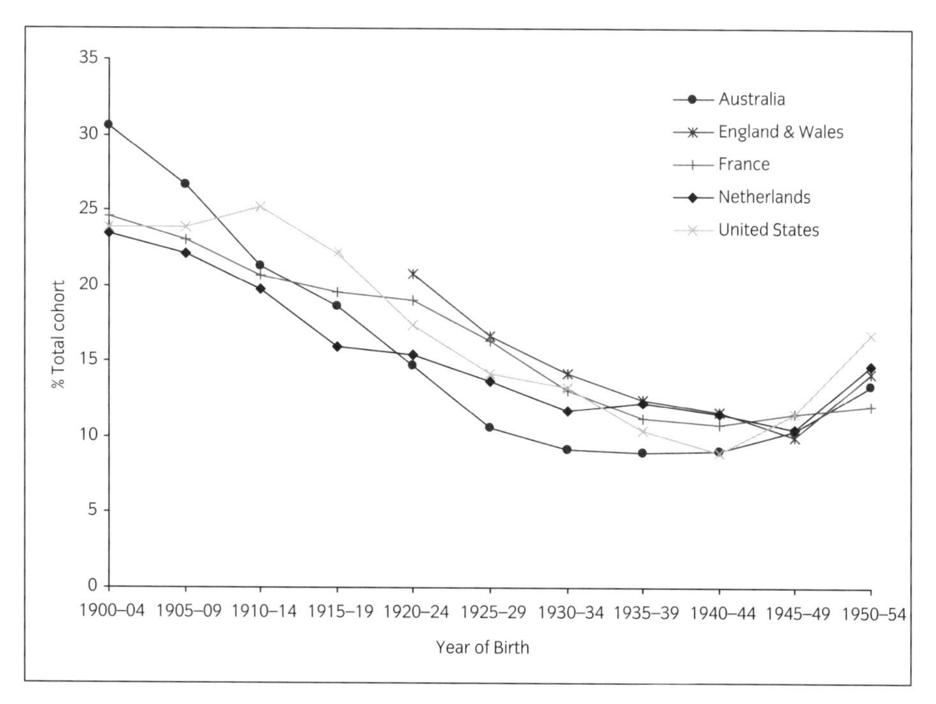

Figure 7.1 Total percentages childless at ages 45–49 in female cohorts born c. 1990–1954, selected countries.

Source: Rowland (2002).

Clearly, in conducting research on the family in Western societies, it is difficult to choose appropriate points of comparison in the past, because changes have been ongoing. Even the twentieth century's high point of supposedly 'traditional' family building behaviour – the baby boom years after the Second World War – was a discontinuity, unrepresentative of previous experience. The increase at that time in the proportions marrying and having children was exceptional, and comparisons between the baby boom years and subsequent changes can give an exaggerated impression of family 'decline'. The second demographic transition provides a basis for considering developments since the baby boom years.

7.2 The second demographic transition

Whereas demographic transition theory (see Chapter 1) addresses the nature of the long term trend from large to small families, other approaches are needed as starting points for interpreting contemporary low fertility and diversity in life cycle experience. One such is van de Kaa's (1987) theory of the *second demographic transition*. Its aim is to describe and explain family building behaviour in post-transition Europe and, by extension, circumstances in a number of other low fertility societies. In the late twentieth century, the United States was one of the few more developed countries with a birth rate sufficient to maintain a balance of births and deaths in the long term.

Whereas the demographic transition anticipated such a balance in the post-transition stage, the birth rate in the majority of developed countries has tended to decline even further, creating a long-term prospect of deaths exceeding births and the population not replacing itself. This situation of *below replacement fertility* is not seen as part of the demographic transition, but as a key distinguishing feature of a new set of circumstances. Hence the name 'the second demographic transition'. In European countries, van de Kaa (1987: 8–11) suggested, there was a typical sequence of events leading to below replacement fertility namely:

- A shift from legal marriage to cohabitation: van de Kaa estimated that ultimately 40 per cent or more of men and women in Western Europe would never be legally married.
- A shift in the focus of the family from children to the adult couple. This brings greater emphasis on childbearing as a means of enriching the lives of the parents, or on childlessness as an alternative basis for achieving personal fulfilment. It marks the end of the reign, during the (first) demographic transition, of the 'child-king', and the succession of the 'king-pair' (van de Kaa 1987; Ariès, 1980: 649).
- A shift from preventive contraception, to contraception to permit self-fulfilling choices. Contraception thus changes from a means of preventing births that could reduce a family's well-being and standard of living, to a means of achieving greater self-fulfilment, even through having no children (van de Kaa 1987: 26).
- A shift from uniform to diversified families and households. The wider spectrum of socially sanctioned choices, together with a higher incidence of divorce, leads to the replacement of the nuclear family household by a complex range of alternatives.

Whether this sequence will be relevant to all European countries is uncertain, but even more important than the above changes are contrasts between the attitudes that are thought to underlie the first and second demographic transitions. During the first transition, fertility control was a response to large family size becoming a social handicap, disadvantaging parents in their goals of giving children better opportunities for education and employment. Norms and attitudes were dominated by concerns for the welfare and future prospects of offspring. At the same time, secularization reduced the influence of traditional religious teachings and made more couples willing to use methods of family planning. Within marriage, the number of children was controlled – quality replaced quantity. Thus society was child-oriented – *altruism* was the underlying motivation in family life (van de Kaa 1987: 5). Similarly, Ariès considered that the historical decline of the birth rate, in the first demographic transition, was unleashed by an enormous sentimental and financial investment in the child. Wise management required reducing family size so that more time and care could be devoted to each child with better results: 'seeing that one's children got ahead in a climate of social mobility was the deep motivation behind birth control.'(Ariès 1980: 647).

In contrast, van de Kaa described the motivation underlying the second demographic transition as *individualism*: norms and attitudes emphasizing the rights and self-fulfilment of individuals. Others have also emphasized the importance of individualism in explaining contemporary low birth rates in more developed countries. Couples and individuals are no longer seen as planning life in terms of the child and his or her future. The child has not disappeared from such plans, but fits into them as one of the options that make it possible for adults to achieve self-fulfilment (Ariès 1980: 650). Thus the child is no longer the essential variable in plans for the future. Whereas people planned their future in terms of *familism* (a family-oriented lifestyle) during the first demographic transition, in the second they plan their future in terms of any combination of familism, consumerism, careerism and other lifestyles. Replacement fertility becomes unattainable when many remain single, or married and childless, or have small families in which the total numbers of children are insufficient to counterbalance the childlessness of others.

Overall, the second demographic transition argues for a turning point in demographic history entailing a shift from altruism to a greater influence of individualism, and a shift from replacement to below replacement fertility. Although there is no consensus that current developments are best described and explained in terms of a 'second demographic transition', it is a significant attempt to find order in a new and diverse situation. As such, it is a valuable starting point for considering the nature of the field within which demographic data and indices on fertility, marriage and the family are employed. D. A. Coleman's (1998) paper on 'Reproduction and survival in an unknown world', for example, discussed aspects of the second demographic transition in the context of a major overview of trends and explanations of contemporary fertility in industrial societies. The detailed explanation of changes in fertility and the family remains a continuing preoccupation of demographic research and theorizing, not least because the maintenance of below replacement fertility is as detrimental as the maintenance of above replacement fertility: 'Excess is not sustainable, deficit brings extinction.' (Coleman 1998: 33).

One initial approach to explanation entails the examination of differential behaviour, comparing the birth rates, or marriage rates, of people with different characteristics. Like mortality differentials, *fertility differentials* are investigated through comparisons of characteristic-specific rates, such as the birth rates of one socio-economic group versus those of other groups. Another very informative initial approach to explanation is through reference to the concept of *proximate determinants*, focusing on the immediate causes of change (Box 7.1).

BOX 7.1 **Module on the proximate determinants of fertility (*Proximate Determinants.xls*)**

Causes of fertility change consist of underlying causes, such as social and economic forces, together with the immediate, or proximate, causes through which the underlying causes operate. For example, changes in age at marriage and the proportions marrying have a direct effect on fertility, whereas factors such as changes in the role and status of women in society contribute to shifts in marriage patterns.

There is much debate about the nature and relative importance of underlying causes, partly because they are difficult to identify and measure, and partly because their influence varies through time and from place to place. Proximate determinants are better known, especially since the ground-breaking work of Bongaarts and Potter (1983), whose research suggested that just four proximate determinants accounted for most fertility variations between countries.

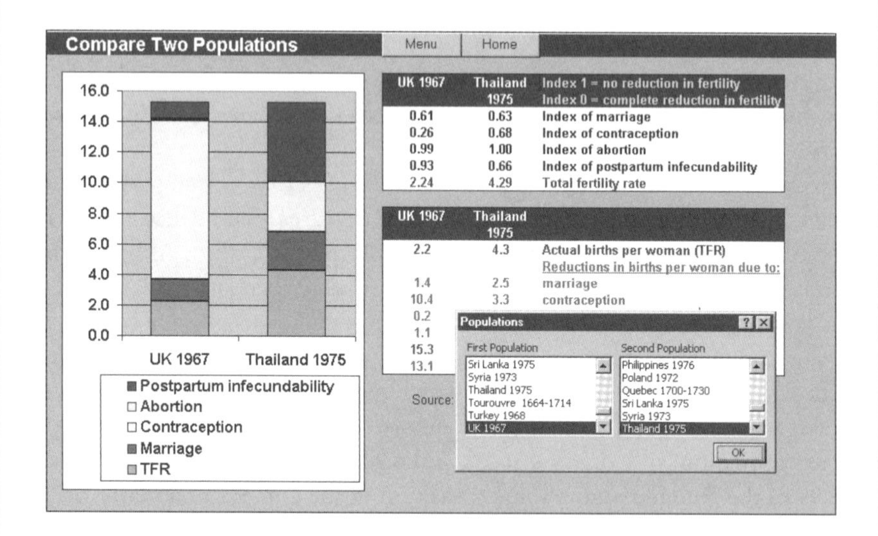

This module, the first version of which was prepared by Lualhati Bost, provides a visual demonstration of the findings of Bongaarts and Potter. The authors attributed variations in the total fertility rate mainly to the influence of four proximate determinants influencing the duration of the reproductive period and the rate of childbearing within it (Bongaarts and Potter 1983: 4) namely: marriage (or first cohabitation), contraception, induced abortion and post-partum infecundability.

1. *The index of marriage* (C_m) refers to the proportion married, taking account of the ages of married females and the relative importance of each age group in childbearing. The index equals 1 if all women of reproductive age are married, and 0 if none are married (Bongaarts and Potter 1983: 81–82).

continued

BOX 7.1 continued

2. *The index of contraception (C_c)* takes account of the prevalence and effectiveness of contraception. The index of contraception equals 1 if contraception is absent or completely inefficient, and 0 if all fecund women use 100 per cent effective contraception.
3. *The index of induced abortion (C_a)* ranges from 1, in the absence of induced abortion, to 0 if all pregnancies are aborted.
4. *The index of postpartum infecundability (C_i)* equals 1 in the absence of breastfeeding and postpartum abstinence, and 0 if the duration of infecundability is infinite. Postpartum infecundability modifies the interval between births; the longer women breastfeed their babies or abstain from intercourse, the greater the spacing of births. In the absence of contraception, induced abortion, breastfeeding and postpartum abstinence, the average birth interval is about 20 months (Bongaarts and Potter 1983: 87). Intensive breastfeeding is about 98 per cent effective in preventing conception in the first 6 months after the birth (Lucas and Meyer 1994).

The four proximate determinants each have potential to reduce the total fertility rate. If they had no limiting effect – as would occur if all women of reproductive age were married, and contraception, abortion and breastfeeding were absent – fertility would rise to an average limit of 15.3 births per woman. Bongaarts and Potter (1983: 87) called this maximum figure *total fecundity* (TF). By multiplying TF by each of the proximate determinants, the total fertility rate (TFR) is obtained:

$$\text{TFR} = C_m \times C_c \times C_a \times C_i \times \text{TF}$$

Thus the equation reveals the components that reduce the maximum potential figure (TF) to the observed fertility level (TFR). Any total fertility rate can be calculated from the varied combinations of the proximate determinants. Hence, 'populations with the same TFR do not necessarily, and in fact only rarely, have the same set of proximate determinants' (Bongaarts and Potter 1983: 98).

The four indexes, with values between 0 and 1, measure the effects of the proximate determinants. When a proximate determinant does not reduce fertility, its corresponding index equals 1. Conversely, the lower the index the greater the reduction in fertility. If any index equalled 0, there would be no births (Bongaarts and Potter 1983: 80). Therefore:

- If no proximate determinant reduces fertility: $\text{TFR} = 1 \times 1 \times 1 \times 1 \times \text{TF} = 15.3$.
- If the influence of marriage is reduced to half: $\text{TFR} = 0.5 \times 1 \times 1 \times 1 \times \text{TF} = 7.7$.
- If the influence of marriage and contraception are both 0.5: $\text{TFR} = 0.5 \times 0.5 \times 1 \times 1 \times \text{TF} = 3.8$.

Equations for estimating the values of the four indices were developed from empirical data, to enable them to be calculated for many different populations. In the authors own data set of 41 populations, the TFRs estimated from the model explained 96 per cent of the variation in the observed total fertility rates (Bongaarts and Potter 1983: 91–92).

continued

BOX 7.1 continued

Instructions

To use the module, insert the CD in the drive, start Excel and open the file *Proximate Determinants.xls*. Enable macros when prompted.

- Click **Data Base** to compare proximate determinants for two selected populations. The data base is a table from Bongaarts and Potter (1983).
- Click **Simulate** to show the effects of varying the proximate determinants.
- Click **Transition** to compare proximate determinants over the demographic transition.
- Click **Menu** to select populations on the later displays.
- Click **Home** to return to the opening display.

See Appendix B for further advice on using the Excel modules.

Illustrative applications

- To compare populations, click **Data Base**
- To change the populations displayed on the graph, click **Menu**.
- The vertical axis of the graph shows the numbers of children, with the various components adding to 15.3, the 'total fecundity' (TF).
- The components of the bar graph are the average number of children per woman (TFR) and the numbers of potential births lost as a result of each of the proximate determinants.

Historical populations

Set the first population to Hutterites, the population with the highest TFR in the data set and, from the graph, identify the reasons why their fertility is less than the maximum.

Set the second population, in turn, to Bavarian villages, Ille de France, Quebec and Werdum. From the graphs and tables, explain why these populations have such low fertility in the absence of contraception. Also, explain the differences in the TFRs of these populations.

Developing countries

Compare the proximate determinants for Bangladesh, Indonesia, Jamaica and Kenya – with particular reference to the role of marriage and postpartum infecundability. Kenya has had one of the highest national fertility levels: to what extent do the proximate determinants help to explain this? What accounts for variations in the total fertility rate of countries making little use of contraception?

Developed countries

Compare the proximate determinants for the United Kingdom, the United States, Denmark, Finland, Hungary, Poland and Yugoslavia. Which countries have similar proximate determinants? Are there varied ways in which countries have obtained low birth rates? How important is the role of contraception compared with that of abortion? Is very low fertility surprising when all four proximate determinants contribute to reductions?

continued

BOX 7.1 continued

Experimental applications

To experiment with varying the proximate determinants, from the opening menu click **Simulate**. This screen displays a selected population and model data to compare with it. Values for the model data are adjusted by moving the controls at the bottom of the screen.

For practice, use Hutterites as the selected population, by clicking **Menu** and choosing from the dialog box, then adjust the controls for the model population to match the Hutterite data – begin by setting the values for contraception and abortion to 1.0.

Now experiment with the following 'What if?' scenarios:

- The prevalence of marriage decreases (e.g. index of marriage = 0.5)
- Contraception is introduced (e.g. index of contraception = 0.25)
- Abortion is introduced (e.g. index of induced abortion = 0.75)
- Breastfeeding among the fertile is extended to 18 months (index of postpartum infecundability = 0.55, see table at the bottom of the screen).

Experimentation will reveal that the total fertility rate is highly susceptible to variations in the four main proximate determinants. Do the model data give any insight into how low-fertility societies might increase their birth rates?

Demographic transition

On the opening menu, click **Transition** to compare selected populations with the average pattern of change in the proximate determinants over the demographic transition. Change the selected population by clicking **Menu**. Bongaarts and Potter (1983: 104) produced the 'synthetic transition' in the proximate determinants, by averaging data for between 4 and 11 countries at each stage. Examination of the graph provides a summary of how the proximate determinants may change through time.

7.3 Period and cohort approaches

There are two main approaches to comparing and explaining changes in fertility. The first is through 'period', or 'cross-sectional', analysis to compare populations in particular years. An example of a period measure is the crude birth rate (see Chapter 1). The crude birth rate is useful for measuring trends through time from minimal data, as in tracing the course of the demographic transition, or for providing ready comparisons of fertility in different countries.

The second approach is through cohort analysis, as in the example of the Massachusetts females discussed previously (see Chapter 4, Figure 4.3). Cohort analysis of fertility seeks to provide a long-run view of family building throughout women's reproductive years. Ideally, cohort analysis of fertility is based on the reproductive history of a group of women who were born or married in a particular period of time, that is a *real cohort* of women. For instance, a cohort analysis might compare the birth rates of cohorts of women born in 1900, 1925 and 1950 at each age from 15–19 to 45–49. This would reveal shifts in the ages of childbearing, as well as in the total numbers of children born to each female cohort. Without recourse to projections, however, it is impossible to describe the entire reproductive histories of younger cohorts in such studies, because they are still of childbearing age.

In practice, therefore, lack of complete information through time on real cohorts means that period data are often used to provide fertility measures for *synthetic* or *hypothetical cohorts*. Statistics for ages between 15 and 49, in a single year, are taken to represent the experience of a cohort moving through each of these ages. In other words, data for 35 single year cohorts (aged 15, 16, . . . 48 and 49) in one year are treated as if they describe 35 years of experience for one cohort. In fact, synthetic cohort measures of fertility, calculated from period data, are more common than measures from real cohort data. This is because there is always great interest in understanding the current situation and the behaviour of people in their twenties and thirties. Real cohort data are always incomplete for the young, but use of synthetic cohort data provides a quick projection of what would happen if the present situation remained unchanged. The ensuing sections on measures of fertility discuss period, synthetic cohort and real cohort measures in turn.

7.4 Period measures of fertility

The period measures of fertility range from crude measures, based on the total population, to more refined measures that relate births to the population 'at risk' of bearing a child. Limiting the denominators of rates to women of childbearing age is an obvious enhancement. Tables 7.1 and 7.2 list the formulas for the basic measures discussed in this chapter, and present worked examples of each.

Total births

Fertility rates and ratios are intended to provide comparable information for different times and places. However, they do not necessarily take precedence over the study of the actual numbers of births, since these have an essential contribution to make in understanding the significance of trends and changes in fertility. In Italy, for example, there were over 900 000 births in 1961 and less than 600 000 in 1991, even though the number of women of reproductive age had increased. Figures for the intervening years are needed to confirm a trend, but the numbers suggest a spectacular decline, with long-range flow-on effects in terms of falling school enrolments and diminishing numbers of young people entering the labour force, as well as the childbearing ages.

Table 7.1 Basic measures of fertility

Measures	Examples
Crude birth rate (CBR) $CBR = \frac{\text{number of live births in a year}}{\text{mid-year population}} \times 1000$ or $CBR = \frac{B}{P} \times 1000$	Egypt 1990 (estimates) Births 1 737 000 Mid-year pop. 52 426 000 CBR = 33.1
General fertility rate (GFR) $GFR = \frac{\text{live births in a year}}{\text{mid-year pop. of females aged } 15-49} \times 1000$ or $GFR = \frac{B}{F_{15-49}} \times 1000$	Egypt 1990 (estimates) Births 1 737 000 Women 15–49 12 423 000 GFR = 140
Child–woman ratio (CWR) $CWR = \frac{\text{number of children aged } 0-4}{\text{number of females aged } 15-49} \times 1000$ or $CWR = \frac{P_{0-4}}{F_{15-49}} \times 1000$	Egypt 1990 (estimates) Children 0–4 7 588 000 Women 15–49 12 423 000 CWR = 611
Age-specific fertility rate (ASFR) ASFR $= \frac{\text{no. of births in a year to women aged } x \text{ to } x+n}{\text{mid-year pop. of women in aged } x \text{ to } x+n} \times 1000$ or, for single years of age $ASFR = \frac{B_x}{F_x} \times 1000$ where x is a single year of age, e.g. age 20 years For grouped ages $ASFR = \frac{B_i}{F_i} \times 1000$ where i is a five year age group, e.g. 20–24 years	Russian Federation 1995 (see below)

Russian Federation 1995

Age	Mid-year pop. females	Births	ASFRs
15–19	5 335 672	238 019	44.6
20–24	4 986 816	561 796	112.7
25–29	4 645 374	309 371	66.6
30–34	5 802 104	171 115	29.5
35–39	6 448 795	68 219	10.6
40–44	6 013 599	13 073	2.2
45–49	4 845 679	578	0.1
		TFR =	1.3

Sources: Shryock and and Siegel (1973); United Nations (1993 & 2000b); US Census Bureau 2001, *Statistical Abstract of the United States 2000.*

Table 7.2 Basic measures of marriage and divorce

Measures	Examples
Crude marriage rate Crude marriage rate $= \frac{\text{marriages in year}}{\text{mid-year population}} \times 1000$ or Crude marriage rate $= \frac{\text{M}}{\text{P}} \times 1000$	United States, 1950 and 1990 Marriages 1950 1 667 000 1990 2 443 000 Mid-year pop. 1950 152 271 000 1990 249 907 000 Crude marriage rate 1950 10.9 1990 9.8
Crude divorce rate Crude divorce rate $= \frac{\text{divorces in year}}{\text{mid-year population}} \times 1000$ or Crude divorce rate $= \frac{D}{P} \times 1000$	United States, 1950 and 1990 Divorces 1950 385 000 1990 1 182 000 Mid-year pop. 1950 152 271 000 1990 249 907 000 Crude divorce rate 1950 2.5 1990 4.7
General marriage rate General marriage rate $= \frac{\text{marriages in year}}{\text{mid-year pop. aged 15 and over}} \times 1000$ or General marriage rate $= \frac{M}{P_{15+}} \times 1000$	United States, 1950 and 1990 Marriages 1950 1 667 000 1990 2 443 000 Mid-year pop. aged 15 and over* 1950 110 214 837 1990 195 142 002 General marriage rate 1950 15.1 1990 12.5 *Data refer to 1 April
General divorce rate General divorce rate $= \frac{\text{divorces in year}}{\text{mid-year population aged 15 and over}} \times 1000$ or General divorce rate $= \frac{D}{P_{15+}} \times 1000$	United States, 1950 and 1990 Divorces 1950 385 000 1990 1 182 000 Mid-year pop. aged 15 and over* 1950 110 214 837 1990 195 142 002 General divorce rate 1950 3.5 1990 6.1 *Data refer to 1 April

Table 7.2 *continued*

Age-specific marriage rate

Age-specific marriage rate (females)

$$= \frac{\text{marriages of females aged } x \text{ to } x+n}{\text{mid-year female pop. aged } x \text{ to } x+n} \times 1000$$

or

$$\text{Age-specific marriage rate (females)} = \frac{M_i^f}{P_i^f} \times 1000$$

where i is a five year age group

Australia 2000

	Mid-year pop. females	Women marrying	Marriage rate
15–19	658 054	3 221	4.9
20–24	665 870	27 332	41.0
25–29	732 731	37 723	51.5
30–34	712 112	19 438	27.3
35–39	749 160	9 953	13.3
40–44	723 665	5 901	8.2

Age-specific divorce rate

Age-specific divorce rate (females)

$$= \frac{\text{divorces of females aged } x \text{ to } x+n}{\text{mid-year female pop. aged } x \text{ to } x+n} \times 1000$$

or

$$\text{Age-specific divorce rate (females)} = \frac{D_i^f}{P_i^f} \times 1000$$

where i is a five year age group

Australia 2000

	Mid-year pop. females	Females divorcing	Divorce rate
15–19	658 054	283	0.4
20–24	665 870	3 973	6.0
25–29	732 731	9 702	13.2
30–34	712 112	10 223	14.4
35–39	749 160	9 328	12.5
40–44	723 665	7 310	10.1

Sources: Shryock and and Siegel (1973); United Nations (1993 & 2000); US Census Bureau 2001, *Statistical Abstract of the United States 2000;* Australian Bureau of Statistics 2001, *Marriages and Divorces, Australia 2000.*

Perhaps the best-known demonstration of the impact of changing numbers of births is the history of the post-Second World War baby boom in the United States (Bouvier and De Vita 1991) and elsewhere. The baby boom successively affected planning for maternity hospitals, primary and secondary schools, universities, and the labour force as well as forward planning for the retired age groups. The number of baby boomers has also been a massive engine of change in the market for housing, goods and services. The analysis of rates extends and enhances the study of the numbers of births, but by no means makes the raw numbers redundant: numbers of births are vital information in the planning decisions of governments, organizations and businesses.

Rate of natural increase

The rate of natural increase (RNI), previously discussed in Chapter 1, is the most widely available indicator of the contribution of vital processes to population growth. Although not strictly a measure of fertility, it places fertility in the essential context of the balance between births and deaths. The rate of natural increase (RNI) is calculated as:

$$\text{Rate of natural increase} = \frac{\text{births} - \text{deaths in a year}}{\text{mid-year population}} \times k$$

where $k = 1000$ or 100.

In the year 2000, the world's population had a rate of natural increase estimated at 14 per thousand, but there were great differences between the rates for countries. Overall, the more developed countries had rates of only 1 per thousand compared with 16 per thousand for the less developed regions (Population Reference Bureau 2001a). Rates of natural increase in excess of 10 per thousand have been a consequence of rapid mortality decline and sustained high fertility, leading to a 'population explosion' in some countries.

Crude birth rate

Also discussed in Chapter 1, the crude birth rate (CBR) is calculated as:

$$\text{Crude birth rate} = \frac{\text{number of live births in a year}}{\text{mid-year population}} \times 1000$$

Over the course of the demographic transition, crude birth rates have varied from a maximum of 35 per thousand or more, to a minimum of about 10. During the 'second demographic transition', however, they may fall further because of the previously unanticipated phenomenon of below-replacement fertility. In the year 2000, Hong Kong, Japan and several European countries, including Germany, Italy and Spain, had CBRs of less than 10.

While the crude birth rate readily identifies variations in the fertility of populations, it completely ignores the concept of risk. It relates births to the total male and female population, not to the portion of the female population capable of bearing children. Crude birth rates, moreover, are not necessarily comparable through time because of the influence of changing age structures. Other things being equal, the greater the proportion of the population of reproductive age, the higher the CBR.

One potential solution is to standardize the crude birth rate by age, through a procedure similar to standardizing the crude death rate. This involves calculating what the crude birth rate would be if the population had the age structure of a selected standard population and its own age-specific fertility rates. The expected births at each age are obtained by multiplying the number of females, at each reproductive age in the standard population, by the corresponding age-specific fertility rate. The total expected births are then divided by the total standard population to obtain the (direct) standardized crude birth rate (see Shryock and

Siegel 1973: 481–482). Standardized crude birth rates are used mainly where the influence of age structure variations are thought to be appreciable. More commonly, there is a preference for fertility measures, such as the general fertility rate, that are based on the population at risk – or an approximation of it.

General fertility rate

The general fertility rate (GFR) represents a first step towards obtaining a more refined measure of fertility than the crude birth rate:

$$\text{General fertility rate} = \frac{\text{live births in a year}}{\text{mid-year population of females aged } 15-49} \times 1000$$

This is a 'general' rate in that it attributes births to all women of reproductive age, irrespective of whether they had a birth. Its potential range is from about 50 to 300 per 1000 (Newell 1988: 39). Like the CBR, it derives partly from birth statistics, and partly from a census count or estimate of the mid-year population. Also, like the crude birth rate, it conceals a great amount of variation by age, since only a small proportion of births occur among the youngest and oldest women in the age range 15–49. Indeed, the age range is sometimes limited to 15–44, if there are few births to older women.

Child-woman ratio

A similar kind of measure, based entirely on census statistics, is the child–woman ratio (CWR):

$$\text{Child–woman ratio} = \frac{\text{number of children aged } 0-4}{\text{number of females aged } 15-49} \times 1000$$

The child–woman ratio enables fertility measurement where no statistics on births are available. This is a useful characteristic which greatly facilitates comparisons of fertility between parts of cities and other small areas. Because of its applications in spatial comparisons, the child–woman ratio is important in geographical studies of population.

A disadvantage is that, like the general fertility rate, the child–woman ratio does not allow for differences in the fertility of women of different ages or marital statuses. The child–woman ratio relates births to all women aged 15–49, irrespective of their likelihood of having a baby. Mortality may also affect the ratio, because the numerator includes only surviving children, whose numbers will be substantially less than the original number of live births if infant mortality is high.

Migration is a further complicating factor, especially since the birth of a child may alter a family's housing needs. If a new housing development has substantial inward movement of couples with young children, its child–woman ratio will not truly reflect the fertility occurring there. This is scarcely a problem, however, if the main concern is where families live, rather than precisely where parents resided when their children were born. Temporary movement to hospitals should have little effect on the data – the place of birth recorded in vital statistics is usually the mother's place of residence at the time, rather than the location of the hospital where the birth occurred.

Age-specific fertility rate

Fertility studies seldom include all possible measures – the availability of data and the aims of the research limit the choices made. Nevertheless, age-specific fertility rates (ASFRs) have a place in many fertility studies, because the likelihood of having a child varies greatly by age. Information on fertility variations according to the ages of mothers is invaluable when seeking to explain family-building patterns and overall levels of fertility. In traditional societies, childbearing begins early and continues through all the reproductive years. Figure 7.2 illustrates this age pattern for a population with 'natural fertility', that is, one not practising contraception or induced abortion. In contrast, many women in contemporary Western societies have their children between the ages of 25 and 34.

The formula for the age-specific fertility rate is much like that for the general fertility rate, except that all the figures relate to a particular age group:

$$\text{Age-specific fertility rate} = \frac{\text{number of births in a year to women in age group } x \text{ to } x+n}{\text{mid-year population of women in aged group } x \text{ to } x+n} \times 1000$$

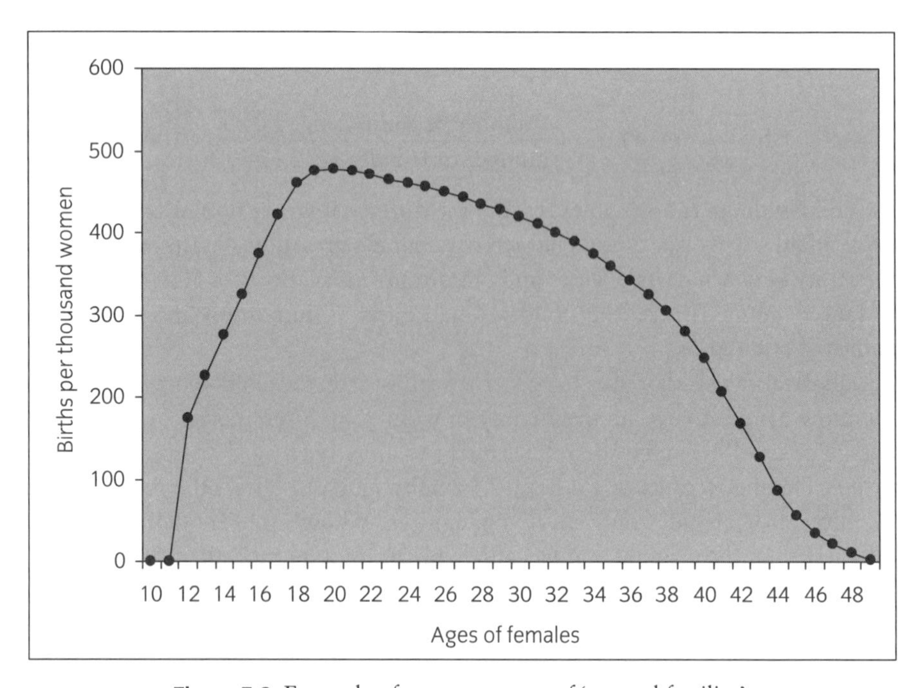

Figure 7.2 Example of an age pattern of 'natural fertility'.

Data source: model estimates from Coale and Trussel (1974).

For example

$$\text{ASFR}_{20\text{–}24} = \frac{\text{number of births to women aged 20–24}}{\text{mid-year population of women aged 20–24}} \times 1000$$

The ASFRs most often refer to five-year age groups (i.e. the seven age groups from 15–19 to 45–49), although detailed investigations may require rates for single years of age (i.e. the 35 single ages from 15 to 49). Five year figures are less cumbersome and published statistics commonly specify mothers' ages in such groupings. Vital statistics publications on births are the data source for tables on births by age of mother, but census figures or estimates are needed to supply the data for the denominators. For completeness, the small number of births to mothers aged less than 15, or more than 49, are normally added to those in the nearest age group. The effect of this is negligible in most societies today – hence there is seldom any need to adjust the denominators. Box 7.2 covers strategies used when, for some births, the ages of mothers are unknown.

Table 7.3 shows the derivation of ASFRs for Italy in 1961 and 1991. For example, the ASFR for Italian women aged 25–29 in 1961 was calculated as: 297782

Table 7.3 Calculation of age-specific fertility rates and total fertility rates for Italy, 1961 and 1991

Age	Births		Women		Age-specific fertility rates per 1000 women	
	1961	1991	1961	1991	1961	1991
A	B	C	D	E	F B/D × 1000	G C/E × 1000
15–19	34848	17156	1860432	2133945	18.73	8.04
20–24	224053	115215	2013149	2317778	111.29	49.71
25–29	297782	214259	1894394	2363387	157.19	90.66
30–34	219059	149252	1943603	2078483	112.71	71.81
35–39	118993	56021	1948647	1912789	61.06	29.29
40–44	32052	10538	1415117	2037106	22.65	5.17
45–49	2871	346	1689975	1715229	1.70	0.20
Total	929658	562787	12765317	14558717	485.34	254.88
Multiply total by 5					2426.69	1274.38
Divide by 1000 (to obtain total fertility rates per woman)					2.43	1.27

Notes: Births to mothers whose age was not stated were distributed proportionately and the resulting figures were rounded to whole numbers. Births to mothers aged less than 15, or more than 49, were added to the adjacent age groups.

Data source: United Nations 2000. *Demographic Yearbook, Historical Supplement 1948–1997* (CD).

BOX 7.2 **Distributing births when the mothers' ages are unknown**

When statistics on births by age of mother include a 'not stated' category, meaning that the ages of some mothers at the birth of their children are unknown, there are two main strategies:

- Either, omit the 'not stated' figures, if the numbers are relatively low.
- Or, distribute the 'not stateds' proportionally through all the mothers' age groups, assigning them to each age group according to its share of the total stated births, as in the example below.

Distribute 200 births, where the ages of the mothers are unknown

Age group	Total births, age of mother stated	Proportion of total births, age of mother stated	Distributed births	Estimated total births in age group
15–29	700	0.7	200 × 0.7 = 140	700 + 140 = 840
30–49	300	0.3	200 × 0.3 = 60	300 + 60 = 360

/1 894 394 × 1000 = 157.19 births per thousand women. It is apparent from the table that even two sets of age-specific fertility rates begin to amount to a sizeable body of data. Hence ways are needed to assist in analysing and presenting such figures. Line graphs are an effective approach to summarizing and comparing sets of ASFRs. Accordingly, Figure 7.3 more clearly illustrates the decline of fertility at all ages in Italy between 1961 and 1991, highlighting a near collapse in the age-specific fertility rates in the prime childbearing ages 20–24 and 25–29. The contrast apparent between the situation in 1961 and 1991 further emphasizes the desirability of having a single figure to summarize individual sets of ASFRs and provide a basis for overall comparisons. The total fertility rate, explained in the next section, fulfils this purpose.

7.5 Synthetic cohort measures of fertility and replacement

As discussed earlier, synthetic 'cohort' measures of fertility are often calculated from period data for a single year. The synthetic cohort approach is pursued for reasons of immediacy, to provide regular assessments of the current situation. Synthetic cohort measures would correspond to real cohort measures only if rates

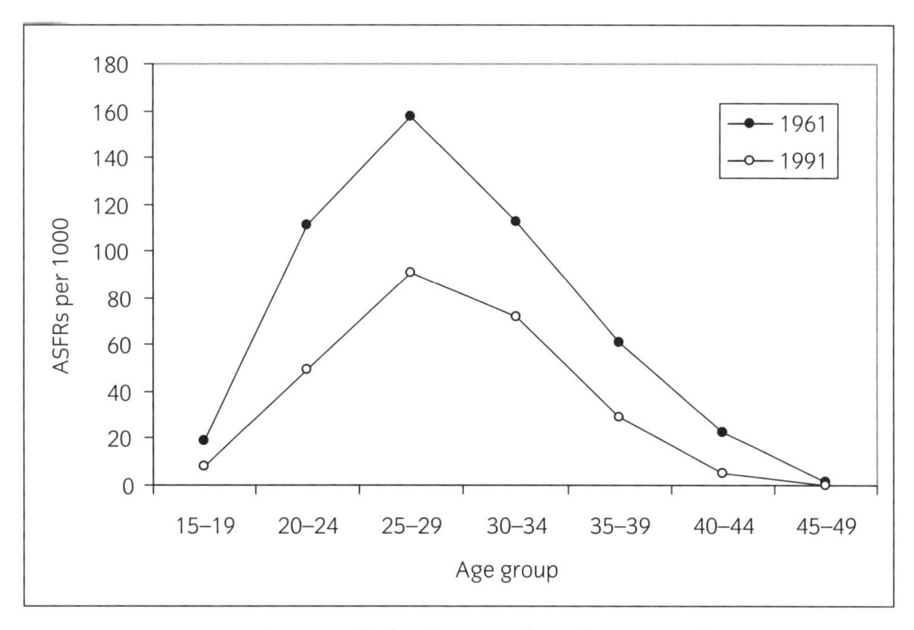

Figure 7.3 Age-specific fertility rates for Italy, 1961 and 1991.

Data source: see Table 7.3.

actually remained constant for a long time. Strictly speaking, synthetic cohort measures are 'period' measures: they use information about many cohorts in a single year, whereas real cohort measures use information about a single cohort over many years.

Total fertility rate

Best known among the synthetic cohort measures is the total fertility rate (TFR). This is calculated from a set of age-specific fertility rates for a single year. A major advantage of the total fertility rate is that it produces a single summary figure from seven or more age-specific fertility rates. In doing so, it measures what used to be called 'average family size'. In the present-day context of complex patterns of relationship formation and breakdown, the total fertility rate is more accurately described as indicating the average number of children per woman.

The total fertility rate is the sum of the single year age-specific fertility rates (per thousand), usually expressed, in its most easily understood form, as a rate per woman:

$$\text{TFR} = \sum_{x=15}^{49} f_x / 1000$$

where f_x is an age-specific fertility rate per thousand for single years; i.e.

$$\text{Total fertility rate (per women)} = \frac{\text{sum of the single year ASFRs}}{1000}$$

Division by 1000 expresses the TFR as a rate per woman. Note the convention that subscript x is used for single year ages in demographic equations, whereas i is used for grouped ages. Since single year figures are voluminous, data for five year age groups are employed more frequently for calculating the total fertility rate. In this case, the sum of the ASFRs must be multiplied by 5:

$$\text{TFR} = 5 \sum_{i=15\text{–}19}^{45\text{–}49} f_i / 1000$$

where f_i is the age-specific fertility rate for grouped ages; i.e.

$$\text{Total fertility rate (per woman)} = \frac{\text{sum of the five year ASFRs} \times 5}{1000}$$

Multiplying by 5 is necessary because the figures refer to events in one year, whereas the women are in each age group for five years. The rate for a five year age group may be interpreted as the average of the single year rates in the age range. This is shown in Table 7.4, which presents hypothetical age-specific rates for single year ages within the range 20–24 years. Adding the single year rates gives a total of 1500. However, calculating the age-specific fertility rate for age group 20–24 from total births, in a single year, and total women produces a rate of 300 per thousand (150/500 × 1000). Multiplying this by 5 produces a figure equal to the sum of the single year rates (300 × 5 = 1500).

Table 7.3 includes the working for the calculation of the total fertility rates for Italy in 1961 and 1991. The steps are as follows:

- Starting with a set of age-specific fertility rates for five-year age groups, add them to obtain a total figure. The rates for Italy in 1961 sum to 485.34 (Table 7.3).

Table 7.4 Comparison of single year and five year age-specific fertility rates

Age	Births	Women	ASFRs per 1000
20	20	100	200
21	25	100	250
22	30	100	300
23	35	100	350
24	40	100	400
Total	150	500	[1500]
20–24	150	500	300

Notes: The bracketed figure in the last column is a total, not an ASFR. The sum of the single year ASFRs = 1500.
Source: hypothetical data.

- Multiply this total by 5 (485.34 × 5 = 2426.69).
- Since the ASFRs were expressed as rates per thousand women, divide the new total by 1000 to express the total fertility rate as a rate per woman (2426.5/1000 = 2.43 children per woman).

The TFRs in Table 7.3 provide an illuminating summary of Italy's fertility, emphasizing the sharp contrast between circumstances in 1961 and 1991. The TFR for 1991 indicates that Italian women would have an average of only 1.3 children each if Italy's age-specific fertility rates remained constant from 1991 into the future. The corresponding figure for the year 2000 was 1.2 (Population Reference Bureau 2001a), which is one of the lowest national figures ever observed.

In contrast, the highest observed total fertility rate is 9.2 children per woman (Bongaarts and Potter 1983). This figure derives from the age-specific fertility in the 1920s of Hutterite women in North Dakota. Women in these Hutterite religious communities typically married early, used no contraception and breast-fed the babies for only a short time.

An even higher figure is obtained when the calculations are based only on the age-specific fertility of married women, to produce the *total marital fertility rate* (TMFR). The Hutterite TMFR for the 1920s was 12.4 children per married woman (Newell 1988: 46). As the maximum figure known for marital fertility, it has been an important basis for comparisons, most notably in Coale's indices of fertility, which obtain, from basic demographic data, ingenious measures of the extent of fertility control (see Newell 1988: 44–49; Pressat 1985: 98–99).

For the Hutterite, and other, populations in the Proximate Determinants module (Box 7.1) the TFR is employed, rather than the TMFR, because the comparisons are between the fertility of all women. Nationally, the TFR for populations with the highest fertility have seldom exceeded 7, for reasons explained in Box 7.3.

The exact meaning of the total fertility rate, derived from period data, is 'the average number of children each woman would have if the population's age-specific fertility rates remained constant'. However, ASFRs continue to vary, even in populations with low birth rates. While the assumption of constant rates is invalid, the TFR at least provides a concise indication of the level of fertility in a given year. Also, from it estimates can be made of the extent to which current fertility is above or below replacement level.

Reference is often made to a TFR of 2.1 as denoting *replacement level fertility*. An average of two children would 'replace' all mothers and fathers, but only if all the children survived to reproductive age. Thus an extra 0.1 is needed to offset the effects of premature mortality, as well as the unbalanced sex ratio of births (see Hinde 1998: 223). Although the figure of 2.1 is a reasonable approximation to the number of children per woman needed to achieve replacement in contemporary more developed countries, it is not a universal benchmark. In pre-transition populations, a TFR of 6 or more children was necessary at times to offset high mortality and ensure the maintenance of population numbers. The TFR denoting

BOX 7.3 **Why are high birth rates so low?**

John Bongaarts (1975) analysed this intriguing question, which arises from the observation that the average number of live births per woman in societies with 'high' fertility is about 7. While there are specific instances of 15 or more births per woman, national average fertility is much lower. Bongaarts and Potter (1983: 8) proposed a theoretical maximum, in the absence of all biological and behavioural constraints on reproduction, of 35 births, not counting multiple births. This envisages one birth per year from age 15 to 49, allowing 9 months per pregnancy together with an infertile interval after each birth. A Russian woman holds the record for the number of children borne, namely 69 from 27 pregnancies (Newell 1988: 35).

The following are the main factors Bongaarts (1975) identified to explain why the total fertility rates of populations are far from the 'theoretical maximum', even in the absence of contraception and induced abortion:

- **Marriage** In high fertility countries, many women spend only 75 per cent of their reproductive span in marriage. Early widowhood contributes to this.
- **Waiting time to conception** Tentative estimates of the monthly probability of conception range from 0.15 to 0.50 for ovulating women who have intercourse regularly. The probabilities are quite low because 'conception is possible only during a very brief period, probably only two days, in the middle of a woman's cycle' (Bongaarts 1975: 294). Consequently, the average waiting time to conception may be several months.
- **Intra-uterine deaths** When an ovum is fertilized, it has a 50 per cent chance of yielding a live birth. Approximately one third of fertilized ova are rejected during the first 2 weeks after conception, evidently because of genetic defects and other abnormalities. On average, two fertilizations are needed to yield one live birth. These intra-uterine deaths inevitably extend the waiting time to a viable conception.
- **Postpartum amenorrhea** The infertile interval after a birth may last from a minimum of about 3 months to 1.5 years, depending particularly on the duration and intensity of breastfeeding.
- **Sterility** When malnutrition and poor health are widespread, women may be sterile (i.e. unable to produce a live birth (Pressat 1985: 214)) for more than 20 per cent of their married life. Greater numbers of women will be infertile (i.e. bearing no children) if some husbands are sterile. Sterility is also age-related, resulting in lower age-specific birth rates at higher ages.

Because of these influences, women in high fertility societies are pregnant for only about one-sixth of their reproductive years. Crude birth rates rarely exceed 50 per thousand in populations taking no deliberate actions to influence fertility (Bongaarts 1975: 295). To summarize the more detailed outcomes, Bongaarts (1975: 294) noted:

A typical thirty-month birth interval in a natural fertility population may be divided as follows: twelve months postpartum amenorrhea, four months waiting time to conception before an intrauterine death, a one-month nonsusceptible period associated with the intrauterine death, another four months' conception waiting time before a live birth, and finally a nine-month full-term pregnancy.

Further insight into why high birth rates are so low may be obtained by examining the proximate determinants of fertility for 'historical populations' (see Box 7.1).

replacement level has drawn close to 2 only as mortality has declined. Thus a population's replacement level fertility depends on its death rate.

Overall, the total fertility rate is an important and widely used measure of fertility, applicable both to synthetic and real cohort data. Nevertheless it has several potential disadvantages:

- When based on period data, it does not strictly measure completed fertility, since it assumes that the age-specific fertility rates for a particular year will remain constant. In reality, period-based total fertility rates may vary appreciably through time, none giving a true indication of long-run trends in completed fertility. For example, short-lived baby booms can produce relatively high total fertility rates. However, if the 'boom' is mainly due to a change in the timing of childbearing, with more couples deciding to have their children within a particular period, there will be no real long-run change in completed fertility. In this case neither the peak in the TFR, nor the subsequent trough, will provide a true indication of trends. Thus variations in period-based TFRs are often due more to changes in the timing of births than to changes in completed fertility.
- It does not measure the average number of children per mother; rather, it measures the number of children per woman, including those who had no children.
- It takes no account of the impact of mortality. This is not a major drawback in long-lived populations, but in populations with high death rates of infants and children, many in the new generation do not survive to become the next generation of parents or potential parents.
- Different sets of age-specific fertility rates will produce the same total fertility rate if they sum to the same total. Therefore it is not possible to construe, from the TFR, the age pattern of fertility.
- It does not, in itself, measure the extent to which fertility is above or below replacement – further calculations, or an alternative index – viz. the *net reproduction rate* (see below) – are needed to determine this.

Gross reproduction rate

While the total fertility rate provides a basis for considering population replacement in terms of total children born, the other two synthetic cohort measures are limited to the female population, namely the *gross reproduction rate* and the *net reproduction rate*. Replacement may also be measured with reference to the male population, but this occurs only in special studies because:

- The reproductive span of women is shorter and more clearly defined than that of men. This offers advantages in assembling statistics and performing calculations; at most, the methods require data for ages 15–49 years.
- Information about demographic and social characteristics of parents is more reliable and more complete for mothers than for fathers, especially if the mothers are single or remarried.

- In contemporary Western countries, women usually have a greater role in decisions about whether and when to have children, because child-bearing has a more profound impact on their lives, including their careers, income and use of time. Therefore the characteristics of women – such as their ages, educational attainments, and occupations – tend to be more influential determinants of fertility than the characteristics of husbands or partners.

The gross reproduction rate resembles the total fertility rate, except that it measures the number of daughters per woman, instead of the total number of children. Therefore, while the method of calculation is the same, the age-specific rates refer only to female births.

The formula for the gross reproduction rate, using single years of age, is:

$$\text{GRR} = \sum_{x=15}^{49} f_x^d / 1000$$

where f_x^{d} is the age-specific fertility rate for single ages, daughters only; i.e.

$$\text{Gross reproduction rate (per woman)} = \frac{\text{sum of the single year ASFRs for daughters}}{1000}$$

For five year age groups, the formula becomes:

$$\text{GRR} = 5 \sum_{i=15\text{–}19}^{45\text{–}19} f_i^d / 1000$$

where f_i^{d} is the age-specific fertility rate for grouped ages, daughters only; i.e.

$$\text{Gross reproduction rate (per woman)} = \frac{\text{sum of the five year ASFRs for daughters} \times 5}{1000}$$

Table 7.5 illustrates the procedure, applying the second formula – since statistics for five year age groups are most widely available. The steps are:

- list total female births (column B);
- list the total numbers of women in the reproductive age groups 15–19 to 45–49 (column C);
- calculate the female age-specific fertility rates, i.e. for daughters only (column D);
- sum the ASFRs and multiply the total by 5, because the women are in each age group for five years;
- Divide by 1000 to express the result as daughters per woman (i.e. GRR = 0.62).

Alternatively, an estimate of the GRR can be found by multiplying the total fertility rate by the proportion of total births in the year that were female (Shryock

Table 7.5 Calculation of gross and net reproduction rates for Italy, 1991

Age	Female births 1991	Total women 1991	Female ASFRs per 1000 1991	Females surviving to age group (from life table, assuming 1000 births per year)	Probability of surviving from birth to age group of mother	Expected survivors per 1000 female births
A	B	C	D B/C × 1000	E	F E/5000	G D × F or (B × F)/C × 1000
15–19	8 383	2 133 945	3.93	4949.45	0.989 89	3.89
20–24	55 864	2 317 778	24.10	4941.32	0.988 26	23.82
25–29	103 541	2 363 387	43.81	4931.44	0.986 29	43.21
30–34	72 567	2 078 483	34.91	4919.19	0.983 84	34.35
35–39	27 145	1 912 789	14.19	4902.65	0.980 53	13.92
40–44	5 145	2 037 106	2.53	4877.75	0.975 55	2.46
45–49	158	1 715 229	0.09	4839.00	0.967 80	0.09
Total	272 803	14 558 717	123.56			121.74
Multiply total by 5			617.82			608.68
Divide by 1000		GRR =	0.62		NRR =	0.61

Notes: Births to mothers whose age was not stated were distributed proportionately and the resulting figures were rounded to whole numbers. Births to mothers aged less than 15, or more than 49, were added to the adjacent age groups.
Data sources: United Nations 2000. *Demographic Yearbook, Historical Supplement 1948–1997* (CD), and Table 7.6.

and Siegel 1973: 524). For Italy in 1991, the TFR was 1.27 (Table 7.3) and the proportion of births that were female was 271 803/562 787 = 0.482 96 (Tables 7.2 and 7.4). Thus the GRR for Italy in 1991 was: 1.27 × 0.482 96 = 0.62 daughters per woman.

Since the sex ratio at birth is normally about 105 : 100, a quick approximation of the GRR is the product of the TFR and 0.487 80 (i.e. 100/205, denoting 100 female births in a total of 205, = 0.487 80); for Italy in 1991 the result is: 1.27 × 0.487 80 = 0.62.

Achieving replacement level fertility depends not only on the number of live births, but also on the number of offspring surviving through the reproductive ages. High mortality populations need a greater number of live births to achieve replacement than low mortality populations, most of whose children will live for seven or eight decades. In contemporary more developed countries, maintenance of a GRR slightly above 1.0 would ensure the replacement of a generation of women by a new generation of daughters. The GRR measures replacement in terms of the average number of daughters there are to succeed each woman. Like the total fertility rate, it is usually calculated from period data, but it is possible for it to become a real cohort measure if data are available on the lifetime experience of actual cohorts. The GRR also has the same disadvantages presented earlier for the TFR. Nevertheless, one of the disadvantages – the lack of recognition of the impact of mortality on population replacement – is readily overcome by including survival estimates. This is the feature that distinguishes the gross reproduction rate from the net reproduction rate.

Net reproduction rate

Calculation of the net reproduction rate requires some basic knowledge of life tables. This section provides the necessary background, which also serves as an initial introduction to the subject matter of the next chapter. Alternatively, this section could be read after Chapter 8.

Much enthusiasm greeted the invention, or re-invention, of the net reproduction rate (NRR) in the 1930s, when Western societies were experiencing unprecedented low fertility, evoking considerable concern about the prospect of population decline. 'This simple and beautiful method', as Carr-Saunders (1936: 123) described the net reproduction rate, provided a clever solution to the problem of ascertaining whether populations were replacing themselves. In the early 1930s, the United Kingdom, the United States, Australia, New Zealand, France, Germany, Austria and the Scandinavian countries all had below-replacement fertility, as established from their net reproduction rates (ibid: Figure 25).

The net reproduction rate estimates the number of daughters who will live to replace their mothers in the future, thereby measuring the replacement of one generation by another. It allows for mortality between birth and the age of the mother at the time of bearing the child. Thus it defines replacement in terms of the numbers of daughters living to their mothers' ages at confinement:

- NRR = 1, signifies exact replacement, or one daughter per woman: women are bearing just sufficient daughters to replace themselves in the future. The replacement level is always an NRR of 1, irrespective of whether the population has high or low mortality.
- NRR < 1, denotes below-replacement fertility, where there are fewer daughters to succeed their mothers' generation. Any value less than one means that the population is not replacing itself.
- NRR > 1, indicates above-replacement fertility – the future generation of potential mothers will be larger than the one that produced them.

The NRR is similar to the GRR, except that it allows for mortality. It is therefore always lower than the GRR. It is especially useful for studying high mortality populations, or for accurate comparisons between past and present levels of replacement. In low mortality populations, there is little difference between the GRR and the NRR.

Using statistics for single years of age, the formula for the net reproduction rate is:

$$\mathrm{NRR} = \sum_{x=15}^{49} \left(f_x^d \times \frac{L_x}{l_0} \right) \Big/ 1000$$

where $f_x{}^d$ is the age-specific fertility rate for single ages, daughters only, and $\frac{L_x}{l_0}$ is the probability of daughters surviving to their mother's age; i.e.

$$\text{Net reproduction rate (per woman)} = \frac{\text{sum of (single year ASFRs for daughters} \times \text{the proportion surviving to their mother's age)}}{1000}$$

Using statistics for five year age groups, the formula becomes:

$$\mathrm{NRR} = \sum_{x=15-19}^{45-49} \left(f_i^d \times \frac{{}_5L_x}{5 \times l_0} \right) \Big/ 1000$$

where f_i^{d} is the age-specific fertility rate for grouped ages, daughters only, and $\frac{{}_5L_x}{5 \times l_0}$ is the probability of daughters surviving to the mid-point of the mother's age group; i.e.

$$\text{Net reproduction rate (per woman)} = \frac{\text{sum of (five year ASFRs for daughters} \times \text{the proportion surviving to the mother's age group)} \times 5}{1000}$$

The only addition to the previous formula for the GRR is the probability of survival of daughters, from birth to their mother's age. Figures for this are obtained from life tables, which show, for each year of age in a given year or period, death rates, proportions surviving since birth, life expectancy and other indices of mortality and survival. National statistical agencies publish life tables at regular intervals, usually after finalizing the latest census figures – which provide the denominators for the mortality rates in the tables (see Chapter 8).

When assembling statistics for the net reproduction rate, the figures on survival are usually taken from a published life table for the same year as the other data, or the nearest year. Otherwise the survival data are taken from a set of model life tables (see Chapter 9), which can substitute for official life tables if necessary.

Table 7.6 is a model life table approximating Italy's female mortality in 1991, when female life expectancy at birth was about 80 years. The only life table figures needed in the calculation of the net reproduction rate are the highlighted values.

- The first highlighted figure is 1000 in column B. This denotes the number of live births occurring in the population each year. (The mathematical notation for this number is l_0.) In a life table the number of live births is set at a convenient

Table 7.6 West model life table, Level 25, females (life expectancy at birth = 80 years)

Age A	Number surviving at exact age x B l_x	Age-specific death rates C ${}_nm_x$	Probability of dying D ${}_nq_x$	Average number alive between exact ages E ${}_nL_x$	Total population aged x and over F T_x	Average life expectancy at exact age x G e_x
0	1000.000	6.04	6.01	995.19	80000	80.00
1	993.989	0.44	1.76	3969.85	79005	79.48
5	992.244	0.18	0.90	4958.76	75035	75.62
10	991.350	0.15	0.75	4954.96	70077	70.69
15	990.605	0.30	1.50	4949.45	65122	65.74
20	989.117	0.36	1.79	4941.32	60172	60.83
25	987.341	0.44	2.22	4931.44	55231	55.94
30	985.147	0.55	2.77	4919.19	50299	51.06
35	982.422	0.80	4.01	4902.65	45380	46.19
40	978.480	1.25	6.24	4877.75	40478	41.37
45	972.374	1.97	9.80	4839.00	35600	36.61
50	962.845	3.06	15.19	4779.12	30761	31.95
55	948.220	4.60	22.73	4689.36	25982	27.40
60	926.664	7.28	35.79	4553.72	21292	22.98
65	893.498	11.90	57.86	4343.41	16739	18.73
70	841.797	21.45	102.02	4002.88	12395	14.72
75	755.921	40.57	184.17	3431.56	8392	11.10
80	616.702	76.10	318.23	2578.93	4961	8.04
85	420.447	134.21	493.17	1545.03	2382	5.67
90	213.096	222.52	679.55	650.76	837	3.93
95	68.287	346.88	833.97	164.17	186	2.73
100	11.338	517.35	1000.00	21.92	22	1.93

Note: the radix of the life table is 1000; the ${}_nm_x$ values are rates per 1000.
Source: Coale and Guo (1990: 31).

number, such as 1000 or 100000, and is assumed to remain constant through time. Other figures in the same column are the numbers alive at subsequent birthdays or 'exact ages'. For example, out of 1000 born, 68 would be alive on their 95th birthday (exact age 95), if the population experienced the age-specific death rates given in column C. The life table assumes that these age-specific death rates also remain constant through time; they would result in the population having a life expectancy at birth of 80 years, as listed in the last column.

- The other highlighted figures are for the reproductive ages in column E, headed 'Average number alive between exact ages'. (The mathematical notation for this is ${}_nL_x$.) Among people moving through a five year age group, such as 20–24, the numbers alive are highest for the youngest age and lowest for the oldest. The figures in column E take this into account, showing the *average* numbers alive in each age group.

For example, the average number alive between exact ages 20 and 25, i.e. in age group 20–24, is 4941. The figure denotes the average number of survivors from an original 5000 live births. Since the life table assumes that 1000 babies are born alive each year, the 20–24 year olds would have numbered 5000 at birth. The average numbers of survivors in each five year age group are also listed in column E of Table 7.5, which shows the working for calculating the NRR as well as the GRR.

Dividing the average number alive in each five year age group by 5000 gives the probability of surviving to each age – the figure is a probability since it is based on the initial population (see Section 1.4).

The procedure for finding the NRR is an extension of the working for the GRR (Table 7.5). The new steps are:

- For each age group list, from the life table, the average numbers surviving between exact ages (column E), then calculate the probabilities of survival to each age (column F). At age group 20–24, 4941.32/5000 = 0.988 26.
- Multiply each ASFR (daughters only) by the corresponding probability of survival, (column G). At ages 15–19, for example, 3.93 × 0.989 89 = 3.89. Alternatively (B × F)/C × 1000 gives the same result: (8383 × 0.989 89)/2 133 945 × 1000 = 3.89.
- Add the figures in column G, multiply by 5, and then divide by 1000 to express the net reproduction rate as the numbers of daughters per woman.

When using an official life table, figures on the average number alive between exact ages usually refer to single years of age, rather than grouped ages. In these circumstances there are two ways to obtain the daughters' probabilities of surviving to the five-year age group of their mothers: either add the single year figures for each age group, and divide by the number of live births occurring during 5 years (e.g. 5000); or take the middle value in each age group and divide by the number of births in one year (e.g. 1000). Using life table data for the middle age group is an approximation, which saves adding sets of values. For the five year age groups between ages 15 and 49, the middle values are those for ages 17, 22, 27, 32, 37, 42 and 47.

Typically, in calculating the NRR, both the fertility data and the life table data refer to a single year. This means that the results denote the experience of a synthetic cohort. As far as the probabilities of survival are concerned, the estimates assume that the figures for a single year can be treated as representing long term trends in survival. Often this assumption is far less restrictive than any assumption that the fertility rates remain constant, since mortality, in more developed countries, is no longer subject to rapid changes at ages 0–49.

The NRR, like the GRR and the TFR, is tied to the fertility data for a single year, which may not be representative of more protracted trends. This is its greatest weakness. Indeed, the foreboding raised by the very low NRRs for the 1930s was somewhat misplaced since, in the long term, the completed fertility of women of reproductive age in the 1930s was higher than anticipated.

7.6 Real cohort measures of fertility

As previously mentioned, the TFR, GRR and NRR can become real cohort measures. This is achieved by replacing the period-based age-specific rates with rates for the complete 35 year reproductive span of individual cohorts. The most commonly used real cohort measures, however, are simply census or survey figures on total children ever born.

Census questions asking women 'How many children have you ever had?' provide information on the *completed fertility* of many birth cohorts, from the middle aged to the oldest. Combining such figures from different censuses can provide a realistic long term view of trends in average family size, without the uncertainties of period variations that are inherent in synthetic cohort measures. The same source also provides details of completed fertility in terms of the numbers of women with no children, one child, two children and so on (Figure 7.4). Cross-tabulating this information with other census or survey figures provides insights into questions such as who remains childless and which social groups have the largest families?

Figure 7.4 illustrates, for the United States, trends through time in the total number of children per woman, measured at ages 45–49, the end of the reproductive span. Subtracting 20 from the census years roughly indicates when the women were in the main ages of family formation: thus those aged 45–49 in 1950 and 1990 were at such ages in 1930 and 1970 respectively. The high proportion childless, or with only one child, is striking in 1950 and 1960 – an important influence was the impact of the Great Depression on family building. In contrast, the figures for 1970 and 1980 owe much to the effects of the baby boom, bringing a wider prevalence of marriage and childbearing. By 1990, families of two children had reached their highest representation.

A problem with fertility statistics for real cohorts is that they are complete only for women of middle age or more. To obtain a full set of time series data on the child-bearing experience of females aged 15–19 in 2001 will mean waiting until the 2030s. Hence the frequent recourse to synthetic cohort data. Nevertheless, cohort comparisons at younger ages, including fertility rates for females in their teens and twenties, are possible and can provide important insights into changing patterns of child-bearing.

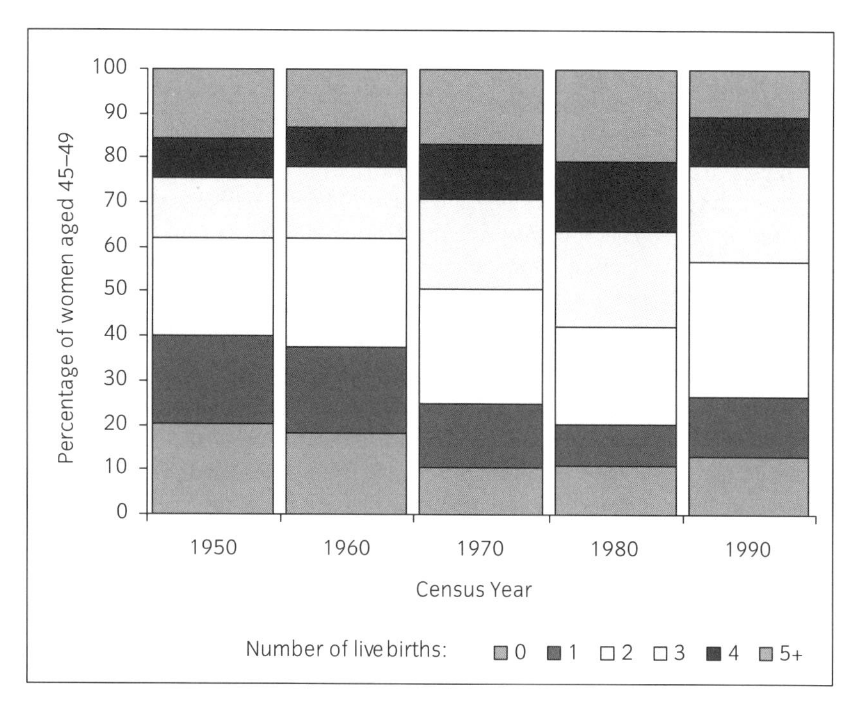

Figure 7.4 Female population aged 45–49 by number of live births, United States 1950–1990.

Data source: United Nations 2000. *Demographic Yearbook, Historical Supplement 1948–1997* (CD).

The discussion has identified a range of issues complicating the analysis of fertility, which raises the question of whether research on fertility is more complex than research on mortality. Although the principal methods for the investigation of both processes are reasonably straightforward, it is often thought that social, economic and cultural influences inevitably make fertility research the more challenging. The question is not clear cut, however, and Box 7.4 provides a comparison to illustrate the arguments.

7.7 Measures of marriage and divorce

The family setting in which child-bearing occurs is a basic consideration in the interpretation of fertility, not least because marriage is one of its main proximate determinants (Box 7.1). Measures of marriage and marriage breakdown help to construct the social context of fertility behaviour, as well as being important for their own sake.

Statistics on marriage and divorce refer only to people entering or ending registered marriages. Yet many people belong to consensual unions, or de facto marriages, which can have similar characteristics to registered marriages in terms of comprising stable relationships in which children are born and reared (Carmichael

BOX 7.4 Fertility analysis versus mortality analysis: which is more complex?

Fertility is often considered a more complex area of study than mortality: fertility is a voluntary, repeatable event, subject to the complexities of human relationships and decision processes, as well as to social, economic and political forces. In contrast, mortality is involuntary, unrepeatable and generally beyond human decision-making. However, in explaining why individuals succumb to particular diseases, there remain many unanswered questions:

'The process of mortality is more complicated than other demographic processes because it is more biologically determined. In addition, age at death can be affected by events occurring over about 70 years of life. Thus the events leading to early death may be widely separated in time from the event. Current knowledge of mortality causation is limited; our methods of gathering appropriate data are even more limited. . . . As theoretical development in mortality proceeds, models of health and mortality will come to incorporate disease environments of childhood, chemical environments throughout life, and health policies and health programs experienced at various points of life.' (Crimmins 1993: 587–588)

Infanticide, euthanasia and other instances of human agency in mortality also introduce extra considerations into mortality analysis. Clearly, both fertility and mortality present considerable challenges. The grid below compares their major features. The first three categories present difficulties for both fertility and mortality analyses, the fourth heightens problems for mortality studies, while the last three do so for fertility analysis.

	Fertility	Mortality
1 **Statistics**	Statistical information is most abundant in countries with low fertility.	Statistical information is most abundant in countries with low mortality.
2 **External influences**	Influenced by norms and values of the society and social groups.	Influenced by environmental factors, norms, values, living standards as well as lifestyles of the society and social groups.
3 **Proportions at risk**	For biological and social reasons, not everyone is at risk of bearing or fathering a child.	Everyone is at risk, but risk varies according to age, sex and other characteristics and circumstances.
4 **Age at occurrence**	Occurs within a 35 year age range for women.	Occurs at all ages.
5 **Origin**	Usually voluntary. Subject to the decisions of individuals and couples.	Usually involuntary. Normally beyond the decisions of individuals.
6 **Frequency**	Repeatable for the same individuals, but with variations in timing, spacing and total number of births.	Occurs only once per individual.
7 **Dependence on relationships**	Normally requires the formation of a continuing relationship between fecund partners.	Does not require the formation of human relationships, but supportive relationships can enhance survival.

1995). Similarly, many breakdowns in marriage-like relationships occur within consensual unions and never enter into official statistics on marriage dissolution. Thus, during the second demographic transition, statistics on marriage and marriage dissolution, based on registration data, have provided incomplete coverage of the formation and dissolution of 'marriages'.

Restricting the definition of the married to those in registered unions does not adequately represent the contemporary concept of marriage, which some sociologists define broadly as entailing a 'commitment' to a union, whether or not formally recognized through a legal or religious ceremony (English and King 1983: 13). Contemporary practice in censuses and family surveys is to seek information on both formal (registered) and informal (consensual) marriage relationships. The measures of marriage and marriage breakdown, while normally based on registration data, are potentially adaptable to measuring trends in the formation and dissolution of informal marriage relationships as well.

Crude rates

The most common measures of marriage and divorce are straightforward and, like the period measures of fertility, are amenable to refinement though seeking to better identify the population 'at risk'. Like other crude rates, the crude marriage rate and the crude divorce rate are based on the total mid-year population (Table 7.2):

$$\text{Crude marriage rate} = \frac{\text{marriages in year}}{\text{mid-year population}} \times 1000$$

$$\text{Crude divorce rate} = \frac{\text{divorces in year}}{\text{mid-year population}} \times 1000$$

The numerators of the rates are the number of marriages or divorces, rather than the number of people experiencing these events. For this reason the rates tend to be low, since at most, there will be only one event in the numerator per two people in the denominator.

While easily calculated, the crude rates have obvious deficiencies through being based on the total population, including children and others not currently at risk of marrying (e.g. the currently married) or divorcing (e.g. the never married, the widowed and the currently divorced). The age structure of the population also affects these crude rates, giving standardization a potential role in interpreting them (see Shryock and Siegel 1973: 571–572).

General rates

General marriage and divorce rates omit children to produce marriage and divorce rates based on the population aged 15 years and over. The denominators approximate the population of marriageable age, even though marriage laws may specify higher minimum ages at marriage.

$$\text{General marriage rate} = \frac{\text{marriages in year}}{\text{mid-year population aged 15 and over}} \times 1000$$

$$\text{General divorce rate} = \frac{\text{divorces in year}}{\text{mid-year population aged 15 and over}} \times 1000$$

These rates may be calculated separately for males and females. Another refinement is to limit the denominator for the marriage rate to persons aged 15 and over who are unmarried (i.e. single, divorced or widowed). A *general first marriage rate* is calculated by restricting the numerator to first marriages and the denominator to the never married. Similarly, the denominator for the divorce rate may be restricted to persons aged 15 and over who are married, and hence potentially vulnerable to becoming divorced (Table 7.2).

Age, sex and duration-specific rates

While the crude and general rates have applications in broad comparative studies, more searching enquiries employ age and sex specific rates, since the timing of marriage is related to the age and sex of individuals. Women have tended to marry at younger ages than men although marriage booms produce younger marriages for both sexes. The timing of marriage is also a basic consideration in explaining fertility trends: early and universal marriage is associated with higher fertility, while late marriage, together with high proportions remaining unmarried, is associated with lower fertility. Early marriage, or pronounced differences in the ages of bride and groom, are also linked with greater risks of divorce.

Age-specific rates, therefore, are valuable for showing whether marriage or divorce is occurring relatively early or late (Table 7.2). The rates are often calculated separately for each sex, because of age differences between spouses; similarly, the denominators are commonly restricted to the unmarried population, to better identify those 'at risk' of marrying:

Age-specific marriage rate (females)

$$= \frac{\text{marriages of females aged } x \text{ to } x+n}{\text{mid-year female population aged } x \text{ to } x+n} \times 1000$$

Age-specific divorce rate (females)

$$= \frac{\text{divorces of females aged } x \text{ to } x+n}{\text{mid-year female population aged } x \text{ to } x+n} \times 1000$$

The marriage rates may also be modified to show first marriage rates or remarriage rates. In societies with high levels of divorce, many marriages are remarriages. For divorce, it is also helpful to have further information on the duration of marriage before divorce, since there is particular concern when divorce occurs early or while there are dependent children to be cared for:

$$\text{Duration-specific divorce rate} = \frac{\text{divorces after } n \text{ years of marriage}}{\text{mid-year population married } n \text{ years}} \times 1000$$

Median age at marriage and divorce

Summary figures on trends in marriage and divorce may be obtained by calculating the median ages at marriage and divorce, using the same method described for deriving the median age of a population (see Chapter 3). For accurate results, the median age at marriage (or remarriage or divorce) should be calculated using single year ages for the main ages at which these events occur (Shryock and Siegel 1973: 563).

7.8 Conclusion

The analysis of fertility has immediate applications in accounting for population growth and age structure evolution, thereby providing a starting point for projecting future developments. Where and when fertility occurs also has wide implications for levels of child and aged dependency in populations, the provision of hospitals and health facilities, the opening and closure of schools, the nature of the demand for housing, and the spatial patterns of markets for goods and services. Many of the uses of the methods covered in this chapter are in practical and applied areas.

The family also figures prominently in the planning decisions of governments, organizations and businesses, since it is the main unit within which incomes are earned, goods are consumed and individuals are nurtured and supported. The future of entire societies depends substantially on the processes of family formation.

Finally, studies of human behaviour in relation to fertility and the family are essential in understanding social changes and discerning appropriate responses. Thorough analysis of fertility and the family involves more than calculating rates and other indices, although they are key indicators. The elucidation of trends and changes requires research on the characteristics of individuals, the experiences in their life histories and the reasoning behind personal decisions. The social context – including cultural practices, laws and policies – is also important in explaining developments, because of its influence on people's lifestyles and life chances.

Study resources

KEY TERMS

Age-specific divorce rate
Age-specific fertility rate
Age-specific marriage rate
Child–woman ratio
Cohort analysis
Completed fertility
Crude birth rate
Crude divorce rate
Crude marriage rate
Duration-specific divorce rate
Familism
Family
Fertility
Fertility differentials
General divorce rate
General fertility rate
General marriage rate
Gross reproduction rate
Marriage
Median age at divorce
Median age at marriage
Natural fertility
Net reproduction rate
Period analysis
Proximate determinants of fertility
Replacement-level fertility
Second demographic transition
Synthetic cohort
Total fertility rate
Total marital fertility rate

FURTHER READING

Campbell, Arthur A. 1983. *Manual of Fertility Analysis*. London: Churchill Livingstone, for the World Health Organization.

Coleman, D. A. 1988. 'Reproduction and Survival in an Unknown World'. *NIDI Hofstee Lecture Series*, 15. The Hague: Netherlands Interdisciplinary Demographic Institute.

Hinde, Andrew. 1998. *Demographic Methods*. London: Arnold, Chapter 8, pp. 95–106.

Lucas, David and Meyer, Paul (editors). 1994. *Beginning Population Studies*. Canberra: National Centre for Development Studies, The Australian National University.

Palmore, James A. and Gardner, Robert W. 1983. *Measuring Mortality, Fertility and Natural Increase: a Self-Teaching Guide to Elementary Measures*. Honolulu: East-West Population Institute., East-West Center, Chapter 3, pp. 59–119.

Pollard, A. H., Yusuf, Farhat, and Pollard, G. N. 1990. *Demographic Techniques* (third edition). Pergamon Press, Sydney, Chapter 6, pp. 81–103.

van de Kaa, Dirk J. 1987. 'Europe's Second Demographic Transition'. *Population Bulletin*, 42 (1).

Weeks, John R. 2002. *Population: an Introduction to Concepts and Issues*. Belmont, CA: Wadsworth, Chapter 5, 'Fertility Concepts and Measurements', pp. 165–202.

Internet resources

Subject	Source and Internet address
Document on *World Population Monitoring*	***United Nations Population Division*** http://www.un.org/esa/population/unpop.htm
National and regional statistics, including some fertility data, for the United Kingdom	***HM Government, National Statistics*** http://www.statsbase.gov.uk/statbase/mainmenu.asp
United States births statistics by state, together with related publications	***US Department of Health and Human Services, National Centre for Health Statistics*** http://www.cdc.gov/nchs/births.htm#Statebirths
Gateway to information about sexual and reproductive health	***International Planned Parenthood Federation*** http://www.ippf.org
Gateway to publications and data from the *Demographic and Health Surveys* – cross-national surveys that are the successors to the *World Fertility Survey*	***Macro International Inc.*** http://www.measuredhs.com
Demographic data for Canada, including fertility statistics	***Statistics Canada*** http://www.statcan.ca/english/Pgdb/People/popula.htm

Note: the location of information on the Internet is subject to change; to find material that has moved, search from the home page of the organization concerned.

EXERCISES

Statistics for fertility calculations, Australia 1996

Age	Total births by age of mother	Female births by age of mother	Total women	Probability of daughters surviving to age of mother
15–19	12509	5988	621542	0.99175
20–24	44837	21807	694273	0.98985
25–29	82782	40278	709746	0.98792
30–34	76435	37227	720453	0.98566
35–39	31864	15359	727555	0.98261
40–44	5113	2470	672182	0.97826
45–49	128	61	640985	0.97152

1 Distinguish between the following measures of fertility:

(a) CBR
(b) CWR
(c) GFR
(d) ASFR
(e) TFR
(f) GRR
(g) NRR

2 Using the statistics in the above table, calculate:

(a) ASFRs for ages 15–19 to 45–49

(b) TFR

(c) GRR

(d) NRR

3 Write a brief account of the level of fertility observed and the differences between the measures. Include a line graph of the ASFRs.

4 What is meant by 'replacement-level fertility'?

5 Is there a single TFR that denotes population replacement?

6 Why is the NRR always lower than the GRR?

7 Why are the GRR, the NRR and the TFR sometimes described as age-standardized fertility rates?

SPREADSHEET EXERCISE 7: MEASURES OF FERTILITY

The main aim of this exercise is to give practice in calculating some of the principal measures of fertility. A secondary aim is to provide initial familiarization with life table construction – since this subject area involves concepts that take time to assimilate. Although life tables are the subject of the next chapter, it is possible to dispel some concerns about the complexity of life table work through early use of several simple formulas required in their construction. This exercise explains all of the steps required to complete the life table in Table 1. Alternatively, to omit the life table work, start at step 4 and, in Table 2, type the probabilities of survival in column E instead of using formulas referring to figures in Table 1.

Calculate a single year life table for females aged 15–49 years and, using data on survival from the life table as necessary, obtain the ASFRs, TFR, GRR and NRR. Plot the ASFRs on a line graph. Use formulas in all the unshaded cells of Tables 1 and 2. (Table 1 is derived from Table 8.2, in the next chapter; however the q_x values here are rounded to 5 decimal places, resulting in slight differences in the other figures.)

	A	B	C	D	E	F
1	Table 1: Life Table for Females Aged 15-49 in a Contemporary Western Population (female life expectancy at birth = 78.3 years)					
2						
3	Age	Number surviving at exact age x	Number of deaths between exact ages	Probability of surviving between exact ages	Probability of dying between exact ages	Average number alive between exact ages
4		l_x	d_x	p_x	q_x	L_x
5	15	98699	31	0.99969	0.00031	98684
6	16	98668	37	0.99962	0.00038	98650
7	17	98631	44	0.99955	0.00045	98609
8	18	98587	47	0.99952	0.00048	98563
9	19	98539	48	0.99951	0.00049	98515
10	20	98491	48	0.99951	0.00049	98467
11	21	98443	48	0.99951	0.00049	98419
12	22	98394	48	0.99951	0.00049	98370
13	23	98346	48	0.99951	0.00049	98322
14	24	98298	48	0.99951	0.00049	98274
15	25	98250	48	0.99951	0.00049	98226
16	26	98202	49	0.99950	0.00050	98177
17	27	98153	49	0.99950	0.00050	98128
18	28	98104	49	0.99950	0.00050	98079
19	29	98054	50	0.99949	0.00051	98029
20	30	98004	51	0.99948	0.00052	97979
21	31	97954	54	0.99945	0.00055	97927
22	32	97900	57	0.99942	0.00058	97871
23	33	97843	61	0.99938	0.00062	97813
24	34	97782	66	0.99933	0.00067	97749
25	35	97717	71	0.99927	0.00073	97681
26	36	97645	78	0.99920	0.00080	97606
27	37	97567	87	0.99911	0.00089	97524
28	38	97480	97	0.99901	0.00099	97432
29	39	97384	107	0.99890	0.00110	97330
30	40	97277	119	0.99878	0.00122	97217
31	41	97158	132	0.99864	0.00136	97092
32	42	97026	147	0.99849	0.00151	96953
33	43	96879	163	0.99832	0.00168	96798
34	44	96717	181	0.99813	0.00187	96626
35	45	96536	200	0.99793	0.00207	96436
36	46	96336	220	0.99772	0.00228	96226
37	47	96116	242	0.99748	0.00252	95995
38	48	95874	266	0.99723	0.00277	95741
39	49	95609	292	0.99695	0.00305	95463
40						

The life table for females aged 15–49

1. Enter the ages for Table 1, the life table, by using the Series feature: i.e. type the first two ages, highlight these two cells, click the mouse on the small square at the bottom right of the cursor and drag the mouse to fill the required cells. Excel will calculate a series based on the interval between the first two numbers you typed. Type the other life table values given in the shaded cells of Table 1. Use the numeric keypad on the right of the keyboard to enter the q_x values (the probability of dying between exact ages or birthdays), together with the first figure in the second column, the number alive at exact age 15 (l_{15}).
2. Complete the rest of the life table using the formulas given below. Note that it is necessary to type only the first formula in each column; they are then copied to the cells for the other ages.
   - p_x (probability of surviving between exact ages or birthdays): $p_x = 1 - q_x$. In Excel, the formula in cell D5 is: =1 – E5.
   - d_x (number of deaths between exact ages): $d_x = l_x \times q_x$. Excel formula in C5: =B5*E5. When this formula is copied down, there will be zeros in the cells until the l_x values in column B are completed.

- l_x (number of survivors, from an original 100 000 live births, at each exact age): $l_{x+1} = l_x \times p_x$. Excel formula in B6: =B5*D5. In the formula, the subscript $x + 1$ refers to age 16 when $x = 15$.
- L_x (average number living between two exact ages): $L_x = l_x - 0.5 \times d_x$. Excel formula in F5: =B5 – 0.5*C5.

3. If you wish to rule lines across the table, use the Borders button on the Formatting Toolbar. Use Edit Repeat to rule other lines. The Edit Repeat command is useful for formatting numbers or text as well as for other operations where you want to repeat steps without going through the menus. It is preferable to rule lines when the spreadsheet is completed to avoid copying unwanted lines.

	A	B	C	D	E	F	G	H
41	Table 2: Fertility Statistics for a Contemporary Western Population (fictitious data)							
42								
43	Age	Total births by age of mother	Female births by age of mother	Total women	Probability of daughters surviving to age of mother	ASFRs per 1000	Female ASFRs /1000 women	Expected survivors of female births /1000 women
44	15-19	17500	8300	632000	0.98609	27.69	13.13	12.95
45	20-24	68300	33300	657000	0.98370	103.96	50.68	49.86
46	25-29	89800	43500	620000	0.98128	144.84	70.16	68.85
47	30-34	48800	23900	606000	0.97871	80.53	39.44	38.60
48	35-39	13400	6600	526000	0.97524	25.48	12.55	12.24
49	40-44	1900	900	423000	0.96953	4.49	2.13	2.06
50	45-49	100	50	364000	0.95995	0.27	0.14	0.13
51	Total	239800		3828000		387.26	188.23	184.69
52	Multiply by 5					1936.28	941.15	923.44
53	Rate per woman:					1.94	0.94	0.92
54	TFR	1.94						
55	GRR	0.94						
56	NRR	0.92						
57								

Calculating the fertility rates

4. Type the headings for Table 2, and then copy the age group labels from your spreadsheet for Exercise 3 or 4 – to save having to type them again and to practise copying between spreadsheets. To do this, open the file for Exercise 3, then choose Window Arrange Tiled to see the two spreadsheets at the same time; highlight the information to be copied, then use the Copy and Paste buttons to transfer the information.

5. Complete Table 2 to obtain values for TFR, GRR and NRR, placing formulas in the unshaded cells. The probabilities in column E of Table 2 are the L_x values in the middle of each five year age group, divided by 100 000 (Table 1). For example the formula in cell E44 in Table 2 is: =F7/100 000 (i.e. 0.986 09). An alternative is to calculate the average of the single year L_xs for each age group and divide by 100 000, as mentioned in the discussion of the NRR in Section 7.5.

6. To view figures in widely separated sections of the spreadsheet simultaneously, split the screen by clicking the mouse pointer on the small box at the top of the vertical scroll bar; hold the mouse button down and drag the box to the point where you want the screen divided. To restore the screen to one window, reverse these steps, or double-click on the dividing line, or click Window Remove Split.

7. Draw a line graph of the ASFRs based on total births. Use the Chart Wizard and follow the instructions on the screen.

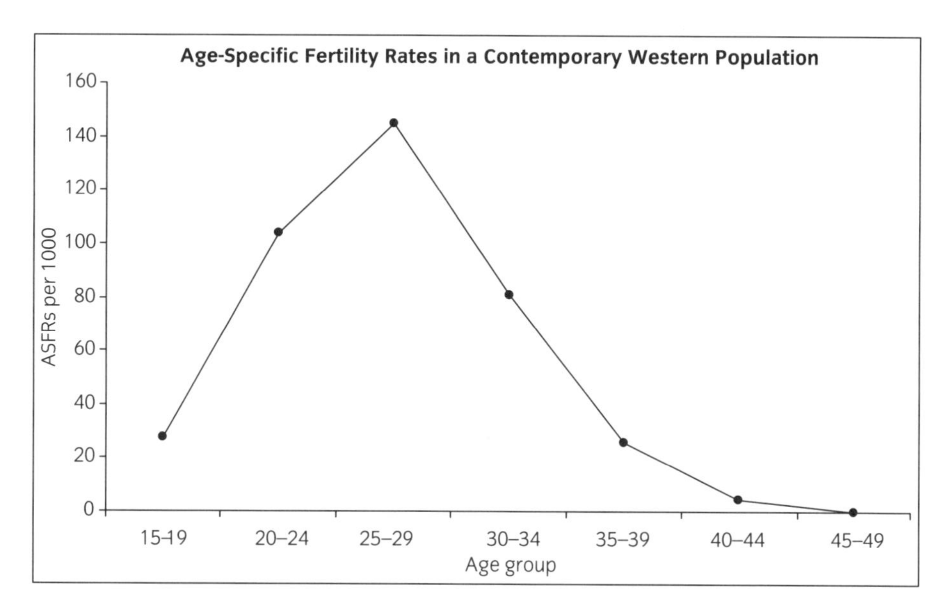

Section IV

Demographic models

This section discusses the demographic models of widest interest in the analysis of populations. The chapters cover the construction and applications of stationary and stable population models. Specific sections include the calculation of single year and abridged life tables, the derivation of intrinsic growth rates and stable age distributions, and the nature of population momentum and population ageing.

Chapter 8 **Life tables**

Chapter 9 **Stable and stationary models**

8 Life tables

Whereas we have found, that of 100 quick Conceptions about 36 of them die before they be six years old, and that perhaps but one surviveth 76, we, having seven *Decads* between six and 76, we sought six mean proportional numbers between 64, the remainder, living at six years, and the one, which survives 76, and finde, that the numbers following are practically near enough to the truth; for men do not die in exact Proportions, nor in Fractions: from whence arises this Table following.

(Graunt 1662: 69)

OUTLINE

8.1 The first life tables

8.2 Types of life tables

8.3 Constructing complete life tables

8.4 Constructing abridged life tables

8.5 Conclusion

Study resources

LEARNING OBJECTIVES

To understand:

- the nature of period and cohort life tables
- the meaning of the main life table functions

To obtain skills in:

- calculating complete life tables
- calculating abridged life tables

COMPUTER APPLICATIONS

Excel module:	Computer-assisted learning exercise on abridged life tables (Box 8.3)
Spreadsheet Exercise 8:	Abridged life tables (see Study Resources)

The life table has been one of demography's most influential discoveries. It examines the toll of mortality, measuring life expectancy and the extent to which death diminishes population numbers as ages increase. Although seemingly preoccupied with morbid concerns, the life table is an important measure of progress, a valid indicator of whether populations are achieving the goal of long life for all. Sometimes viewed as a 'death table', it is equally concerned with survival and length of life.

How many survive to successive ages is important information for the life insurance industry, superannuation funds and pension schemes, since the viability of financial arrangements depends on knowing the likelihood that clients will live to older ages. The life table is an essential foundation for estimates and projections in these enterprises. It is also widely used in the study of the survival of plant, animal and insect populations, as diverse as grasses, forest trees, mice and fruit flies. The versatility and relevance of the life table in measuring the survival or endurance of animate, and even inanimate, objects – such as engines and machine parts – has assured it an enduring role in many areas of inquiry outside demography. For this reason the life table is one of the best-known and most appreciated inventions of demography.

Skills in life table construction are also a valuable asset for people who work with population statistics. They not only permit the calculation of life tables and life table figures, such as for estimates and projections, but also facilitate understanding of important measures based on life table techniques. Research and applied work commonly need to employ life tables, or life table methods, to supply assumptions or answer questions. Life tables have been important in demography from its earliest beginnings, and today they have a central role in the main areas of inquiry, including studies of fertility, mortality and migration, as well as population growth and structure. This chapter covers the essentials of life table construction and includes a self-teaching computer module to facilitate learning the methods of calculating abridged life tables.

8.1 The first life tables

For those who have read the earlier chapters in this book, it is probably no surprise to learn that the concept of a life table was the brainchild of John Graunt (1620–74), although the term 'life table' itself was a later innovation. Graunt's original table, reproduced in Table 8.1, shows the numbers surviving at successive ages out of 100 'quick conceptions' or live births. According to his figures, only 25 per cent were living at age 26, and 1 per cent at 76.

The importance of Graunt's table derives not from the statistics themselves, but from the concept of a table using mortality data to obtain the proportions surviving at each age – which Graunt evidently originated. Regarding the statistics in the table, however, there has been controversy over their authorship and how they

Table 8.1 Graunt's life table for London

From whence it follows, that of the said 100 conceived there remains alive at six years end 64.	
At Sixteen years end.......40	At Fifty six......................6
At Twenty six 25	At Sixty six 3
At Tirty six 16	At Seventy six 1
At Fourty six 10	At Eighty 0

Note: original spelling used.
Source: Graunt (1662: 69–70).

were calculated, some having attributed them to William Petty (1623–87), an eminent friend of Graunt (Willcox 1939: xi–xii; Glass 1950).

The methods of life table construction needed further development but, within two decades of Graunt's death, Edmund Halley (1656–1742) produced a more rigorous mathematical approach foreshadowing methods used today. Halley is best remembered for his prediction of the return of the comet that bears his name, but he contributed to many areas of scientific inquiry. Through his life table work, based on the Bills of Mortality for the town of Breslau in Germany, Halley is credited, along with Graunt and Malthus, as one of demography's earliest pioneers. Halley was also among the first to attribute the authorship of the whole of Graunt's *Observations* to William Petty (Willcox 1939: iv). After Graunt's death in 1674, Petty had published a new and enlarged edition of Graunt's 1662 monograph, which may explain much of the confusion about authorship (Hewins 1896, cited in Willcox 1939: vi).

8.2 Types of life tables

Period life tables

In constructing his life table, Halley made the key assumption that the size of the population was stationary, or constant. He noted that the recorded numbers of births and deaths in Breslau were approximately equal and that migration was unimportant. Much later it was realized that real populations are rarely stationary (Cox 1970: 299), but Halley's assumption has persisted as a foundation for life table construction. There are three reasons for this: (i) it simplifies the calculation of important measures of mortality and survival; (ii) it permits comparisons between life tables for different places without having to make allowances for differences in age structure (Pressat 1985: 124); (iii) the resulting figures are readily understood and adequate for many purposes.

Today, the concept of a *stationary population* describes a population with the following characteristics:

1. *Constant size* The total numbers do not change because the number of births is equal to the number of deaths: balancing every 100 000 births, there are 100 000 deaths. It follows that the population's crude birth and death rates are also constant and equal, and the growth rate is zero. A population experiencing ZPG (zero population growth) will be a stationary population, unless there is a complex balance between natural increase and net migration, such as if net immigration offsets natural decrease.
2. *Constant age structure* The numbers and percentages are unchanging at every age. Although the shape of the age structure is constant, different stationary populations have different age structures: their shape depends on their mortality rates. In a stationary population where very many die young, the age structure is triangular, because relatively few survive to older ages. Conversely, in stationary populations where long-life is typical, the stationary population is fairly rectangular in form. There is no single age structure that is characteristic of a stationary population or of a population experiencing zero population growth (see Chapter 9, Box 9.1).
3. *Closed to migration* The population experiences no inward or outward migration, and it is therefore maintained solely through natural increase. A population with no migration arrivals or departures is described as a *closed population*. In contrast, an *open population* is one experiencing inward and outward migration. Technically, a stationary population need not be closed, provided inward and outward migration at each age balance exactly, which is unlikely.

The concept of a stationary population is the basis for *period life tables*, which are derived from the age-specific mortality rates for one year, or average figures for a longer period, such as three years. A period life table represents the experience of a *synthetic or hypothetical cohort* located in a single year: the age-specific mortality rates for one year represent the experience of the cohort. The synthetic cohort's mortality rates are a combination of rates from many different real birth cohorts during the year.

If a population is stationary, the numbers born and the numbers alive at each age remain constant. This is illustrated in Figure 8.1 – a Lexis diagram (see Chapter 4) for the youngest ages – which shows the numbers alive at each 'exact' age in the year 2000, including exactly 0 (i.e. birth), exactly 1 (the first birthday), exactly 2 (the second birthday) and so on. For brevity and mathematical convenience in life table work, the numbers alive at each exact age are referred to by the notation l_x. 'Little l_x', or 'lower case l_x', refers to the numbers alive at exact ages, while the subscript x denotes the age. Thus l_0, ('little l zero') is the number alive at birth, l_1 is the number alive at the first birthday and l_2 is the number alive at the second birthday. The l_xs derive from age-specific death rates for the population in a particular period. Thus, in Figure 8.1, the l_xs reflect the mortality experience in 2000 of cohorts born in many different years: the 1 year olds were born in 1999, the 2 year olds in 1998, and the 100 year olds in 1900 (see Table 8.2).

The numbers alive at each exact age appear on the horizontal lines of Figure 8.1. In the year 2000, we assume that there were 100 000 live births. Mortality rates for the year 2000 imply that, out of each 100 000 born, there will be 99 092 alive on their first birthday (i.e. about 99 per cent), 99 008 on their second birthday and so on. Since the population is stationary, we can project numbers forward in time by inserting identical figures for the years 2001–2004. As discussed later, the actual age structure of the stationary population is represented by the average number alive between exact ages (${}_nL_x$), rather than by the numbers at exact ages.

Although the assumption of a stationary population is unrealistic, the life table provides a valuable means of assessing, in detail, the nature of mortality in a given period. Also, being able to make calculations on the basis of a quick projection of mortality is very useful; it need not strain the bounds of credibility if change is fairly slow, or if the aim is to demonstrate long run implications of holding a given situation constant.

Cohort life tables

Life tables based on the observed mortality of *real cohorts* are also possible. These are termed *cohort life tables*, or *generational life tables*, since they describe the mortality through time of actual cohorts rather than mortality in one period. Often the data for such tables are incomplete, and necessarily so for cohorts that are still alive.

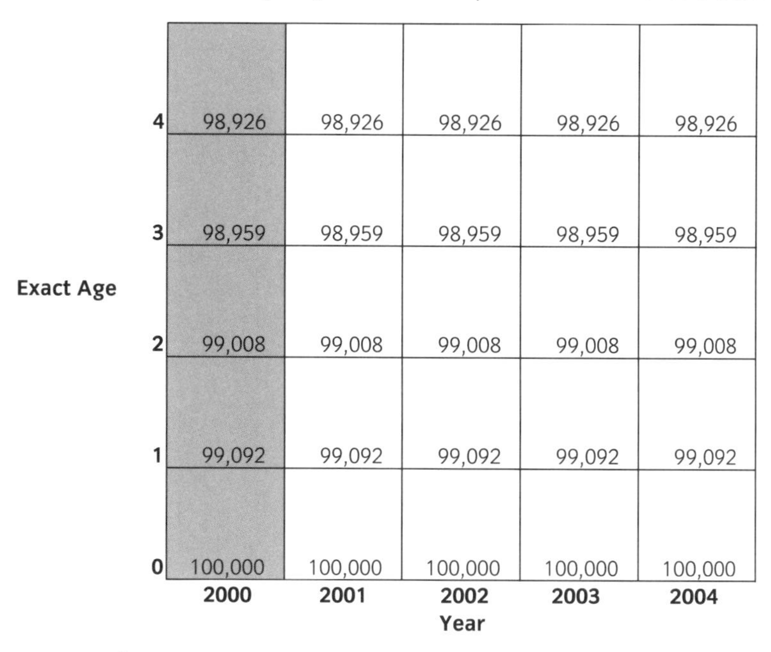

Figure 8.1 Lexis diagram showing numbers alive at exact ages in a stationary population.

Note: shaded cells based on the observed data for the year 2000.
Data source: Table 8.2.

Consequently, projections of mortality are required to complete the cohort life tables. Figure 8.2 compares cross-sectional (period) and real cohort life expectancies for females in Asia. The latter are based on observed data through time, together with projections. Life expectancy at birth is the average length of life. While both the cross-sectional and real cohort figures show a pronounced improvement in life expectancy, the real cohort measures are appreciably higher, because they reflect actual and anticipated improvements in survival during people's lifetimes.

Complete, or unabridged, life tables provide information on single years of age, whereas abridged life tables provide a summary for selected ages. The next two sections explain the methods of constructing both complete and abridged life tables. The Excel module on the epidemiologic transition (Chapter 6, Box 6.1) presents graphs that help to clarify the nature of the *life table functions* – the measures of mortality and survival in the life table – and the differences between them (see also Chapter 9, Box 9.1 on *West models*).

8.3 Constructing complete life tables

Actuaries and mathematicians use complex methods of life table construction, but straightforward methods can produce similar results that are adequate for general use. Starting with mortality rates from an actuary's life table, the example in Table 8.2

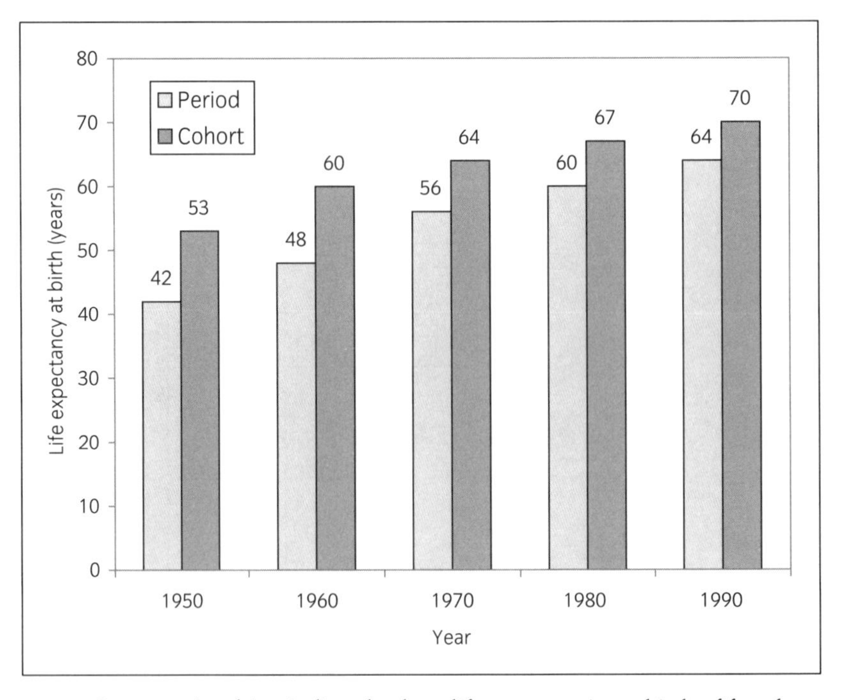

Figure 8.2 Cross-sectional (period) and cohort life expectancies at birth of females in Asia.

Data source: Rowland (1994b: 29).

Table 8.2 A complete (single year) life table for a Western female population, 2000

Age	$_nM_x$ per 1000	q_x	p_x	l_x	d_x	L_x	T_x	e_x
0	9.12160	0.00908	0.99092	100000	908	99364	7826402	78.26
1	0.84807	0.00085	0.99915	99092	84	99042	7727038	77.98
2	0.49502	0.00049	0.99951	99008	49	98983	7627996	77.04
3	0.33352	0.00033	0.99967	98959	33	98942	7529013	76.08
4	0.27296	0.00027	0.99973	98926	27	98912	7430070	75.11
5	0.23258	0.00023	0.99977	98899	23	98887	7331158	74.13
6	0.20229	0.00020	0.99980	98876	20	98866	7232271	73.14
7	0.19221	0.00019	0.99981	98856	19	98846	7133405	72.16
8	0.19225	0.00019	0.99981	98837	19	98827	7034558	71.17
9	0.18216	0.00018	0.99982	98818	18	98809	6935731	70.19
10	0.18219	0.00018	0.99982	98800	18	98791	6836922	69.20
11	0.18223	0.00018	0.99982	98782	18	98773	6738131	68.21
12	0.19239	0.00019	0.99981	98764	19	98754	6639358	67.22
13	0.21268	0.00021	0.99979	98745	21	98734	6540603	66.24
14	0.25325	0.00025	0.99975	98724	25	98711	6441869	65.25
15	0.31411	0.00031	0.99969	98699	31	98683	6343157	64.27
16	0.38518	0.00039	0.99961	98668	38	98649	6244474	63.29
17	0.44618	0.00045	0.99955	98630	44	98608	6145825	62.31
18	0.47682	0.00048	0.99952	98586	47	98563	6047217	61.34
19	0.48721	0.00049	0.99951	98539	48	98515	5948654	60.37
20	0.48744	0.00049	0.99951	98491	48	98467	5850139	59.40
21	0.48768	0.00049	0.99951	98443	48	98419	5751672	58.43
22	0.48792	0.00049	0.99951	98395	48	98371	5653253	57.45
23	0.48816	0.00049	0.99951	98347	48	98323	5554882	56.48
24	0.48840	0.00049	0.99951	98299	48	98275	5456559	55.51
25	0.48864	0.00049	0.99951	98251	48	98227	5358284	54.54
26	0.49907	0.00050	0.99950	98203	49	98179	5260057	53.56
27	0.49932	0.00050	0.99950	98154	49	98130	5161879	52.59
28	0.49957	0.00050	0.99950	98105	49	98081	5063749	51.62
29	0.51002	0.00051	0.99949	98056	50	98031	4965669	50.64
30	0.52049	0.00052	0.99948	98006	51	97981	4867637	49.67
31	0.55140	0.00055	0.99945	97955	54	97928	4769657	48.69
32	0.58236	0.00058	0.99942	97901	57	97873	4671729	47.72
33	0.62360	0.00062	0.99938	97844	61	97814	4573856	46.75
34	0.67515	0.00067	0.99933	97783	66	97750	4476043	45.78

Table 8.2 *continued*

Age	${}_nM_x$ per 1000	q_x	p_x	l_x	d_x	L_x	T_x	e_x
35	0.72681	0.00073	0.99927	97717	71	97682	4378293	44.81
36	0.79907	0.00080	0.99920	97646	78	97607	4280611	43.84
37	0.89203	0.00089	0.99911	97568	87	97525	4183004	42.87
38	0.99549	0.00099	0.99901	97481	97	97433	4085480	41.91
39	1.09927	0.00110	0.99890	97384	107	97331	3988047	40.95
40	1.22398	0.00122	0.99878	97277	119	97218	3890717	40.00
41	1.35944	0.00136	0.99864	97158	132	97092	3793499	39.04
42	1.50578	0.00150	0.99850	97026	146	96953	3696407	38.10
43	1.68380	0.00168	0.99832	96880	163	96799	3599454	37.15
44	1.87305	0.00187	0.99813	96717	181	96627	3502655	36.22
45	2.07374	0.00207	0.99793	96536	200	96436	3406029	35.28
46	2.28609	0.00228	0.99772	96336	220	96226	3309592	34.35
47	2.52075	0.00252	0.99748	96116	242	95995	3213366	33.43
48	2.76762	0.00276	0.99724	95874	265	95742	3117371	32.52
49	3.04801	0.00304	0.99696	95609	291	95464	3021629	31.60
50	3.34149	0.00334	0.99666	95318	318	95159	2926166	30.70
51	3.64844	0.00364	0.99636	95000	346	94827	2831006	29.80
52	3.99052	0.00398	0.99602	94654	377	94466	2736179	28.91
53	4.35799	0.00435	0.99565	94277	410	94072	2641713	28.02
54	4.76231	0.00475	0.99525	93867	446	93644	2547641	27.14
55	5.19386	0.00518	0.99482	93421	484	93179	2453997	26.27
56	5.68611	0.00567	0.99433	92937	527	92674	2360817	25.40
57	6.20842	0.00619	0.99381	92410	572	92124	2268143	24.54
58	6.79514	0.00677	0.99323	91839	622	91528	2176019	23.69
59	7.42672	0.00740	0.99260	91217	675	90879	2084491	22.85
60	8.12774	0.00809	0.99191	90542	733	90175	1993612	22.02
61	8.89053	0.00885	0.99115	89809	795	89411	1903437	21.19
62	9.74129	0.00969	0.99031	89014	863	88582	1814026	20.38
63	10.68500	0.01063	0.98937	88151	937	87682	1725443	19.57
64	11.73947	0.01167	0.98833	87214	1018	86705	1637761	18.78
65	12.91226	0.01283	0.98717	86196	1106	85643	1551056	17.99
66	14.22468	0.01412	0.98588	85090	1202	84489	1465413	17.22
67	15.71228	0.01559	0.98441	83888	1308	83235	1380923	16.46
68	17.35403	0.01720	0.98280	82581	1421	81870	1297689	15.71
69	19.16595	0.01898	0.98102	81160	1541	80390	1215818	14.98
70	21.20612	0.02098	0.97902	79619	1671	78784	1135429	14.26

Table 8.2 *continued*

Age	$_nM_x$ per 1000	q_x	p_x	l_x	d_x	L_x	T_x	e_x
71	23.436 28	0.023 16	0.976 84	77 948	1 806	77 046	1 056 645	13.56
72	25.963 66	0.025 63	0.974 37	76 143	1 952	75 167	979 599	12.87
73	28.830 38	0.028 42	0.971 58	74 191	2 109	73 137	904 432	12.19
74	32.102 59	0.031 60	0.968 40	72 083	2 277	70 944	831 296	11.53
75	35.834 56	0.035 20	0.964 80	69 805	2 457	68 576	760 352	10.89
76	40.096 91	0.039 31	0.960 69	67 348	2 647	66 024	691 775	10.27
77	44.964 77	0.043 98	0.956 02	64 700	2 845	63 278	625 751	9.67
78	50.473 92	0.049 23	0.950 77	61 855	3 045	60 332	562 474	9.09
79	56.711 30	0.055 15	0.944 85	58 810	3 243	57 188	502 141	8.54
80	63.736 96	0.061 77	0.938 23	55 567	3 432	53 851	444 953	8.01
81	71.611 61	0.069 14	0.930 86	52 134	3 604	50 332	391 102	7.50
82	80.388 33	0.077 28	0.922 72	48 530	3 750	46 655	340 770	7.02
83	90.151 69	0.086 26	0.913 74	44 780	3 863	42 848	294 115	6.57
84	100.870 32	0.096 03	0.903 97	40 917	3 929	38 952	251 267	6.14
85	112.564 62	0.106 57	0.893 43	36 988	3 942	35 017	212 315	5.74
86	125.257 33	0.117 87	0.882 13	33 046	3 895	31 098	177 298	5.37
87	138.929 67	0.129 91	0.870 09	29 151	3 787	27 257	146 200	5.02
88	153.574 92	0.142 62	0.857 38	25 364	3 617	23 555	118 943	4.69
89	169.229 23	0.156 03	0.843 97	21 746	3 393	20 050	95 388	4.39
90	185.871 83	0.170 07	0.829 93	18 353	3 121	16 793	75 338	4.10
91	203.418 06	0.184 64	0.815 36	15 232	2 812	13 826	58 545	3.84
92	222.053 03	0.199 86	0.800 14	12 420	2 482	11 179	44 719	3.60
93	241.698 67	0.215 64	0.784 36	9 937	2 143	8 866	33 541	3.38
94	262.240 30	0.231 84	0.768 16	7 795	1 807	6 891	24 675	3.17
95	283.832 79	0.248 56	0.751 44	5 987	1 488	5 243	17 784	2.97
96	306.410 26	0.265 70	0.734 30	4 499	1 195	3 901	12 541	2.79
97	329.809 73	0.283 12	0.716 88	3 304	935	2 836	8 639	2.61
98	354.166 67	0.300 88	0.699 12	2 368	713	2 012	5 803	2.45
99	379.656 16	0.319 08	0.680 92	1 656	528	1 392	3 791	2.29
100	406.150 58	0.337 59	0.662 41	1 127	381	937	2 400	2.13
101	434.571 89	0.357 00	0.643 00	747	267	614	1 462	1.96
102	462.121 21	0.375 38	0.624 62	480	180	390	849	1.77
103	491.869 92	0.394 78	0.605 22	300	118	241	459	1.53
104	832.500 00	1.000 00	0.000 00	182	182	218	218	1.20
Total					100 000	7 826 402		

Notes: Shaded cells denote values or formulas different from those in the rest of the column; the last age interval is open-ended, i.e. ages 104 and over.

calculated the same life table using the simple formulas to be described in this section. The formulas produced almost identical figures on life expectancy at every age under 98. The differences, however, would have been greater if the new table had been based on actual mortality rates, rather than the smoothed mortality data that the actuary used. This illustrates that results may vary more because of the nature of the input data than because of the methods of calculation.

At first sight, life tables appear complicated because of the unfamiliar notation for labelling different measures, such as q_x and L_x. These are no more difficult to master than half a dozen words in a foreign language, but to remember them it is necessary to understand what they mean and how they differ from each other. Table 8.3 lists definitions of the life table functions.

Life table construction mainly entails repeating some simple arithmetical operations many times. Without the aid of a computer, preparing a life table for single years of age is a daunting prospect. There are many hundreds of calculations to perform, and the chances of inaccuracy are high – one mistake can create a chain of errors through the rest of the table. Yet when using a spreadsheet, the labour required to construct a two-sex life table for a hundred or more single years of age is minimal – provided the mortality data are adequate – and the opportunities to avoid or detect errors are increased.

Table 8.3 Definitions of life table functions

Function	
l_x	The number alive at exact age x, out of the original number of births.
${}_nq_x$	The probability of dying between exact ages x and $x + n$.
${}_np_x$	The probability of surviving from exact age x to exact age $x + n$.
${}_nd_x$	The number of deaths between exact ages x and $x + n$.
${}_nL_x$	The average number alive in the interval between exact ages x and $x + n$, i.e. the age distribution of the stationary population. It also denotes the number of person-years lived in the interval between exact ages x and $x + n$.
T_x	The total population aged x and over, or the total number of person-years lived from exact age x.
e_x	The expectation of life at exact age x, i.e. the average number of years lived by a person from exact age x.

Notes:
1. n is the number of years in the age interval; when $n = 1$, the initial subscript is usually omitted.
2. Exact age = age on a birthday.
3. *Person-years* are the sum of all the years members of a population have lived, during a fixed interval or over their whole lives. If a population has 1000 people alive on their fifth birthday, and all survived to their tenth birthday, they would live 5000 person-years between exact ages 5 and 10. Thus, person-years lived are calculated as the product of the number of people and the number of years they have lived (see Palmore and Gardner 1983: 3).

A useful strategy in constructing a new life table, or any spreadsheet, is to replicate results from a textbook example before entering your own data. Recreating previous work is one of the best ways to ensure accurate results. Another benefit of the spreadsheet approach is that once an accurate spreadsheet is completed, further life tables can be obtained quickly – by copying the spreadsheet and typing in a new set of age-specific death rates, since the whole spreadsheet depends on these initial values. Having created a life table for females, for example, a male life table is made by copying the spreadsheet, pasting it in another location, and replacing the female age-specific death rates with figures for males: the rest of the life table updates automatically.

Much of life table construction merely entails typing in a formula for one age, then copying it to all, or most of, the cells for the other ages. The exceptions are some of the formulas for the youngest and oldest ages. Recognizing where the few exceptions occur – that is in the shaded cells in Table 8.2 – is one of the main prerequisites for accurate results.

The rest of this section provides a step-by-step approach to understanding life table construction, highlighting the few difficulties. An effective way to understand the material is to create a spreadsheet of Table 8.2, adding columns as each section is read. Alternatively, if a computer is not available, recreate the first and last five lines (ages 0 to 4 and 100 to 104+) with a calculator. Since spreadsheets retain many decimal places in intermediate calculations, rounding errors will cause slightly different results when using a calculator.

Age-specific death rate (M_x)

Separate life tables are usually calculated for males and females, because there are marked differences in mortality patterns between the sexes. The starting point for life table construction is to obtain information on the number of male and female deaths in each single year of age, and calculate age-sex-specific death rates. Where possible, the denominator for the rates should be a census year, to avoid the inaccuracies of deriving rates from estimates. Also, to reduce the effects of short-term fluctuations in mortality or deficiencies in the data, life tables are preferably based on the average number of deaths in each age over a period longer than a year, such as the three years centred on a census year.

In life table calculations, the letter M denotes the age-specific death rate; it is capitalized when the rate derives from observed data for the population, while a lower case m denotes an age-specific death rate calculated from the life table itself, as discussed later. Capitals are also used in the formula for M_x to denote observed deaths (D) and observed population size (P). The subscript x stands for an age, such as 80 years. Accordingly, the age-specific death rate is defined as:

$$M_x = \frac{D_x}{P_x} \times k \quad \text{e.g.} \quad M_{80} = \frac{D_{80}}{P_{80}} \times 1000$$

where:

M_x = age-specific death rate of persons aged x
D_x = deaths in a year at age x
P_x = mid-year population aged x
k = 1 or 1000.

Probability of dying between exact ages (q_x)

The bridge between the observed data on mortality and the life table itself is provided by the q_x function. Based on the age-specific death rates, q_x denotes the probability of dying between pairs of exact ages. Thus, q_{80} denotes the probability of dying between exact age 80 and exact age 81.

The probability of dying is the first life table function calculated, and all the others follow from it. Using the age-specific death rates, life table probabilities of dying can be obtained as:

$$q_x = \frac{2M_x}{2 + M_x}$$

where:

q_x = the probability of dying, between exact ages x and $x + 1$
M_x = the age-specific death rate of persons aged x last birthday.

For example

$$q_{80} = \frac{2M_{80}}{2 + M_{80}}$$

In Table 8.2,

$$q_{80} = \frac{2(63.73696/1000)}{2+(63.73696/1000)} = 0.06177$$

More refined approaches to calculating q_x seek to take account of the uneven distribution of deaths between ages, especially for infants and the very old (Woods 1979: 51; Newell 1988: 69; Preston et al. 2001: 38ff). One alternative for q_0 is to use the infant mortality rate instead of the above formula, since it is based on the initial population – the number of live births (Palmore and Gardner 1983: 126). Note that the probability of dying in an open-ended age interval is always 1; in other words $_{\text{infinit}}q_x$ (or $_\infty q_x$) = 1. Thus $_\infty q_{104}$ is shaded in Table 8.2, to denote an exception in the derivation of the q_x column.

Although M_x (per person) and q_x have similar values where death rates are low, they are different concepts. M_x is a rate based on the mid-year population, while q_x is a probability based on the initial population. The formula for q_x is explained further in Box 8.1.

Probability of surviving from one exact age to the next (p_x)

The complement of the probability of dying is the probability of surviving (p_x). Together they sum to 1 in each age interval, because everyone must either survive or die between birthdays. Since the probability of dying in an open-ended age

BOX 8.1 **Derivation of the formulas for q_x and $_\infty L_x$**

The probability of dying between exact ages x and $x + 1$ (q_x)

$$q_x = \frac{D_x}{P_x + 0.5D_x}$$

where:

D_x = observed number of deaths of persons aged x last birthday in a year
P_x = observed mid-year population

i.e. the probability of dying is equal to the number of deaths divided by the initial population aged x. (Adding half the deaths to the mid-year population provides an estimate of the size of the initial population aged x.)
Divide the numerator and denominator on the right-hand side by P_x:

$$q_x = \frac{D_x/P_x}{1 + 0.5D_x/P_x}$$

Since the age-specific death rate (M_x) is defined as:

$$M_x = \frac{D_x}{P_x}$$

therefore, substituting M_x for (D_x/P_x):

$$q_x = \frac{M_x}{1 + 0.5M_x}$$

Multiplying each term by 2:

$$q_x = \frac{2M_x}{2 + M_x}$$

Average number alive in an open-ended age group ($_\infty L_x$)

The life table age – specific death rate (m_x) is defined as:

$$m_x = \frac{d_x}{L_x}$$

Hence:

$$_\infty m_x = \frac{_\infty d_x}{_\infty L_x} \quad \text{and} \quad _\infty L_x = \frac{_\infty d_x}{_\infty m_x}$$

Since m_x is approximately equal to M_x, and $l_x = {}_\infty d_x$

$$_\infty L_x = \frac{l_x}{_\infty M_x}$$

References: Pollard et al. (1990: 28); Newell (1988: 68–69, 76–77); Palmore and Gardner (1983: 125–127).

range is always 1, the probability of surviving ($_{\infty}p_x$) is always 0. The formula for the probability of surviving is one minus the probability of dying:

$$p_x = 1 - q_x$$

Numbers surviving at exact ages (l_x)

Derived from p_x is an important indicator of the impact of mortality on a population, namely the proportions surviving at each exact age. They are one of the clearest ways of summarizing the toll of mortality at successive ages: for example, if 80 per cent survive to age 85, individuals can look forward with some confidence to a long life, but governments will need to make financial arrangements to support a long-lived population. Figures on the proportions surviving are of great interest in relation to management of life insurance and pension arrangements.

The life table function that provides this information is l_x, representing the number of survivors at exact age x from an original 100 000 live births. Thus l_0, the number alive at exact age zero, is 100 000. This figure is called the *radix* of the life table; it is written into the first cell in the l_x column (shaded in Table 8.2). While this is the most common radix, 1000 or 10 000 are alternatives. In mathematics, the word radix refers to the base of a number system – 10 is the radix of the decimal system. In life table work, the subscript $x + 1$ refers to the age that is one year older than age x: if x is 40 years, $x + 1$ is age 41.

Function l_x is calculated as:

$$l_{x+1} = l_x \times p_x \quad \text{e.g.} \quad l_{41} = l_{40} \times p_{40}$$

In Table 8.2, $l_0 = 100\,000$
and $l_{41} = 97\,277 \times 0.99878 = 97\,158$

If the numbers at each age are survivors from an original 100 000, the percentages surviving at each age are obtained through dividing the l_x values by 1000. For instance, if the number of females alive at exact age 40 (l_{40}) is 25 432, we can say that only 25 per cent of females will live to their 40th birthday – if death rates remained constant through time. This was probably the situation in the most primitive populations, but in long-lived contemporary populations l_{40} for females is around 98 per cent.

It is important to note that the number of survivors in the l_x column of the life table has no relationship to the actual number of people in the population whose death rates were used (Newell 1988: 73). Basing each life table on 100 000 births is intended to make the tables easier to interpret, and to facilitate comparisons between mortality patterns in different places and times.

Finally, l_xs refer to exact ages and are expressed as whole numbers, whereas some other functions refer to intervals and/or are expressed as decimal numbers. These distinctions and conventions are listed in Box 8.2.

BOX 8.2 **Life table functions: types of measures and formatting conventions**

Point and interval measures

Functions referring to exact age x	Functions referring to the interval between exact ages x to $x+n$
l_x	${}_nq_x$
T_x	${}_np_x$
e_x	${}_nd_x$
	${}_nL_x$

Note: when n = 1 year, the left-hand subscript is usually omitted.

Formatting conventions

- Whole numbers for persons (l_x, d_x, L_x and T_x)
- Five decimal places for rates and probabilities (M_x, q_x, p_x)
- Two decimal places for life expectancy (e_x)

Number of deaths between exact ages (d_x)

Just as q_x and p_x complement each other, so too do l_x and the next life table function, d_x. Whereas l_x is the number alive at an exact age, d_x is the number of deaths between exact ages. It is obtained by multiplying l_x, the number alive at the start of the age interval, by the probability of dying:

$$d_x = l_x \times q_x \quad \text{e.g.} \quad d_{60} = l_{60} \times q_{60}$$

In Table 8.2, $d_{60} = 90\,542 \times 0.00809 = 733$.

The number dying in an age interval can also be expressed as the difference between two adjacent l_x values:

$$d_x = l_x - l_{x+1} \quad \text{e.g.} \quad d_{60} = l_{60} - l_{61}$$

For a final open-ended age interval, the number dying is the same as the number alive at the start of the interval:

$${}_\infty d_x = l_x$$

For example, deaths at ages 104 years and over: ${}_\infty d_{104} = l_{104}$.

In Table 8.2, the sum of the d_x column is 100 000. This is because, in a stationary population, the number of births is equal to the number of deaths. A useful check, therefore, is to ensure that the sum of the d_x column is equal to the radix of the life table.

Average number alive between exact ages (L_x)

Although l_x is a valuable measure of mortality patterns, for some purposes it is limited in that it refers to exact ages rather than age last birthday. Censuses and surveys, which are the main sources of statistics on the age distribution of populations, refer to age last birthday. At any point in time, only a minority of people are celebrating their birthday, the rest of the population are between birthdays. Hence, if we require information about the age structure of the life table population, for instance to make estimates using census figures (as in net migration estimates), or even just to plot an age–sex pyramid for the life table population, a life table concept is needed that is closer to the notion of age last birthday.

The relevant life table function here is L_x ('big L_x' or 'capital L_x'), which is defined as the average number alive between exact ages, or the number of person-years lived between exact ages (see Table 8.3). L_x is calculated as the average of the l_x ('little l_x') values at the start and end of the interval:

$$L_x = 0.5(l_x + l_{x+1}) \quad \text{e.g.} \quad L_{60} = 0.5(l_{60} + l_{61})$$

In Table 8.2, $L_{60} = 0.5(90542 + 89809) = 90175.$

The above formula is an approximation, since it assumes that deaths are distributed evenly over the age interval so that the numbers surviving between exact ages follow a straight line rather than a curve. While fairly reasonable for most age intervals, the assumption of linear change is least satisfactory for the very young, since the younger the infant, the greater the risk of death. For example, among the newborn, most deaths occur in the first days and weeks of life, and there is much better survival in the post-neonatal period. To address this problem, different equations are used for L_0 and L_1 (see shaded cells in Table 8.2).

In the absence of empirical data on the relative weighting of the l_x values in the equations, various authors have recommended the following approximations for general use (Barclay 1966: 104; Pollard et al. 1990: 38; Shryock and Siegel 1973: 443; Palmore and Gardner 1983: 43):

$$L_0 = 0.3l_0 + 0.7l_1$$

In Table 8.2, $L_0 = 0.3(100000) + 0.7(99092) = 99364;$
and

$$L_1 = 0.4l_1 + 0.6l_2$$

In Table 8.2, $L_1 = 0.4(99092) + 0.6(99008) = 99042.$

Concerning these equations, Barclay (1966: 104) noted that 'a considerable degree of inaccuracy can be tolerated here without serious distortion'. A more refined approach utilizes the life table function a_x, which denotes the average number of person-years lived by people dying in an age interval (see Hinde 1998: 31). Values of a_x, however, are often unavailable.

Another exception is the equation for L_x when the final age interval is open-ended (i.e. ${}_{\infty}L_x$):

$$ {}_{\infty}L_x = \frac{l_x}{{}_{\infty}M_x} \quad \text{e.g.} \quad {}_{\infty}L_{104} = \frac{l_{104}}{{}_{\infty}M_{104}} $$

In Table 8.2,

$$ {}_{\infty}L_{104} = \frac{182}{(832.5/1000)} = 218 $$

The derivation of this equation is explained in Box 8.1.

Total population aged x and over (T_x)

From the L_x values are obtained the T_x values, defined as the total population aged x years and over. For example, the total life table population is T_0, while the total aged 65 and over is T_{65}. The T_x function has applications in calculating life expectancy, one of the main summary measures of mortality, as well as other estimates based on total population numbers.

Another interpretation of the T_x values (see Table 8.3) is illuminating from the point of view of measuring how many person-years of life are lived from each age. The T_0 value in Table 8.2 shows, for example, that 100 000 births have more than 7.8 million years of living to do! At age 65, they still have over 1.5 million years of living ahead (T_{65}). The latter figure underlines the astronomical nature of the numbers entering into estimates of funding requirements for income support and health care in later ages.

T_x is calculated as the sum of all the L_x values from age x to the highest age in the life table. Thus T_{65} is the sum of the L_x values from L_{65} to the last L_x value in the life table:

$$ T_x = \sum_{t=0}^{\infty} L_{x+t} $$

For example,

$$ T_{65} = \sum_{t=0}^{\infty} L_{65+t} \quad \text{or} \quad T_{65} = L_{65} + L_{66} + L_{67} + \ldots + L_{103} + {}_{\infty}L_{104} $$

In Table 8.2,

$$T_{100} = L_{100} + L_{101} + L_{102} + L_{103} + {}_{\infty}L_{104}$$
$$= 937 + 614 + 390 + 241 + 218 = 2400$$

Where there is a final open-ended age interval, the last T_x is equal to ${}_{\infty}L_x$ (see Table 8.2):

$$T_x = {}_{\infty}L_x \quad \text{e.g.} \quad T_{104} = {}_{\infty}L_{104}$$

In Table 8.2, $T_{104} = 218$.

When the last T_x is known, other T_xs can then be obtained by working backwards from the end of the life table:

$$T_x = T_{x+1} + L_x \quad \text{e.g.} \quad T_{103} = T_{104} + L_{103}$$

In Table 8.2, $T_{103} = 218 + 214 = 459$.

This method is useful, for instance, when manual calculation is necessary.

Table 8.4 Life expectancy at birth in selected regions and countries, c. 2001

Region/country	Males	Females
More developed countries	72	79
Less developed countries	63	66
Sub-Saharan Africa	49	52
Brazil	65	72
Canada	76	81
China	69	73
France	75	83
Germany	74	81
India	60	61
Italy	76	82
Japan	77	84
United Kingdom	75	80
United States	74	80

Source: Population Reference Bureau 2001a.

Expectation of life (e_x)

As foreshadowed, an important application of T_x is to calculate the final life table function to be considered here, namely life expectancy. This is a well-known summary measure of mortality, widely employed in comparisons through time and between countries, especially in relation to life expectancy at birth. The life table function e_x refers to life expectancy from exact age x. For example, e_{60} is life expectancy at exact age 60, or the average number of years lived from the 60th birthday.

The equation for e_x shows that it is equal to the total person years of life lived after exact age x (T_x) divided by the total persons alive at exact age x (l_x). This represents the average number of years lived by people aged x:

$$e_x = \frac{T_x}{l_x} \quad \text{e.g.} \quad e_{65} = \frac{T_{65}}{l_{65}}$$

In Table 8.2,

$$e_{65} = \frac{1551056}{86196} = 17.99$$

Populations in the pre-transition and middle stages of the demographic transition exhibit what is known as *the paradox of the life table*. This refers to a situation where life expectancy at birth is lower than life expectancy at age one, and sometimes even at age five. The paradox of an increase in life expectancy with age reflects the high rates of infant and child mortality in traditional cultures; those who survive the high-risk period have better prospects. Beyond the early years, however, life expectancy declines at every age (see graphs in the module described in Box 6.1).

Table 8.4 illustrates the range of national life expectancies at birth observed in 2001, and Table 8.5 summarizes the formulas for constructing complete life tables.

Other life table rates

While the above steps constitute the main procedures for constructing single year, or complete, life tables, there are several other basic rates derived from the life table, including the life table age-specific death rate, and the crude birth and death rates of the life table population.

Life table age-specific death rates may be expressed as:

$$m_x = \frac{d_x}{L_x}$$

This function approximates M_x: the lower case m denotes that the figures are from the life table, rather than from observed data.

The crude birth rate of the stationary population is equal to the total births divided by the total population:

Table 8.5 Formulas for single year life tables

1 Age-specific death rate

$$M_x = \frac{D_x}{P_x} \times k$$

$$m_x = \frac{d_x}{L_x}$$

2 Probability of dying between exact ages x and $x + 1$

$$q_x = \frac{2M_x}{2 + M_x}$$

$$q_x = \frac{d_x}{l_x}$$

$$q_x = \frac{d_x}{L_x + \frac{1}{2} d_x}$$

Also, for ages x and over:

$${}_{\infty}q_x = 1$$

3 Probability of surviving from exact age x to exact age $x + 1$

$$p_x = 1 - q_x$$

$$p_x = \frac{l_{x+1}}{l_x}$$

Also, for ages x and over:

$${}_{\infty}p_x = 0$$

4 Number of survivors at exact age x

$$l_{x+1} = l_x \times p_x$$

$$l_{x+1} = l_x - d_x$$

5 Number of deaths between exact ages x and $x + 1$

$$d_x = l_x \times q_x$$

$$d_x = l_x - l_{x+1}$$

6 Average number living between exact ages x and $x + 1$

$$L_x = 0.5(l_x + l_{x+1})$$

$$L_x = l_x - 0.5d_x$$

$$L_0 = 0.3l_0 + 0.7l_1$$

$$L_1 = 0.4l_1 + 0.6l_2$$

Also, for and open-ended age interval:

$${}_{\infty}L_x = \frac{l_x}{{}_{\infty}M_x}$$

$${}_{\infty}L_x = T_x$$

7 Total population at exact age x and over

$$T_x = \sum_{t=0}^{\infty} L_{x+t}$$

$$T_x = T_{x+1} + L_x$$

8 Expectation of life from exact age x

$$e_x = \frac{T_x}{l_x}$$

$$\text{CBR} = \frac{l_0}{T_0} \times 1000$$

In Table 8.2,

$$\text{CBR} = 100000/7826402 \times 1000 = 12.78$$

In a stationary population, the crude birth rate is also equal to the crude death rate, since the total deaths, the sum of the d_x column, is equal to l_0, the total live births. Table 8.5 includes some further equations for calculating life table functions. These may be more convenient to use, depending on the context.

8.4 Constructing abridged life tables

For many purposes, the full detail of a single year of age life table is unmanageable or unnecessary. This is so when the need is for summary comparisons between different populations, or for projections and estimates for five year age groups. A summary approach is also advisable if the data are deficient and best used in an aggregated form, such as when mortality rates have to be based on estimates, rather than on census figures. In such contexts *abridged life tables* come into their own.

An abridged life table is a summary life table. It normally has rows for ages 0, 1 and 5, then for every fifth year of age. Age 1 is invariably included because of the mortality peak in infancy. Since certain functions in a life table refer to exact ages (see Box 8.2), the rows of an abridged life table are usually labelled as single years, 0, 1, 5, 10, 15, . . . , rather than as age groups 5–9, 10–14, 15–19. . . . When using a spreadsheet, an extra column showing the number of years in each age interval is a useful addition to the life table, avoiding the need to write different formulas for different intervals (see Table 8.6). As necessary, formulas can refer to the cells in this extra column instead.

Table 8.6 An abridged life table for a Western female population, 2000

Age	n	${}_nM_x$per1000	${}_nq_x$	${}_np_x$	l_x	${}_nd_x$	${}_nL_x$	T_x	e_x
0	1	9.121 60	0.009 08	0.990 92	100 000	908	99 364	7 821 185	78.21
1	4	0.487 52	0.001 95	0.998 05	99 092	193	395 982	7 721 820	77.93
5	5	0.200 30	0.001 00	0.999 00	98 899	99	494 247	7 325 838	74.07
10	5	0.204 54	0.001 02	0.998 98	98 800	101	493 747	6 831 591	69.15
15	5	0.421 86	0.002 11	0.997 89	98 699	208	492 975	6 337 844	64.21
20	5	0.487 92	0.002 44	0.997 56	98 491	240	491 855	5 844 869	59.34
25	5	0.499 32	0.002 49	0.997 51	98 251	245	490 642	5 353 014	54.48
30	5	0.590 55	0.002 95	0.997 05	98 006	289	489 308	4 862 372	49.61
35	5	0.902 36	0.004 50	0.995 50	97 717	440	487 485	4 373 064	44.75
40	5	1.528 71	0.007 61	0.992 39	97 277	741	484 534	3 885 579	39.94
45	5	2.538 01	0.012 61	0.987 39	96 536	1 217	479 639	3 401 045	35.23
50	5	4.017 30	0.019 89	0.980 11	95 319	1 896	471 857	2 921 406	30.65
55	5	6.255 07	0.030 79	0.969 21	93 424	2 877	459 925	2 449 549	26.22
60	5	9.819 09	0.047 92	0.952 08	90 547	4 339	441 886	1 989 624	21.97
65	5	15.824 30	0.076 11	0.923 89	86 208	6 561	414 635	1 547 738	17.95
70	5	26.165 33	0.122 79	0.877 21	79 646	9 780	373 782	1 133 102	14.23
75	5	45.144 32	0.202 83	0.797 17	69 866	14 171	313 904	759 321	10.87
80	5	79.862 71	0.332 86	0.667 14	55 695	18 539	232 130	445 417	8.00
85	infinity	174.210 76	1.000 00	0.000 00	37 157	37 157	213 286	213 286	5.74
Totals						100 000	7 821 185		

Note: Shaded cells denote values or formulas different from those in the rest of the column.

With knowledge of single year life tables, the construction of abridged life tables is reasonably straightforward. Table 8.7 lists all of the formulas required. They appear more complicated, mainly because of the presence of subscripts indicating the size of the age interval. Subscripts are used in the labelling of life table functions mostly when the age interval is greater than one year.

The rest of this section provides a concise commentary on the construction of the life table shown in Table 8.6. As before, calculating some or all of the figures in each column, after reading each subsection, will consolidate understanding of the methods. Another way of studying the same techniques, in conjunction with reading the text, is to work with the computer module on abridged life tables (Box 8.3). The module provides a brief account of each step, an opportunity to test skills by entering formulas into the spreadsheet, and immediate feedback on whether the formulas are correct. Using the module, most people can study the methods required and construct an accurate abridged life table in about 40 or 50 minutes. It is often preferable to work with someone else on the spreadsheet exercise, enabling discussion and clarification of the steps involved.

Table 8.7 Formulas for abridged life tables

1 Age-specific death rate

$$ {}_nM_x = \frac{{}_nD_x}{{}_nP_x} \times k $$

$$ {}_\infty M_x = \frac{{}_\infty D_x}{{}_\infty P_x} \times k $$

2 Probability of dying between ages x and $x + n$

$$ {}_nq_x = \frac{2n \times {}_nM_x}{2 + n \times {}_nM_x} $$

$$ {}_nq_x = \frac{{}_nd_x}{l_x} $$

3 Probability of surviving from exact age x to exact age $x + n$

$$ {}_np_x = 1 - {}_nq_x $$

4 Number of survivors at exact age x

$$ l_{x+n} = l_x \times {}_np_x $$

$$ l_{x+n} = l_x - {}_nd_x $$

5 Number of deaths between ages x and $x + n$

$$ {}_nd_x = l_x \times {}_nq_x $$

$$ {}_nd_x = l_x - l_{x+n} $$

6 Average number living between ages x and $x + n$

$$ {}_nL_x = \frac{n}{2}(l_x + l_{x+n}) $$

$$ {}_nL_x = T_x - T_{x+n} $$

$$ L_0 = 0.3l_0 + 0.7l_1 $$

$$ {}_4L_1 = \frac{4}{2}(l_1 + l_5) $$

$$ {}_\infty L_x = \frac{l_x}{{}_\infty M_x} $$

7 Total population aged x and over

Working from the start of the life table:

$$ T_x = \sum_{i=x}^{\infty} {}_nL_i $$

Working from the end of the life table:

$$ T_x = T_{x+n} + {}_nL_x $$

For a final open-ended interval:

$$ T_x = {}_\infty L_x $$

8 Expectation of life from age x

$$ e_x = \frac{T_x}{l_x} $$

BOX 8.3 Computer assisted learning exercise on abridged life tables (*Abridged Life Table Exercise.xls*)

This Excel module provides an opportunity to learn about the construction of abridged life tables, using the computer to check the answers as you proceed. On-screen information provides a summary of the essential background arising from Chapter 8. The exercise entails entering Excel formulas (*not* numbers) in each column of life table functions, starting with $_nq_x$. The cells for data entry are highlighted in yellow. Ticks and crosses in the adjacent column (headed 'Check') indicate whether the entered result is correct. As each function is completed accurately, scroll down to start on the next one.

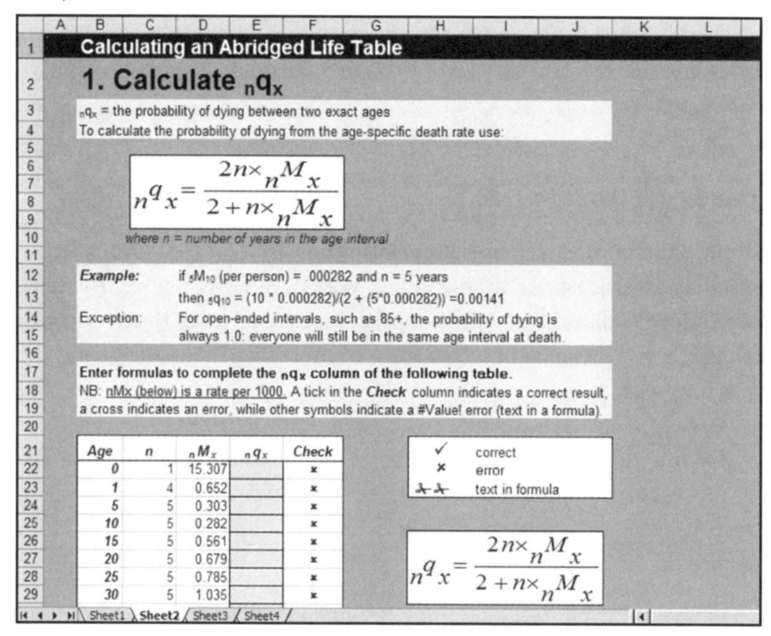

Instructions

To use the module, insert the CD in the drive, start Excel and open the file *Abridged Life Table Exercise.xls*. This workbook does not contain macros.

- To start, click the tab for **Sheet1**, and read the introductory material.
- Next click **Sheet2** to work on the exercise.
- **Sheet3** provides a summary of the notes for printing, if required.
- **Sheet4** is the master spreadsheet with the answers. It is recommended that this be referred to only as a last resort.
- **NB** To avoid scrambling the text boxes in the spreadsheet and deleting background formulas, it is important to type answers *only in the yellow cells* and to use the vertical scroll bar to move through the spreadsheet. Altering other cells, especially in the 'Check' columns, will corrupt the spreadsheet, making it necessary to start again.

See Appendix B for further advice on using the Excel modules.

Applications

This module is an alternative to formal instruction on abridged life tables, showing that the essentials can be learnt fairly readily. The module is intended to be used in conjunction with Section 8.4 of the text, after the material on single year life tables is understood. Completing the module is recommended to consolidate understanding of Section 8.4 and as preparation for attempting the exercises for Chapter 8 (see Study resources).

To permit comparisons, the single year life table in Table 8.2, and the abridged life table in Table 8.6, employ the same base data. The differences between the corresponding l_x and e_x values in the two tables are small and are due mainly to the use of less detailed information in calculating Table 8.6. Utilizing longer age intervals inevitably means that the assumption of a linear distribution of deaths between exact ages becomes less tenable, such as when calculating ${}_nL_x$. Also, adopting a lower limit for the final open-ended age interval – 85 years and over – further reduces the likelihood of an exact match between the two tables.

In the ensuing discussion, worked examples are presented from Table 8.6; minor differences arise when repeating the working with a calculator, because of rounding errors – the full detail employed in the intermediate steps is not shown in the table.

Age-specific death rate (${}_nM_x$)

To begin the calculations, age-specific death rates are required that refer to an age group, such as 10–14, rather than a single year of age. As before, the death rates are calculated from data for a twelve month period, or from the average for three years. Hence they represent the experience of mortality in an age group during an interval of one year.

The age-specific death rate for grouped ages is written ${}_nM_x$ where n is the number of years in the age interval. Thus ${}_5M_{10}$ = deaths for ages 10 to 10 + 5, or deaths in the age group 10–14. Since the age interval is 5 years, the age range is actually from 10.000 to 14.999, that is the whole time span from the tenth birthday to just before the fifteenth.

The formula for the age-specific death rate for grouped ages is:

$$ {}_nM_x = \frac{{}_nD_x}{{}_nP_x} \times k \quad \text{e.g.} \quad {}_5M_{70} = \frac{{}_5D_{70}}{{}_5P_{70}} \times 1000 $$

$$ {}_\infty M_x = \frac{{}_\infty D_x}{{}_\infty P_x} \times k \quad \text{e.g.} \quad {}_\infty M_{85} = \frac{{}_\infty D_{85}}{{}_\infty P_{85}} \times 1000 $$

where

n = the number of years in the age interval
${}_nP_x$ = mid-year population aged x to $x + n$
${}_nD_x$ = number of deaths during the year to persons aged x to $x + n$ (or an annual average).
Set $k = 1$ to express as a rate per person, or $k = 1000$ to express as a rate per thousand.

The former example above is the age-specific death rate for ages 70–74, while the latter is for ages 85 years and over. As before, the infinity symbol denotes an open-ended age interval.

Probability of dying between exact ages ($_nq_x$)

To calculate the probability of dying from the age-specific death rate, use the following formula for grouped data:

$$_nq_x = \frac{2n \times {_nM_x}}{2 + n \times {_nM_x}} \quad \text{e.g.} \quad {_5q_{70}} = \frac{2 \times 5 \times {_5M_{70}}}{2 + 5 \times {_5M_{70}}}$$

where n is the number of years in the age interval.

The formula measures the probability that people in a particular age interval will die while they are in that age interval – that is, between two exact ages. When using the formula, it is important to observe the correct order of operations (see Appendix A): multiplication precedes addition. Also, since the $_nM_x$s in the table are rates per thousand, it is necessary to divide them by 1000 when calculating $_nq_x$. Thus, in Table 8.6,

$$_5q_{70} = \frac{2 \times 5 \times 0.02616533}{2 + 5 \times 0.02616533} = 0.12279$$

As in the discussion of complete life tables, for open-ended ages such as 85 years and over, the probability of dying is always 1.0. All who reach an open age interval will still be in the same age interval at death, although their deaths could be spread over many years.

Probability of surviving from one exact age to the next ($_np_x$)

The calculation of the probability of surviving follows the same pattern as before, except that the interval between exact ages is greater. Also, the probability of surviving in a final open-ended age interval is always zero.

$$_np_x = 1 - {_nq_x}$$

In Table 8.6, $_5p_{50} = 1 - {_5q_{50}} = 1 - 0.01989 = 0.98011$.

Number surviving at exact ages (l_x)

As in a single year life table, the l_x values in an abridged life table represent the number of survivors at exact ages from an original 100 000 births. The l_x values may be calculated using $_np_x$:

$$l_{x+n} = l_x \times {_np_x}$$

In Table 8.6, $l_{65} = l_{60} \times {_5p_{60}} = 90\,547 \times 0.95208 = 86\,208$.

Table 8.7 provides an alternative formula using $_nd_x$.

Deaths between exact ages (${}_nd_x$)

The ${}_nd_x$ function is obtained by multiplying the number alive at exact age x, by the probability of dying:

$$ {}_nd_x = l_x \times {}_nq_x $$

In Table 8.6, ${}_5d_{10} = l_{10} \times {}_5q_{10} = 98\,800 \times 0.00102 = 101$.

As in the single year life table, the ${}_nd_x$ column sums to 100 000, since the number of deaths occurring over all the age intervals equals the number of births.

Average number alive between exact ages (${}_nL_x$)

Apart from ${}_nq_x$, the ${}_nL_x$ is the only function for abridged life tables that differs appreciably from those for complete life tables:

$$ {}_nL_x = \frac{n}{2}(l_x + l_{x+n}) \quad \text{e.g.} \quad {}_5L_{65} = \frac{5}{2}(l_{65} + l_{70}) $$

Since n is the number of years in the age interval, the formula may also be written as:

$$ {}_nL_x = n \times \frac{1}{2}(l_x + l_{x+n}) \quad \text{e.g.} \quad {}_5L_{65} = 5 \times \frac{1}{2}(l_{65} + l_{70}) $$

In Table 8.6, ${}_5L_{65} = 5 \times \frac{1}{2}(86\,208 + 79\,646) = 414\,635$.

In the above, the average number alive between two exact ages (out of 100 000 births) is multiplied by 5 in recognition that in the age interval 65–69 there are five single year cohorts, representing the survivors from births during a five year period. Thus if the radix of the life table is 100 000, in a 5 year age group the average number alive will be nearly 5 times 100 000 until mortality takes its toll.

As mentioned earlier, abridged life tables typically include separate figures for the first year of life, given that the risks of mortality are high for infants. Thus we use the same single year life table function as before for L_0:

$$ L_0 = 0.3l_0 + 0.7l_1 $$

In Table 8.6, $0.3(100\,000) + 0.7(99\,092) = 99\,364$.

Subsequent age-related variations in the timing of mortality have less impact and, for the second age group, 1–4 years, the general formula may be followed:

$$ {}_nL_x = \frac{n}{2}(l_x + l_{x+n}) $$

Therefore

$$ {}_4L_1 = \frac{4}{2}(l_1 + l_5). $$

In Table 8.6, ${}_4L_1 = \frac{4}{2}(99\,092 + 98\,899) = 395\,982.$

Since this formula is not exceptional, the cell for ${}_4L_1$ is not shaded in Table 8.6, but the shaded values of n, in the second column, have to be taken into account. The last cell in the ${}_nL_x$ column, however, is shaded because a different formula is needed for the open-ended interval 85 years and over. This is the same formula as used in the complete life table in Table 8.2:

$$ {}_\infty L_x = \frac{l_x}{{}_\infty M_x} \quad \text{e.g.} \quad {}_\infty L_{85} = \frac{l_{85}}{{}_\infty M_{85}} $$

In Table 8.6,

$$ {}_\infty L_{85} = \frac{37157}{174.21076/1000} = 213\,286 $$

Total population aged x and over (T_x)

Like l_x and e_x, T_x is a life table function that refers to exact ages – hence, it requires no special allowance for variations in the age intervals. T_x is the sum of the ${}_nL_x$ values from age x, and T_0 is equal to the sum of the ${}_nL_x$ column, e.g.

$$ T_{65} = {}_5L_{65} + {}_5L_{70} + {}_5L_{75} + \ldots + {}_5L_{80} + {}_\infty L_{85} $$

In Table 8.6, $T_{75} = {}_5L_{75} + {}_5L_{80} + {}_\infty L_{85} = 313\,904 + 232\,130 + 213\,286 = 759\,321.$
For a final open ended age interval:

$$ T_x = {}_\infty L_x $$

In Table 8.6, $T_{85} = {}_\infty L_{85} = 213\,286$

Expectation of life (e_x)

Life expectancy at each age is also calculated in the same way as for single year life tables:

$$e_x = \frac{T_x}{l_x} \qquad \text{e.g.} \quad e_{10} = \frac{T_{10}}{l_{10}}$$

In Table 8.6,

$$e_{10} = \frac{6831477}{98798} = 69.15$$

8.5 Conclusion

The Excel module described in Box 8.3 provides practice with the formulas, as well as with spreadsheets for life table work. Following the conventions for the layout of life tables (Box 8.2) will make the table easier to check and more intelligible to others. Life table calculations, however, sometimes start from an existing life table, rather than calling for a completely new set of calculations. For example, abridged life table functions may be required when a reliable single year life table is available. For functions referring to exact ages, every fifth value can be extracted as needed. Also, ${}_5L_x$ values are readily obtained by adding up the single year figures; e.g.

$${}_5L_{20} = L_{20} + L_{21} + L_{22} + L_{23} + L_{24}$$

Such figures are often wanted for calculating survival ratios (see Chapter 12) for population estimates, including estimates of net migration. Where possible, working from a published official life table is usually preferable to constructing a table from raw data, not least because it provides consistency with earlier research. Box 8.4

BOX 8.4 **Life tables for both sexes combined**

Sometimes combined life table figures for both sexes are required when separate life tables are available for each sex. Rather than calculating a new life table based on mortality rates for the total population, the combined figures may be obtained by weighting figures in the l_x, d_x, L_x and T_x columns of the male life table by 1.05 (i.e. weighting the life table functions referring to numbers of persons). This assumes that the sex ratio at birth is 105 males per 100 females. For example, l_x and T_x for total persons are obtained as:

$$l_x = \frac{1.05 l_x^{m} + l_x^{f}}{2.05}$$

$$T_x = \frac{1.05 T_x^{m} + T_x^{f}}{2.05}$$

The superscripts, m and f, denote figures from the male and female life tables. Dividing the numerator by 2.05 produces life table values weighted by the sex ratio at birth.

describes the derivation of life table figures for both sexes combined, using existing life tables for males and females.

Demography has a considerable literature on life table construction; discussions of more advanced methods are presented, for instance, in Hinde (1998), Barclay (1966), Smith (1992) and Namboodiri (1991). Also, demographers have produced model life tables that may be appropriate to use where mortality statistics are limited, instead of attempting to construct a new life table from defective data. The next chapter discusses model life tables, as well as model stable populations and the methods of constructing stable population models from raw data.

Study resources

KEY TERMS

Abridged life table
Age-specific death rate
Closed population
Cohort life table
Complete life table
Exact age
Life expectancy
Life table
Life table functions
Open population
Paradox of the life table
Period life table
Person-years
Probability of dying
Probability of surviving
Stationary population
Life table functions:
l_x
${}_nq_x$
${}_np_x$
${}_nd_x$
${}_nL_x$
T_x
e_x

FURTHER READING

Barclay, George W. 1966. *Techniques of Population Analysis*. New York: John Wiley, Chapter 4, 'The Life Table', pp. 93–122.

Cox, Peter, R. 1970. *Demography*. Cambridge: Cambridge University Press, Chapter 11, 'Life Tables', pp. 181–205.

Haupt, Arthur. 1986. 'Halley's Other Comet'. *Population Today*, 14 (1): 3 and 10–11.

Hinde, Andrew. 1998. *Demographic Methods*. London: Arnold, Chapter 4, 'The Life Table'.

Newell, C. 1988. *Methods and Models in Demography*. London, Frances Pinter, Chapter 6, 'Mortality and Life Tables', pp. 63–81.

Pollard, A. H., Yusuf, Farhat, and Pollard, G. N. 1990. *Demographic Techniques* (third edition). Sydney: Pergamon Press, Chapter 3, 'The Life Table', pp. 30–48.

Shryock, Henry S. and Siegel, Jacob S. 1973. *The Methods and Materials of Demography* (two volumes). US Bureau of the Census, US Government Printing Office, Washington, Chapter 15, 'The Life Table', pp. 429–461.

INTERNET RESOURCES

Subject	Source and Internet address
US Decennial Life Tables, using national, state and cause-eliminated data	***National Center for Health Statistics*** http://www.cdc.gov/nchs/products/pubs/pubd/lftbls/decenn/1991-89.htm
Berkeley Mortality Data Base: life tables for national populations and raw data used in constructing them	***Department of Demography, University of California Berkeley*** http://www.demog.berkeley.edu/wilmoth/mortality/index.html
Life tables for Japan	***Ministry of Health, Labour and Welfare, Japanese Government*** http://www.mhlw.go.jp/english/database/db-hw/lifetb00/
Life tables for Israel	***Central Bureau of Statistics, Israel*** http://www.cbs.gov.il/publications/mortality/mort_e.htm
Paper on Finnish life tables since 1751	***Demographic Research (Online Journal) Max Planck Institute for Demographic Research*** http://www.demographic-research.org/Volumes/vol1/1/1.pdf
Life tables for the United Kingdom and constituent countries	***Government Actuary's Department, United Kingdom*** http://www.gad.gov.uk/

Note: the location of information on the Internet is subject to change; to find material that has moved, search from the home page of the organization concerned.

EXERCISES

1 From Table 8.2 find:

(a) The life expectancy of females aged 60.

(b) The percentage of females alive at exact age 85.

(c) The total person-years lived between exact ages 18 and 19.

(d) The total person-years lived from exact age 90.

(e) The probability of dying at age 49.

(f) The average number of females alive between 100 and 101.

(g) The total life table population aged 100 or more.

(h) The total deaths between exact ages 0 and 1.

(i) The probability of surviving in the final open-ended age group.

(j) The first death rate, in the adult ages, that exceeds the death rate at age 0.

2 Define the following terms:

(a) q_0

(b) q_x

(c) p_{21}

(d) p_x

(e) M_x

(f) e_0

(g) e_{41}

(h) e_{84}

(i) e_x

(j) m_x

3 Represent the following l_x values for real cohorts by a Lexis diagram:

	2001	2002	2003	2004
l_0	1000	1000	1000	1000
l_1	980	985	990	
l_2	975	980		
l_3	970			

4 Complete the following table:

Age	Pop.	deaths	M_x	q_x	p_x	l_x	d_x	L_x	T_x	e_x
0	117 724	1769								
1	121 143	143								
2	125 743	101								
3	129 258	88								
4	137 559	69						98 209	6 562 207	

5 Provide brief answers to the following questions:

(a) Is there an age structure – 'young', 'mature' or 'old' – that is characteristic of a stationary population? Explain your answer.

(b) Why is life expectancy at 60 not equal to e_0 minus 60?

(c) What is the sum of the L_x column of a life table?

(d) What is the sum of the d_x column of a life table?

(e) What is $l_0/T_0 \times 1000$?

(f) Why is there no such function as ${}_nl_x$?

(g) In an abridged life table, why is T_x, which is calculated from ${}_nL_x$, never accompanied by a subscript denoting the size of the age interval?

6 Obtain the most recent official life table for your own country and make a summary table comparing the l_x and e_x functions for ages 0, 1, 20, 40, 60, and 80. Write a brief statement explaining the meaning of the life table functions and the differences between the figures for males and females.

SPREADSHEET EXERCISE 8: ABRIDGED LIFE TABLES

When people have to rely on calculators for life table work, the task of constructing a two-sex life table is time-consuming and subject to many sources of error. The background in Excel obtained in previous exercises is now more than sufficient to enable the use of a spreadsheet to accomplish the same task with speed and accuracy. Refer to Table 8.6 for an example of the layout of the work, and to Table 8.7 for a summary of the formulas employed in constructing abridged life tables.

Calculate an abridged life table from the data below, providing separate tables for males and females. Note that the last age group is open ended (95 years and over). The full answers are given at the end.

Table 1 ${}_nM_x$ values (per 1000) for males and females

Age	Females	Males
0	32.19	43.03
1	1.99	2.47
5	0.69	0.97
10	0.56	0.78
15	0.89	1.38
20	1.23	1.87
25	1.50	1.92
30	1.83	2.23
35	2.36	2.87
40	3.22	4.11
45	4.69	6.32
50	7.10	10.04
55	10.41	15.46
60	16.49	24.03
65	27.26	37.32
70	46.18	58.78
75	78.03	93.40
80	125.59	145.56
85	194.11	218.33
90	288.06	315.16
95+	423.16	447.87

Procedure

1. First, construct a life table, replicating Table 8.6 or another example. This is to check that the formulas are correct. Save the spreadsheet at each stage of the work, in order to recover quickly from any unforeseen problems.
2. When finished, copy the table to another worksheet and edit it as necessary to show all of the ages required for Exercise 8.
3. Use a radix of 100 000 and replace the original ${}_nM_x$ values with those for the female population, as shown in the table above. Upon completion, the table should then have updated automatically to show the life table values corresponding to the new ${}_nM_x$ s.
4. Check that the ${}_nd_x$ column sums to 100 000 and that the sum of the ${}_nL_x$ column equals T_0. Also check that the other values are logically consistent. Errors tend to arise if formulas are retyped, instead of being copied downwards wherever possible.
5. When the life table for the females is finished, copy the entire table and paste the copy below the female life table. Edit the titles, and replace the female ${}_nM_x$ values

with those for the males. If the first table is accurate, the second one should be too, unless there were absolute references in the formulas.

Table 2 Answers to Spreadsheet Exercise 8

Females

Age	n	${}_nM_x$	${}_nq_x$	${}_np_x$	l_x	${}_nd_x$	${}_nL_x$	T_x	e_x
0	1	32.190	0.031 68	0.968 32	100 000	3 168	97 782	6 997 475	69.97
1	4	1.990	0.007 93	0.992 07	96 832	768	385 793	6 899 692	71.25
5	5	0.690	0.003 44	0.996 56	96 064	331	479 494	6 513 900	67.81
10	5	0.560	0.002 80	0.997 20	95 733	268	477 998	6 034 406	63.03
15	5	0.890	0.004 44	0.995 56	95 466	424	476 269	5 556 408	58.20
20	5	1.230	0.006 13	0.993 87	95 042	583	473 752	5 080 139	53.45
25	5	1.500	0.007 47	0.992 53	94 459	706	470 531	4 606 386	48.77
30	5	1.830	0.009 11	0.990 89	93 753	854	466 632	4 135 855	44.11
35	5	2.360	0.011 73	0.988 27	92 899	1 090	461 773	3 669 223	39.50
40	5	3.220	0.015 97	0.984 03	91 810	1 466	455 382	3 207 451	34.94
45	5	4.690	0.023 18	0.976 82	90 343	2 094	446 481	2 752 068	30.46
50	5	7.100	0.034 88	0.965 12	88 249	3 078	433 551	2 305 587	26.13
55	5	10.410	0.050 73	0.949 27	85 171	4 321	415 054	1 872 036	21.98
60	5	16.490	0.079 19	0.920 81	80 850	6 402	388 246	1 456 982	18.02
65	5	27.260	0.127 60	0.872 40	74 448	9 500	348 491	1 068 736	14.36
70	5	46.180	0.207 00	0.793 00	64 948	13 444	291 131	720 245	11.09
75	5	78.030	0.326 46	0.673 54	51 504	16 814	215 484	429 114	8.33
80	5	125.590	0.477 90	0.522 10	34 690	16 578	132 003	213 630	6.16
85	5	194.110	0.653 45	0.346 55	18 111	11 835	60 970	81 627	4.51
90	5	288.060	0.837 31	0.162 69	6 277	5 255	18 244	20 657	3.29
95	infinity	423.160	1.000 00	0.000 00	1 021	1 021	2 413	2 413	2.36
Totals						100 000	6 997 475		

Males

Age	n	${}_nM_x$	${}_nq_x$	${}_np_x$	l_x	${}_nd_x$	${}_nL_x$	T_x	e_x
0	1	43.030	0.04212	0.95788	100000	4212	97051	6596649	65.97
1	4	2.470	0.00983	0.99017	95788	942	381267	6499598	67.85
5	5	0.970	0.00484	0.99516	94846	459	473082	6118331	64.51
10	5	0.780	0.00389	0.99611	94387	367	471017	5645249	59.81
15	5	1.380	0.00688	0.99312	94020	647	468482	5174232	55.03
20	5	1.870	0.00931	0.99069	93373	869	464693	4705750	50.40
25	5	1.920	0.00955	0.99045	92504	884	460311	4241057	45.85
30	5	2.230	0.01109	0.98891	91620	1016	455562	3780746	41.27
35	5	2.870	0.01425	0.98575	90604	1291	449795	3325184	36.70
40	5	4.110	0.02034	0.97966	89314	1817	442026	2875389	32.19
45	5	6.320	0.03111	0.96889	87497	2722	430679	2433363	27.81
50	5	10.040	0.04897	0.95103	84775	4151	413496	2002684	23.62
55	5	15.460	0.07442	0.92558	80623	6000	388116	1589188	19.71
60	5	24.030	0.11334	0.88666	74623	8458	351971	1201072	16.10
65	5	37.320	0.17068	0.82932	66165	11293	302594	849101	12.83
70	5	58.780	0.25624	0.74376	54872	14061	239210	546507	9.96
75	5	93.400	0.37860	0.62140	40812	15451	165430	307296	7.53
80	5	145.560	0.53362	0.46638	25360	13533	92970	141866	5.59
85	5	218.330	0.70619	0.29381	11828	8353	38257	48896	4.13
90	5	315.160	0.88137	0.11863	3475	3063	9718	10639	3.06
95	infinity	447.870	1.00000	0.00000	412	412	920	920	2.23
Totals						100000	6596649		

9 Stable and stationary models

. . . about one third of all that were ever quick die under five years old . . .

(Graunt 1662: 9)

OUTLINE

9.1 Stable populations
9.2 Constructing stable population models
9.3 Model life tables
9.4 Model stable populations
9.5 Population momentum
9.6 Population ageing
9.7 Further applications of stable and stationary models
9.8 Conclusion
Study resources

LEARNING OBJECTIVES

To understand:

- the concept of a stable population
- how to construct stable population models
- the nature and uses of model life tables and model stable populations
- the nature of population momentum and population ageing

To become aware of:

- applications of the stable and stationary models in preparing demographic estimates

COMPUTER APPLICATIONS

Excel modules: West Models.xls (Box 9.1)
Relational Models.xls (Box 9.2)
Momentum.xls (Box 9.3)

Spreadsheet exercise: Stable population models (See Study resources)

Demography has a long-established interest in generalizations about the characteristics of stages of demographic evolution, such as the level of birth and death rates and the shape of the age profile, before, during and after the demographic transition. Pre-transition and post-transition populations have been likened to stationary populations with zero growth and constant age structures. Similarly, those in the early part of the demographic transition have been likened to *stable populations*, with constant age structures and constant rates of growth.

Comparisons between demographic stages and demographic models are merely broad generalizations that never duplicate reality, but the similarities can be sufficient to permit the application of stable and stationary models in diverse areas of demographic analysis and estimation. The usefulness of the concepts of stable and stationary models depends not on achieving an exact correspondence with actual circumstances, but on achieving a sufficient correspondence to permit analyses that would otherwise be impossible. For these and other reasons, demographers have described stationary population models as 'the most versatile and most useful of the demographer's tools' (Pollard et al. 1990: 30) and stable population models as 'unquestionably one of the main achievements of mathematical demography' (Pressat 1985: 211).

The previous chapter discussed the nature and computation of stationary populations. This chapter does the same for stable populations. It then describes published sources of model life tables and model stable populations and discusses some applications of the models in demographic estimation, as well as in understanding population momentum and population ageing.

The chapter includes four computer applications arising directly from the subject matter. The Excel module *West Models.xls* (see Box 9.1) displays graphs of life table functions for a wide range of different mortality levels, to aid understanding of life tables and mortality trends. It also shows the age structures of stable populations corresponding to many different combinations of mortality and fertility levels, together with examples of stable population models for selected countries and regions. The module entitled *Relational Models.xls* (see Box 9.2) enables the instant calculation of 'families' of life tables from user-defined criteria. The third Excel module, *Momentum.xls* (see Box 9.3) illustrates the nature of population momentum in the world's age structure, as well as in more developed and less developed regions and 17 countries. Finally, the spreadsheet exercise provides practice in constructing stable population models.

9.1 Stable populations

A stable population is a model or hypothetical population with the following characteristics:

- Constant birth and death rates and a constant growth rate.
- A constant age–sex structure. While the absolute numbers in age groups change through time, the percentages in each age group remain constant.
- The population is closed to migration; in other words there is no inward or outward migration.

The population is 'stable' in the sense that its growth rate and age structure are unchanging. A population would develop a stable age structure after 100 years if fertility and mortality remained constant and there was no inward and outward migration. The age structure changes occurring after 70 years are normally minor (Weller and Bouvier 1981: 338). A stationary population is a special case of a stable population, because it has the same characteristics, except that its growth rate is zero and, therefore, the numbers in each age group remain constant. Stable population models can illustrate the long-run effects on a population's age structure of given levels of fertility and mortality, that is, the situation pertaining after eliminating the effects on the age structure of short-lived events, such as baby booms, famines, epidemics, wars and natural disasters.

The shape of a stable population's age structure depends on the level of natural increase. If natural increase is high, with births greatly exceeding deaths, the age profile will be triangular or 'young', as in the stable population model for Australia in 1971 (Figure 9.1). If natural increase is low, the age profile will be rectangular or 'old'. An excess of deaths over births produces an age profile that tapers downwards. A stable population with such an age profile is inherently on a pathway to extinction as the consequences of a constant rate of decline unfold. The 1996 stable population for Australia in 1996 is an 'old' age profile with some tapering evident in the smaller sizes of younger cohorts (Figure 9.1).

The growth rate of a stable population, termed the *intrinsic rate of natural increase*, is based only on the population's birth and death rates, and is therefore independent of its current age structure. Such a model for Russia in 2001 would use the country's vital rates, but the model's age structure would be independent of the country's age structure in 2001. Accordingly, stable population models can provide a means of obtaining a long-range projection of the shape of the population's age structure, without the detailed work normally required in calculating projections. The strict assumptions of the stable model, however, mean that its main use for projections is to show the long-range implications of particular levels of births and deaths.

Contrasting age distributions will evolve to identical stable age distributions if the fertility and mortality levels are the same. This is illustrated in the 'desert island' scenario in the computer module *Age Structure Similations.xls* (See Box 3.2), where an initially unusual population develops the age structure of a stationary population with a female life expectancy of 30 years. It is also demonstrated through the regional and country examples in the *West Models.xls* module (Box 9.1). Alfred Lotka, one of the pioneers of stable population theory, proved this principle mathematically in 1907 concluding:

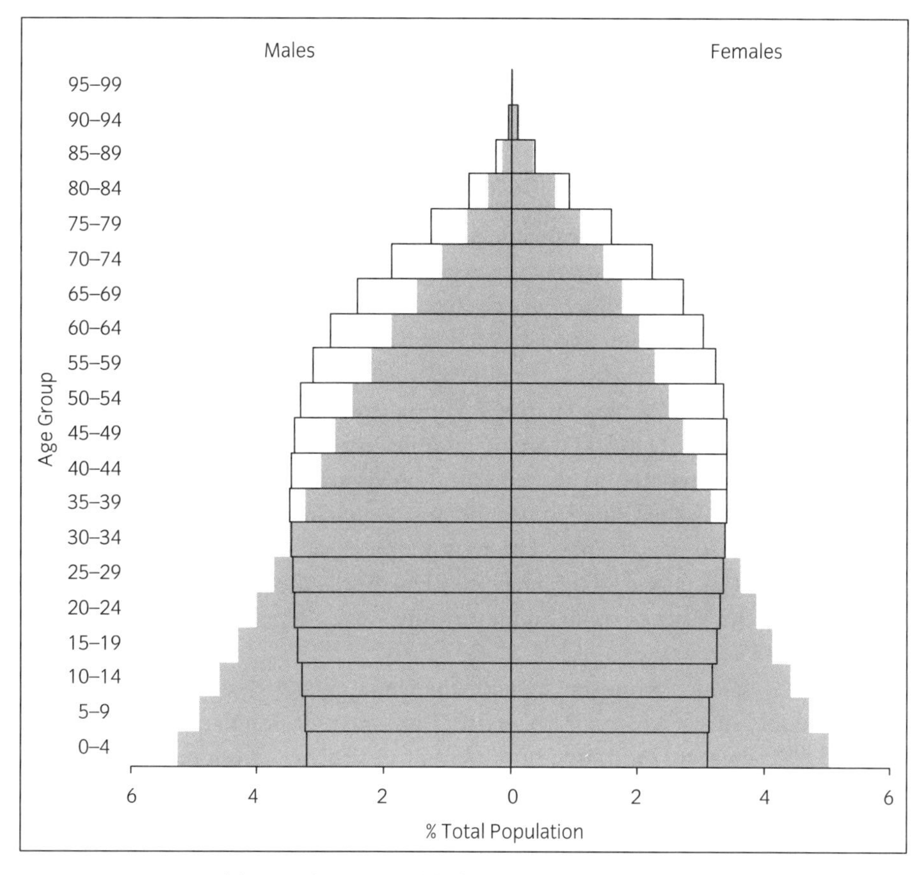

Figure 9.1 Stable population models for Australia, 1971 (shaded) and 1996.

A population that has been exposed indefinitely to unchanging conditions of overall fertility and age-specific mortality tends toward a population with an unchanging age distribution and growth rate. They depend on the fertility and mortality conditions but not on the initial age distribution. This population at the limit is characterized as stable. (quoted in Henry 1976: 216).

9.2 Constructing stable population models

The previous chapter showed that mortality data alone are required to construct a stationary model or life table, in which the number of births (l_0) equals the number of deaths (the sum of the d_x column). In comparison, the stable model entails additional procedures because of the need to take account of the effects of fertility rates as well, although the task is simplified if a stable population model employs mortality data (the L_x column) from an existing life table. When a suitable life table is available, the steps in constructing a stable population model reduce to two: calculating the population's intrinsic rate of natural increase – from fertility rates and the life table – and deriving its age structure.

The intrinsic rate of natural increase (*r*)

A forerunner of the growth rate of a stable population was T. R. Malthus's concept of the *natural power of mankind to increase*. This he developed after the publication of the first edition of his *Essay on the Principle of Population* (1798) from studying population growth as recorded in the early censuses of the United States. Malthus was interested in discovering the rate at which a population would grow in the absence of constraining factors. Potential constraints included limits on the amount of agricultural land, and an inequitable distribution of food to the labouring class. Malthus expected that the natural power of mankind to increase would be most apparent in the United States, because of its abundance of good land and because he believed that the population as a whole was adequately fed. He wrote that: 'The greatest check to the increase of plants and animals, we know from experience, is the want of room and nourishment. . . .' (Malthus 1830, in Demko et al. 1970: 45).

Malthus concluded from statistics for the United States that a 25 year interval between doublings represented 'the natural progress' of population when not restrained by premature mortality. In other words, unchecked population increases in geometric progression to double itself every 25 years (ibid: 52). Malthus's rate of increase is equivalent to a constant annual growth rate of 2.8 per cent, a rate actually exceeded during the demographic transition of some developing countries.

In modern demographic writing, the focus is on a range of inherent growth rates, rather than the maximum. The key concept is the *intrinsic rate of natural increase* (or the *intrinsic growth rate*), which is the constant growth rate (r) of a stable population. The rate is 'intrinsic' to, or inherent within, the population concerned because it is based solely on its age-specific fertility and mortality in the year of observation.

Population momentum (see Section 9.5) has a considerable impact on observed growth rates of contemporary populations, but the intrinsic rate of natural increase is independent of this influence. For example, many European populations with below-replacement fertility were still growing at the turn of the twenty-first century because of the population momentum built into their age structures, but their intrinsic growth rates were negative.

As well as excluding the effects of the initial age structure, stable models exclude the effects of migration, whereas in the United States, Canada, Australia and New Zealand, for example, migration augments population growth appreciably. Even sustained below-replacement fertility need not result in population decline in countries with long-standing experience of, and commitment to, accepting migrants. However, migration is not a simple countermeasure to population decline in countries unaccustomed to managing migration and settlement programs (see United Nations 2000a).

The steps in calculating the intrinsic growth rate are straightforward, even though more complex mathematics underlies them. The discussion here focuses on practical procedures; accounts of the mathematics are available in other texts (see the 'Further reading' section in Chapter 1).

Table 9.1 presents an example of the calculation of the intrinsic growth rate for a growing Western population with a gross reproduction rate of 1.2 and female and male life expectancies at birth of 75.0 and 71.2 years respectively (West model life table, Level 23, from Coale and Demeny 1983: 53). The specified gross reproduction rate implies a total fertility rate of about 2.5 births per woman (1.2 × (205/100) = 2.46). For brevity, stable models are usually calculated from data for five year age groups. Although the example employs hypothetical data, figures for any national or regional population may be used, as in the *West Models.xls* module; the results show the long-run implications of fertility and mortality rates, in the absence of migration.

Table 9.1 Calculation of the intrinsic growth rate of a growing Western population, year 2000 (GRR = 1.2, ${}^{f}e_0 = 75$)

Age group A	Age group mid-point y B	Female ASFRs per woman C	Probability of Survival D	R_0 working E (C × D)	R_1 working F (B × E)
15–19	17.5	0.01326	0.97914	0.01298	0.22721
20–24	22.5	0.04324	0.97703	0.04225	0.95055
25–29	27.5	0.07812	0.97421	0.07611	2.09290
30–34	32.5	0.07113	0.97061	0.06904	2.24378
35–39	37.5	0.02906	0.96577	0.02807	1.05245
40–44	42.5	0.00506	0.95870	0.00485	0.20617
45–49	47.5	0.00013	0.94751	0.00012	0.00585
Total		0.24000		0.23341	6.77891
Total times 5:		**GRR = 1.20000**		**R_0 = 1.16707**	**R_1 = 33.89456**
Working:	$\ln R_0$ =	0.15450			
	R_1/R_0 =	29.04237 (mean length of a generation)			
	r =	**0.00534**			
	r =	0.534% per annum			

Note: the probabilities of survival were calculated as $5L_x/500000$ from figures in Table 9.3.
Data source: West level 23 model life table (Coale & Demeny 1983: 53).

Coale's formula provides a straightforward, yet accurate, method for calculating the intrinsic growth rate, r (Woods 1982a: 203; Pollard et al. 1990: 108). It requires only two terms, labelled R_0, which is the net reproduction rate, and R_1 (where R_1/R_0 is the *mean length of a generation*, as discussed later):

$$r = \frac{\ln R_0}{\frac{R_1}{R_0} - 0.7 \ln R_0}$$

where $\ln R_0$ is the natural logarithm of R_0.

The first step in using Coale's formula is to calculate the population's net reproduction rate (see Chapter 7) as follows:

1. List the reproductive age groups for females (column A) and the central age or mid-point (labelled y, i.e. letter y, italicized) of each age group (column B). For any five year age interval, spanning ages x to $x + 5$, the mid-point is equal to $x + 2.5$. Thus, for age group 20–24, the mid-point is 22.5.
2. List the age-specific fertility rates for female births (column C). In the example, the set of female age-specific fertility rates produce a GRR of 1.2. The figures in column C are expressed as rates per woman.
3. List the proportion of females surviving from birth to the mid-point of the mother's age group (column D). (From the life table for the year of observation, obtain ${}_5L_x/5 \times l_0$, or L_x/l_0; in the latter case, use the middle L_x value in each five year age range.) The life table usually refers to the same year as the fertility data, to reflect demographic conditions at a single point in time – unless a particular combination of birth and death rates is wanted. If a life table is not available for the same year, use one for the nearest year available or a model life table (see Section 9.3).
4. In column E, multiply each ASFR (column C) by the corresponding probability of survival (column D); e.g. at ages 15–19, $0.01326 \times 0.97914 = 0.01298$.
5. Sum the figures in column E and multiply the total by 5 to obtain R_0. The total is multiplied by 5, since the statistics are for five year age groups. R_0 is the net reproduction rate, sometimes described as the ratio of the number in one generation to the number in the preceding generation (Pollard et al. 1990: 108).

Having obtained R_0 (NRR) in the usual way, it is necessary to add only one more column of working to Table 9.1 to calculate R_1. The figures in the working column for R_1 (column F) are calculated as the sum of the central age times the expected survivors from female births (cells in column B times corresponding cells in column E). Thus, for age group 15–19: $17.5 \times 0.01298 = 0.22715$ (this result is not entirely consistent with manual calculation, because the spreadsheet for Table 9.1 retained more than 5 decimal places in column E).

Table 9.1 includes additional figures for $\ln R_0$ and R_1/R_0, so that the formula for r can refer to the cells containing these values. From the values in the table, the intrinsic growth rate is:

$$r = \frac{0.15450}{29.04230 - 0.7 \times 0.15450} = 0.00534$$

Thus the population was growing at a rate of just over 0.5 per cent annually, which would double the population in about 130 years. Having obtained the intrinsic rate of natural increase, it may now be applied in obtaining the age distribution of the stable population.

The age distribution of a stable population

Deriving the age distribution of a stable population begins with the age distribution of the life table (the L_x column) used in obtaining the intrinsic rate of natural increase. Calculation of the age distribution of the stable population entails adjusting this age distribution – using a set of multipliers – to recognize that each cohort in a stable population originates from annual births that are growing, or declining, at a constant rate through time. By contrast, in a stationary population, each cohort of males or females originates from a constant number of births equal to the radix of the life table. If the radix is 100 000 ($l_0 = 100\,000$), the totals at every age represent the survivors from an original 100 000. Before discussing the derivation of the age distribution, it is necessary to explain the nature of the multipliers used to adjust the age structure of the stationary population.

Figure 9.2 shows the number of births in a population growing at a constant annual rate of 0.534 per cent through time, as in Table 9.1. If, for example, there were 58 709 live births in 1900, there would be 100 000 live births in 2000. In a growing stable population the youngest people are the survivors from the largest numbers of births. Reading the graph chronologically, or forward through time, it is obvious that the figures increase. Conversely, if we read the graph in the opposite direction, the numbers of births decrease. The latter way of thinking about the numbers of births, i.e. looking back through time, is the basis of the multipliers.

Included in Table 9.2 are examples of the formulas for calculating the numbers of births plotted in Figure 9.2. Working forwards in time, it is assumed that the oldest cohort, born in 1900, numbered 58 709. The numbers at birth in the other cohorts are obtained by multiplying 58 709 by $(1 + r)^y$, where r is the population's intrinsic growth rate, and y is the number of years over which the growth rate has operated. This is an application of the geometric growth rate formula for finding the end of period population (see Table 2.3, formula 1). Thus the babies born in the year 2000 numbered 100 000 ($58\,709 \times (1 + r)^{100} = 100\,000$), since the intrinsic growth rate ($r = 0.005\,34$) had been operating for 100 years since the 1900 cohort was born.

When calculating stable populations, however, it is necessary to work retrospectively, because the population has been growing (or declining) over the lifetimes of everyone alive in the base year, or year of observation. Thus, in the second half of Table 9.2, the formulas show the calculation of the numbers in each cohort, assuming that there were 100 000 births in the year 2000. Since the base year is 2000,

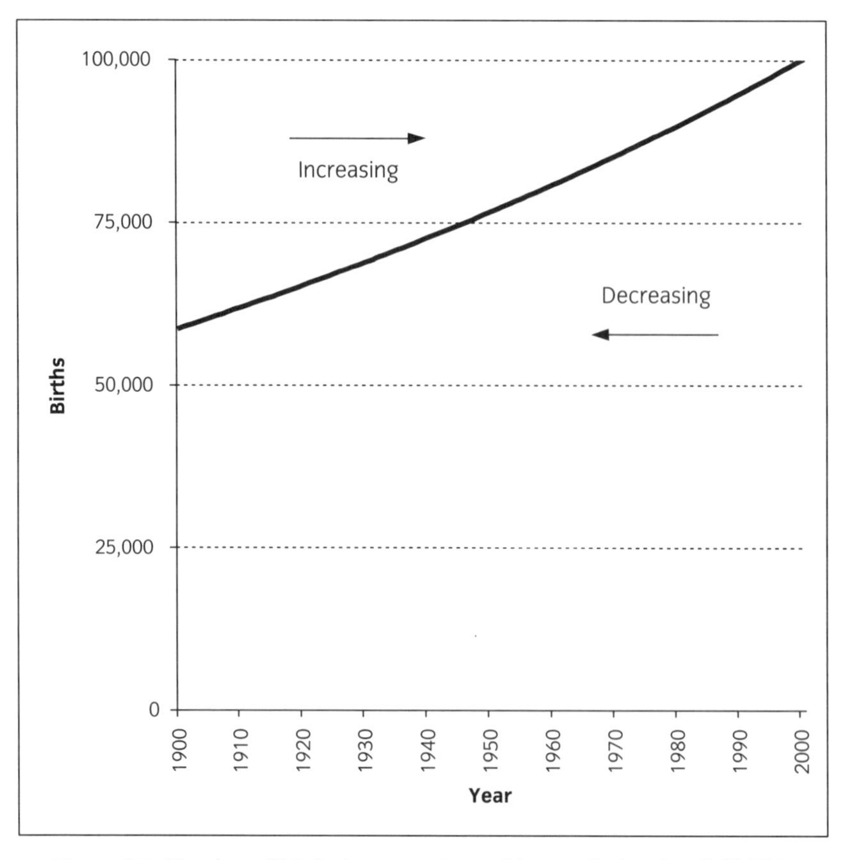

Figure 9.2 Number of births in a growing stable population ($r = 0.534\%$).

the number born three years earlier, in 1997, is equal to $100\,000(1 + \mathrm{r})^{-3}$. This is an application of the geometric growth rate for finding the initial population when the population at a more recent date is known (see Table 2.3).

The formula given in Table 2.3 was:

$$P_0 = \frac{P_n}{(1+r)^n}$$

When the power of a number is moved from the denominator of a fraction to the numerator (or vice versa), the sign of the power is changed. Thus the above can also be written as:

$$P_0 = P_n(1+r)^{-n}$$

It follows that:

$$\text{Births}_{1997} = \text{Births}_{2000}(1+r)^{-3} \qquad \text{and} \qquad \text{Births}_{1900} = \text{Births}_{2000}(1+r)^{-100}$$

The multipliers, or correction factors, for modifying the age distribution of the stationary population are therefore equal to:

Table 9.2 Multipliers for calculating cohort size at birth

Intrinsic growth rate: 0.005 34			
Age in 2000 A	Birth year B	Cohort size at birth C	Births D
Working forwards though time			
100	1900	58 709	58 709
99	1901	$58\,709\,(1+r)$	59 023
98	1902	$58\,709\,(1+r)^2$	59 338
97	1903	$58\,709\,(1+r)^3$	59 655
3	1997	$58\,709\,(1+r)^{97}$	98 415
2	1998	$58\,709\,(1+r)^{98}$	98 941
1	1999	$58\,709\,(1+r)^{99}$	99 469
0	2000	$58\,709\,(1+r)^{100}$	100 000
Working backwards though time			
0	2000	100 000	100 000
1	1999	$100\,000\,(1+r)^{-1}$	99 469
2	1998	$100\,000\,(1+r)^{-2}$	98 940
3	1997	$100\,000\,(1+r)^{-3}$	98 415
97	1903	$100\,000\,(1+r)^{-97}$	59 654
98	1902	$100\,000\,(1+r)^{-98}$	59 338
99	1901	$100\,000\,(1+r)^{-99}$	59 022
100	1900	$100\,000\,(1+r)^{-100}$	58 709

Source: after Pressat (1972: 318–319).

$$(1+r)^{-y}$$

where

r is the intrinsic growth rate and
y is the number of years since each cohort's birth.

The multipliers take account of the number of years that the intrinsic growth rate has operated since the birth year of each cohort. As noted earlier, y is the mid-point of each age group. For a final open-ended interval, such as ages 85 and over, the mid-point may be set at $x + e_x$, that is the lower limit of the age group (85) plus the life expectancy of persons aged 85 years and over (see Woods 1982a: 206).

In a growing population, the older the cohort, the smaller will be their numbers at birth, because we are looking back in time. However, in the calculation of the age distribution of a stable population, we apply the multipliers not to figures on births (as in Table 9.2 above), but to the number of survivors at each age in the stationary population. The multipliers adjust the size of each age group consistently, according to the number of years since their birth:

$$ {}_5L_x \times (1+r)^{-y} = \text{adjusted cohort size} $$

The resulting absolute numbers at each age are of interest mainly for indicating the relative size of each age group: hence they are converted to percentages to show the age distribution in a more comparable form. The percentage distribution denotes the constant age structure of the stable population. In a growing stable population, the numbers in each age group increase, but the percentages remain constant.

Table 9.3 provides a worked example of the calculation of the age distribution, which proceeds as follows:

1. List the age groups (column A).
2. List the central ages (y) (column B).
3. List the age distribution of the stationary population, that is the ${}_5L_x$ values for males and females. Use the same life table as before for calculating R_0 and R_1, since the stable population must reflect only one mortality pattern. In the example, each five year cohort originated from five single-year cohorts numbering 100 000 at birth (columns C and D).
4. Calculate the multipliers for each age group (column E). For age group 0–4, the multiplier is $(1.005\,34)^{-2.5} = 0.986\,77$, while for age group 50–54 it is $(1.005\,34)^{-52.5} = 0.756\,08$.
5. To obtain the age distribution of males in the stable population, take the number of males in the stationary population, multiply it by the corresponding multiplier, then by 1.05, to allow for there being about 105 male births per 100 female births from year to year. This will give the approximate representation of males relative to females in the stable population. Thus, for age group 0–4, the working is $488\,600 \times 0.986\,77 \times 1.05 = 506\,243$, and for 95–99 it is $1876 \times 0.594\,96 \times 1.05 = 1172$.
6. The age distribution of females is found in the same way, but without any further adjustment for the sex ratio of births; e.g. for ages 95–99, $4000 \times 0.594\,96 = 2380$. The ${}_5L_{95}$ value, 4000, is the number of survivors expected from 100 000 births, whereas 2380 is the number of survivors expected in the same age group from a smaller number of births in the past.
7. Convert the resulting age distribution to percentages to make the figures more easily comparable: e.g. for males aged 0–4, $506\,234/(6\,155\,528 + 6\,131\,135) \times 100 = 4.1$. The figures are sometimes converted to sum to other convenient totals such as 100 000.

Table 9.3 Stable population model of a growing Western population, year 2000

Intrinsic growth rate: 0.005 34[1]								
Age group	Age group mid-point	Life table Pop. males	Life table Pop. females	Multiplier[1] $(1+r)^{-y}$	Stable population numbers		Stable population percentages	
	y	$5L_x$	$5L_x$		Males	Females	Males	Females
A	B	C	D	E	F C × E × 1.05	G D × E	H	I
0–4	2.5	488 600	491 897	0.986 77	506 243	485 389	4.1	4.0
5–9	7.5	486 890	490 802	0.960 84	491 215	471 582	4.0	3.8
10–14	12.5	485 948	490 266	0.935 60	477 386	458 693	3.9	3.7
15–19	17.5	484 591	489 572	0.911 01	463 541	446 005	3.8	3.6
20–24	22.5	482 443	488 516	0.887 07	449 359	433 348	3.7	3.5
25–29	27.5	479 960	487 105	0.863 76	435 299	420 742	3.5	3.4
30–34	32.5	477 360	485 305	0.841 06	421 563	408 171	3.4	3.3
35–39	37.5	474 200	482 886	0.818 96	407 768	395 464	3.3	3.2
40–44	42.5	469 679	479 348	0.797 44	393 268	382 251	3.2	3.1
45–49	47.5	462 323	473 755	0.776 49	376 939	367 866	3.1	3.0
50–54	52.5	450 108	464 942	0.756 08	357 334	351 533	2.9	2.9
55–59	57.5	430 119	451 366	0.736 21	332 491	332 300	2.7	2.7
60–64	62.5	399 032	430 305	0.716 87	300 357	308 473	2.4	2.5
65–69	67.5	353 417	396 621	0.698 03	259 030	276 853	2.1	2.3
70–74	72.5	290 250	343 578	0.679 69	207 144	233 527	1.7	1.9
75–79	77.5	209 498	265 435	0.661 83	145 585	175 673	1.2	1.4
80–84	82.5	124 792	172 978	0.644 44	84 442	111 474	0.7	0.9
85–89	87.5	54 522	84 734	0.627 50	35 923	53 171	0.3	0.4
90–94	92.5	14 691	26 392	0.611 01	9 425	16 126	0.1	0.1
95–99	97.5	1 876	4 000	0.594 96	1 172	2 380	0.0	0.0
100+[2]	101.3	77	197	0.583 04	47	115	0.0	0.0
Total		7 120 376	7 500 000		6 155 528	6 131 135	50.1	49.9

[1] Rounded to 5 decimal places, to produce results consistent with manual calculation. Unless figures are rounded, spreadsheets retain more decimal places than those displayed.
[2] Central age for ages 100 and over = $x + e_{100}$ = 101.3 for both sexes.
Data source: West level 23 model life table (Coale & Demeny 1983: 53).

Other characteristics of stable populations

Apart from the intrinsic growth rate and the constant age distribution, there are other measures describing the characteristics of the stable population, including the *intrinsic birth and death rates* and the *mean length of a generation*.

The intrinsic birth rate, the constant crude birth rate in a stable population, may be calculated for the base year in the usual way as total births/total population × 1000. The model assumes 105 000 male births and 100 000 female births in the base year, 2000, while the total population is the sum of the male and female populations in the base year (Table 9.3, columns F and G). Therefore the population's crude birth rate is 205 000/(6 155 528 + 6 131 135) = 0.016 68, or 16.68 per thousand (see Barclay 1966: 222; Shryock and Siegel 1973: 529–530).

Also, just as the crude death rate is equal to CBR – RNI, the intrinsic death rate (the constant crude death rate in a stable population) is equal to the intrinsic birth rate minus the intrinsic rate of natural increase. Thus, in the same example, the intrinsic death rate = 0.016 68 – 0.005 34 = 0.011 34, or 11.34 per thousand.

Other basic measures, such as dependency ratios and the median age, could also be used to describe a stable population, but there is one final measure to mention that is a by-product of the calculation of the intrinsic rate of natural increase, namely *the mean length of a generation*, which closely approximates the mean age of mothers at the birth of their daughters – in other words *the mean age at child-bearing* (Hinde 1998: 172–3). The mean length of a generation is calculated as R_1/R_0 (Pollard et al. 1990: 108). The working in Table 9.1 shows this figure to be 29.04 years. Figures of this magnitude occur in many populations, and demographic estimates often assume that the mean length of a generation is 29 years.

9.3 Model life tables

So far, the discussion of stable and stationary models has assumed that data are available for their calculation, which is not always the case. When adequate data are unavailable, for historical studies or even contemporary research on poorly documented populations, model life tables and model stable populations can serve in their stead.

Model life tables are sets of hypothetical life tables spanning the full range of life expectancies as well as different patterns of age-specific mortality. They permit an educated reconstruction of mortality conditions in all kinds of populations, from primitive peoples, to ancient civilizations, pre-industrial communities, contemporary populations of developing countries and even future populations. This section explains the nature of the two types of model life tables, empirical and relational, both of which provide important and versatile resources for demographic studies.

Empirical model life tables

There are two series of empirical model life tables, one first produced by the United Nations in 1955 and revised in 1982 (United Nations 1982a), the other originating from the Office of Population Research at Princeton University in 1966, with revisions in 1983 (Coale and Demeny 1983) and 1989 (Coale and Guo 1989 and 1990). The two series serve somewhat different purposes, the United Nations series focusing on mortality patterns in developing countries, the Princeton series on mortality patterns – past, present and future – in the developed countries.

The *Regional Model Life Tables* (Coale and Demeny 1983), from Princeton University, provide four sets of abridged life tables for females and males. The female life expectancies at birth in each set range, in steps of 2.5 years, from 20 to 80 years in the 1983 edition (Coale and Demeny 1983), and to 85 in the later revision.

The sets denote four regions, each with different patterns of age-specific mortality, as described in Table 9.4. The tables are based almost entirely on North American and European life tables from the late nineteenth century through the twentieth century. The West region life tables draw upon life tables recording the historical experience of mortality transition in many Western countries. The other regional patterns derive from smaller and more limited data sets, essentially referring to the experience of mortality change in North, South and East regions of Europe.

Ever since their first publication, the *Regional Model Life Tables*, and those for the West region in particular, have been a mainstay for demographic estimation and reconstruction in many countries and societies including Australian Aborigines and Ancient Egyptians. The West model life tables have been recommended for use when the age pattern of mortality is unknown (Newell 1988:138). Table

Table 9.4 Characteristics of the regional model life tables

Region	Country life tables	Age pattern of mortality
North	Norway, Sweden, Iceland	Relatively low mortality in infancy and after age 50.
South	Southern Europe, including Spain, Portugal and southern Italy	High mortality under age 5, low mortality at ages 40–60 and high mortality over age 65.
East	Mainly central Europe, including Germany, Austria and Poland	Relatively high mortality in infancy and after age 50.
West	Mainly western Europe and other countries of European settlement, including the United States, Canada, Australia and New Zealand Also includes Japan, Taiwan and Israel.	An 'average' mortality pattern.

Source: Coale and Demeny (1983: 11–12).

9.5 is an example of one of the life tables; the module *West Models.xls* includes a display illustrating the characteristics of the West region model life tables (see Box 9.1).

Since the *Regional Model Life Tables* represent the historical experience of the more developed countries, they have been less relevant to describing the experience of contemporary developing countries. Here the United Nations series of life tables, especially the 1982 edition, came into their own. They made substantial use of the body of fairly reliable data available for less developed countries and have been an important resource for demographic research in such areas. Like the *Regional Model Life Tables*, the *Model Life Tables for Developing Countries* (United Nations 1982a) contain four series of regional tables, representing different age patterns of mortality, together with a *General* or average pattern similar to the West region in the Princeton series.

Table 9.5 Example of a regional model life table: West mortality level 25 (female life expectancy 80 years)

Female						
Age	l_x	${}_nm_x$	${}_nq_x$	${}_nL_x$	T_x	e_x
0	1000.000	6.04	6.01	995.19	80 000	80.00
1	993.989	0.44	1.76	3969.85	79 005	79.48
5	992.244	0.18	0.90	4958.76	75 035	75.62
10	991.350	0.15	0.75	4954.96	70 077	70.69
15	990.605	0.30	1.50	4949.45	65 122	65.74
20	989.117	0.36	1.79	4941.32	60 172	60.83
25	987.341	0.44	2.22	4931.44	55 231	55.94
30	985.147	0.55	2.77	4919.19	50 299	51.06
35	982.422	0.80	4.01	4902.65	45 380	46.19
40	978.480	1.25	6.24	4877.75	40 478	41.37
45	972.374	1.97	9.80	4839.00	35 600	36.61
50	962.845	3.06	15.19	4779.12	30 761	31.95
55	948.220	4.60	22.73	4689.36	25 982	27.40
60	926.664	7.28	35.79	4553.72	21 292	22.98
65	893.498	11.90	57.86	4343.41	16 739	18.73
70	841.797	21.45	102.02	4002.88	12 395	14.72
75	755.921	40.57	184.17	3431.56	8 392	11.10
80	616.702	76.10	318.23	2578.93	4 961	8.04
85	420.447	134.21	493.17	1545.03	2 382	5.67
90	213.096	222.52	679.55	650.76	837	3.93
95	68.287	346.88	833.97	164.17	186	2.73
100	11.338	517.35	1000.00	21.92	22	1.93

Table 9.5 *continued*

Male						
Age	l_x	${}_nm_x$	${}_nq_x$	${}_nL_x$	T_x	e_x
0	1000.000	7.72	7.67	993.86	73 880	73.88
1	992.327	0.55	2.18	3961.72	72 886	73.45
5	990.159	0.27	1.37	4947.07	68 924	69.61
10	988.806	0.25	1.24	4941.09	63 977	64.70
15	987.582	0.77	3.82	4928.85	59 036	59.78
20	983.807	1.04	5.18	4906.81	54 107	55.00
25	978.712	1.14	5.68	4880.21	49 200	50.27
30	973.149	1.13	5.64	4852.57	44 320	45.54
35	967.658	1.48	7.37	4821.18	39 468	40.79
40	960.531	2.32	11.52	4776.10	34 646	36.07
45	949.468	3.73	18.50	4705.18	29 870	31.46
50	931.900	6.25	30.76	4590.69	25 165	27.00
55	903.231	9.78	47.75	4412.63	20 575	22.78
60	860.097	15.52	74.82	4146.03	16 162	18.79
65	795.741	25.01	117.98	3753.39	12 016	15.10
70	701.859	41.05	186.83	3194.59	8 262	11.77
75	570.731	69.52	296.13	2431.12	5 068	8.88
80	401.719	112.71	436.36	1555.21	2 637	6.56
85	226.424	176.18	596.00	765.96	1 082	4.78
90	91.475	265.50	750.92	258.73	316	3.45
95	22.784	385.72	873.37	51.59	57	2.50
100	2.885	547.55	1000.00	5.27	5	1.83

Note: the radix of the life table is 1000; the ${}_nm_x$ values are rates per 1000.
Source: Coale and Guo (1990: 31).

When selecting a model life table to represent mortality conditions at a particular time and place, the first step is to identify the region whose pattern of age-specific mortality best represents the assumed conditions. From this set a selection is then made according to the level of mortality that seems most appropriate, such as by referring to the figures on the probability of dying (e.g. in infancy), or the proportions surviving at particular ages, or life expectancy.

As 'models', the model life tables are intended to correspond closely to real conditions, but the relevance of a selected model life table to a particular situation depends on making an informed choice – on the basis of comparisons with whatever data are available for the population in question. This requires careful evaluation of evidence available for the particular society, or for similar societies. Where sufficient mortality statistics are available to calculate life tables, comparisons with

BOX 9.1 **West model life tables and stable populations (*West Models.xls*)**

This module provides a visual introduction to the regional model life tables and stable populations devised by Coale and Demeny (1983) and updated by Coale and Guo (1989 and 1990). It also presents population pyramids depicting stable population models calculated from data for 20 countries and regions. The three main displays in the module consist of:

- *Model life tables*: shows age and sex-specific data for the main life table functions, for instance the survival curves (l_x values) for populations at various levels of mortality. The mortality level is selected using a scroll bar on the screen. The information displayed here is the same as that for the *Epidemiologic Transition* module (see Box 6.1), but the applications focus on understanding life tables.
- *Model stable populations*: presents age structures of stable populations corresponding to any level of mortality and fertility, selected using scroll bars.
- *Country/region examples*: presents the age structures of a country or region in 2000, superimposed on the stable age structure corresponding to chosen levels of mortality and fertility, selected using scroll bars. Moving the scroll bar for the 'current year' projects the population forward from 2000 until the population develops the stable age structure.

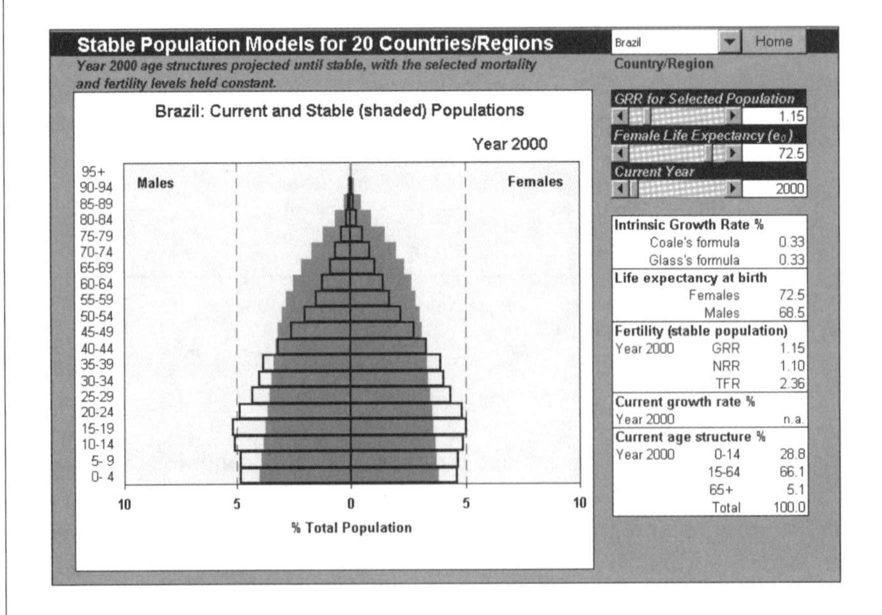

The life table and stable population data presented approximate figures in Coale and Demeny (1983), but the module is not intended to replicate the original tables. The base data were sets of l_x values from the West region model life tables (see Section 9.3). From these, the module calculates life tables and stable populations that correspond to the original models sufficiently for illustrative purposes.

continued

BOX 9.1 continued

Instructions

To use the module, insert the CD in the drive, start Excel and open the file *West Models.xls*. Enable macros when prompted.

- Click **Model Life Tables** to display graphs of life table functions at various mortality levels.
- Click **Model Stable Populations** to display the age structures of stable population at various levels of mortality and fertility.
- Click **Countries & Regions** to project national and regional age structures until they become stable at the chosen levels of mortality and fertility.
- Use the scroll bars and menus to change each display.
- Click **Home** to return to the opening display.

See Appendix B for further advice on using the Excel modules.

Illustrative applications

1 Life tables

The following activities illustrate age and sex-specific trends through time in the main life table functions. For some questions, sketching 'expected' results before using the computer helps to test understanding, both of the life table functions and of trends in mortality through time.

The figures refer to abridged life tables. The youngest ages depicted in the graph are 0 and 1, followed by ages in 5 year steps from 5 onwards. The details for the youngest ages reveal the impact of infant mortality. Although the module allows only female mortality levels to be set, it automatically chooses the male mortality corresponding to the same level in Coale and Demeny's (1983) model life tables. Female life expectancy changes in increments of 2.5 years from a minimum of 20 years.

(a) Set female life expectancy to 20 and select the d_x function. Increase the level of life expectancy gradually to illustrate changes in the pattern of deaths in populations experiencing the demographic transition.

(b) Choose the l_x function and observe changes associated with rising life expectancy, especially the substantial improvements in survival at younger ages and, ultimately, the rectangularization of the survival curve.

(c) Using the e_x function, examine whether life expectancy always declines with age. Test this by comparing age-specific life expectancy for various mortality levels. What do the results indicate in relation to 'the paradox of the life table'?

(d) With female life expectancy at birth set at 70 years, compare the graphs of all the life table functions. Why are there similarities between the distribution of the m_x and q_x values? Why do the graphs for l_x and L_x differ so much in these examples? Why are there similarities here between the graphs for e_x and T_x? Why is the pattern for d_x different from the others?

continued

BOX 9.1 continued

2 Stable populations

The display of stable populations includes calculations of the intrinsic rate of natural increase from Coale's and Glass's formulas – the latter is more complex but more precise (for Glass's formula, see Shryock and Siegel 1973: 527). The gross reproduction rate increases in increments of 0.05 daughters per woman.

(e) *Effects of fertility decline on the age structure* For female life expectancies of 20 and 80, observe the extent to which varying the level of fertility affects the age profile of the stable population (e.g. set female e_0 to 20, then use the gross reproduction rate scroll bar to change the level of fertility).

(f) *Effects of mortality decline on the age structure* For GRRs of 3.0, 1.0 and 0.5, observe the extent to which varying the level of mortality affects the age profile of the stable population (e.g. set the GRR to 3.0, then use the female life expectancy scroll bar to change the level of mortality).

(g) *Population ageing* Use the module to view a selection of the age profiles of the stable populations depicted in Figure 9.5 – by setting life expectancy and GRR to the levels indicated.

(h) *Intrinsic growth rate* Use the module to examine the relationship between the intrinsic growth rate and the levels of fertility and mortality. Is the shape of the age profile an indicator of the intrinsic growth rate?

3 Country and regional examples

Click the **Countries & Regions** button on the opening display, and choose a country from the menu at the top right. Set the gross reproduction rate (GRR) and female life expectancy to the required values. Representative values may be taken from the table below, but the display also permits experimentation with a wide range of different levels of fertility and mortality. The chart shows the current age structure, e.g. in the year 2000, compared with the age structure of the stable population corresponding to the selected levels of fertility and mortality. Moving the 'current year' scroll bar shows the projected population converging to a stable age structure. The projection can extend to 200 years, but most of the changes are completed within 70 to 100 years.

(i) *Populations forget their past* Set GRR and e_0 to any levels that seem interesting and, keeping these constant, observe how different national populations converge to the same stable age distribution (i.e. choose different populations from the drop-down menu and run a projection for each using the 'current year' scroll bar.). This is a demonstration of Lotka's theorem, referred to in the quotation at the end of Section 9.1.

(j) *Current and stable age distributions* Compare the age structures of selected populations in 2000 with the stable age distributions that would result from holding constant the gross reproduction rates and the life expectancies shown in the table below (i.e. set the current year to 2000 and GRR and e_0 to the values in the table). Comment on consequences potentially ensuing from such changes in the shape of the age structure.

continued

BOX 9.1 continued

(k) *Responding to below-replacement fertility* Choose a country with below-replacement fertility in 2001, such as China or Germany and, adjusting the GRR until the NRR in the right-hand table is around 1.0, note the changes in the age profile that would result from a return to replacement-level fertility. What is the intrinsic growth rate when NRR is 1.0?

(l) *Effects of changes in life expectancy* Using the example of South Africa or India or Pakistan, examine the effects on the age structure of higher or lower levels of life expectancy than those shown in the table below (keep GRR constant at the level shown in the table). Do changes in mortality have an appreciable effect on the age structure? Does an understanding of population ageing help to explain the results?

Measures of fertility and mortality for selected countries and regions, 2001

	Region/Country	TFR	GRR	Female e_0
1	World	2.8	1.37	69
2	More developed regions	1.6	0.78	79
3	Less developed regions	3.2	1.56	66
4	Australia	1.7	0.83	82
5	Brazil	2.4	1.17	72
6	Canada	1.4	0.68	81
7	China	1.8	0.88	73
8	Egypt	3.5	1.71	68
9	France	1.9	0.93	83
10	Germany	1.3	0.63	81
11	India	3.2	1.56	61
12	Indonesia	2.7	1.32	70
13	Italy	1.3	0.63	82
14	Japan	1.3	0.63	84
15	Mexico	2.8	1.37	78
16	New Zealand	2.0	0.98	80
17	Pakistan	5.6	2.73	61
18	South Africa	2.9	1.41	54
19	United Kingdom	1.7	0.83	80
20	United States of America	2.1	1.02	80

Source of TFR and e_0: Population Reference Bureau 2001a: *World Population Data Sheet 2001*.

model life tables can alert researchers to deficiencies or idiosyncrasies in the observed figures.

Relational model life tables

While the Princeton and United Nations model life tables derive from empirical data, William Brass's (1971) relational model life table system is an inventive alternative approach to modelling a population's mortality. Instead of choosing a table from a published set, the relational approach enables researchers to calculate a life table, or a whole 'family' of related life tables, from a 'standard' set of l_x values.

Thus a life table for Sweden in 2001 could be used as a standard from which to generate a series of life tables representing Sweden's mortality at various dates in the future, for instance to take account of anticipated increases in life expectancies. These life tables could then serve to provide the mortality assumptions for cohort component projections, which require age-specific survival data for each step in the projection (see Chapter 12).

Brass's model life tables are *relational* models, because they refer to the relationship between the l_x columns of two life tables – the 'standard' life table selected as the base for the calculations, and the 'predicted' life table obtained from it. The module on relational models, included in this chapter, enables the calculation of life tables using Brass's method (see Box 9.2). The steps in using the module entail selecting a standard population, and setting two parameters, alpha (α) and beta (β).

BOX 9.2 **Module on relational model life tables (Relational Models.xls)**

The relational model life table system is an imaginative demographic invention of William Brass, formerly Professor of Medical Demography at the London School of Hygiene and Tropical Medicine. Relational models enable the production of 'families' of related life tables by choosing a standard life table and varying two parameters, alpha and beta (see Section 9.3). This module uses graphs and animations to demonstrate the nature of the relational models, and enables life tables to be viewed and printed. It is intended that experimentation with the module will convey much more about the construction and uses of relational models than is immediately accessible from a verbal description.

Life Table

Standard: West, female, e0=80

Age	mx/1000	lx	dx	qx	px	Lx	Tx	ex
0	0.60	100000	60	0.00060	0.99940	99958	9003579	90.04
1	0.01	99940	5	0.00005	0.99995	399749	8903622	89.09
5	0.01	99935	3	0.00003	0.99997	499667	8503872	85.09
10	0.01	99932	3	0.00003	0.99997	499652	8004205	80.10
15	0.01	99929	5	0.00005	0.99995	499632	7504553	75.10
20	0.02	99924	8	0.00008	0.99992	499600	7004921	70.10
25	0.02	99916	11	0.00011	0.99989	499554	6505320	65.11
30	0.03	99906	15	0.00015	0.99985	499491	6005766	60.11
35	0.05	99891	24	0.00024	0.99976	499393	5506276	55.12
40	0.09	99867	45	0.00045	0.99955	499220	5006883	50.14
45	0.18	99822	92	0.00092	0.99908	498878	4507662	45.16
50	0.32	99730	160	0.00161	0.99839	498248	4008784	40.20
55	0.57	99569	284	0.00286	0.99714	497136	3510536	35.26
60	1.01	99285	501	0.00505	0.99495	495172	3013400	30.35
65	2.10	98784	1031	0.01044	0.98956	491341	2518228	25.49
70	4.43	97753	2142	0.02191	0.97809	483408	2026888	20.73
75	10.05	95611	4688	0.04903	0.95097	466333	1543480	16.14
80	22.93	90922	9859	0.10843	0.89157	429966	1077147	11.85
85	59.44	81064	20974	0.25874	0.74126	352883	647182	7.98
90	159.99	60089	34335	0.57140	0.42860	214609	294299	4.90
95	309.43	25754	22466	0.87233	0.12767	72606	79690	3.09
100	464.14	3288	3288	1.00000	0.00000	7084	7084	2.15

Standard Population
West, female, e0=80

Alpha -1.00
Range -1.5 to 0.8.

Beta 1.00
Range 0.6 to 1.4.

Home

continued

BOX 9.2 continued

Instructions

The buttons on the opening display link to the following operations:

- **Functions**: presents graphs of life table functions for selected standard populations and any values of alpha and beta.
- **Life Tables**: displays abridged life tables calculated from selected standard populations and any values of alpha and beta.
- **Logits**: displays a scatter diagram of standard and predicted logits of l_x, for selected standard populations and any values of alpha and beta.
- **Functions & Logits**: a combined display.
- Click **Home** to return to the opening display.
- The **Example** tab, at the bottom of the screen, is a spreadsheet with a worked example of the calculation of 'predicted' l_x values, and an option to vary alpha and beta.

Use the menus and scroll bars to alter the characteristics of the models.

Illustrative applications

Note that values of alpha represent the level of mortality and range from about −1.5 to 0.8 (see Section 9.3). Also, beta represents the relationship between childhood and adult mortality, and ranges in value from about 0.6 to 1.4.

1. Choose the **Functions** display, set alpha = 0 and beta = 1 then compare the characteristics of the standard populations in terms of d_x, l_x and e_x.
2. On the **Functions** display, set the standard population to 'West, female, e_0 = 40' and observe changes in d_x arising when alpha and beta are increased or decreased. Repeat using the l_x function. Note that unrealistic results may occur for values of alpha and beta outside the ranges specified above.
3. On the **Life Tables** display, set beta = 1 and the standard to 'West, female, e_0 = 60', then adjust alpha to produce a life table with a life expectancy at birth of 70 years. Also, observe that many different life tables can be produced by changing the standard life table and varying alpha and beta.
4. On the **Logits** display, observe how the intercept and slope of the regression line change as alpha and beta increase or decrease.
5. On the **Functions and Logits** display, set the function to l_x and alpha to 0, then observe the changes in the survival curve as the intercept and slope of the regression line are altered. Again, note that unrealistic results may occur for values of alpha and beta outside the ranges specified above.

Alpha denotes the level of mortality; the value of alpha determines whether life expectancy in the new life table is higher, the same or lower than in the standard life table. Beta denotes the relationship between childhood and adult mortality (Newell 1988: 156). In effect, relational model life tables permit the production of a wide range of life tables consistent with any chosen standard, but give the user latitude to determine the extent to which the new life table differs from the standard.

When corresponding pairs of l_x values from two life tables are plotted on a scatter diagram, the points will lie on a curve (Figure 9.3). The l_x values for the

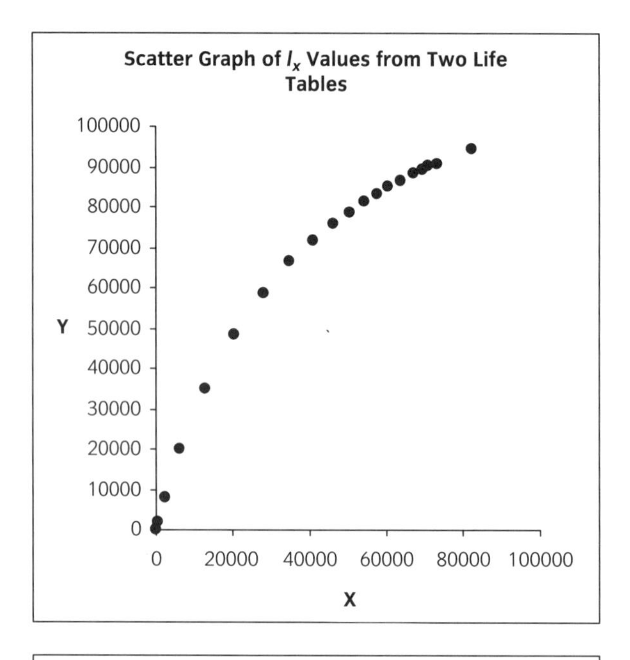

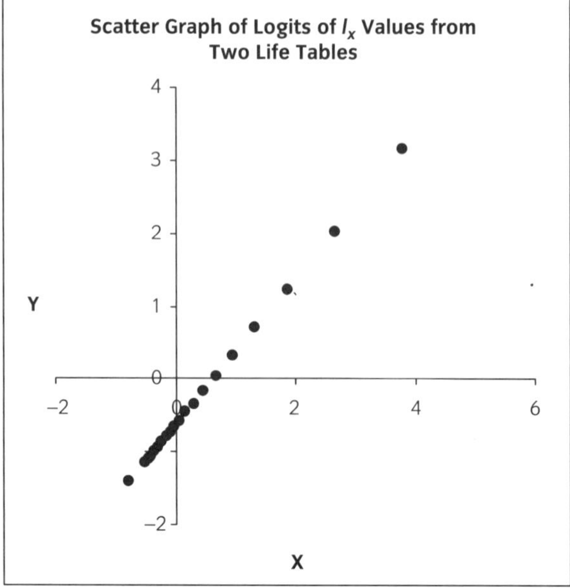

Figure 9.3 Scatter graphs of l_x values and logits of l_x values from two life tables.

Source: Spreadsheet Exercise 13.

'standard' life table are plotted on the horizontal axis, and the l_x values for a second life table on the vertical axis. Since curvilinear relationships are more complex to describe mathematically than linear relationships, Brass transformed the l_x values into logits: data transformation entails changing the scale of measurement of a variable, such as from an arithmetic scale to a logarithmic scale, or from proportions to logits (see Armitage and Berry 1987: Chapter 11).

If a life table has a radix of 100 000, l_x values are divided by 100 000 to express them as proportions. The logit of the proportion aged x may then be calculated as follows (Newell 1988: 153):

$$\text{logit}(p_x) = 0.5\ln\left(\frac{1-p_x}{p_x}\right)$$

where ln denotes the natural logarithm and p_x is a proportion >0 and <1. Thus if $p_x = 0.25$, the logit = 0.549 31. There is no logit for the proportion surviving at age zero, since the proportion is 1. In Excel, the formula is written: logit = 0.5 *LN((1 − p_x)/p_x), where LN( ) is the natural logarithm function. The spreadsheet exercise in Chapter 13 provides practice in using data transformation, including converting proportions to logits.

On a scatter diagram, the logits of corresponding pairs of l_x values tend to be arranged in a straight line, although the points are not necessarily as perfectly aligned as in Figure 9.3. Thus the advantage of the logit transformation is that it enables the relationship between two life tables to be expressed in terms of a straight line (a linear regression line), which can be defined by two parameters or measures: the *intercept* (alpha) where it crosses the Y axis, and the *slope* (beta) of the line. (Detailed accounts of linear regression are presented in statistics texts.)

Accordingly, a relational model life table is calculated from a standard life table by making assumptions about the relationship between the standard and the new life table. The relationship is described by a linear regression line. Any point (y) on the linear regression line is equal to $\alpha + \beta x$ since the equation for a straight line is: $y = \alpha + \beta x$

In relational models:

y represents the logits of the values in the 'predicted' life table,
x represents the logits of the corresponding values in the 'standard' life table,
α (alpha) = intercept,
β (beta) = slope.

By varying the *alpha coefficient* (the intercept of the regression line) and the *beta coefficient* (the slope of the regression line), a wide range of life tables can be generated with varying life expectancies and different patterns of age-specific mortality (see Box 9.2).

For example, if in the standard population l_{10} = 90 000, converting this to a proportion gives 0.900 00. Converting the proportion to a logit gives −1.098 61. If we assume that $\alpha = -0.5$ and $\beta = 1$, then $y = -0.5 + (-1.09861) = -1.59861$.

To convert this back to a proportion, use (Newell 1988: 162):

$$\text{antilogit of logit }(p_x) = \frac{1}{1+e^{2y}}$$

where y is the logit of the 'predicted' life table value, and e is the base for natural logarithms, i.e. approximately 2.718 282.

Using Excel, antilogit = 1/(1 + EXP(2 * y)). In the above example, logit (p_{10}) = −1.598 71, therefore the antilogit = $1/(1 + e^{2\times-1.598\,71}) = 0.960\,74$. Multiplying by 100 000, the predicted value of l_{10} in the new life table is 96 074.

If $\alpha = 0$ and $\beta = 1$ the 'predicted' value of l_{10} will be the same as the original 'standard' value: $y = 0 + (-1.098\,61) = -1.098\,61$; antilogit ($p_{10}$) = 0.900 00; therefore l_{10} = 90 000.

From the l_x values, the other functions in the new relational life table are calculated. Thus by changing the values of alpha or beta, or both, a set of life tables can be calculated from the same standard life table. A worked example of this can be viewed in the relational model life tables module by selecting the spreadsheet tab labelled 'Example' (Box 9.2).

Values of alpha represent the level of mortality, and range in value from about −1.5 to 0.8 (Newell 1988: 157):

−1.5	0.0	0.8
Higher life expectancy	Life expectancy the same as the standard	Lower life expectancy

Values of beta represent the relationship between childhood and adult mortality, and range in value from about 0.6 to 1.4:

0.6	1.0	1.4
High infant and child mortality, and low adult mortality, relative to the standard	The relationship between adult and child mortality remains the same as in the standard life table	Low infant and child mortality, and high adult mortality, relative to the standard

Leaving beta at 1.0 would result in a simpler (one parameter) system based on the alpha values alone. Conversely, others have devised more complex (four parameter) systems to provide a better correspondence with certain, less common, mortality patterns (Newell 1988: 157 and 163–164).

The standard life table may be the most recent official life table for the population in question, for instance if the aim is to produce further life tables showing the likely mortality of that population in future years. If no suitable life table is available for the population in question, a model life table could be used instead. The standard should have a mortality pattern similar to that of the population being studied, and should reflect conditions expected to prevail. This means that if epidemic diseases were prevalent in a country's past, but no longer, the standard life table for deriving future mortality patterns should exclude the impact of such diseases.

Referring to the relational model life table system, Coale and Demeny (1983: 1) noted:

This system for generating model life tables has the advantage of permitting variation in two parameters for each 'family' of tables considered, but the set of tables covered is nevertheless determined by the choice of a 'standard' life table. The logit transformation can closely approximate all members of any one of the four families of the regional tables presented in this book by assigning a given table (for instance, that with a life expectancy of 30 years) as the standard, and by a suitable choice of α and β for the other tables in the family. However, it is not possible to give a close approximation of a life table in one family as a logit transformation of a standard table in another family.

9.4 Model stable populations

Accompanying both sets of empirical model life tables are corresponding sets of model stable populations (United Nations 1982b; Coale and Demeny 1983); the discussion here refers to examples from the latter publication. In the *Regional Model Life Tables*, Coale and Demeny present the age distributions and other attributes of the stable populations with intrinsic rates of natural increase between −1 and +5 per cent, at intervals of 0.5 per cent. Thus the model stable populations cover an extensive range of mortality patterns, mortality levels and intrinsic rates of natural increase.

The authors also provided a second set of stable populations with gross reproduction rates (GRR) ranging from 0.8 to 6.0, as illustrated in Table 9.6, which presents selected data from a much more detailed tabulation. The table refers to West region females with a life expectancy at birth of 60 years. It shows stable age distributions ranging from a very young profile when GRR = 6 daughters per woman, to old age structures for the lowest values of GRR; the figures correspond to percentages of total females in each age group. The intrinsic growth rates of the stable populations range between −1.4 and 5.9 per cent (in the table, r is expressed as a rate per 1000). Corresponding stable populations for males are also included in the volume. The GRRs in the tables assume a mean age of maternity, or paternity, of 29 years (Coale and Demeny 1983: 35). The module *West Models.xls* illustrates the model stable populations by generating the age–sex pyramids for stable populations corresponding to a wide range of life expectancies and fertility levels, based on data approximating the life tables and stable populations in the volume by Coale and Demeny (1983).

Clearly, the model stable populations represent a vast archive of information covering very many of the stable populations that researchers may wish to use. The model stable populations also serve to illustrate the interrelationships between demographic variables, including the effects on age structures of different levels

Table 9.6 Stable populations at given rates of the female GRR, females, West mortality level 17 (life expectancy 60 years)

GRR(29) =	0.80	1.00	1.25	1.50	1.75	2.00	2.25	2.50	3.00	3.50	4.00	5.00	6.00
Population at age (x)													
0–1	0.93	1.26	1.68	2.07	2.44	2.79	3.11	3.42	3.97	4.46	4.90	5.66	6.31
1–4	3.68	4.92	6.40	7.78	9.05	10.22	11.29	12.28	14.03	15.55	16.87	19.09	20.88
5–9	4.80	6.20	7.79	9.21	10.45	11.55	12.52	13.39	14.85	16.04	17.02	18.55	19.67
10–14	5.09	6.33	7.66	8.76	9.68	10.45	11.09	11.63	12.48	13.11	13.57	14.18	14.53
15–19	5.40	6.46	7.52	8.33	8.95	9.43	9.81	10.09	10.48	10.69	10.80	10.82	10.71
20–24	5.69	6.56	7.34	7.88	8.25	8.49	8.64	8.72	8.76	8.69	8.56	8.23	7.87
25–29	5.99	6.64	7.15	7.44	7.57	7.61	7.58	7.51	7.30	7.04	6.76	6.24	5.76
30–34	6.28	6.70	6.95	7.00	6.93	6.80	6.63	6.45	6.06	5.68	5.33	4.71	4.21
35–39	6.56	6.74	6.73	6.56	6.32	6.06	5.79	5.52	5.02	4.57	4.18	3.55	3.06
40–44	6.83	6.76	6.48	6.12	5.75	5.38	5.03	4.71	4.14	3.67	3.27	2.66	2.22
45–49	7.07	6.73	6.21	5.68	5.19	4.74	4.34	3.99	3.39	2.92	2.54	1.98	1.60
50–54	7.23	6.62	5.88	5.21	4.63	4.13	3.71	3.34	2.75	2.30	1.95	1.46	1.14
55–59	7.26	6.40	5.47	4.70	4.06	3.54	3.11	2.75	2.19	1.78	1.47	1.06	0.79
60–64	7.07	6.01	4.94	4.11	3.45	2.94	2.53	2.19	1.69	1.33	1.08	0.74	0.54
65–69	6.56	5.36	4.24	3.41	2.80	2.32	1.95	1.66	1.24	0.95	0.75	0.50	0.35
70–74	5.60	4.41	3.35	2.62	2.09	1.69	1.39	1.16	0.84	0.63	0.48	0.31	0.21
75–79	4.15	3.15	2.30	1.74	1.35	1.07	0.86	0.71	0.49	0.36	0.27	0.16	0.11
80–84	2.48	1.81	1.28	0.94	0.71	0.55	0.43	0.35	0.24	0.17	0.12	0.07	0.05
85–89	1.05	0.74	0.50	0.36	0.26	0.20	0.15	0.12	0.08	0.06	0.04	0.02	0.01
90–94	0.26	0.18	0.12	0.08	0.06	0.04	0.03	0.03	0.02	0.01	0.01	0.00	0.00
95+	0.03	0.02	0.01	0.01	0.01	0.00	0.00	0.00	0.00	0.00	0.00	0.00	0.00
Birth rate	9.71	13.28	17.69	21.92	25.92	29.69	33.22	36.54	42.61	48.02	52.89	61.35	68.52
Death rate	23.38	19.32	16.01	13.86	12.42	11.43	10.74	10.26	9.69	9.44	9.36	9.47	9.74
Growth rate	−13.67	−6.04	1.68	8.06	13.50	18.26	22.48	26.28	32.92	38.58	43.53	51.88	58.78
NRR (29)	0.672	0.840	1.050	1.259	1.469	1.679	1.889	2.099	2.519	2.939	3.358	4.198	5.038

Note: GRR = gross reproduction rate; NRR = net reproduction rate.
Source: Selected data from Coale and Demeny (1983: 96).

of fertility and mortality, a topic examined further in Section 9.6 on 'Population Ageing'.

Model stable populations have had many applications in demographic analysis and estimation. As Coale and Demeny (1983: 29) stated:

> The scheme of tabulation is designed with two considerations in mind: (a) to provide stable age distributions for the full range of mortality and fertility levels likely to be relevant either in fitting the tables to observed data, or in analysing the hypothetical consequences of fertility and mortality schedules; and (b) to tabulate values at small enough intervals so that linear interpolation is sufficiently exact.

9.5 Population momentum

As well as the modules on life tables and stable populations, this chapter includes a module illustrating the phenomenon of *population momentum*, through statistics for 20 countries and regions, together with hypothetical data (Box 9.3). Population momentum denotes the long-range potential for change in the overall size of a population, inherent in the different numbers in younger and older cohorts. It is sometimes known also as 'the momentum of growth' (Keyfitz 1971) or simply 'growth potential'. Population momentum is the potential for growth (or decline) that is inherent within an age structure. If (i) fertility is at replacement level, (ii) mortality is constant and (iii) net migration is zero, the only force for growth potentially remaining is population momentum.

Thus population momentum is the potential for change in total population numbers, measured as the difference in size between the present population and the future stationary population, obtained from a projection holding mortality constant at the present level and fertility constant at replacement level. If the difference is positive, *positive momentum* (momentum of growth) is present. If the difference is negative (i.e. the present population is larger than the projected stationary population), *negative momentum* (momentum of decline) is present.

Positive momentum was illustrated earlier in the *Age Structure Simulations* module (see Box 3.2), where the 'Population ageing' example demonstrates that if the United States population experienced replacement level fertility from 1970 onwards, as well as constant mortality and zero migration, it would increase in size by 32 per cent in the next 100 years (from 203 212 000 in 1970 to 268 482 000 in 2070). All of the growth would be due to the movement of larger cohorts into older ages. Comparing the age structures at the start and end of the simulation shows that most of the growth occurs in the middle and older ages. The 32 per cent increase is due to population momentum.

A necessary distinction, however, is between momentum and self-reinforcing growth, whereby larger generations of parents beget still larger generations of

BOX 9.3 **Module on population momentum (*Momentum.xls*)**

This module presents graphical displays of the phenomenon of population momentum. It projects populations, with constant mortality and replacement fertility, until they become stationary, measuring momentum at each step by projection and, for comparison, by the method of estimation described in Section 9.5. The module uses both hypothetical and real data to provide a diverse range of examples. There are two main displays in the module:

- *Population momentum, hypothetical data*: projects populations, with stable age structures in year zero, until they become stationary. Changing the gross reproduction rate affects the shape of the initial population only. Changing e_0 affects the initial age structure as well as determining the replacement fertility for the projection.
- *Population momentum in 20 countries & regions*: projects national and regional populations for the year 2000 until they become stationary at the chosen level of mortality and the corresponding replacement fertility. Momentum figures are shown for each step in the projection.

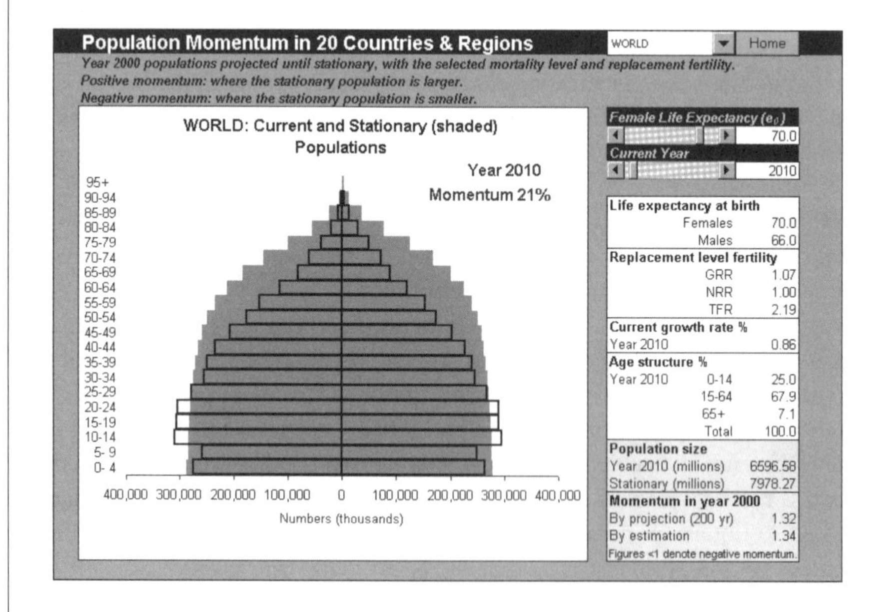

Both displays compare the age structure of the current population with that of the projected stationary population (after 200 years). Differences between the numbers in the two age profiles denote population momentum.

Where the stationary profile is larger, positive momentum (momentum of growth) is present. Conversely, where the stationary population is smaller, negative momentum (momentum of decline) is present. Often there is positive momentum in some age groups (those set to expand in size) and negative momentum in others (those that will contract in size, such as the child age groups when above-replacement fertility falls to replacement level).

continued

BOX 9.3 continued

As well as showing the initial and stationary populations, populations at intermediate steps in the projection can also be displayed, by moving the scroll bar for selecting the 'current year'. Differences between the size of the initial and stationary populations, or between the size of intermediate and stationary populations, indicate the current growth potential.

Instructions

To use the module, insert the CD in the drive, start Excel and open the file *Momentum.xls*. Enable macros when prompted.

- Click **Hypothetical Populations** to study projections and momentum measures beginning, in year zero, with a stable population at any selected level of fertility and mortality. The base data for the display are from Coale and Demeny (1983) and Coale and Guo (1990).
- Click **Countries & Regions** to begin the projections with a particular population in the year 2000. The data source was: Population Division of the Department of Economic and Social Affairs of the United Nations Secretariat 2001: *The Sex and Age Distribution of Populations*, United Nations publications, ST/ESA/Ser.A/199. New York: United Nations.
- Use the scroll bars and menus to change each display.
- Click **Home** to return to the opening display.

See Appendix B for further advice on using the Excel modules.

Illustrative applications

1 Hypothetical populations

This spreadsheet calculates the momentum inherent in a wide range of age structures and illustrates how momentum declines to zero as the population approaches a stationary state, assuming replacement level fertility from the initial year. The negative momentum inherent in populations with below-replacement fertility can also be illustrated, by setting the gross reproduction rate of the initial population to a low figure.

The table beside the graph includes data on the initial size of the population compared with its ultimate size, when stationary. A 100 year projection (at constant mortality and replacement level fertility) suffices to achieve zero momentum and a stationary age profile, but small perturbations can continue longer in some circumstances. The module projects the ultimate (stationary) population to 200 years so that longer-range projections can also be displayed, if required.

The table includes measures of population momentum obtained by projection and, in the starting year, by the short-cut formula described in Section 9.5. It also compares the intrinsic rate of natural increase from Coale's approximation and Glass's more exact formula (see Shryock and Siegel 1973: 527).

By providing a visual display of how momentum varies, the spreadsheet aims to clarify the concept, in effect through a series of 'movies' produced by changing values with the scroll bars. The scroll bars permit the selection of both likely and unlikely example populations – such as one with a life expectancy of 85 years and a GRR of 6.0. Many permutations and

continued

BOX 9.3 continued

combinations are within reach and worth exploring. Some more conventional scenarios to view are:

(a) *High fertility, high mortality* Set the current year to zero, GRR to 3.00 and e_0 to 20. Move the current year scroll bar to observe the transition to zero momentum and a stationary state (the initial and stationary populations will be almost identical at year 0; hence momentum is close to zero). Setting GRR to 3.15 will result in zero momentum in year 0.

(b) *High fertility, medium mortality* Repeat the projection, with initial settings of year = 0, GRR = 3, $e_0 = 50$.

(c) *Low fertility, low mortality* Again, run a projection with initial settings of year = 0, GRR = 1, $e_0 = 80$ (the initial and stationary populations will be almost identical at year 0; hence momentum remains low).

(d) *Very low fertility, low mortality* Repeat, with initial settings of year = 0, GRR = 0.5, $e_0 = 80$.

2 Country and regional data

This spreadsheet calculates population momentum for 20 countries and regions in 2000, and shows projected changes in their age structures as momentum approaches zero, that is as their age structures became stationary under the influence of replacement fertility. Mortality levels for the start of the twenty-first century can be set from the table included in Box 9.1. The display also permits experimentation with other mortality levels. The current age structure is superimposed on the stationary age structure to show the population's ultimate destination in the projection.

Some topics for exploration from this module are (1) the nature of 'unstoppable growth', due to the positive momentum in the age structure of the world's population, as well as that of developing countries (e.g. Mexico, India and Pakistan); (2) the problem of 'the momentum of decline' evident in countries with long-standing fertility levels below replacement, such as Italy and Germany. Although the projections to a stationary population (and replacement level fertility) always show momentum converging to zero, in reality some countries with below-replacement fertility will experience increasing levels of negative momentum through time.

children, as in the 'Population explosion' simulation (See Box 3.2). Momentum is measured in relation to fertility at replacement level, whereas self-reinforcing growth is associated with fertility above replacement level (see also Preston 1986: 350; Preston et al. 2001: 165–166).

Over the course of the classical demographic transition, momentum rises and falls producing a bell-shaped momentum curve (Figure 9.4). Initially, falling death rates add to cohort size and increase momentum. Later, falling birth rates decrease cohort size and reduce momentum. Stationary age structures with *zero momentum* characterize the theoretical starting point in the pre-transition stage, as well as the theoretical end point in the post-transition stage.

Populations experiencing replacement level fertility (and constant mortality and zero net migration) indefinitely, will eventually run out of momentum as the

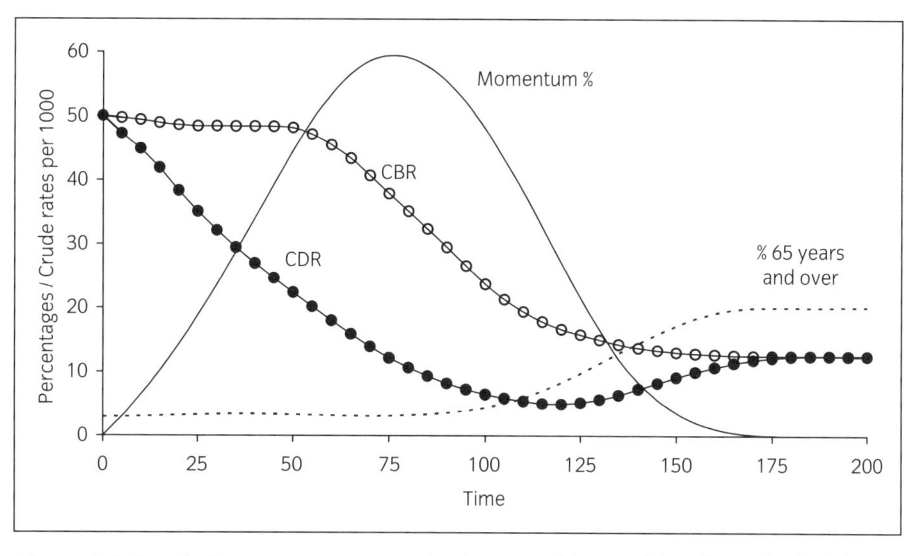

Figure 9.4 Population momentum over the demographic transition (hypothetical data).

growth rate falls to zero and the age structure becomes stationary, as in the United States ('Population ageing') simulation. Projections demonstrate that the shift to zero growth and zero momentum is largely completed after 70 years of constant birth and death rates, although minor departures from a stationary age structure may still be evident. *Negative momentum* (or the *momentum of decline*) will develop if a population experiences below-replacement fertility in the long run. Negative momentum is inherent within age structures that taper downwards, as illustrated by the simulations for Germany and Italy in the *Momentum* module (Box 9.3).

Calculating population momentum

There are various methods of calculating population momentum (Keyfitz 1971; Preston 1986; Kim and Schoen 1997) among which the most accurate is the cohort-component projection method (see Chapter 12). A country's population is projected for 100 years assuming replacement fertility, constant mortality at the level for the initial year and no migration – as in the 'Population ageing' simulation. After a century, the population will be stationary. Comparing the size of the stationary population with the size of the initial population reveals the latter's inherent level of momentum.

Momentum is then calculated as the ratio of the size of the stationary population to that of the initial population:

$$\text{Momentum} = \frac{\text{size of the stationary population}}{\text{size of the initial population}}$$

In the United States example mentioned above,

$$\text{Momentum} = \frac{\text{size of the population in 2070 (Year 100)}}{\text{size of the population in 1970 (Year 0)}} = \frac{268\,482\,000}{203\,212\,000} = 1.32$$

For brevity, the projection method is sometimes based only on the female population. Population momentum, calculated by projection for the year 2000, showed the world's population had an inherent potential for growth of 30 per cent in that year, while the figure for sub-Saharan Africa was 50 per cent (Bongaarts and Bulato 1999; Bongaarts and Bulato 2000: 27–29). Such indices draw attention to the fact that the world's population growth is unstoppable for decades, and that momentum in a majority of countries makes continuing increases inevitable. Even if countries with high birth rates achieved replacement fertility immediately, the inherent growth potential in their age structures would still cause great increases. Illustrations of this include the examples of Egypt, India, and Mexico in the *Momentum* module (Box 9.3).

Alternative methods of calculating momentum avoid the need to prepare projections, mainly through formulas employing data for the female population. One of the simplest methods of estimation yields results close to those obtained by projection (Bongaarts and Bulatao 1999: 523; Bogue 1993: volume 5, page 19-57, equation 9):

$$\text{Momentum} = \frac{\text{proportion of females under age 30 in the initial population}}{\text{proportion of females under age 30 in the stationary populatior}}$$

The formula states that the momentum of a population is close to the ratio of (a) the proportion under its mean age at childbearing to (b) the proportion of the life table population under that age. The numerator refers to the observed population in the initial year, the denominator to the life table describing mortality in the same year. In a life table, the proportion of females under 30 is calculated as $(T_0 - T_{30})/T_0$. If the proportion of females under age 30 is 0.422 in the initial population and 0.365 in the life table, the momentum will be 0.422/0.365 = 1.16. This indicates that momentum alone has the potential to increase total numbers by 16 per cent (see Bogue 1993: volume 5, pages 19-57 to 19-58).

9.6 Population ageing

Population ageing, or *demographic ageing*, entails an increase in the percentage of the population in older ages, often taken as 65 years and over. In the year 2000, about 7 per cent of the world's population of 6.1 billion were aged 65 or more. This percentage might double by 2050 and treble by 2100 (World Bank 1994). Population ageing and population momentum are interrelated phenomena. In the classical demographic transition, rising momentum is due to lower death rates of infants and children, which make populations younger, while the later decline in population momentum is associated with lower birth rates, which make populations older (Figure 9.4). Heightened levels of ageing result if below-replacement fertility and negative momentum emerge.

The greatest improvements in survival through time occur among infants and children. If high birth rates persist while death rates decline, populations stay demographically young – because each new cohort of children is larger than the one before it. This situation creates substantial population momentum arising from there being larger birth cohorts poised to reach successive ages in the future. The numbers in older age groups expand throughout the demographic transition because of the long-run, flow-on effects of lower death rates among the young.

Nevertheless, the percentages in older age groups remain low during the demographic transition until birth rates start to decline (Figure 9.4). In the context of low mortality, falling birth rates reduce differences between the size of child, parent and grandparent generations, as families averaging six or more children give way to families averaging about two children.

Thus, lower birth rates constitute the principal cause of the expansion of the percentages in older ages, from less than 3 per cent (pre-transition) to around 20 per cent (post-transition). In already long-lived populations, however, a further decline in death rates, at older ages, contributes noticeably to ageing, because such changes have an immediate impact in sustaining numbers and augmenting percentages of older people.

Stable population models clarify this phenomenon; indeed, the explanation of how population ageing occurs is one of their best-known experimental applications. Figure 9.5 is based on 35 female stable population models, each

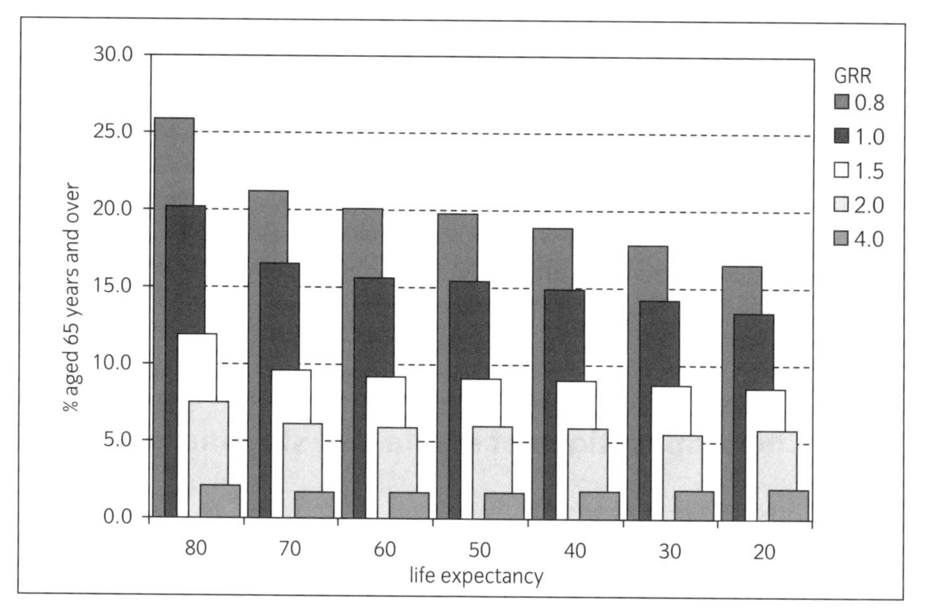

Figure 9.5 Percentages aged 65 years and over in various stable populations.

Source: Coale and Demeny (1983: 31).

showing the implications for the age structure of a particular combination of birth and death rates. The diagram uses life expectancy as the measure of mortality, and the gross reproduction rate as the measure of fertility. It illustrates the following main points:

- Lower fertility makes populations older. For a population with a female life expectancy of 70 years, the percentages aged 65 and over range from 2 to 21 per cent for GRRs between 4.0 and 0.8.
- In populations with a life expectancy of less than about 70 years, better survival has a relatively small impact on demographic ageing, because most of the improvements augment the younger age groups.
- Further rises in life expectancy, to 80 years, boost the representation of the older ages. This phenomenon is most obvious when fertility is low. It reflects that, in long-lived populations with rectangular survival curves, much of the improvement in survival necessarily occurs among middle aged or older people, augmenting older groups rather than the young. This gives mortality decline considerable prominence as a cause of further ageing in developed country populations with below-replacement fertility.

Apart from this last exception, over the demographic transition the main effect of mortality decline is an increase in the *numbers* in older ages, whereas the main effect of fertility decline is an increase in the *percentages* in older ages.

Finally, demographic ageing is a concept that emphasizes percentages, which are important because of their bearing on the question of dependency, sometimes measured as the ratio of the elderly to other adults or the ratio of pensioners to taxpayers. Nevertheless, the numbers of older people are also important in the demographic study of the aged, because major concerns arise from how many will require support in terms of pensions, housing, and health and welfare services, even in countries still demographically 'young', such as India. By 2050, India could have far more older people than Europe and the United States combined. In India, the population aged 65 and over numbered 11 million in 1950 (3.3 per cent of the total population), around 52 million in 2000 (5.2 per cent) and might reach 237 million in 2050 (14.6 per cent) (World Bank 1994).

9.7 Further applications of stable and stationary models

Projections and estimates

Familiarity with stable and stationary models is relevant not only because they have valuable applications in themselves, but also because other demographic procedures employ them. Life tables, for example, are the main source of mortality assumptions in demographic analyses and projections. For projections, model life tables can provide a range of mortality patterns and levels from which to select assumptions about a population's survival and life expectancy at successive years into the future (Coale and Trussell 1996: 480–481). Alternatively, relational model life tables, based on a country's national life table for the initial year, or an alternative standard, may be calculated to provide mortality data for each year in the projection.

Life tables are also the foundation for a wide range of demographic estimates, as well as estimates in other fields. Such applications include estimates of:

- wastage and turnover in the labour force;
- contraceptive effectiveness, where the 'survivors' are women who did not become pregnant;
- the proportions surviving at intervals after contracting a disease or undergoing a course of medical treatment – to provide evidence of the effectiveness or otherwise of medical interventions;
- the numbers likely to live to pensionable ages and hence become eligible for a publicly funded retirement income;
- the number of person-years lived in receipt of a pension or other government benefit – to reveal the long-run financial implications of economic dependency in a population.

Pollard et al (1990: 49–59) provide many examples of the uses of life table techniques in preparing estimates for human populations, non-human populations and even inanimate objects. Also, the following quotation from Nathan Keyfitz (1993: 538) illustrates the versatility of life tables in such applications:

> Biologists, operations researchers, quality control engineers, insurance mathematicians, as well as demographers all think about mortality and survivorship in essentially the same way. Formally the survival of a human being subject to multiple causes of sickness and death is no different from the survival of an item of equipment containing multiple parts, any of which can fail, and the failure of certain parts, or certain combinations of parts, puts the equipment out of service for a period of time or ends its usefulness altogether. The analogy to the sickness and death of a human being is complete, though one would not know it from the terms used . . . our life expectancy is the engineer's mean time to failure . . . Lotka saw a fleet of trucks as being subject to a mortality table, and during World War II the stock of engines needed to be stored in the South Pacific was calculated using a life table for those engines operated under active service conditions.

Estimates from incomplete data

Demographic models, and other bases for estimation, have also helped to bridge many gaps in statistical coverage of historical and developing country populations for which demographic statistics are limited or defective. Model stable populations, for example, have enabled the provision of quite detailed estimates of fertility and mortality rates in such contexts.

Until the 1970s, the absence of censuses, or of adequate censuses, left a void in demographic statistics for many developing countries. This challenged demographers' ingenuity to use limited sources, such as administrative or medical records and incomplete vital registrations, to make demographic estimates. A particular concern has been to derive figures which, in developed countries, are normally obtained from vital statistics, since vital registration systems in less developed countries have commonly been absent or unreliable.

Extension of census collections has improved the availability of the basic information for such estimates. Moreover, relatively low-cost sample surveys have provided complementary data on a greater range of population changes. Considerable advances in the availability of data for demographic estimation have come from major cross-national studies such as the World Fertility Survey and the Demographic and Health Surveys (see Lucas and Meyer 1994). Today, many of the estimation techniques applied to developing country populations rely on information from censuses and large scale surveys, rather than from other sources (Brass 1996: 453).

Table 9.7 lists examples of the types of estimates sought and the data employed, and illustrates the major interest in estimates of vital rates from censuses and surveys (Brass 1996: 453 ff). Coverage of techniques for these and other purposes is available in the sources listed under 'Further reading' at the end of this chapter.

Table 9.7 Examples of types of estimates derived from incomplete data

Type of estimate	Source data
Crude birth and death rates and age-specific mortality rates	Age–sex distributions from two censuses
Crude birth and death rates	Age–sex distribution from one census or survey
Crude birth and death rates	Age distribution of deaths
Infant and child mortality	Census/survey data on the number of children born and the number still alive.
Age-specific fertility rates and fertility differentials	Census/survey data on mothers and children in households ('own children method')
Age-specific fertility rates	Census/survey data on women's ages and the number of children born
Age-specific fertility rates	Survey data on women's ages and the number of children born in the previous 12 months
Cohort and period age-specific fertility rates and parity progression ratios	Survey data on birth histories
Survival to successive stages of the life course	Census/survey data on the survival of relatives
Emigration	Survey data on the place of residence of relatives

9.8 Conclusion

This chapter and the preceding one have introduced the concepts of stationary and stable population models, methods of calculation and some of their main applications. Other texts cover the mathematics as well as more advanced topics (see references under 'Further reading' in Chapter 1). Without additional mathematical background, however, knowledge of stable and stationary models can be applied in major areas of interest in planning and policy-making, such as estimates of migration and population projections. The remaining two sections of the book – on *Spatial Patterns and Processes* (Section 5) and *Applied Demography* (Section 6) – include coverage of these topics.

Study resources

KEY TERMS

- Intrinsic birth rate
- Intrinsic death rate
- Intrinsic rate of natural increase
- Mean length of a generation
- Mean age at childbearing
- Model life tables
- Model stable populations
- Negative momentum
- Population ageing
- Population momentum
- Positive momentum
- Relational model life tables
- Stable population

FURTHER READING

Hinde, Andrew. 1998. *Demographic Methods*. London: Arnold, Chapter 13, 'Models of Population Structure' and Chapter 14, 'Applications of Stable Population Theory'.

Kinsella, Kevin and Velkoff, Victoria A. 2001. *An Aging World: 2001*. US Census Bureau, Series P95/01-1. Washington, DC: US Government Printing Office.

Pollard, A.H., Yusuf, Farhat, and Pollard, G.N. 1990. *Demographic Techniques*. Sydney: Pergamon Press, Chapter 4 'Applications of Stationary Population Models'.

Woods, Robert. 1982. *Population Analysis in Geography*. London: Longman, Chapter 8 'Models of Population Structure and Change'.

Reference works

Coale, Ansley, J. and Demeny, Paul. 1983. *Regional Model Life Tables and Stable Populations* (second edition). New York: Academic Press.

Coale, Ansley and Guo, Guang. 1990. 'New Regional Model Life Tables at High Expectation of Life'. *Population Index*, 56: 26–41.

United Nations. 1982a. *Model Life Tables for Developing Countries*. New York: United Nations Department of International Social and Economic Affairs.

United Nations. 1982b. *Stable Populations Corresponding to the New Model Life Tables for Developing Countries*. New York: United Nations Department of International Social and Economic Affairs.

Indirect estimation

Hill, K. and Zlotnik, H. 1982. 'Indirect Estimation of Fertility and Mortality'. In J.A. Ross (editor), *International Encyclopaedia of Population*. New York: Free Press, pp. 324–34.

Preston Samuel H., Heuveline, Patrick, and Guillot, Michel. 2001: *Demography: Measuring and Modeling Population Processes*. Oxford: Blackwell, Chapter 11, 'Indirect Estimation Methods'.

Pollard, A.H., Yusuf, Farhat, and Pollard, G.N. 1990. *Demographic Techniques*. Sydney: Pergamon Press, Chapter 12, 'Estimating Demographic Measures from Incomplete Data'.

United Nations. 1983. *Manual X: Indirect Techniques for Demographic Estimation*. New York: Department of International Economic and Social Affairs.

INTERNET RESOURCES

Subject	Source and Internet address
Population analysis spreadsheets, including a downloadable Excel spreadsheet for calculating a stable population	***US Census Bureau*** http://www.census.gov/ipc/www/pas.html
Home page of a major United States organization concerned with promoting research, training and information dissemination on ageing.	***National Institute on Aging (NIA)*** http://www.nih.gov/nia/
Home page of a major British organization concerned with the welfare of older people.	***Age Concern*** http://www.ageconcern.org.uk/
Discussion paper on a model life table system for the WHO. Includes a review of previous model life tables	***World Health Organization*** http://www3.who.int/whosis/discussion_papers/htm/paper08.htm
An informative introduction to the concept of population momentum: 'Deconstructing Population Momentum' by John Knodel	***Population Reference Bureau*** http://www.prb.org/Content/NavigationMenu/PRB/AboutPRB/Population_Today1/march99_pt.pdf
Report on 'Population Momentum and Population Aging in Asia and Near-Asia Countries'	***East–West Centre, University of Hawaii*** http://www.eastwestcenter.org/res-pr-detail.asp?resproj_ID=45

Note: the location of information on the Internet is subject to change; to find material that has moved, search from the home page of the organization concerned.

EXERCISES

1 Provide brief answers to the following:

(a) Is a stationary population also a stable population? Explain your answer.

(b) What is population momentum?

(c) Is there any way of decreasing a population's inherent momentum of growth?

(d) What is population ageing?

(e) Explain the nature and purpose of a model life table.

(f) Why is the mid-point of an age group not necessarily the same as the median age of people in the age group?

(g) What is the intrinsic growth rate of a population?

(h) How can a population have a positive growth rate and a negative intrinsic growth rate?

(i) Is rapid population growth in the early part of the demographic transition due to population momentum?

2 At the start of the twenty-first century, China had an estimated R_0 of 0.812 97 and an R_1 of 23.528 50. Calculate the mean length of a generation in China and the population's intrinsic rate of natural increase.

3 Using the life table in Table 9.5 and the female age-specific birth rates given below, construct a stable population model for the United States in the year 2000. Assume that the mid-point of the last age interval is 101.9 years. Include the following measures:

- the gross reproduction rate
- the net reproduction rate
- the mean length of a generation
- the intrinsic rate of natural increase
- the age distribution of the stable population; present figures for both sexes expressed as percentages of the total stable population.

Age-specific birth rates (of daughters) per woman, United States 2000

15–19	0.023 66
20–24	0.054 78
25–29	0.059 22
30–34	0.045 90
35–39	0.019 71
40–44	0.003 85
45–49	0.000 24

Estimated from: National Centre for Health Statistics 2002: *National Vital Statistics Report*, 50 (5), Table 4.

4 At the start of the twenty-first century, Egypt's total population was 67 884 000 and its total population momentum was 47 per cent. Using these figures, calculate the country's projected stationary population size.

5 From the life tables in Table 9.5, estimate the population momentum in female populations with a life expectancy at birth of 80 years and the following percentages aged less than 30: (a) 50, (b) 40, and (c) 30.

6 Using the *West Models* module (Box 9.1), complete the missing cells in the following table showing the characteristics (for both sexes) of stable populations with particular life expectancies and gross reproduction rates. (In the module, use the 'Stable Population Models' display.) Comment on the variations in the percentages aged 65 and over.

	% Total population aged 65 years and over			
Female e_0	GRR			
	0.5	1.0	2.0	4.0
20				
40				
60				
80				

7 Compare the results of calculations of population momentum, for the year 2000, in the world's population and in five countries or regions of your own choice. Use the module *Momentum.xls*, and the information on life expectancy and GRR in countries and regions in the table in Box 9.1.

SPREADSHEET EXERCISE 9: STABLE POPULATION MODELS

Complete Tables 1, 2 and 3 below to calculate intrinsic growth rates and the age structure of a stable population.

The intrinsic rate of natural increase

Calculate the intrinsic rate of growth in the two examples (Tables 1 and 2), using formulas to obtain the figures in bold.

1. Type the row and column headings for Table 1 and the age-specific data in columns B to D.
2. Calculate the total for column C and the gross reproduction rate (GRR).
3. Enter formulas in columns E and F to obtain the terms needed to calculate the intrinsic growth rate.
4. Calculate the two terms needed in the intrinsic growth rate formula (i.e. the cells labelled 'Working' below the table). LN is the natural logarithm function, the formula for which is written =LN(number).
5. Enter an Excel formula to calculate the intrinsic growth rate using Coale's approximation. The only values needed are in the two 'working' cells.
6. Create Table 2 by copying the whole of Table 1, editing the title, and entering new age-specific values in columns C and D. The values in the rest of the table should update to accurate results automatically.

Table 1 Calculation of the intrinsic rate of natural increase, Example 1

Age group A	Central age B	Female ASFRs C	Probability of survival D	R_0 working E (C × D)	R_1 working F (B × E)
15–19	17.5	0.010 70	0.986 12	**0.01**	**0.18**
20–24	22.5	0.043 57	0.983 75	**0.[illegible]**	**0.96**
25–29	27.5	0.069 65	0.981 34	**0.07**	**1.88**
30–34	32.5	0.043 09	0.978 76	**0.04**	**1.37**
35–39	37.5	0.013 12	0.975 29	**0.01**	**0.48**
40–44	42.5	0.002 14	0.969 57	**0.00**	**0.09**
45–49	47.5			**0.00**	**0.00**
Total		**0.182 27**		**0.18**	**4.97**
		GRR =		R_0 =	R_1 =
Total times 5		**0.911 35**		**0.89**	**24.84**

$$r = \frac{\ln R_0}{\frac{R_1}{R_0} - 0.7 \ln R_0}$$

Working:

R_1/R_0 = **27.78**

LN R_0 = **−0.11**

r = **−0.004 02**

−0.40% per annum

Table 2 Calculation of the intrinsic rate of natural increase, Example 2

Age group A	Central age B	Female ASFRs C	Probability of survival D	R_0 working E (C × D)	R_1 working F (B × E)
15–19	17.5	0.012 89	0.986 15	**0.01**	**0.22**
20–24	22.5	0.050 07	0.983 76	**0.05**	**1.11**
25–29	27.5	0.071 20	0.981 34	**0.07**	**1.92**
30–34	32.5	0.039 47	0.978 77	**0.04**	**1.26**
35–39	37.5	0.012 05	0.975 30	**0.01**	**0.44**
40–44	42.5	0.002 15	0.969 60	**0.00**	**0.09**
45–49	47.5	0.000 12	0.960 03	**0.00**	**0.01**
Total		**0.187 95**		**0.18**	**5.04**
		GRR =		R_0 =	R_1 =
Total times 5		**0.939 75**		**0.92**	**25.21**

$$r = \frac{\ln R_0}{\frac{R_1}{R_0} - 0.7 \ln R_0}$$

Working:

R_1/R_0 = **27.34**

LN R_0 = **−0.08**

r = **−0.002 96**

−0.30% per annum

The age structure of a stable population

Assuming that a stable population has an intrinsic growth rate of 0.01269 and a sex ratio at birth of 1.05, complete Table 3 to calculate its age structure, using formulas to obtain the figures in bold.

Table 3 Age distribution of a stable population
Intrinsic growth rate = 0.01269

Age group	Central age	Life table pop.		Multiplier	Stable population		Percentages	
		Males $_5L_x$	Females $_5L_x$		Males	Females	Males	Females
0–4	2.5	489914	492185	**0.96897**	**498446**	**476911**	**5.3**	**5.1**
5–9	7.5	487702	490520	**0.90976**	**465876**	**446255**	**4.9**	**4.7**
10–14	12.5	486708	489846	**0.85417**	**436517**	**418411**	**4.6**	**4.4**
15–19	17.5	484620	488804	**0.80198**	**408086**	**392009**	**4.3**	**4.2**
20–24	22.5	479960	487220	**0.75297**	**379466**	**366863**	**4.0**	**3.9**
25–29	27.5	476217	485704	**0.70696**	**353500**	**343374**	**3.7**	**3.6**
30–34	32.5	472810	483842	**0.66376**	**329525**	**321156**	**3.5**	**3.4**
35–39	37.5	468512	481086	**0.62320**	**306578**	**299815**	**3.2**	**3.2**
40–44	42.5	462009	476746	**0.58512**	**283849**	**278955**	**3.0**	**3.0**
45–49	47.5	451331	469875	**0.54937**	**260345**	**258135**	**2.8**	**2.7**
50–54	52.5	433803	459222	**0.51580**	**234944**	**236867**	**2.5**	**2.5**
55–59	57.5	406040	443590	**0.48428**	**206471**	**214823**	**2.2**	**2.3**
60–64	62.5	364418	420594	**0.45469**	**173983**	**191241**	**1.8**	**2.0**
65–69	67.5	306832	386805	**0.42691**	**137539**	**165130**	**1.5**	**1.8**
70–74	72.5	234908	335998	**0.40082**	**98864**	**134676**	**1.0**	**1.4**
75+	80.0	284859	568492	**0.36465**	**109068**	**207302**	**1.2**	**2.2**
Total		**6790643**	**7460529**		**4683056**	**4751923**	**49.6**	**50.4**

Section V

Spatial patterns and processes

Approaches to the study of spatial aspects of population are of considerable interest to many users of demographic information. The chapters in this section cover the main methods of measuring and mapping population distribution and migration, and techniques for estimating migration from basic demographic data.

Chapter 10 **Population distribution**

Chapter 11 **Migration**

10 Population distribution

I conceive . . . That *London*, the *Metropolis* of *England*, is perhaps a Head too big for the Body, and possibly too strong: That this Head grows three times as fast as the Body unto which it belongs, that is, It doubles its People in a third part of the time. . . .

(Graunt 1662: 4)

OUTLINE

10.1 Where people live

10.2 Types of spatial units

10.3 Measures of population distribution

10.4 Housing

10.5 Population mapping and GIS

10.6 Types of population maps

10.7 Conclusion

Study resources

LEARNING OBJECTIVES

To understand:

- the significance of population distribution
- the nature of the main types of spatial units, and the problems in defining urban and rural areas
- how to calculate basic measures of population distribution
- the relevance of housing and housing-based measures in the study of population distribution

To become aware of:

- applications of population mapping and the main types of population maps

COMPUTER APPLICATIONS

Excel module: Population Distribution.xls (Box 10.1)

Spreadsheet exercise: A descriptive statistics module (See Study *resources*)

Since national statistics conceal many variations, questions about population composition and processes of change may call for comparative information on regions or communities. Studying population at a small scale deepens understanding of society-wide developments, especially where the national situation arises from a mosaic of trends and characteristics. Hence, many users of demographic methods need to be familiar with a wide range of techniques that include some with specific relevance to sub-national studies. This is all the more relevant today, since a distinguishing feature of contemporary *applied demography* is an emphasis on the analysis of population data for small areas.

Although population distribution is an important subject area, along with migration it has received little mention in some demographic texts. This can create a misleading impression of the relevance of demographic methods, for many with valuable applications at the national level are equally useful in sub-national studies. Growth rates, birth rates, death rates, standardization techniques and life tables, for instance, are commonly employed in studies of local and regional developments, as well as those at a larger scale. While academic demographers have had a particular interest in national trends and comparisons between social groups, sub-national research has become more common in demography, as well as being a core interest in geography and sociology. It is also noteworthy that the most celebrated demographic studies include some making substantial use of sub-national data, including Hajnal (1974), on age at marriage in Europe, and the Princeton European Fertility Project, on the historical decline of fertility in Europe (Coale and Watkins 1986).

While methods covered in previous sections have applications in the study of regional and small area populations, this section on *Spatial Patterns and Processes*, and the next on *Applied Demography* discuss methods and concepts that are especially relevant to demographic work at the sub-national level, often for practical purposes. This chapter introduces spatial population data, basic measures of population distribution, geographic information systems and population maps. A basic concern is with approaches that facilitate understanding of where people live, a key influence on which is the type and distribution of housing. The next chapter, on migration, introduces further concepts and methods that are important in explaining population distribution and its dynamics.

10.1 Where people live

Where people live is a vital consideration in planning and policy-making. Planning work in government departments and private organizations regularly refers to population at the sub-national level. Where people live is also essential to understanding the nature of communities and variations within and between them in relation to welfare needs, service use and consumer behaviour. This reflects that many population characteristics vary spatially, including age structures, incomes and occupations. So too do diverse aspects of behaviour from purchasing of goods to childbearing and voting preferences in elections. Formerly neglected aspects of population distribution, such as spatial variations in the incidence of crime and diseases, today receive considerable attention as inputs to policy-making. Planning of social surveys, the location of business enterprises and marketing strategies all rely heavily on information about the distribution of population.

How governments define where people live also has significant outcomes. Governments' budget allocations can depend on the type of area in which people reside. In the United States, the US Office of Management and Budget is responsible for the standards for defining metropolitan areas. Whether communities are inside or outside a metropolitan area affects their federal and state budget allocations (*Anon.* 1998).

Research on where people live requires attention to the details of the definition of areas and their boundaries. Even at the national level, compiling comparable figures through time for Germany or states in the Balkans or the former Soviet Union requires careful identification of component areas and the definition of borders. Similarly, boundary adjustment can be needed in time series data for states, provinces, cities and census tracts. For small areas, the simplest approach to obtaining comparable units through time is to aggregate areas with boundary changes in common. For studies at a single date there are few such complications, but informed decisions are still needed on the appropriate units of analysis, such as whether to use large or small areas. Other things being equal, the smaller the areas the greater the apparent diversity of the population, since statistics for larger areas are likely to conceal internal variations.

10.2 Types of spatial units

Legal and statistical areas

Government organizations compile census figures, vital statistics and other social and economic data for two main types of spatial units:

- *Legal/administrative areas* Examples include states, provinces, electoral districts, cities, municipalities, counties and shires. Statistics on such areas are needed for administrative purposes. They denote the area of jurisdiction of different levels of government – national, state and local – together with electoral divisions and postal districts.
- *Statistical/geographical areas* These range in size from some of the largest to the smallest areas for which demographic data are available, including regions, metropolitan areas, urban areas, census tracts and street blocks.

Figure 10.1 illustrates the hierarchy of legal and geographical areas in the United States. Information for the smallest statistical areas, which can consist of

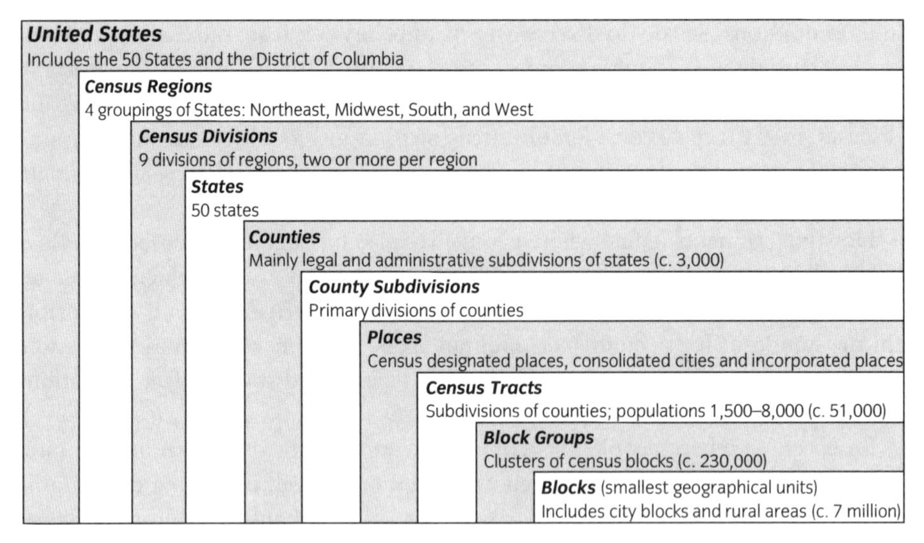

Figure 10.1 Summary of legal and geographical areas for the tabulation of population statistics in the United States, 2000 census.

Source: United States Census Bureau, 2000 Census, *Geographic Terms and Concepts*.

individual street blocks, reveals detailed variations in population characteristics. Census atlases, which contain maps of population characteristics, commonly plot information for the smallest areas. Also, by aggregating small area data, researchers are better able to define the region or community they are investigating. This is useful for studies of immigrant communities, areas of high or low socio-economic status, or other districts and social groups that either occupy a small part of an administrative district, or straddle the boundaries of different ones.

Urban areas and urban regions

Urban areas are commonly employed units of analysis, because they are intended to define the boundaries of entire towns and cities. Definitions of what is urban vary from country to country, but population size is the main criterion, often combined with the additional criteria such as population density and the percentage in non-agricultural employment. Cross-national studies of urban populations sometimes adopt a definition based on a uniform minimum size, such as centres with 1000 or 5000 inhabitants. In developing countries, however, this can include places where a high proportion of the population is engaged in agriculture. An alternative is to compare settlement hierarchies rather than total urban populations, that is, identifying populations living in cities and towns of different sizes. The notion of a *settlement hierarchy* refers to places sorted into categories according to their total population size. The nature of the settlement hierarchy is highly relevant in comparisons, for example, when two countries with the same urban percentage include one where a majority of the population live in a single large city, while in the other the population is distributed through many centres of different sizes.

Sometimes settlement patterns can be described only through considering urban regions, rather than discrete urban areas. Probably the first to anticipate the development of modern urban regions was the author of *The Time Machine*, H. G. Wells. Writing in 1902, he predicted:

railway-begotten 'giant cities' . . . are destined to such a process of dissection and diffusion as to amount almost to obliteration . . . these coming cities will not be, in the old sense, cities at all; they will present a new and entirely different phase of human distribution. . . . We are in the early phase of a great development of centrifugal possibilities . . . A city of pedestrians is inexorably limited by a radius of about four miles . . . a horse-using city may grow out to seven or eight . . . is it too much . . . to expect that the available area for even the common daily toilers of the great city of the year 2000 . . . will have a radius of over one hundred miles? What will be the forces acting upon the prosperous household . . . ? The passion for nature . . . the allied charm of cultivation . . . and that craving for a little private imperium are the chief centrifugal inducements. . . . The city will diffuse itself until it has taken up considerable areas and many of the characteristics of what is now country . . . The country will take itself many of the qualities of the city. The old antithesis will . . . cease, the boundary lines will altogether disappear. (quoted in Berry 1974: 36–7)

Wells' vision predicts that the urban region will supersede the urban area – that clearly bounded concentrations will give way to larger, but less dense and less compact, agglomerations of centres. In the United States, for example, many cities and towns are clustered within larger metropolitan regions, bound together through commuting and other social and economic linkages. Vast regions have a dense network of towns and cities situated in close proximity to one another. Arguably much of the north-eastern seaboard of the United States forms a huge 'megalopolis' of coalescing cities, rather than many separate entities (Gottmann 1961). Processes of coalescence inevitably create major difficulties in determining boundaries between metropolitan areas and United States census concepts are designed to address this complexity (Table 10.1). The urban region, rather than the urban centre, defines the true setting of the lives of a high proportion of the American population.

Uniform and nodal regions

Although government agencies publish statistics for urban regions, other types of regions and communities, they are not necessarily suited to the needs of particular studies. In such circumstances, researchers may aggregate small areas, such as census tracts or blocks (see Figure 10.1), to define regions or communities more appropriately. This can produce *uniform regions*: areas that are relatively homogenous in terms of the phenomena in question (Abler et al. 1972: 85).

Commonly, however, the aim is to identify *nodal regions*, consisting of central places and their hinterlands (Morrill 1974: 218). Whereas uniform regions are defined in terms of phenomena exhibiting relatively little variation, nodal – or

Table 10.1 Summary of urban concepts in the United States, 2000 census

Concept / Definition
Metropolitan Area (MA)
A large population nucleus, together with adjacent communities that have a high degree of economic and social integration with that nucleus
Metropolitan Statistical Area (MSA)
An MA not closely associated with other MAs
Primary Metropolitan Statistical Area (PMSA)
A part of a large MA, with more than 1 million persons, that has very strong internal economic and social links.
Consolidated Metropolitan Statistical Area (CMSA)
An MA containing PMSAs.
Urbanized Area (UA)
Densely settled territory that contains 50 000 or more people; delineated to provide a better separation of urban and rural territory, population and housing in the vicinity of large places.
Urbanized Area Central Place
One or more dominant centres in a UA
Urban Cluster
Densely settled territory with at least 2500 people, but fewer than 50 000.

Source: United States Census Bureau, 2000 Census, *Geographic Terms and Concepts*.

functional – regions 'depend on phenomena in motion, and significant variations in intensity of flow within the region are usually the rule' (Abler et al. 1972: 85). Nodal regions are delimited with reference to spatial interactions, through identifying places that are linked together economically, socially or politically, such as through trade, commuting, shopping or use of medical and educational services. For example, the proportion of workers commuting to a central town or city is likely to vary across a region according to distance and accessibility by transport, yet the ties with the node are important criteria identifying interrelationships or common bonds between places. Where the zones of influence of competing centres overlap, the strength of interactions is a basis for defining regional boundaries (Morrill 1974: 218). Clearly, the definition of regions is seldom clear-cut; nonetheless nodal regions can provide useful, though imperfect, generalizations about the spatial organization of human activities (Morrill 1974: 220).

10.3 Measures of population distribution

The nature and definition of spatial units are matters particular to each country, but the measures of population distribution are applicable in many settings. This section introduces the most commonly needed methods for analysing population distribution. Chapter 13 discusses some further techniques that are particularly relevant to the analysis of spatial variations in population composition, including measures of population diversity, segregation and inequality.

Numbers and percentages

Information on population distribution is often presented in tables and maps showing absolute numbers living in different places, or the percentage of the total population in each place. Although rudimentary, such statistics are sometimes the most needed and most informative. Both absolute numbers and percentages are vital for comparisons between places as well as through time. The absolute numbers of people are also a necessary consideration in planning, because the viability of a service or enterprise depends on the numbers of clients or customers.

Table 10.2 includes statistics on the absolute and percentage distribution of population in regions of Africa in 1950 and 2000. These show that during the half century, the total population of Africa more than trebled from 221 million to 799 million; the magnitude of the absolute increase is an initial indicator of the pressures of change in African societies. It is also apparent that regional differences in population growth were not very marked during the 50 years, since the main change in the percentage of the total population of Africa living in each of the five regions was a small increase in the proportions in the Western and Eastern regions. Interpretation of these changes, however, can really only begin when reference is made to a map of the regions and their constituent countries. Maps are indispensable to interpreting population distribution and to communicating findings to others.

Table 10.2 Population distribution in regions of Africa, and calculation of the index of redistribution, 1950–2000

Region	1950 Population		2000 Population		Absolute
	millions	%	millions	%	differences |C – E|
A	B	C	D	E	F
Western	62	28.1	234	29.3	1.2
Eastern	62	28.1	246	30.8	2.7
Northern	52	23.5	173	21.7	1.8
Middle	29	13.1	96	12.0	1.1
Southern	16	7.2	50	6.3	0.9
Total	221	100.0	799	100.0	7.8
Index of redistribution (half the total)					**3.9**

Sources: United Nations *Demographic Yearbook*; Population Reference Bureau *World Population Data Sheet 2000*.

Population density

The population density of an area – the number of people per square mile or square kilometre – can convey an initial impression of crowding or emptiness, the pressure of population on resources and the environment, as well as the abundance or scarcity of demand for goods and services:

$$\text{Population density} = \frac{\text{total population}}{\text{total area}}$$

Population density is employed in comparisons between countries as well as within them. Statistics on the land area of countries is published in the United Nations *Demographic Yearbook*, while the land area of places within countries is published in conjunction with census data and other official statistics. Table 10.3 presents examples of the calculation of population density. The world's population density in 2000 was about 45 people per square kilometre of land, that is, the ratio of population to area was: 6067000000/134133811 = 45.

The simplicity of population density, as the ratio of population to land area, is an advantage from the point of view of calculation, but a disadvantage from the ease with which potentially misleading information may be compiled. A classic example of mismatching population and area occupied is in the population density

Table 10.3 Calculation of population density in selected countries, 2000

Country A	Population (millions) B	Area square miles C	Area square kilometres ($C \times 2.58998$)[1] D	Density per square mile[2] ($B \times 10^6/C$) E	Density per square kilometre ($B \times 10^6/D$) F
Bangladesh	128.1	55598	143998	2304	890
Taiwan	22.3	13969	36179	1596	616
Netherlands	15.9	15768	40839	1008	389
United Kingdom	59.8	94548	244877	632	244
France	59.4	212934	551495	279	108
United States	275.6	3717796	9629017	74	29
Russia	145.2	6592819	17075269	22	9
Canada	30.8	3849670	9970568	8	3
Australia	19.2	2988888	7741160	6	2
World	6067.0	51789516	134133811	117	45

Notes: [1] 1 square mile = 2.589975 square kilometres (1 mile = 1.60934 kilometers).
[2] Some of the densities differ slightly from the figures in the source publication, because they are based on the rounded figures in column B.
Source: Population Reference Bureau 2000, *World Population Data Sheet 2000.*

for Australia – two persons per square kilometre (Table 10.3). Australia has a land mass similar in size to that of the United States without Alaska, but its total population is a little less than that of the American state of Texas. Australia's overall population density takes no account of the fact that about two-thirds of the country is uninhabited and largely uninhabitable, and gives rise to the misconception that Australia is an 'empty' continent. Similar problems arise over the population density of countries such as Saudi Arabia, China and Canada.

Comparisons of population density in entire countries are rarely satisfactory because they merely compare total populations with total areas, irrespective of whether or not particular tracts of land are occupied. The 'land' includes mountains, deserts, swamps, forests, national parks and wilderness reserves, although rivers and lakes are usually excluded. Conversely, seeking to limit the estimate to 'occupied' lands could conceal the greater 'footprint' of the population in terms of its use of resources and effect on the environment. Most populations rely on international trade for part of their livelihood and sustenance and, consequently, have impacts far beyond their own national boundaries. Other problems of population density include:

1. Creating a false impression of similarities between places with the same density. Localities with contrasting characteristics may have the same density, such as one with its houses on separate blocks of land, and another of similar size with a high rise apartment building and a golf course.
2. Creating a false impression of homogeneity within areas in terms of population density. New York City has a population density of over 9000 per square kilometre, while its borough of Manhattan has a density of 26 000 per square kilometre.
3. Within cities and towns, the presence of non-residential land uses commonly affects population densities. For the United States, according to Downs (1994, 144), a reasonable assumption is that 50 per cent of the land in suburbs is residential, while in large cities the figure drops to 25 per cent. This has led to alternative definitions of urban population density, some of which include only residential land in the calculations (Baldassare 1979; Magri, 1994; Richardson et al. 1998).
4. Decisions about where to draw administrative and statistical boundaries greatly affect the apparent density of areas. The density of a town or city can vary substantially according to whether boundaries include or exclude lands on the rural–urban fringe.

Location quotients

Percentages are the basis for calculating many other measures of population distribution including location quotients and various indices of dissimilarity. Location quotients compare two percentages or proportions. A quotient is simply the result of the division of one number by another. Thus the formula for the location quotient (LQ) for a particular population characteristic (i) is:

$$LQ_i = \frac{X_i}{Y_i}$$

where

X_i is the percentage for the first population
Y_i is the percentage for the second population.

Quotients greater than 1.0 denote an over-representation of a particular characteristic (i) in the first population, those less than one denote an under-representation – in comparison with the second population. The second population is selected according to its relevance as a basis for comparisons; it is often the total population of the area, region or nation. As a means of facilitating comparisons, the location quotient comes into its own when many populations are compared with the same base population, as in Table 10.4.

The location quotients in the table measure the extent to which the provinces and territories of Canada had an under-representation or over-representation of

Table 10.4 Calculation of location quotients for the population of French origin in Canada, 1901 and 1991

Province/Territory	Percentage of French origin		Location quotients	
	1901	1991	1901 (B/Total)	1991 (C/Total)
A	B	C	D	E
Total (Canada)	**30.7**	**24.5**	1.0	1.0
Newfoundland	–	1.9		0.1
Prince Edward Island	13.4	10.0	0.4	0.4
Nova Scotia	9.8	7.5	0.3	0.3
New Brunswick	24.2	34.3	0.8	1.4
Quebec	80.2	76.6	2.6	3.1
Ontario	7.3	6.7	0.2	0.3
Manitoba	6.3	7.7	0.2	0.3
Saskatchewan	2.9	6.0	0.1	0.2
Alberta	6.2	5.3	0.2	0.2
British Columbia	2.6	3.6	0.1	0.1
Yukon	6.5	5.6	0.2	0.2
Northwest Territories	0.2	4.3	0.0	0.2

Data source: McVey and Kalbach (1995: 356).

people of French origin, compared with the figure for the total population of Canada. The base population for the location quotients was the percentage of French origin in the whole of Canada in the specified year. Thus for New Brunswick in 1991, the LQ was calculated as 34.3/24.5 = 1.4. The quotients for 1901 and 1991 show clearly the persistence of the relative over-representation of people of French origin in Quebec and their marked under-representation in almost all other parts of Canada. The percentages for Quebec were 2.6 and 3.1 times the national figures for 1901 and 1991 respectively. Other examples of location quotients are available in the *Population Distribution* module (Box 10.1).

BOX 10.1 **Module on measures of population distribution (Population Distribution.xls)**

This module illustrates the calculation and mapping of location quotients and measures based on the index of dissimilarity, such as the index of redistribution and the index of concentration. The controls in the module can generate many different percentage distributions to illustrate the range of values possible for each measure.

Maps of the location quotients and the two percentage distributions employed in the index of dissimilarity enable the information to be visualized and interpreted in a spatial context.

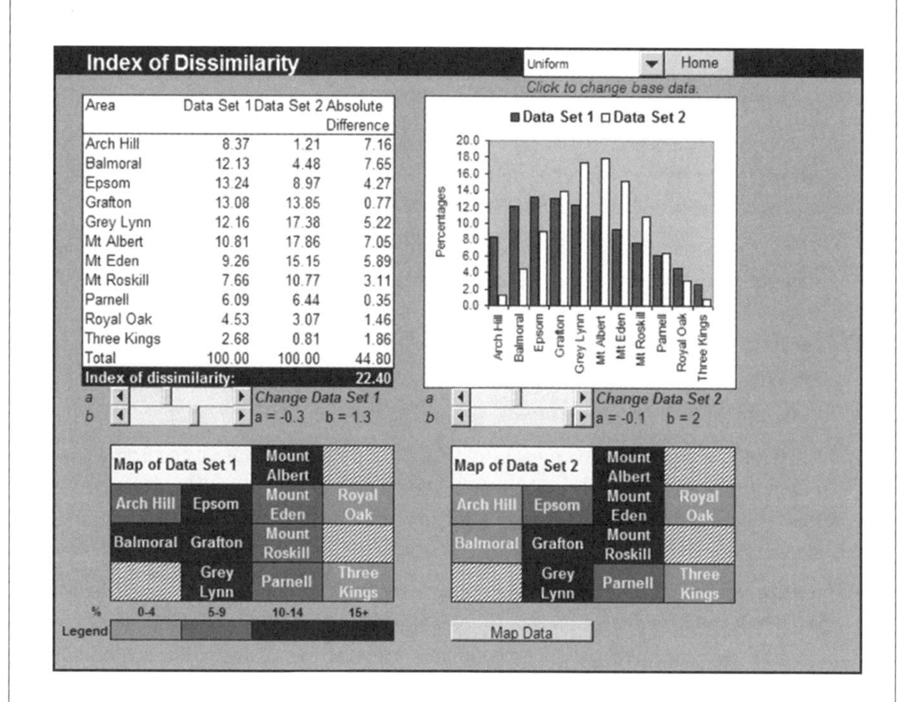

continued

BOX 10.1 continued

The two main displays in the module consist of:

- *Location quotients*: calculates location quotients for eleven 'suburbs', from two sets of simulated percentage distributions. The results are presented as a table, a graph and a map. The map, like the source data, is hypothetical. A 'location quotient calculator' on the same display can be used to show the theoretical range of values of the location quotient.
- *Index of dissimilarity*: calculates the index of dissimilarity from two sets of simulated percentage distributions for eleven 'suburbs'. The percentage distributions may be read as referring to two populations, two variables (e.g. population and area), or the same population at two dates, depending whether the primary interest is the index of dissimilarity or its restatement as the index of concentration or the index of redistribution. A table presents the source data and the index of dissimilarity, while a graph and two maps facilitate comparisons between the two percentage distributions.

Instructions

To use the module, insert the CD in the drive, start Excel and open the file *Population Distribution.xls*. Enable macros when prompted.

- Click **Location Quotients** to work with these statistics.
- Click **Index of Dissimilarity** to work with this measure or the index of redistribution, or the index of concentration.
- Use the **scroll bars** to change one or both of the two percentage distributions. There are two scroll bars for each percentage distribution, showing values of 'a' and 'b', which are displayed alongside.
  1. The scroll bar labelled 'a' affects whether the values, as displayed in the table, are skewed to the left ($a = -1$) or the right ($a = 1$). To illustrate this, set $b = 1$ and change the values of 'a' (use the 'uniform' Base Data).
  2. The scroll bar labelled 'b' affects whether the values are clustered in the first and last categories ($b = 0$) (i.e. the 'suburbs' of Arch Hill and Three Kings) or more clustered in the middle categories ($b = 2$). To illustrate this set $a = 0$ and change the values of 'b' (use the 'uniform' Base Data).
  3. When $a = 0$ and $b = 1$, the values match the distribution of the **Base Data** used to generate the two data sets (i.e. $y = x$, see next paragraph).

  The presence of two scroll bars reflects that the simulated data are produced from a linear regression model: $y = a + bx$, where y is the simulated or 'predicted' value, x is the corresponding value in the Base Data, a is the y intercept and b is the slope of the regression line.
- The **Base Data** menu, on the top right, permits the selection of different base data from which the percentage distributions are generated. Using the 'uniform' distribution will suit most purposes, but 'bimodal', 'unimodal' and 'irregular' distributions are available for experimentation.

continued

BOX 10.1 continued

- Click **Map Data** to plot maps of the location quotients or the current data sets for the index of dissimilarity.
- Click **Home** to return to the opening display.

See Appendix B for further advice on using the Excel modules.

Illustrative applications

1 Location quotients

Using the 'Location Quotient Calculator', on the bottom right of the screen, examine the following (answers at the end):

(a) What are the maximum and minimum location quotients?

(b) When do the highest values of the LQ occur?

(c) Can different combinations of numerator and denominator values produce the same location quotient?

Referring to the map on the location quotients screen:

(d) Maps are a valuable means of revealing patterned variations in the distribution of population. Use the following setup:

Base Data: uniform.
Data Set 1, $a = 1$, $b = 2$.
Data Set 2, $a = 1$, $b = 2$.
Click **Map Data**.

The map shows a uniform pattern, since the LQ for each area is 1.0.

Now change Data Set 1 to $a = -1$, leaving b unchanged.
Click **Map Data**.

The new map reveals a marked contrast in the distribution of the location quotients, with high values in the western suburbs. The middle suburb in the series, Mt Albert, is now the only area where the percentages are the same in both data sets. The eastern suburbs have relatively low location quotients, since figures below the 'norm' must fall within the range 0 to 0.99. Conversely, since the range above the norm is open-ended, variations in the west are much greater. The mapping categories conceal much of this variation – hence reference needs to be made to the location quotients themselves to describe the true situation.

(e) Minor differences between location quotients can also produce very misleading contrasts when the data are mapped. To illustrate this, set the following initial values:

Base Data: uniform.
Data Set 1, $a = 0$, $b = 1$.
Data Set 2, $a = 0$, $b = 1$.
Click **Map Data**.

continued

BOX 10.1 continued

The map will show a uniform pattern.

Now change Data Set 1 to $a = -0.01$, leaving b unchanged.

Click **Map Data**.

A west–east (left–right) contrast is again apparent on the map, but the range of LQs is only 0.036. On a choropleth map, minor differences can appear as sharp contrasts, depending on the categories selected for mapping.

2 Index of dissimilarity (I_D)

Use the Index of Dissimilarity display to examine the examples in the following table; map the data for each example.

Data Set 1		Data Set 2		I_D	Comments
a	*b*	*a*	*b*		
1	0	0	2	98.0	Very high index due to spatial concentration of the first population/data set.
–1	2	1	2	75.1	Very high index due to one distribution being the mirror image of the other.
0	0.75	0	2	40.1	High index arising when data set 1 is bimodal while data set 2 peaks in the 'middle' suburbs.
0.5	1	0.5	1.5	19.8	Medium index where about a third of the dissimilarity arises in just one suburb (Three Kings). Differences between the two maps provide some evidence of this.
1	0	1	0	0	Lowest possible index due to highly concentrated, but identical distributions.
0.01	1	0.01	1	0	Lowest possible index arising from identical, even distributions.

Overall, the results emphasize that indices of dissimilarity, redistribution and concentration, as well as maps of population distribution, are best interpreted with reference to the original data, since summary approaches can conceal important features.

Answers

(a) 100 and 0.

(b) When the denominator is low.

(c) Yes, for instance because any numerator that is less than the denominator will produce LQs in the narrow range of 0.0 to 0.9. In principle, location quotients in this range can be associated with any denominator. Also, when the numerator is 0, any denominator will produce LQ = 0.0.

The principal advantage of the location quotient is that it establishes an easily understood benchmark – the number 1.0 – with which the quotients for all areas can be compared and interpreted as being similar or relatively high or low. The main problems are:

- the results will vary according to the base population chosen;
- other things being equal, more detailed data for smaller areas will disclose greater differences than aggregated data for larger areas;
- figures below the 'norm' can range only from zero to 0.99, whereas the range above the 'norm' is open-ended;
- the quotients permit comparisons of only one characteristic at a time, rather than multiple characteristics of areas. Similarities and differences in terms of several features may be important.

Index of redistribution

Besides location quotients and population densities, there are some valuable summary measures of population distribution based on the index of dissimilarity (see Chapter 3, Section 3.4 and Chapter 13, Section 13.4). The first for discussion here is the index of redistribution. This provides a single number summarizing the extent of shifts in population distribution apparent from percentages, such as those for 1950 and 2000 in Table 10.2.

As discussed in Chapter 3, the index of dissimilarity measures the extent of non-overlap between two percentage distributions. In the case of the index of redistribution, the two sets of percentages denote the spatial distribution of the same population at two dates. Thus the formula for the index of redistribution is:

$$I_R = 0.5\sum_{i=1}^{n}|x_i - y_i|$$

where:

x is the percentage distribution of the population at the first date
y is the percentage distribution of the population at the second date
i is a date category, such as a region or area
n is the number of categories.

Reading the formula, the index of redistribution (I_R) is equal to half the sum of the absolute differences between corresponding figures in two columns of percentages. Table 10.2 illustrates the calculation of the index of redistribution for the regions of Africa; column F contains the absolute differences, denoted by vertical rules, between the percentages for 1950 and 2000. (In Excel, use the ABS (absolute) function to calculate absolute differences between pairs of numbers. For example: = ABS(C6 – E6) returns the difference between the numbers in cells C6 and E6, omitting the sign.)

The steps in obtaining the index of redistribution are:

- Calculate the percentage distribution of the population at two dates.
- Find the absolute difference between corresponding percentages for each date.
- Sum the differences.
- Halve the sum of the differences to obtain the index.

The index of redistribution for the five regions of Africa 1950–2000 was 3.9 per cent, denoting that only this small percentage of the population would need to be in a different region in 2000 to produce a distribution of population identical to that for 1950. Although this low figure could be anticipated from the earlier discussion of population distribution in Africa, the index offers the advantages of quantifying differences, facilitating comparisons between different periods of time (e.g. population redistribution in different decades) and producing a single summary figure from data for any number of areas within a country, region or city.

Comparisons of population redistribution in different places, however, are not possible using the index, because the results depend on the number and characteristics of the areas within each place. Other things being equal, a city divided into 30 statistical areas would exhibit a higher index of redistribution than the same city divided into 10 areas.

Index of concentration

Another measure of population distribution, with similar advantages and disadvantages to the index of redistribution, is the index of concentration (Timms 1965: 241). It is a version of the index of dissimilarity focusing on the relationship between population and land area. It compares the percentage of the population in each region or place, with the percentage of the country's total land in each region.

Thus, the index of concentration measures the degree of correspondence between population and area as follows:

$$I_C = 0.5\sum_{i=1}^{n}|x_i - y_i|$$

where:

x is the percentage of the population in each area
y is the percentage of the total land in each area
i is a data category, such as a region or area
n is the number of categories.

Referring to the African data again, the last two columns of Table 10.5 show the working for the index of concentration in 1950 and 2000. These columns indicate the percentage overlap between population and the relative size of the regions. The indices for both years are fairly low, signifying that the population of Africa was not concentrated in any one region: there was a reasonable correspondence

Table 10.5 Calculation of the index of concentration for regions of Africa, 1950 and 2000

Region	1950 Population		2000 Population		Area			Absolute	
	millions	%	millions	%	square miles	square kilometres (F × 2.58998)	% of total	Differences 1950 |C − H|	2000 |E − H|
A	B	C	D	E	F	G	H	I	J
Western	62	28.1	234	29.3	2370015	6138291	20.3	7.8	9.0
Eastern	62	28.1	246	30.8	2456184	6361467	21.0	7.1	9.8
Northern	52	23.5	173	21.7	3286031	8510755	28.1	4.6	6.4
Middle	29	13.1	96	12.0	2553151	6612610	21.8	8.7	9.8
Southern	16	7.2	50	6.3	1032730	2674750	8.8	1.6	2.6
Total	221	100.0	799.0	100.0	11698111	30297874	100.0	29.9	37.6
Index of concentration (half the total)								**14.9**	**18.8**

Sources: United Nations *Demographic Yearbook*; Population Reference Bureau, *World Population Data Sheet 2000.*

between population and area. During the 50 years, however, there was an increase in population concentration, rising from 15 to 19. The latter figure implies that 19 per cent of the population of Africa would need to be in a different region to produce an exact correspondence between population size and land area at this scale.

An index of concentration of zero would denote that each region contained a proportion of the population equal to its percentage of the total land. Conversely an index of concentration near 100 could indicate that the population was concentrated in one region only. The maximum value of the index is 100 minus the percentage of the total area within the one populated region (Timms 1965: 242).

10.4 Housing

Housing is a key influence on population distribution and density within localities. Although censuses count people wherever they happened to be at the census date, most people are counted at home. Thus census figures mainly show population distribution at night time when the majority are at their places of residence, rather than the daytime distribution, when people are out at work, or attending school, or away for other reasons. Housing types and densities therefore have a major role in explaining population distribution.

An important distinction is between dwellings and households, since dwellings sometimes contain more than one household. A dwelling is a residential unit, such as a house, an apartment, a flat or a caravan. A household is commonly understood to be either a person living alone or two or more individuals who live and eat together, such as a family or a group of unrelated individuals. Statistical collections in the United States identify a *family household* as having at least two members related by blood, marriage, or adoption. A *non-family household* is either a person living alone or a householder who shares the housing unit with non-relatives only, such as boarders (Fields and Casper 2000: 1). The term 'householder' is discussed in the last chapter (see Section 13.3).

A further distinction is between private and non-private dwellings. Private dwellings are owned or rented by the householders. Non-private dwellings usually have communal eating facilities and include hotels, motels, hospitals, nursing homes and hostels for the aged. In ageing Western populations, accommodation in non-private dwellings is assuming increasing importance, but so too are retirement villages and apartment blocks with corporate provision of support services. In some censuses, these are not identified separately, but are classified as private dwellings: as a result, little information may come to light from the census on this major shift towards assisted living.

Housing is significant as well because accommodation is one of the largest items of expenditure for most households and families, and because dwellings are the units within which many goods and services are consumed. Although dwellings may contain more than one family or household, their numbers serve as a first approximation to income-earning and consumption units. Dwellings further represent major units to be provided for in terms of land allocation, construction of roads and provision of a wide range of facilities – public transport, schools, hospitals, shops and places of employment, worship and recreation.

This section introduces two basic housing-related measures that have applications in explaining population distribution and its implications for planning, namely dwelling occupancy rates and housing density. These are pertinent to questions concerning crowding and under-utilization, population density and dispersal, the viability of amenities and businesses, as well as the costs of supplying and maintaining utilities – electricity, gas, water, sewerage, and telecommunications. Higher population densities make public transport more economic and encourage the development of diversity in retail provision. Conversely, low housing densities – leading to low population densities – discourage the development of neighbourhood shops, at the same time potentially raising the supply and maintenance costs of municipal services because of the greater distances between homes.

Occupancy rates

Occupancy rates measure the extent to which the dwelling stock is being used, potentially highlighting under- and over-utilization:

$$\text{Dwelling occupancy rate} = \frac{\text{total population in occupied private dwellings}}{\text{total number of occupied private dwellings}}$$

The dwelling occupancy rate measures the average number of persons per dwelling, preferably limited to occupied dwellings since the inclusion of empty dwellings can greatly distort the rate. It differs from average household size, because a dwelling may accommodate more than one household. Comparisons between the number of households and the number of dwellings, therefore, indicates the extent to which people are sharing accommodation. The dwelling occupancy rate does not take account, however, of the use of non-private accommodation.

A related measure, which gives some recognition to differences in dwelling size, is the average *room occupancy rate* for each type of dwelling, or number of persons per room (Neutze 1977: 151):

$$\text{Room occupancy rate} = \frac{\text{total population}}{\text{total number of rooms}}$$

Assessments of the crowding of dwellings also need to refer to other considerations, such as the number of bedrooms, the number of married couples present and the age and sex of children. For a nuclear family with a married couple, a son, a daughter and a grandparent, a four bedroom house would have 1.25 persons per bedroom, and no apparent overcrowding – assuming that the couple shared a bedroom while each of the other residents had their own.

Table 10.6 illustrates the calculation of dwelling occupancy rates. The results show that, in the United States, the average number of persons per dwelling in 1990 was 2.4; only in the smallest settlements was the rate appreciably lower. It should be noted that the data in the table refer to 'urban places', which differentiate between older central cities and newer cities and towns in the so-called 'urban fringe'. Hence the settlement size distribution in the table does not show the population of broader built-up regions, such as metropolitan areas (see Table 10.1). Furthermore, the average figures disguise many trends and variations, such as the increase through time in dwellings with just one resident. In the table, the data on population numbers refer to the resident population while the housing units include both occupied and unoccupied dwellings.

Housing density

Changes in housing densities are at the centre of much discussion of the future of cities, some key features being (Peter Harrison, quoted in Troy 1995: 72–3):

- Non-residential space demands are increasing at a greater rate than population and are the major contributors to urban spread.
- As Western cities increase in size, their overall population density falls because they encroach on more broadacre land uses such as conservation areas and airports.
- Since housing consumes only between a half and a third of the land area of a modern city, increasing its density will not result in limiting the spread of the city to any great extent.

Table 10.6 Population and housing densities, and housing occupancy rates, by settlement size, United States 1990

Size of place	Number of places	Population	Housing units	Area km²	Population density per km² (C/E)	Housing density per km² (D/E)	Housing occupancy Rate (C/D)
A	B	C	D	E	F	G	H
1 000 000 or more	8	19 952 631	8 133 674	6330	3152	1285	2.5
500 000 to 999 999	15	10 107 184	4 214 279	6 891	1467	612	2.4
250 000 to 499 999	41	14 585 006	6 351 594	12 138	1202	523	2.3
100 000 to 249 999[a]	136	19 702 834	8 108 840	17 070	1154	475	2.4
50 000 to 99 999	355	24 027 445	9 653 190	21 511	1117	449	2.5
25 000 to 49 999[b]	775	26 362 252	10 580 962	29 261	901	362	2.5
10 000 to 24 999	1 819	28 389 859	11 323 472	38 806	732	292	2.5
5 000 to 9 999	2 335	16 589 440	6 800 990	32 953	503	206	2.4
2 500 to 4 999	3 026	10 701 219	4 532 894	31 366	341	145	2.4
2 000 to 2 499[c]	1 093	2 437 517	1 067 583	10 677	228	100	2.3
1 500 to 1 999[c]	1 537	2 657 965	1 180 502	12 291	216	96	2.3
1 000 to 1 499[c]	2 307	2 834 902	1 246 019	14 539	195	86	2.3
less than 1 000[c]	9 988	4 000 428	1 825 082	50 731	79	36	2.2
Other urban	–	15 556 714	6 030 845	27 402	568	220	2.6
Other rural[c]	–	50 804 477	21 213 752	8 846 995	6	2	2.4
Total	23 435	248 709 873	102 263 678	9 158 960	27	11	2.4

Notes: [a] Includes 100 000+; [b] Includes < 50 000 and 25 000+; [c] Rural. The land area data omit water surfaces; dwellings include unoccupied as well as occupied.
Source: United States Bureau of the Census, *1990 Census of Population and Housing: Population and Housing Unit Counts, United States*. 1990 CPH-2-1.

Compared with population density, data on housing densities are relatively rare in the literature. This is despite the fact that 'In the United States, developers, city planners, and residents are more accustomed to thinking about dwelling units per acre than about persons per square mile' (Downs 1994: 142–4). The conventional measure of housing density is:

$$\text{Housing density} = \frac{\text{total dwellings}}{\text{total land area}}$$

Excluding non-residential land from the denominator is a potential refinement, depending on the availability of information. Results will vary further according to whether the calculations refer to all dwelling units, or to a subset of these, such as occupied dwellings or private dwellings. Use of total dwellings offers the advantage of including the housing belonging to persons temporarily absent, who are usually included in estimates of the resident population. Total dwellings is also particularly relevant to questions concerning the number of houses requiring the supply of urban services, or the total stock of dwellings available to meet accommodation needs.

Table 10.6 illustrates the calculation of housing densities across the range of settlement sizes in the United States. The results show that the density of dwellings parallels the broad pattern of population densities. The biggest American cities have housing densities of 1285 per square kilometre, while the figures are dramatically lower in smaller settlements, reaching 86 per square kilometre in the smallest towns. Like population densities, however, housing densities can vary greatly according to whether urban boundaries enclose just the built-up area and whether they include non-residential lands.

10.5 Population mapping and GIS

When analysing and presenting information on population distribution, maps are an invaluable aid to understanding. Indeed, important discoveries have arisen from visualizing information in map form. An early and famous example was John Snow's mapping of the places of residence of people infected in a cholera epidemic in London in 1854. From the map the author deduced that one water pump – used locally for drinking water – was the source of the infection (see Tufte 1997: 28–37). Other outstanding examples of information mapping and graphing appear in Edward R. Tufte's books, including *The Visual Display of Quantitative Information* (1983) and *Envisioning Information* (1990).

One point that Tufte emphasizes is the need to represent statistics faithfully, without distortion, through the selection of appropriate graphic techniques, data categories and labelling. Mapping and graphing today are largely automated processes using computers, which can mean that some aspects of design and presentation may not be discretionary. By default, a computer may produce a map in contrasting colours, depicting a dozen or more mapping categories, but the map's value in communicating information will be negligible. Yet more sophisticated computer programs also enable experimentation with different map designs so that, through a series of drafts, ways can be found to best convey the meaning of the data.

Computer mapping initiated a revolution in the study of population distribution. Until about the late 1960s, maps required manual drawing, which limited the use of small area data as well as creating high production costs. Plotters and line printers on mainframe computers subsequently enabled more researchers to produce maps from computer files. Since the late 1980s, mapping programs for desktop computers have transformed the quality, versatility and accessibility of mapping as a research tool. One of the least expensive mapping programs is

POPMAP, produced by the United Nations Fund for Population Activities and the United Nations Statistical Division.

The technological revolution that has underlain developments in computer mapping has also transformed access to demographic data generally. The sets of demographic data accessed through computers are sometimes described today as 'demographic information systems' or, more commonly, *geographic information systems* (GIS). A GIS consists of a data base of spatially referenced information, together with the procedures for storing, retrieving, analysing and displaying it. National population census data are now widely available as computer data bases with associated programs which, collectively, constitute geographic information systems (Plane and Rogerson 1994: 345–346). Increasing use is also being made also of *geocoding*, whereby the locations of areas or addresses are expressed as co-ordinates on a standard reference grid, to facilitate accurate tabulation and mapping.

Commercial interests quickly realized the practical applications of GIS and population mapping, such that they now have a major role in market analysis and research consultancies, as well as in public administration and planning. This has led to the adoption, in business demography, of the term *geodemographics*, to refer to demographic data for spatial units, especially small areas, and the analysis thereof for business applications (see Chapter 13, Section 13.5). Another product of the revolution in data access and mapping technology has been the production of census atlases, mapping many variables for vast numbers of small areas. Census atlases have become valued reference works, through making great amounts of information accessible in considerable detail.

10.6 Types of population maps

It is interesting that from among the many examples of excellence in graphs and maps that Tufte (1983) examined, he nominated as 'the best statistical graphic ever drawn' a work similar to a population flow map used today to illustrate migration patterns (Figure 10.2). Dating from 1861, the map depicts the fortunes of Napoleon's army in his Russian campaign of 1812–13 (Tufte 1983: 40–41). The thickness of the flow line represents the size of the army at each stage, beginning with 422 000 men and ending with 10 000. The flow line, in conjunction with additional information on dates, and temperatures during the retreat from Moscow, dramatically portrays the losses suffered as a result of battles, famine, the Russian winter and many other reversals. The map conveys an unusual amount of detail about the fortunes of an army moving through space and time.

Population maps facilitate the analysis of information on varied aspects of population distribution, as well as enabling effective communication of results. Statistics on population distribution can be presented in table form, but it is

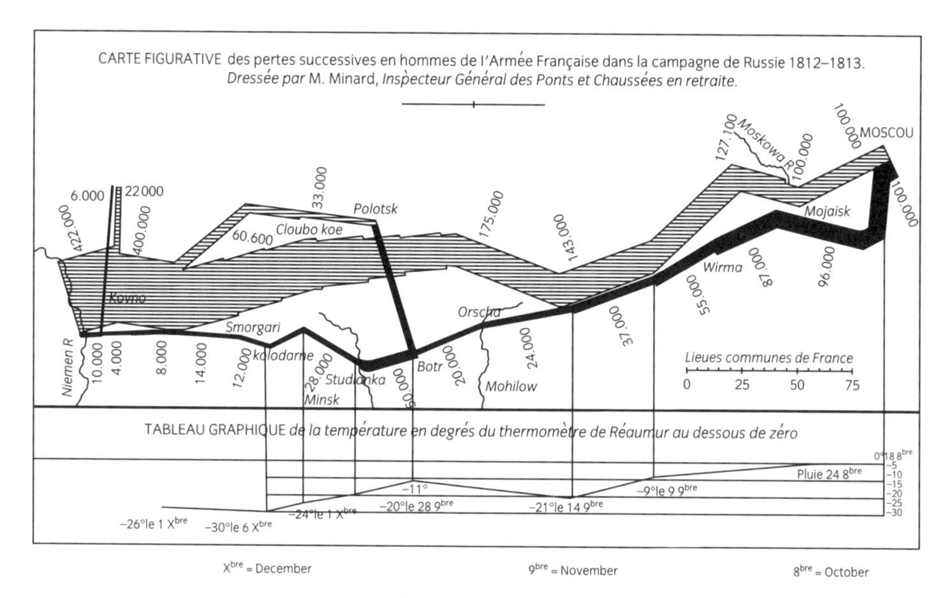

Figure 10.2 Flow map depicting Napoleon's army in Russia, 1812–1813.
Source: Minard's 1861 map, redrawn from Tufte (1983: 41).

often difficult to analyse such tables and it is even more difficult to communicate findings to people unfamiliar with the names and locations of areas.

This section briefly reviews the characteristics of the main types of population maps, including dot maps, choropleth maps, flow maps and maps with proportional symbols. While computers can now accomplish most of the drawing, the production of effective maps still depends on an appreciation of map design, together with the advantages and disadvantages of different mapping techniques.

Dot maps

Population distribution patterns can be represented effectively by dots placed within the boundaries of areas on a map. Each dot may represent individuals or aggregates such as 100 or 1000 people, or even percentages of the total population (see Coale and King 1969: 196). The size of the dots depends on the space available, as well as appearances. The dots should not coalesce into a solid mass. Appropriate choices of dot size can convey an initial impression of relative densities as well as overall distribution patterns.

In principle, dot maps consist of symbols of uniform size. In practice, it may not possible to depict a particular spatial distribution of population using only dots of the same size. Therefore, dots are often used in conjunction with proportional circles representing nuclear settlements or dense concentrations. This overcomes the problem of representing both large and small numbers together, such as when mapping town populations in conjunction with a dispersed rural population. To maintain a true visual impression of the numbers represented, dots and circles of different sizes on the same map are drawn proportional to the numbers they represent, as in Figure 10.3. This is discussed further in the later section on maps with proportional symbols.

Topographical maps, aerial photographs, satellite imagery, street directories and fieldwork enable the identification of unoccupied lands and the most appropriate placement of the dots in occupied areas. In the absence of information on the location of population, the dots may be evenly distributed within areas, such as census tracts within cities. However, in maps of regional and national population distribution it is grossly inaccurate to manufacture dot maps without reference to land uses and the location of urban areas, for instance showing the population of Alaska spread evenly across the state, including the mountains and the tundra. Overall, dots and proportional circles can be effective ways of portraying population distribution in terms of absolute numbers. Dot size and their placement within areas are the main considerations.

Choropleth maps

By far the most common type of population map is the *choropleth map*, depicting the distribution of population densities, growth rates and all kinds of population

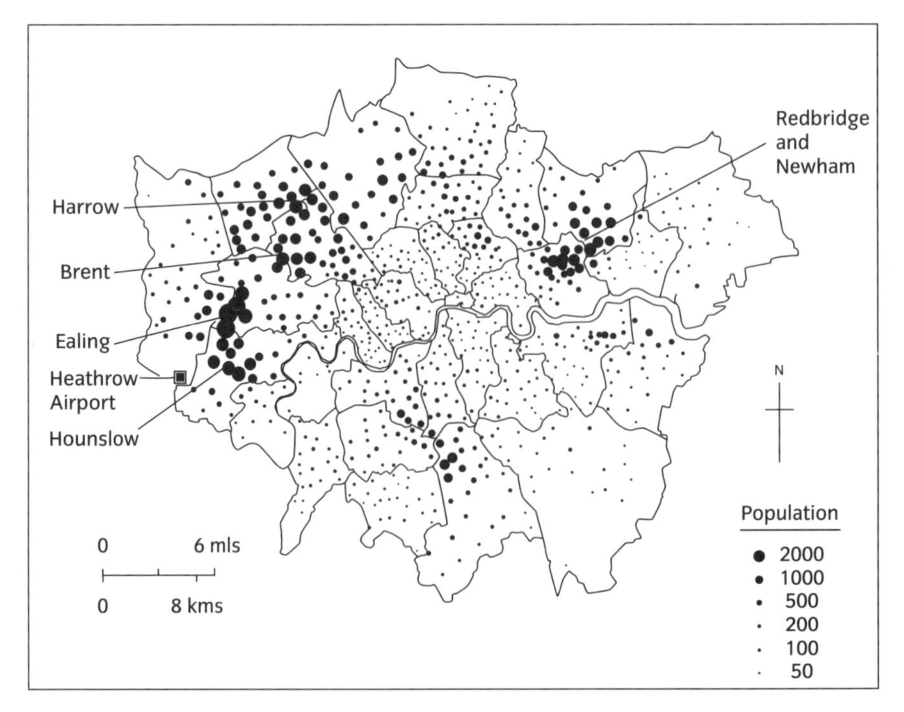

Figure 10.3 Map of the Indian settlement pattern in Greater London wards, 1991.
Source: (Peach 1999: 332).

characteristics. Whereas dot maps present data for absolute numbers, choropleth maps refer to percentages, rates and ratios, rather than absolute numbers. The name choropleth derives from the Greek words *choros* (place) and *plethos* (amount); choropleth maps employ shading of statistical areas to portray spatial variations. They provide an immediate impression of distribution patterns. They also highlight differences between spatial distributions, such as when comparing maps of income and education or patterns at different points in time. Figure 10.4 is an example of a choropleth map; the *Population Distribution* module also presents choropleth 'maps' of simulated data (Box 10.1).

Ideally, the shading categories form a graded density series, from light to dark, representing data categories from lowest to highest. This is achieved most effectively with shading in the same colour. Extra colours may create discontinuities and destroy the visual impression of a gradation of values. Two colours, however, can usefully represent positive and negative values if such a distinction is fundamental, as in maps distinguishing between areas of growth and decline.

While the researcher may gain insights from studying quite detailed maps, the final number of mapping categories is normally kept to the minimum needed to convey the information to other readers. This is because it is difficult to comprehend maps with many levels of shading: it is preferable to use no more than five or six categories. The map may need to go through a number of drafts to attain a final version that is true to the data and comprehensible to others.

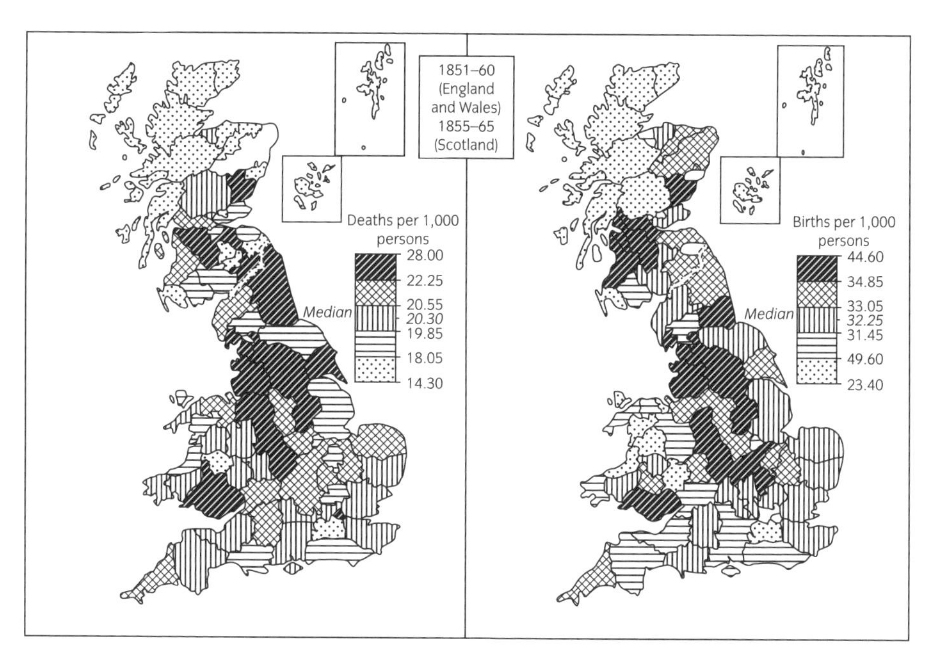

Figure 10.4 Choropleth maps of death and birth rates in mid-nineteenth century Britain.
Source: Lawton (1986: 21).

To enhance the clarity of maps, it may also be preferable to use mapping categories with constant class intervals, such as percentages in rented accommodation of 0–9, 10–19 and 20–29 rather than 0–4, 5–14 and 15–29. Otherwise it can become difficult to gauge the relative differences between areas. Also, the categories should not overlap so that it becomes unclear where a particular value is located; a common instance of this is using the intervals 1–10, 10–20 and 20–30 – here the values of 10 and 20 belong to two categories. Mapping software may not automatically generate classes that are mutually exclusive. Statistical measures, such as *quartiles* and *deciles*, sometimes provide a suitable basis for identifying mapping categories. Whereas the median divides a statistical distribution into two equal parts, quartiles divide it into four; similarly, deciles divide statistical distributions into ten equal parts. These measures are discussed in texts on descriptive statistics.

Particular disadvantages of choropleth mapping relate to the size and nature of the areas to which the data refer, namely:

- values seem to change abruptly at boundaries;
- shading creates the impression that areas are homogeneous in relation to the phenomenon portrayed.

These disadvantages are generally addressed, where practicable, through using smaller rather than larger statistical areas as mapping units. Figure 10.5 is a visually effective choropleth map based on data for small areas.

Flow maps

Whereas choropleth maps portray the characteristics of areas, flow maps illustrate linkages between them (Figure 10.6). Their main demographic application is in the study of migration. Flow maps bring migration patterns to life, with arrows of different widths depicting origins and destinations, together with the volume of movement. The choice of a scale for the width of the arrows depends on the size of the largest and smallest flows to be displayed and the amount of space on the map. Simple linear scales only are needed, such as 1 mm of arrow width per 1000 migrants. If it is difficult to accommodate the flow lines on the map, alternatives are to represent areas diagrammatically, with identifying labels or as a *cartogram*, as discussed later.

Maps with proportional symbols and superimposed graphs

The discussion of dot maps included mention of the use of proportional circles to represent population numbers. Other proportional symbols for population mapping are spheres, squares and cubes. If mapping packages do not generate these, it may be possible to copy them across from other software. Magnitudes are more easily comprehended from two-dimensional symbols, but three-

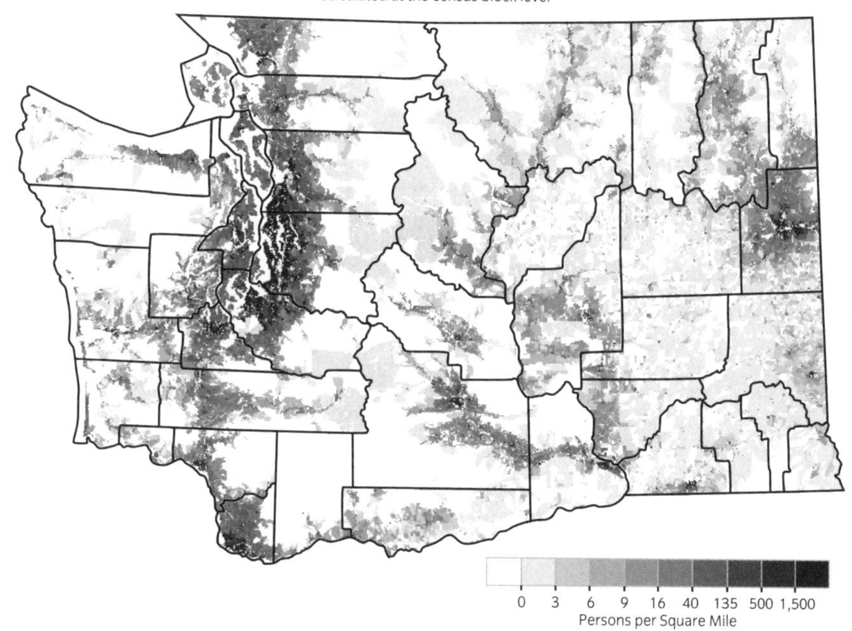

Figure 10.5 Choropleth map of population density in census blocks, Washington state, 2000 census.

Source: Carol Jenner and Mike Mohrman, Office of Financial Management, State of Washington (2002). http://www.ofm.wa.gov/popden/graymap.htm.

dimensional symbols – spheres and cubes – can depict relatively large numbers without occupying excessive space on a map. They may also add visual interest.

Proportional circles vary in size according to the numbers they represent. The area of a circle is equal to:

$$\pi r^2$$

where

π is a constant (3.1416)
r is the radius of the circle.

Pi (π) is a mathematical constant equal to the ratio of the circumference of a circle to its diameter. Since π is a constant, the radius of each circle on a population map is made proportional, at a suitable scale, to the square root of the totals to be depicted. For example, two totals of 2500 and 10 000 have square roots of 50 and 100 respectively; if 1 mm is a suitable length to represent each 10 units of radius on the map, the two circles will have radii of 5 mm and 10 mm:

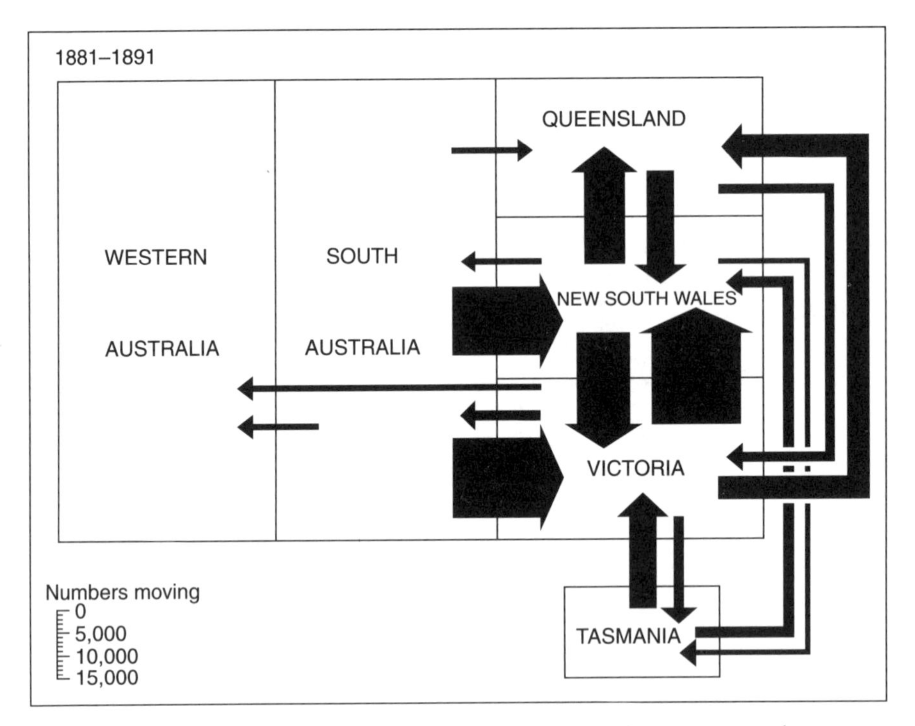

Figure 10.6 Estimated migration of Australian-born persons between Australian colonies, 1881–1891.

Source: Rowland (1980).

Value	Square root	Radius (mm)
10 000	100	10
5 000	71	7
2 500	50	5
1 000	32	3

An extension of the use of proportional circles is to turn the circles into pie charts showing further aspects of population composition, such as the percentage of each area's population in particular age groups (Figure 10.7).

Proportional spheres have a radius proportional to the cube root of the total represented. This is because the volume of a sphere is calculated as:

$$\frac{4}{3}\pi r^3$$

Examples below show the calculation of the radius of spheres to represent some large population totals, again using 1 mm to represent 10 units of radius.

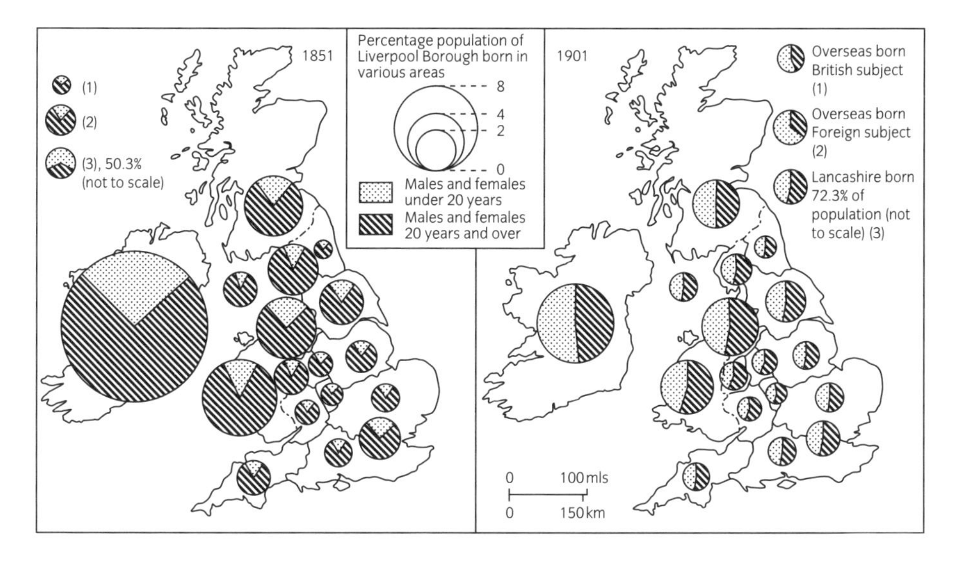

Figure 10.7 Maps with proportional pie charts showing ages and origins of immigrants to Liverpool borough, 1851 and 1901.

Source: Lawton (1986: 29).

Value	Cube root	Radius (mm)
500 000	79.4	8
100 000	46.4	5
50 000	36.8	4
10 000	21.5	2

A similar procedure is followed in drawing proportional squares and cubes. The sides of squares are drawn proportional to the square root of the total represented, while the sides of cubes are proportional to the cube root of the total. This is because the area of a square is equal to the length of one side to the power of two (length^2), while the volume of a cube is equal to the length of one side to the power of three (length^3).

Perhaps the ultimate map employing proportional symbols is the *cartogram*. In a cartogram, states, regions or other areas are represented by squares, rectangles, irregular shapes or circles. These symbols are drawn proportional to the population sizes, densities, or other characteristics they represent. Alternatively, the original shape of areas may be scaled up or down in proportion to their population totals (Figure 10.8). Cartograms provide a means of emphasizing larger populations or the most relevant information. Extensive, sparsely populated areas no longer dominate the picture. An important consideration in designing cartograms is to ensure that the country or region remains recognizable. This requires adequate labelling or maintenance of a semblance of the shape, location and

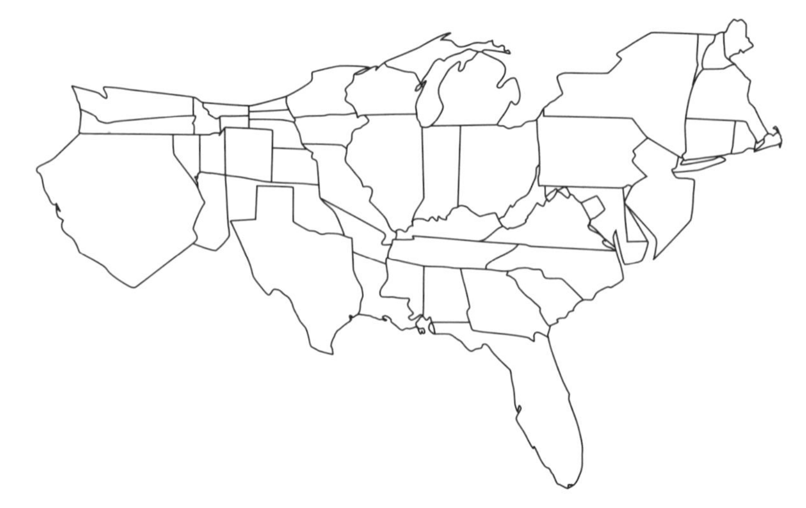

Figure 10.8 United States population cartogram, 1996 (states resized relative to their populations).

Source: Texas A & M University, Visualization Laboratory 2002. http://www-viz.tamu.edu/faculty/house/cartograms/1996Cartogram.html.

contiguity of areas. Figure 10.8 is a cartogram representing the relative sizes of state populations in the United States; for a cartogram of the world's population in 2000 see Weeks (2002: 22–23).

Finally, an effective way of maximizing the amount of information contained on a population map is to place a graph within each area. These may be population pyramids, bar charts or other types of graphs plotting statistics for each place (Figure 10.7). Superimposed graphs can convey much information effectively, provided that there is sufficient space.

10.7 Conclusion

Applications of demography in public administration and business include analyses of population in a spatial setting: a vital concern is where the people live. This chapter has reviewed the principal types of maps used in presenting information on population distribution. It has also discussed the basic methods employed to summarize and compare aspects of population distribution. Additional information on these subjects is available from the references listed below under the heading 'Further reading'. The ensuing chapter on migration extends the consideration of population dynamics in a spatial setting. The final chapters in the *Applied Demography* section also complement the study of population distribution, presenting further methods and concepts needed in understanding, and planning for, people according to their places of residence.

Study Resources

KEY TERMS

Cartogram
Choropleth map
Dot map
Dwelling occupancy rate
Flow map
Geocoding
Geographic information systems (GIS)
Housing density
Housing occupancy rate
Index of concentration
Index or redistribution
Location quotient
Nodal regions
Population density
Room occupancy rate
Settlement hierarchy
Uniform regions

FURTHER READING

Bailey, T.C. and Gattrell, A.C. 1995. *Interactive Spatial Data Analysis*. Harlow: Longman.

Dent, Borden D. 1985. *Cartography:Thematic Map Design*. Dubuque, IA: William C. Brown.

Dorling, D. 1996. 'Area Cartograms: Their Use and Creation'. *Concepts and Techniques in Modern Geography*, 59: 1–68.

Dorling, D. 1995. *A New Social Atlas of Britain*. London: John Wiley.

Plane David A. and Rogerson, Peter A. 1994. *The Geographical Analysis of Population: With Applications to Planning and Business*. New York: John Wiley.

Smith, David M. 1977. *Patterns in Human Geography*. Harmondsworth: Penguin Books.

United Nations (biennial). *World Urbanization Prospects*. New York: United Nations Department of Economic and Social Affairs.

Woods, Robert. 1982. *Population Analysis in Geography*. Longman, Harlow.

INTERNET RESOURCES

Subject	Source and Internet address
Full text of a major journal on population geography.	**International Journal of Population Geography** http://www3.interscience.wiley.com/cgi-bin/jtoc?ID=15448
Links to sites of interest to population geographers, provided by the Population Geography Research Group	**Department of Geography, University of Liverpool** http://www.liv.ac.uk/geography/pgrg/hotlink.htm
Data on the population of cities throughout the world	**United Nations Statistics Division** http://www.un.org/Depts/unsd/demog/
A site providing access to Census Bureau maps and cartographic resources	**US Census Bureau** http://www.census.gov/geo/www/maps/
Statistics from the 2000 US census, including data for each state	**US Census Bureau** http://www.census.gov/prod/cen2000/index.html
Population analysis spreadsheets, including downloadable Excel spreadsheets for calculating indices of urbanization and population distribution	**US Census Bureau** http://www.census.gov/ipc/www/pas.html

Note: the location of information on the Internet is subject to change; to find material that has moved, search from the home page of the organization concerned.

EXERCISES

1 Explain the meaning of the following:

(a) $LQ > 1$

(b) $LQ = 1$

(c) $LQ < 1$

2 Would it be appropriate to use the index of redistribution to compare changes in the distribution of population in Canada and the United States during the interval 1950–2000, working from statistics for provinces and states respectively? Explain your answer.

3 When studying the populations of places that have been affected by boundary changes, how can demographic data be made more comparable through time?

4 From the statistics in Table 10.5, calculate the density of population in regions of Africa in 2000. Express the answers either in terms of population per square kilometre or per square mile. Comment briefly on the results.

5 Identify three disadvantages of the index of redistribution.

6 Calculate the index of redistribution 1990–2000, and the index of concentration in 2000, from the following table:

Resident population in regions of a city, 1990 and 2000

Region	Population 1990	Population 2000	Area sq. km
Central	432 861	471 950	145.65
Northern	172 418	169 350	88.94
Southern	744 582	634 500	700.56
Eastern	191 300	233 900	4 005.92
Western	769 200	876 200	5 930.68
North-Eastern	765 400	667 800	900.64
South-Western	140 300	159 750	2 551.86
Total	3 216 061	3 213 450	14 324.25

7 Find examples of the main kinds of population maps and evaluate their effectiveness in conveying information about population distribution.

8 Using materials from the Internet, describe three contrasting examples of applications of location quotients in the study of population and/or other phenomena. Acknowledge sources by providing Internet addresses for each example.

SPREADSHEET EXERCISE 10: A DESCRIPTIVE STATISTICS MODULE

In Chapter 6, the exercise entitled 'A Graphical Data Base' provided a means of viewing, in graphical form, the contents of a data base – as an aid to becoming familiar with a set of statistics. Exercise 10 shows how a more versatile data base can be presented in graphical form with accompanying descriptive statistics to assist the analysis. Only a limited range of descriptive statistics are included in the present module, but others could be added as needed.

The main aims of the exercise are to introduce the use of statistical functions and to provide further practice in constructing graphical data bases. The module displays graphs and descriptive statistics for 96 sets of data (12 countries/regions by eight variables). Knowledge of previous spreadsheet exercises is assumed.

	A	B	C	D	E	F	G	H	I	J	K	L	M	N	O	P	Q
1	**Menu Lists, Cell Links and Selected Data**																
2																	
3	**Cell Links**			**List of Countries**			**List of Variables**										
4	Selected country		1	S. Eastern Asi		1	Average annual growth rate %										
5	11		2	Brunei		2	Crude birth rate per 1000										
6	Selected variable		3	Cambodia		3	Crude death rate per 1000										
7	6		4	East Timor		4	Total fertility rate per woman										
8			5	Indonesia		5	Life expectancy at birth, males										
9			6	Lao		6	Life expectancy at birth, females										
10			7	Malaysia		7	Life expectancy at birth, both sexes										
11			8	Myanmar		8	Infant mortality rate per 1000 live births										
12			9	Philippines													
13			10	Singapore													
14			11	Thailand													
15			12	Viet Nam													
16																	
17	**Selected row (ie the row number of the selected statistics in the Data spreadsheet)**																
18	**86**																
19																	
20	=(A5-1)*8+A7																
21																	
22	**Selected Data**																
23	Country	**Thailand** ←		=INDEX(D4:D15,$A$5)													
24	Variable	**Life expectancy at birth, females**					←	=INDEX(G4:G11,$A$7)									
25	Estimates									Medium Variant Projections							
26	1950	1955	1960	1965	1970	1975	1980	1985	1990	1995	2000	2005	2010	2015	2020	Mean	Median
27	**49.1**	**52.9**	**56.1**	**58.9**	**61.6**	**63.2**	**66.7**	**69.9**	**71.9**	**73.4**	**74.6**	**75.8**	**76.9**	**77.9**	**78.7**	**67.2**	**69.9**
28																	
29	=INDEX(Data!C1:C96,$A$18)														=INDEX(Data!Q1:Q96,$A$18)		
30															=AVERAGE(A27:O27)		
31															=MEDIAN(A27:O27)		
32																	

The 'Data' spreadsheet

Open the file ***Asia.xls*** provided on the CD. Save the file under a new name (e.g. *Exercise 10.xls*). Right click on the sheet tab for the spreadsheet (the default labelling is 'Sheet1') and rename it **Data**.

The file contains demographic statistics for the United Nations region of south-eastern Asia and the eleven countries in the region extracted from the following publication: United Nations 1993. *World Population Prospects: the 1992 Revision*. New York: United Nations, Department of Economic and Social Information and Policy Analysis.

Note that there are no column headings at the start of the table. The headings have been placed at the end so that the rows of data in the spreadsheet remain numbered from one. Retaining this numbering simplifies the selection of data for analysis. The figures are estimates from 1950 to 1990, and 'medium variant' projections from 1995 to 2020.

The 'Main' spreadsheet

The next step is to create a spreadsheet containing (1) information for drop-down menus, (2) the indices (cell links) that control the selection of data and (3) the row of data selected for plotting and statistical analysis.

- Right click on the tab for Sheet2, and rename it **Main**.
- Now type the information for the Main spreadsheet as shown in the picture. Enter formulas in the cells containing bold numbering. Arrows indicate examples of the formulas.

In cell B23, the index function chooses the selected country from the list of countries (column D) according to the value in cell A5. Similarly, the index function chooses the selected variable from the list of variables (column G) according to the value in cell A7.

The index function is also used to select information from the Data spreadsheet. A formula containing the index function is typed in cell A27 and then copied across as far as cell O27. The statistics in row 27 are selected according to the country number and the variable number. The row number, showing the location of the selected statistics in the 'Data' spreadsheet, is calculated in cell A18. Thus the selection that the index function makes in row 27 depends on the value in cell A18.

In cell P27 type the formula for the mean; in cell Q27 type the formula for the median (see section headed 'Functions'). The mean and median were discussed in Chapter 3.

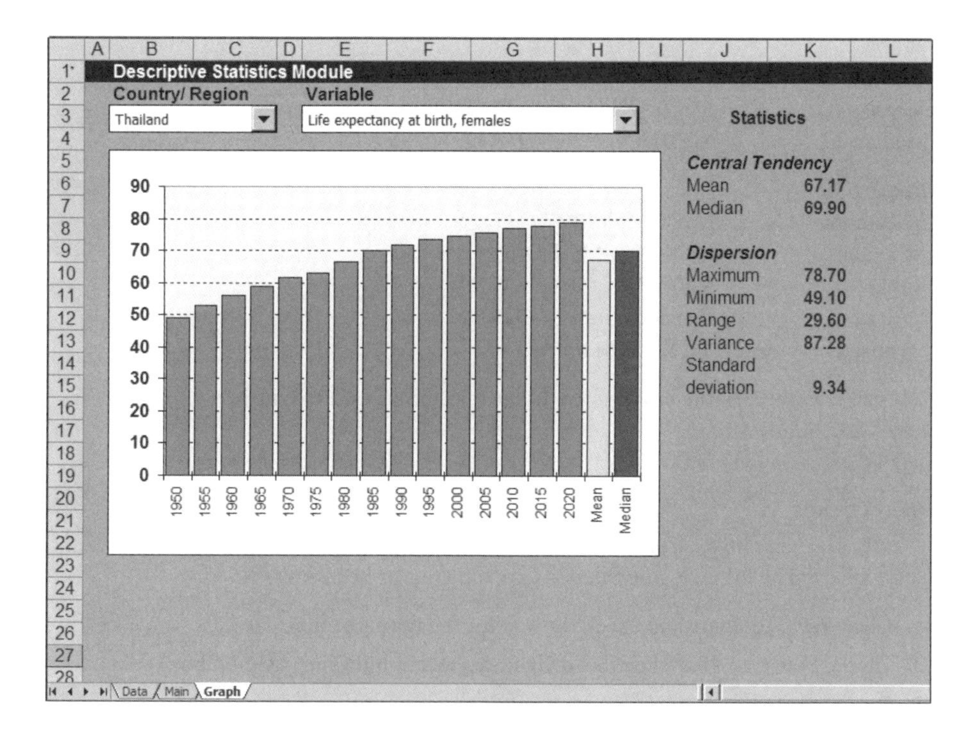

The 'Graph' spreadsheet

- Right click on the tab for Sheet3, and rename it **Graph.**
- On the Main spreadsheet, highlight the headings and selected data in rows 26 and 27, including the mean and median.
- Click the Chart Wizard button, draw a column graph and save it on the Graph spreadsheet.

The colours of the mean and median columns can be changed as follows:

1. click in the middle of the mean bar (all columns will be selected)
2. click on the outer border of the mean column (that column will be selected)
3. double click on the mean column to bring up the editing menu
4. repeat these steps for the median column

Functions

The next step is to construct a table beside the graph to present further descriptive statistics for the selected data. The formulas for the descriptive statistics can be entered using the **Function Wizard** button (if present) on the standard toolbar, or by clicking **Insert Function** on the standard toolbar. All of the functions in this exercise are listed under Function Category: Statistical.

Explanations of the meaning of the statistical functions can be obtained by selecting the function, clicking Help and printing the topic. The Excel formulas used in the table on the Graph spreadsheet are:

Central Tendency
Mean = AVERAGE(Main!A27:O27)
Median = MEDIAN(Main!A27:O27)

Dispersion
Maximum = MAX(Main!A27:O27)
Minimum = MIN(Main!A27:O27)
Range = K10-K11
Variance = VARP(Main!A27:O27)
Standard deviation = STDEVP(Main!A27:O27)

The cell references in the brackets are to the selected statistics in the Main spreadsheet, row 27, columns A to O.

Menu boxes

Create two menu boxes from the list of countries and list of variables on the Main spreadsheet. The following instructions are the same as in Exercise 6.

1. Click View Toolbars and check the box for the Forms toolbar
2. On the Forms toolbar click the 'drop down menu' button or 'combo box'.
3. Draw a rectangle on the screen the size of the required menu box

Editing the menu boxes

1. On the Forms toolbar click the Control Properties button; the Format Object dialog box will appear; alternatively, right click on the menu box you want to edit.
2. Click the mouse in the Input Range box.
3. On the Main spreadsheet, highlight the list of countries; if the Format Object box is in the way, move it by clicking the mouse on the title bar and shifting the box while holding down the left mouse button.
4. Click in the Cell Link box.
5. On the Main spreadsheet click the cell under the heading 'Selected country'.

6. Press Enter, or click OK in the Format Object dialog box.
7. Click in a blank cell to deselect the drop-down menu.
8. Repeat the above steps to create a menu box from the list of variables, but this time the cell link will be to the cell under the heading 'Selected variable'.

If you later wish to change the control properties of the drop-down menu, right click on it and choose 'Format control' from the list. (In older versions of Excel you may need to add the 'drawing selection' button to one of the toolbars, such as the drawing toolbar. To do this click View Toolbars Customize; under 'Category' choose Drawing, then drag the button with the white arrow on to one of the toolbars at the top of the screen.)

Finally, add headings and labels as shown in the picture of the module. Through use of the menus, the module can now display any row from the Data spreadsheet, together with accompanying summary statistics.

For discussion

Describe another data set that could be examined using the same approach. Suggest possible modifications or enhancements of the Descriptive Statistics module.

11 Migration

There come yearly to dwell at *London* about 6000 strangers out of the Country, which swells the burials about 200 *per Annum*.

(Graunt 1662: 15)

OUTLINE

11.1 Concepts and theories
11.2 Migration statistics
11.3 Migration rates
11.4 Migration effectiveness
11.5 Migration expectancy
11.6 Estimating migration
11.7 Calculating survival ratios
11.8 Conclusion
Study resources

LEARNING OBJECTIVES

To understand:

- the concepts of migration and mobility
- migration rates
- the principal measures of migration
- methods of estimating migration
- the calculation of survival ratios for migration estimates

To become aware of:

- applications of classifications of migration
- aggregate and individual perspectives in migration studies
- theories concerning migration and the demographic transition

COMPUTER APPLICATIONS

Excel modules: Migration in the United States.xls (Box 11.2)

Spreadsheet exercise 11: Migration estimates (see Study resources)

Individuals have an impact upon the size of populations only once through birth and once through death but, in all likelihood, many times through migration. Migration is the most repeated demographic event and the least predictable aspect of population change. In Western societies it is commonplace for people to regard a dozen or more dwellings as 'home' at different times. The frequency of migration denotes strong impulses for migration at particular stages of life, as well as ease of movement and, for some, tenuous attachments to communities.

Migration is often the main component of population growth or decline at the community and regional levels. Associated changes, such as urbanization and economic restructuring, have long been major interests of migration studies. Migration itself is a symptom of social and economic processes, such as the decline of employment in manufacturing and changes in lifestyles favouring rural living and retirement to coastal areas. Some further population movements, including refugee flows, are symptoms of the effects of wars, natural disasters and oppressive governments, although it is often difficult for receiving countries to distinguish between political refugees, fleeing persecution, and 'economic refugees', escaping economic adversity.

While migration can denote upheaval and transformation, it is also true that much migration within Western countries is absorbed in exchanges of people who have similar characteristics, such as in terms of their ages, occupations and educational attainments (Richmond 1969; Rowland 1979). Such flows serve more to maintain and renew, rather than alter, the size and demographic composition of community populations. Exchanges of population, as in migration streams between cities, reflect the inherent dynamics of contemporary societies arising from the activities of businesses and governments, as well as from the progress of individuals through their life cycles.

Accordingly, when interpreting migration statistics, economic and social changes are important, but so too are the everyday dynamics arising from the need for organizations to train new staff or replace retirees, as well as from the personal decisions of individuals about their education, employment, marriage, living arrangements and housing. Communities can experience substantial population inflows and outflows which, while altering the membership of the population, have far less impact on its total numbers and overall demographic composition. Thus a large share of all migrations may be absorbed in *population replacement* rather than in *net growth* (see Chapter 2). Over the long run, the net effect of population movements usually entails community change of some kind, but the extent of change can be small compared with the great volume of two-way migration. The index of *migration effectiveness*, discussed later, seeks to measure the efficiency with which inward and outward movements produce a net redistribution of population.

Demographers commonly omit migration from demographic modelling, partly to show the effects of other processes more clearly, and partly because of the complexity of migration. Any population with no inward or outward migration is termed a *closed population*. The opposite is an *open population*, that is, one open to the effects of migration. While closed populations are useful in demographic models, real-world examples of them are now rare: migration is pervasive.

Overall, migration is significant both as a process of change and as a process of demographic maintenance and renewal. Being one of the main components of population growth and decline, it is an essential process to consider when analysing population size and composition. Beyond the study of migration for its own sake, the concepts and methods in the study of migration have applications in other areas of demographic inquiry, including population projections, population distribution, age structure, and spatial aspects of population composition.

This chapter introduces the study of migration through a summary of selected concepts and theories, together with a discussion of sources of statistics on migration. The chapter then introduces migration rates and other widely used measures. Later sections focus on methods of estimating migration, to complete the coverage of the main techniques for the demographic analysis of migration. Further measures and models, of particular interest in geographical studies of migration, are covered in other texts (e.g. Plane and Rogerson 1994).

11.1 Concepts and theories

Mobility and migration

In defining migration, demographers draw initial distinctions between *international (or external) migration*, *internal migration* and *local migration* (or 'residential mobility'), all of which are deemed to entail a change in an individual's usual place of residence (see Table 11.1). Thus, international migration refers to residential moves across national boundaries; internal migration to residential moves between communities within the same country (e.g. cities, towns or villages); while local migration consists of residential moves within a community, such as a city, town or village. The principal concern in the study of all three types of migration is permanent movement, commonly defined as a relocation of a person's usual place of residence lasting at least six or twelve months.

Although these definitions have practical value and are widely accepted, in some contexts they may have a decisive and adverse impact on what is considered relevant and what is discovered about population movement. Comparability with other studies is desirable, but definitions must be appropriate to the situation, not artificial constructs that are culturally inappropriate. In the past, a formidable obstacle to progress in migration research in developing countries was the adoption, in data collections, of textbook definitions of migration that had been derived from the experience of Western countries. Such approaches were intended to yield 'comparable' data, but the comparisons were misleading. Far from capturing the main features of population movements in developing countries, research based on definitions of migration, such as 'a permanent change in place of residence',

Table 11.1 Key terms in the study of population movement

Term	Definition
Circular mobility (circulation)	Temporary population movement where, from the outset, the movers intend to return to their original places of residence.
Closed population	A population experiencing no inward or outward migration.
Fixed period migration	Migration occurring in a specific interval, such as between two censuses.
Gross migration	The sum of arrivals and departures in an area.
Internal migration	Migration between communities in the same country.
International (external) migration	Migration between countries.
Lifetime migration	Migration occurring between birth and the time of a census or survey (Shryock and Siegel 1973: 618).
Local migration	Migration within communities. Often referred to as *residential 'mobility'*.
Migration	Population movement entailing a change in the usual place of residence.
Migration interval (or period)	The interval during which migration is observed, such as between two censuses.
Migration rate	*Either*: The ratio of the number of migrants to the number of people at risk of moving;
	Or: The ratio of the number of movers to the size of the population at risk of sending or receiving the migrants (Shryock and Siegel 1973: 606).
Migration stream	A group of migrants with a common origin and destination.
Mobility	All forms of population movement, whether temporary or permanent (Kosinski and Prothero 1975: 3).
Net migration	The difference between the number of arrivals and departures in an area.
Open population	A population experiencing inward and outward migration.
Return migration	Migration back to the original place of residence.

failed to bring to light the importance of temporary moves that are of great social significance, and often surpass the frequency of 'permanent' moves.

In Indonesia, for example, conventional sources of migration data used before the 1980s – censuses and large-scale surveys – created the impression that most of the population of Indonesia, and Java especially, lived all their lives in the same village. Later detailed studies of villages in West Java revealed that only a third of all the moves associated with work and formal education would have been classified as migration in the census, which focused on 'permanent' changes in place of residence, that is, staying at the destination at least six months. The greater numbers of moves were circular, entailing a return to the place of origin. Some circular migrants were daily commuters, but others were absent for days, weeks or months. The longer absences were frequently seasonal in nature, when there was no employment in the village in planting or harvesting. All of the circular migrants regarded the village as their permanent home and ultimately returned there. The study therefore disputed the view that social change and economic development arose only from permanent population movement. Many Indonesians worked in one place but consumed, spent, and invested their earnings in another. Circular migration denoted a long-term commitment to bilocality (Hugo 1982: 59–83).

Realization of the importance of non-permanent population movements has led, since the 1970s, to a shift in emphasis in research from a focus on migration to a focus on the broader concept of *population mobility*. This is especially true in research on less developed countries, but the concept of mobility has provided a valuable context for considering the diversity of population movements in other countries as well. A pioneering study here was Gould and Prothero's (1975) 'space-time typology', developed for use in Africa.

Classifications of population movements

Gould and Prothero proposed a classification of population movements at a variety of scales of time and space. It included movements unidentified in censuses, and previously overlooked. The classification distinguished between migrations on the two dimensions of space and time (Figure 11.1). The spatial dimension included distance and direction, but difficulties in specifying distance, for instance whether to use physical distance (kilometres) or economic distance (travel costs), led to an emphasis on direction – specified in terms of movements between urban and rural origins and destinations (Figure 11.1).

The temporal dimension distinguished between movements according to the time involved, from shifts lasting a few hours to those that were permanent. 'Migra-

Space	Time					
	Circulation				Migration	
	Daily	Periodic	Seasonal	Long-term	Irregular	Permanent
Rural-rural Rural-urban Urban-rural Urban-urban						

Figure 11.1 Space-time typology of population mobility in Africa.

Source: Gould and Prothero (1975: 42).

tion' was a permanent change of residence, whereas 'circulation' included a great variety of movements, all lacking a declared intention of leaving permanently.

The space-time approach, emphasizing a continuum of mobility from brief to permanent moves, has served as a foundation for many subsequent migration studies. The typology identifies movements that are very important socially and economically, as well as in terms of population distribution, but are likely to be overlooked in studies of migration per se. It provides a more comprehensive framework within which to begin migration research in developing countries. The distinction between circulation and migration is not absolute, however, since some movers do not know whether they will return, or how long they will stay before moving elsewhere.

Classifications serve as frameworks for interpreting migration statistics, since they draw attention to diversity in the nature and consequences of different population movements. As in Gould and Prothero's (1975) work, they can overcome the limitations of conventional definitions. There are a number of other classifications that employ different criteria to refine understanding of the nature of population movements, together with their determinants and consequences (see Lewis 1982).

One example, with applications in the study of residential moves within urban areas, is Roseman's (1971) distinction between *partial and total displacement migrations*, each reflecting different considerations in migration decisions, as well as different consequences for the lives of the movers. In *total displacement migration*, movers severed links with all the locations they regularly visited on a daily or weekly basis, such as places of work, shopping, schooling and recreation. In contrast, *partial displacement migration* was more conservative in that there was, at most, only a partial change in the places visited regularly. Housing was the main consideration in partial displacement migrations. Total displacement required greater adaptations on the part of people moving, as well as entailing greater impacts on the receiving community.

Another informative classification, used in explaining the process of international migration to the United States and other countries, refers to a sequence of stages involved in *chain migration* (Price 1963). Chain migration is the process whereby immigrants encourage and assist relatives and friends to join them. It has a considerable influence on the origins of immigrants as well as their destinations, through establishing links between villages or localities in the country of origin and specific locations at the destination. Thus it becomes misleading to equate origins simply with the country of origin. Migrations that have the outward appearance of a mass movement of people between countries, may in fact arise from the influence of social networks linking specific people and places. Whereas the two classifications described earlier distinguish between different forms of movement, chain migration focuses on stages in a sequence of movements through time (Table 11.2). Chain migration characterized the movement of eastern and southern European peoples in the past, and it continues to be influential in migration patterns and decisions, such as among emigrants from Asian countries.

Table 11.2 The chain migration sequence

Stage	Characteristics
1	Arrival of pioneers, mainly males; highly mobile; movement may be erratic and opportunistic.
2	Pioneers persuade relatives and friends to follow, often giving help with fares, jobs and accommodation. Group remains male dominated with high mobility among new arrivals. Migrants begin to congregate in some areas of settlement.
3	Arrival of womenfolk. Possibly an increase in the scale of migration. Stable settlements formed and mobility declines. New arrivals persuade others to join them through letters or visits. Moves of longer-established settlers often oriented towards finding more congenial surroundings, more profitable areas or places nearer friends and compatriots.
4	Second generation of immigrants reaches maturity. Their patterns of movement may approach that of the host population.
5	Third generation matures – mobility characteristics probably similar to those of stage 4.

Source: Price (1963).

Aggregate and individual perspectives

The decision-making processes of individuals have been of central interest in sociological research on international migration and immigrant settlement – the concept of chain migration has been one of the foundations for such work. Demographic and geographical studies of internal migration long tended to emphasize an aggregate approach to migration studies, relying especially on census materials or large-scale surveys.

E. G. Ravenstein (1885), propounder of certain 'laws' of migration, was one of the earliest writers on the aggregate approach. Everett Lee's (1966) paper entitled 'A theory of migration' built on Ravenstein's observations, stating a series of generalizations about the volume and pattern of migration, as well as the characteristics of movers. The generalizations included: migration takes place mainly in well-defined migration streams (for instance, because of the location of opportunities) and a counter-stream accompanies every major migration stream. Lee's work is one of many examples of the aggregate approach to migration studies. In demography and geography, it is characterized by an emphasis on statistical regularities and models, and interpreting migration as a response to spatial and social organization.

During the late 1960s, geographers began developing a 'behavioural approach' to the geographical study of migration. It arose out of a quest for more searching explanations of population movement than could be obtained from aggregate data. Like the classic studies of chain migration (e.g. Thomas and Znaniecki 1958; Price 1963), the behavioural approach focuses on individuals – their perceptions, decision-making processes and migration behaviour – but it employs its own concepts and terminology (see Lewis 1982 and Woods 1982b). Because of their different emphasis, findings from behavioural studies function as a valuable complement to migration research utilizing aggregate statistics from censuses, administrative records and other sources.

Migration and the demographic transition

Migration was for long the missing component in discussions of the demographic transition. One question, for example, was whether the decline of the birth rate in Europe would have been faster in the absence of mass migration to the United States and elsewhere. There are two main hypotheses about migration and the demographic transition: one by Dov Friedlander, represents the more 'classical' style of approach in which fertility, mortality and migration are seen as closely interrelated. The other, by Wilbur Zelinsky, focuses mainly on population mobility for its own sake, without emphasis on causal links between migration and fertility and mortality.

Friedlander (1969) examined interrelationships between migration, fertility and population growth, especially the way migration influenced the timing of fertility decline in the demographic transition. He hypothesized that the timing of fertility decline depended on whether there were opportunities for internal and/or external migration. Consequently, he interpreted the amount of growth from natural increase occurring in European countries during the transition as related to opportunities for migration.

In contrast, Zelinsky (1971) described changes in mobility believed to accompany the demographic transition in industrialized societies. He proposed that there are patterned changes through time in rates of different types of population movement, such as rural to urban, urban to urban and international. He saw these changes paralleling the stages of the demographic transition. Zelinsky's mobility transition draws attention to the fact that migration and mobility are both mechanisms and symptoms of changes taking place in societies. Zelinsky's model, like Gould and Prothero's space-time typology, has attracted great interest although there have been several areas of debate, such as concerning its relevance to less developed countries, and the notion that there is only one sequence of changes for all societies (Nam 1994: 239; Skeldon 1990).

11.2 Migration statistics

International migration is measured through information collected at borders and airports. People intending to settle permanently, however, commonly comprise only a small proportion of the arrivals. The overwhelming volume of many flows – including tourists and commuters – hinders the maintenance of detailed records, while long land borders aid illegal migration. Freedom of movement between member countries of the European Union has also reduced imperatives to collect information on arrivals and departures. Accordingly, there are many discrepancies between the international migration statistics of different countries.

In this context, census statistics on the foreign-born are often the most complete sources of information on the net outcome of immigration, although some, such as illegal migrants who wish to conceal their foreign origins, will be undercounted. Similarly, census statistics on the foreign-born are the principal basis for determining the consequences of international migration, such as in terms of impacts on the composition of the labour force.

For statistics on internal migration and local migration, censuses are usually the most complete source if a suitable question is included, for instance asking where people lived one year ago or at the time of the previous census. Although most censuses are conducted on a de facto basis, counting people wherever they were at the time of the census, they also often include a question about where each person usually lives. The internal migration statistics are then coded with reference to places of usual residence. This eliminates the effects of temporary movement for work, holidays or other reasons, although it does not address the issues of circular mobility and bilocality discussed earlier.

Other sources of data on movement within countries are surveys (including labour force surveys, which may include questions about movement), and a wide range of administrative records such as those arising from the maintenance of electoral rolls and telephone directories. An inherent limitation is that coverage may be restricted to particular subgroups, such as members of the labour force, persons eligible to vote or telephone subscribers. All of these sources provide what are known as 'direct counts' of migration, whereas the techniques explained in the second half of the chapter provide indirect estimates of migration. The latter derive from sources other than actual counts of the numbers and characteristics of migrants.

Although internal migration is usually defined as permanent movement between communities, people may be uncertain whether their own migration will be permanent, even in terms of the working definition of 'permanent' as six or twelve months residence. Some movers are 'testing the waters' to decide, on the basis of experience, whether the new location will suit their needs. Also, the poor may not be settled in any location for long, while the wealthy may own homes in two or more places, living in each at different times.

Definitions of community may similarly create uncertainties. In internal migration studies, communities are spatially defined units, consisting of an entire city, town, village or rural administrative district. When interpreting the data, it is important to know where the community boundaries lie. If a town is built on two sides of a river, or astride a state or provincial border, the data may identify it as two communities instead of one. Similarly, if the boundaries of a metropolitan region include towns and small settlements on the edge of a city, shifts between the city and the nearby towns may be identified in the data as local, rather than internal, migrations.

Issues concerning boundaries and the definition of communities explain why it is difficult or impossible to obtain comparable data on rates of internal migration for different countries, or even for the same country at different dates. One solution is simply to compare the total changes in places of residence – for internal and local migrations combined – rather than attempting to differentiate between the two (Long 1970). The sum of such figures can reveal high overall levels of population movement which, in a twelve month period in the mid-1990s, ranged from 10 per cent of the population moving in Great Britain, to 16 per cent in Canada and the United States, 19 per cent in Australia and 23 per cent in New Zealand (Bell and Hugo 2000: 27). Box 11.1 identifies a number of factors conducive to high rates of migration.

BOX 11.1 **Factors conducive to high migration rates in developed countries**

1. *Concentration of population in cities* Because of the size and diversity of the housing stock, there are greater opportunities for housing-related moves in cities than in the countryside. Residential movement within cities is the principal cause of high mobility in modern societies.
2. *Housing* Important influences on migration are affordable housing and a varied housing market suited to particular needs at different stages of life, such as rented flats for young couples, single family dwellings for couples with children, and retirement accommodation for the aged.
3. *Life cycle changes* Migration often accompanies changes such as commencing tertiary education, entering the labour force, marriage, childbearing, retirement, widowhood and frailty.
4. *Family migration* Migration of families, rather than lone individuals, produces high migration rates. Marriage and childbearing are concentrated in the young adult ages, where propensities to move for education, employment and housing are highest.
5. *Occupations* More developed countries have a high percentage of their labour force in migration-prone professional, administrative and technical occupations, as well as in service occupations for which there is a national employment market. Relatively immobile farmers and farm workers now comprise only a low proportion of the labour force.
6. *Employers* In more developed countries, a high proportion of the labour force works for governments and large firms. Their staffing policies generate many transfers of personnel, as well as facilitating such moves through housing assistance and financial inducements. Self-employed persons, in contrast, have low mobility and comprise a declining proportion of the labour force.
7. *Modern technology and affluence* These have reduced the financial constraints to movement, as well as the physical obstacles and information barriers.
8. *Returns to the original home* Departures sometimes pre-suppose a subsequent return. The original home community or region may remain the preferred place to live.

A related problem is that internal migration statistics allow great variations in the distances moved, simply because of differences in the size and shape of areas. Consequently, moving house just a short distance across a community border, such as a road, will be recorded as an internal migration, whereas moving a considerable distance within a rural administrative area will not – because the individual did not cross a migration-defining boundary (Figure 11.2).

Finally, as discussed earlier, in some contexts temporary population movements have a considerable effect on the distribution of population, making the de facto population numbers a necessary consideration in planning. Information on the origins and duration of stay of temporary migrants can then become an important adjunct to information on permanent moves and changes in the usual resident population.

11.3 Migration rates

The module on migration in the United States (Box 11.2) illustrates the types of migration statistics available from official collections, as well as differences in the age composition of groups such as:

- total *movers* and *stayers* (i.e. those who migrated and those who did not migrate during the period in question);
- movers from abroad compared with movers within the United States;
- movers between states compared with movers within states;
- movers between counties, regions or divisions compared with movers within them.

The module also provides the option of examining the data in terms of rates of migration as well as raw numbers and percentages. Like rates of mortality and fertility, migration rates are an essential basis for comparative studies, and for gauging the likelihood of events (migrations) in different times, places and social groups.

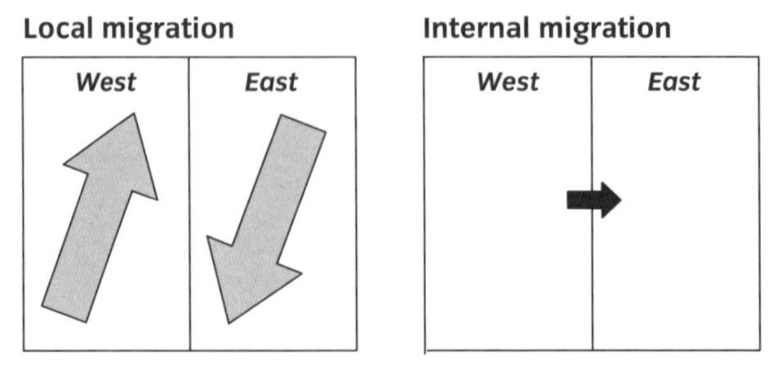

Figure 11.2 Migration within and between two rural areas.

BOX 11.2 **Module on migration in the United States, March 1999–March 2000 (*Migration in the United States.xls*)**

This module presents statistics on migration in the United States by region, sex and age during the 12 months ending March 2000. The data sets include movement within and between counties, states, divisions and regions.

The main purpose of the module is to provide access to a wide range of information on migration, in a way that facilitates comparisons between regions and different types of movement. Most of the module was constructed using techniques introduced in Spreadsheet Exercises 6 and 10, and demonstrates their application in presenting a substantial volume of data in graphical form.

Data source

United States Census Bureau 2001. 'Geographical Mobility March 1999 to March 2000, Detailed Tables', *Current Population Reports*, P20-538.

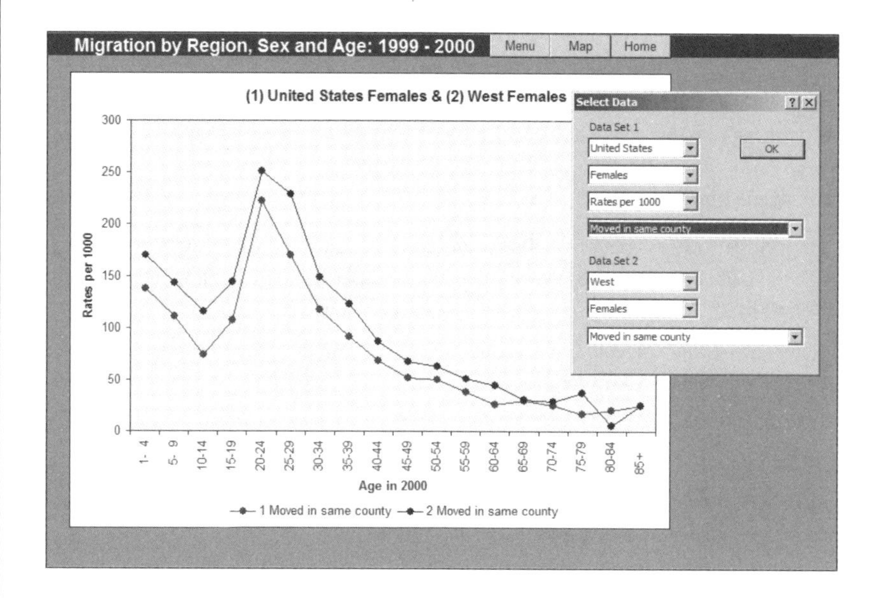

The graph displays data by age, as well as figures for all ages combined, and the menus enable selection of regions and other items, including whether the data are displayed as raw numbers, percentages of the total population or migration rates (based on the end-of-period population). The database contains the following sets of mover and stayer populations:

Totals

1. Total population
2. Total movers
3. Total non-movers

continued

BOX 11.2 continued

Moved within states

4. Moved in same county
5. Moved from different county, same state
6. Total moved in same state

Moved between states

7. Moved in same division
8. Moved from different division, same region
9. Moved from different region
10. Total moved between states

Moved from another country

11. Moved from abroad

Instructions

- Click **Start** to go to the data base display.
- Click **Menu** to change the information displayed in the graph.
- Click **Map** to view a map of US census regions and divisions.
- Click **Home** to return to the opening display.

Illustrative applications

The module illustrates variations in the volume and age–sex composition of different types of migration flows. Answering the following questions will permit study of many of the variations in the data:

1. How many people moved in the 12 months? (For Data Set 1, select United States, Total Population, Numbers, and Total Movers.)
2. How many moved in the same county? Why is short-distance movement so prevalent?
3. Which region(s) have relatively high migration rates?
4. Do males and females have similar rates of migration? (Compare rates for total movers, movers within counties and movers between states.)
5. Are there substantial differences in the age composition of interstate flows compared with other types of movement (e.g. moves within counties and moves from abroad)?
6. Are there peaks in migration rates evident in the later ages in any migration flows (e.g. a 'retirement peak')?
7. Prepare a brief statement summarizing the main variations observed in the migration data.
8. Explaining migration: Obtain a copy of the *Current Population Report* by Jason Schachter on 'Geographical Mobility' (e.g. from http://www.census.gov/population/www/socdemo/migrate/p20-538.html). Consider what other characteristics of movers, apart from age and sex, are important in explaining migration. See also Jason Schachter 'Why People Move: Exploring the March 2000 Current Population Survey', *Current Population Reports* P23-204 (http://www.census.gov/prod/2001pubs/p23-204.pdf).

Basic measures

For consistency with other overall measures of population change, namely crude rates of birth, death and natural increase, migration rates for total populations are usually defined as the number of events divided by the mid-period population. Thus the principal migration rates are:

$$\text{Rate of inward migration} = \frac{\text{arrivals}}{\text{mid-period population}} \times 1000$$

$$\text{Rate of outward migration} = \frac{\text{departures}}{\text{mid-period population}} \times 1000$$

$$\text{Rate of net migration} = \frac{\text{arrivals} - \text{departures}}{\text{mid-period population}} \times 1000$$

$$\text{Rate of gross migration} = \frac{\text{arrivals} + \text{departures}}{\text{mid-period population}} \times 1000$$

These generally produce satisfactory figures for cross-national studies of international migration, and for comparing the occurrence of internal or local migration in different parts of the same country. The figures may be calculated for calendar years, or they may be expressed as rates for longer intervals, such as an intercensal period. Multiplying the figures by 1000 is consistent with the convention for crude birth and death rates, thereby facilitating comparisons of the magnitude of different demographic processes. Rates per cent are also commonly used in migration studies because percentage figures are easily comprehended.

The 'risk' of migration

The rate of outward migration provides a clear illustration of a rate based on the concept of the population at risk of the event. The formula above shows that the outward migration rate is the number of departures from a place divided by its mid-period population. Everyone living in a place could potentially leave it. Thus basing the rate on the area's total population is realistic. The mid-period population is the conventional measure of the average population at risk. Overall, the outward migration rate corresponds to the concept of a rate as the number of events divided by the population at risk.

The rate of inward migration, however, is not at all comparable. The mid-period population by no means includes the population at risk of inward migration. This is apparent from the question: What is the population 'at risk' of moving to Paris? Nobody knows the answer, but it can scarcely be said to be the average population already living in Paris. The same applies to net and gross migration rates (see Table 11.1), because their population at risk includes the numbers at risk of inward migration. Thus, from this viewpoint, the rates of inward, net and gross migration are not demographic rates in the strict sense of being based on the population at risk of the event. Nor are they even comparable to other crude rates in having denominators that include the population at risk. Rather, they are simply ratios expressing the number of movers in comparison with another convenient number, the mid-period population.

A solution to the inconsistencies between the meaning of inward and outward migration rates is to define risk somewhat differently in migration studies, especially since there is considerable interest in the places affected. Accordingly, the population at risk may be redefined as *the population at risk of sending or receiving migrants* during the migration interval (Shryock and Siegel 1973: 606). From this alternative viewpoint, all four of the above rates are comparable with each other, and it is justifiable to describe them as rates rather than ratios. The alternative concept also provides a clearer basis for defining the nature of net and gross migration rates as the population at risk of experiencing net or gross migration.

Choosing denominators

Since migration strictly denotes a change in place of residence, an appropriate base population for migration rates is an area's resident population. If a study is concerned with population mobility more broadly, including both temporary and permanent population shifts, then the de facto population is likely to be a more suitable denominator.

Another question concerning choice of a denominator is whether to use the mid-period population, or an alternative. Adopting the mid-period population has the advantage of consistency with the usual approach to defining the average population at risk. Nevertheless, the initial and end-of-period populations are also frequently employed as denominators for migration rates.

The initial population is often the preferred denominator for migration rates based on data for a fixed period, such as those for a five or ten year census interval. *Fixed period migration statistics* (see Table 11.1) refer only to the population that was alive at both the start and the end of the migration interval (e.g. 1996–2001). They therefore exclude the migration of recently born children (e.g. aged 0–4 in 2001), who were not in the initial population; they also exclude the migration of people who died in the interval. The denominator for the rates is obtained as the sum of the movers and stayers in each area at the start, the 'stayers' being non-movers who were still alive at the end of the period. Accordingly, for fixed period migration statistics, the initial population of movers plus stayers represents the population at risk of being changed through migration.

The end-of-period population, however, is likely to be chosen if the migration data refer to people's places of previous residence at a range of different dates in the past. Also, if the focus is on the migration rates of groups, such as occupational groups, census or survey data may be available only to calculate the rates with reference to their most recent characteristics. Particular interest in current characteristics, such as the ages of people who have moved, also calls for the calculation of rates from end-of-period data. Finally, where there are practical difficulties in assembling statistics for a large number of initial populations, information on the end-of-period population may be employed because it is more readily available (Shryock and Siegel 1973: 619).

Age-specific rates

As just indicated, age-specific migration rates comprise instances where the end-of-period population is the usual denominator. This not only achieves a focus on the most recent characteristics, but also avoids potential confusion arising from mixing references to ages at the start and end of a period:

$$\text{Age-specific migration rate} = \frac{\text{number of migrants aged } x \text{ at the end of the period}}{\text{end-of-period population aged } x}$$

Age-specific data are particularly helpful in understanding how the incidence of migration varies over the life cycle. Figure 11.3 illustrates age-specific migration rates in the United States for the 12 month interval 1999–2000. The graph shows peak migration rates in the child and young-adult ages, when movement related to housing, education and employment are most common. Age groups in the graph include extra detail for people in their late teens and early sixties, when post-school education, initial employment or retirement influence migration decisions.

Figure 11.3 Age-specific migration rates, total movers (including movers from abroad) in the United States, 1999–2000.

Source: US Census Bureau 2001. 'Geographical Mobility March 1999 to March 2000, Detailed Tables', *Current Population Reports* P20-538 (http://www.census.gov/population/www.socdemo/migrate/p20-538.html).

Analyses of the age selectivity of migration sometimes focus not on migration rates but on the numbers of movers in each age group, or the overall age structure of migration streams between particular origins and destinations. Figure 11.4 provides an example of an *origin–destination matrix* of migration depicting the age composition of flows between two cities and two adjacent counties. The diagram summarizes information about 'stayers' in four regions and 'movers' in twelve inter-regional migration streams. Grouping age structures of movers and stayers into similar types is one strategy for summarizing the considerable volume of statistics typically produced in migration studies (see Rowland 1979; Lewis 1982).

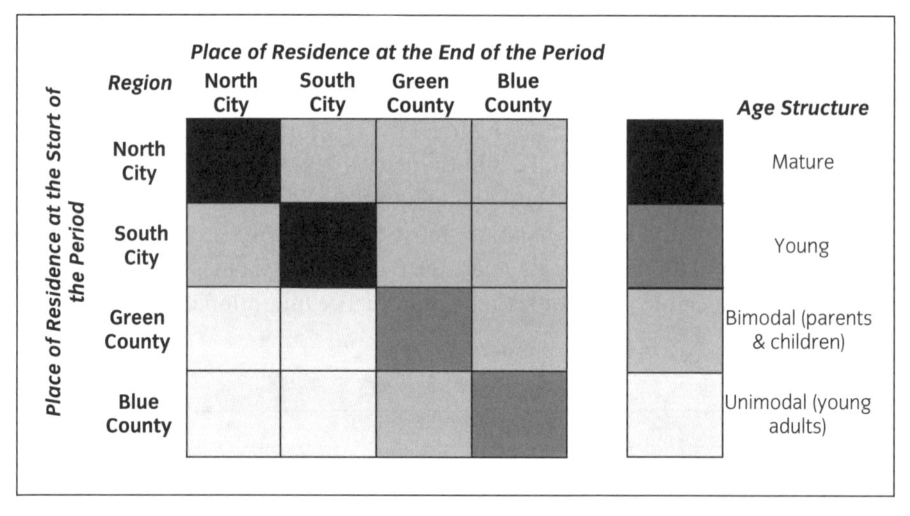

Figure 11.4 Origin–destination matrix showing age-structures of interregional migrations.

11.4 Migration effectiveness

After migration rates, a widely utilized summary measure of internal migration is the *migration effectiveness ratio*. It usually derives from census or survey data on internal migration, but it also has applications in the study of international population flows. Migration effectiveness is the ratio of net migration to gross migration; the lower the ratio the less the effectiveness of migration as a process of population redistribution (Shyrock and Siegal 1973: 656):

$$\text{Migration effectiveness ratio} = \frac{\text{arrivals minus departures (net migration)}}{\text{arrivals plus departures (gross migration)}} \times 100$$

The chief benefit of the ratio is that it provides a clear summary of the extent to which an area's migration consists of a two-way exchange of migrants with other places, or a one-way shift of population to or from the area. In the first diagram in

Figure 11.5, East and West both have low migration effectiveness, because they each attract about as many migrants as they lose. In the second diagram, the East has 'effective' outward migration, while the West has 'effective' inward migration. The key point is that the ratio places net migration gains and losses in the context of gross migration, without which it is impossible to discern whether net shifts are actually occurring in association with an exchange of migrants. The presence of a pattern of exchange has a decisive bearing on the meaning of the statistics.

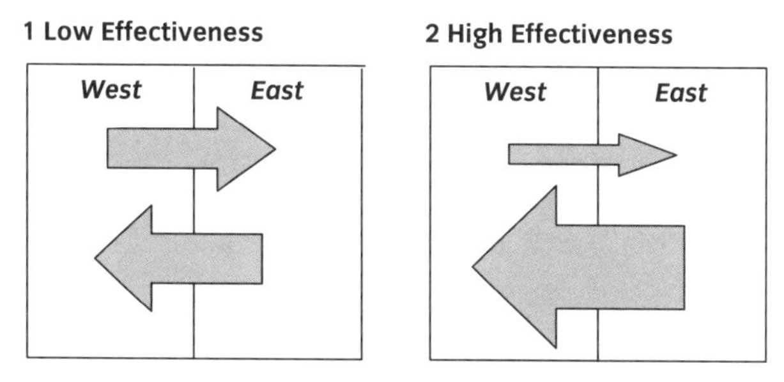

Figure 11.5 Migration effectiveness.

11.5 Migration expectancy

Another important issue for many migration studies is the extent to which mobility varies over the life cycle. Although age-specific migration rates provide a basis for investigating, at points in time, the association between migration and stages of life, there is much less information available directly on the number of migrations individuals make over their lifetimes.

In the absence of complete migration data for real cohorts, life table methods offer a means of estimating how many times people move house during their lives. The result is a measure of *migration expectancy* referring to the average number of moves individuals may be expected to make during their lives (Wilber 1963). As in life tables, the figures refer to the experience of a synthetic cohort, based on a cross-section of the whole population at a single date. Any changes through time in probabilities of moving are necessarily ignored. This is one cause of discrepancies between migration expectancies and survey figures on people's own migration histories (Taeuber 1966: 417). Difficulties in recalling accurately the details of past moves – in childhood or long ago – also make such comparisons uncertain.

Table 11.3 illustrates the computation of migration expectancy. The steps in calculating the figures are:

1. Obtain age-specific migration rates per person (column D) for the area in question and the corresponding l_x and L_x values from a life table for the same place and time. The migration data should preferably refer to a twelve month interval, to minimize the effect of multiple moves that become more likely the

Table 11.3 Calculation of migration expectancy for the United States (both sexes), 1997–98

Age A	Total population[1] March '98 B	Movers 1997–98[1,2] C	Migration rate (C/B) D	l_x E	L_x F	Expected number of moves (D × F) G	Cumulative number of moves H	Migration expectancy I
1–4	15802	3552	0.22478	99268	396721	89176	1048938	10.6
5–9	20453	3526	0.17240	99118	495329	85392	959762	9.7
10–14	19663	2755	0.14011	99022	494883	69338	874370	8.8
15–19	19466	2864	0.14713	98905	493650	72630	805031	8.1
20–24	17613	5607	0.31834	98519	491362	156422	732401	7.4
25–29	18996	5442	0.28648	98020	488766	140022	575979	5.9
30–34	20358	4267	0.20960	97487	485746	101811	435957	4.5
35–39	22691	3568	0.15724	96795	481820	75763	334145	3.5
40–44	21771	2570	0.11805	95881	476549	56255	258382	2.7
45–49	18634	1798	0.09649	94651	469305	45283	202127	2.1
50–54	15424	1242	0.08052	92946	458779	36943	156844	1.7
55–59	12190	916	0.07514	90406	443132	33299	119901	1.3
60–64	10065	574	0.05703	86630	419530	23926	86603	1.0
65–69	9361	439	0.04690	80870	385659	18086	62677	0.8
70–74	8512	389	0.04570	73056	339620	15521	44591	0.6
75–79	6898	276	0.04001	62422	280047	11205	29070	0.5
80–84	4383	186	0.04244	49276	207474	8805	17865	0.4
85+	2928	130	0.04440	33629	204073	9061	9061	0.3
Total	265208	40101	0.15121			1048938		

Notes:
[1] Figures in thousands.
[2] Excluding movers from abroad.

Sources:
United States Census Bureau, *Current Population Reports*, P20-520, 'Geographical Mobility: March 1997 to March 1998' (http://www.census.gov/prod/www/abs/mobility.html).
National Centre for Health Statistics, *National Vital Statistics Reports*, Volume 47, Number 13, p. 5. 'United States Abridged Life Tables, 1996' (http://www.cdc.gov/nchs/data/nvs47_13.pdf).

longer the period of time. The calculations assume no more than one move per person in the twelve months, and no migration of people whose address was the same at the start and the end (Wilber 1963: 445). The migration of infants under one year of age is omitted from the fixed interval migration data in the example.

2. Multiply each L_x by the corresponding migration rate (per person); the result is the expected number of moves in each age interval (column G).
3. Calculate the cumulative number of moves from each age (column H). This shows the expected number of moves the cohort will make during the rest of their lives. It is obtained in the same way as the T_x function in life tables (see Chapter 8). For example, the expected number of moves at ages 20 and over is the sum of the expected number of moves in all the age groups from 20–24 to 85 years and over.
4. Divide these cumulative figures by the corresponding l_x value to derive migration expectancy at each age. This parallels the calculation of life expectancy as T_x/l_x.

Because only one move per year is counted, migration expectancy strictly refers to the number of years in which moves occurred, rather than the total number of moves (Long 1988, cited by Bell 1996: 103). In the example, migration expectancy in the United States is just under 11 moves after the first birthday. Most of the shifts are concentrated in the younger adult ages where education, employment, family formation and housing needs have a major influence. To facilitate a more detailed exploration of cohort mobility, these types of calculations may be extended to population subgroups, as Wilber (1963) demonstrated with reference to sex, type of move (e.g. intrastate and interstate moves), employment and marital status. Bell (1996) developed this approach further.

11.6 Estimating migration

Whereas migration expectancy calculations provide estimates of migration from a lifetime perspective, other techniques produce estimates of net migration for geographical areas. At the sub-national level, migration has such a pervasive impact on population size and composition that estimates of migration are essential aids to explaining population changes in the absence of direct counts from a census or survey.

The methods of estimation discussed in this section are mostly limited in supplying statistics only on net migration (arrivals minus departures). They also omit details on inward and outward flows, as well as origins and destinations. The birthplace method is the only indirect procedure that overcomes some of these disadvantages. However, it requires less widely available data, on place or region of birth within the country, and it does not estimate the movement of the foreign-born.

Nevertheless, the overall benefits of net migration estimates are considerable:

- they are applicable at all geographical scales – to local, regional and national populations;
- they can supply comparable information through time, including information for periods and places where no other migration statistics exist;
- some of the methods provide estimates by age and sex.

As in all demographic analyses, the quality of the results depends on the accuracy of the input data; comparing estimates derived from different methods is a valuable way of validating findings.

The vital statistics method

The simplest, and potentially most accurate, method of estimating net migration is to use the vital statistics (VS) method. Since population change is equal to natural increase plus net migration, this 'demographic balancing equation' may be rewritten making net migration the subject. Thus net migration is equal to population change minus natural increase. Table 11.4 illustrates the method, showing that net migration is obtained for each region by subtracting natural increase from the total population change.

Accurate estimates of net migration by the VS method require high quality vital statistics on births and deaths, together with reliable census statistics, or other data on population change. In more developed countries, registration of births and deaths is virtually complete, and considerable efforts are made to ensure that census totals are as accurate as possible. In these circumstances, the vital statistics method yields the best estimates of net migration.

Table 11.4 Vital statistics method of estimating net migration in four regions, 1995–2000

Region A	Births 1995–2000 B	Deaths 1995–2000 C	Natural Increase 1995–2000 (B – C) D	Total Population Change 1995–2000 E	Net Migration 1995–2000 (E – D) F
North	252 344	126 941	125 403	265 621	140 218
South	9 440	8 317	1 123	26 211	25 088
East	37 750	19 510	18 240	26 820	8 580
West	23 059	8 682	14 377	27 520	13 143
Total	322 593	163 450	159 143	346 172	187 029

One complication is that vital statistics registration is normally according to the place of usual residence of mothers of the newborn and the last place of usual residence of the deceased. Hence, the changes attributed to natural increase relate to changes in the resident or de jure population of the place in question. To produce estimates of the net migration of the resident population, the figures on population change also need to be de jure based.

Alternatively, using figures on de facto population changes, in conjunction with de jure vital statistics, will include in the migration estimates the net effect of temporary movement of visitors to the area. This may influence the findings for some places considerably, for instance if hotels and other short-term accommodation were fully occupied at the first census, and only half full at the second. Vital statistics estimates based on de facto population figures denote the combined net movement of the resident and non-resident population of the area, and need to be interpreted accordingly.

Net migration calculated from vital statistics is also subject to inaccuracies arising from the method's handling of the migration of the new-born and deaths of migrants:

- child out-migrants, born to mothers who were resident in the area but subsequently left, are included in the area's births;
- migrants of any age who took up residence in the area, but died, are included in the area's deaths.

These influences may tend to cancel each other, but the longer the interval, the greater their potential effect on the net migration estimate.

Missing from vital statistics figures is any distinction between the contributions of internal and external migration – only the combined net effect of both types of movement will be measured. Researchers most often calculate net migration estimates in studies of internal migration and population redistribution, but international migration can be the predominant process in some places. Missing also are details of the age and sex composition of the net movement. Here the survival ratio methods come into their own, since they provide net migration estimates by age and sex.

Survival ratio methods

Survival ratio methods of estimating net migration entail two main steps, as illustrated in Figure 11.6. First, calculate how many in a birth cohort survive from one census (A) to the next (B). Second, subtract the number of survivors from the cohort's size at the second census (C). The difference is the intercensal net migration (D). These calculations are usually made separately for males and females, on account of age and sex related variations in probabilities of moving.

The data needed to utilize the survival ratio methods are widely available. The requirements are a set of age- and sex-specific survival ratios, and statistics on the age and sex of the population at two consecutive censuses, preferably five or ten

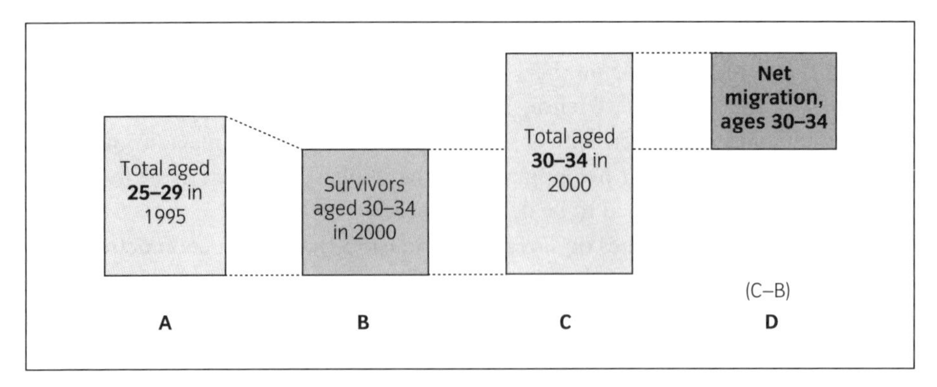

Figure 11.6 Calculation of age-specific net migration.

years apart. Five and ten year intervals both permit the use of data for five year age groups. Other intervals between censuses, such as seven or nine years, require data for single years of age in the initial population. Survivors from single year birth cohorts are grouped into five-year brackets at the end of the interval. Single years of age could be used throughout, but presenting such detailed estimates would create a false impression of their reliability.

The most common approach to survival ratio estimates of net migration is to calculate the figures chronologically, working forward through time. This is known as *forward survival*. An alternative is to work backwards through time from a more recent date to an earlier one; this entails *reverse survival*. The ensuing sections explain the two approaches in turn.

Forward survival method

Figure 11.7a provides a diagrammatic representation of forward survival, referring to a cohort aged 60–64 at the start of an interval and, hence, aged 65–69 five years later. The brace on the left-hand side represents their numbers at the start. Over the five-year interval, the size of the original cohort falls on account of mortality. The brace labelled 'estimated survivors' denotes their numbers after five years, at ages 65–69. To estimate the net migration of people aged 65–69 at the end of the period, the estimated survivors (from the original 60–64 year olds) are subtracted from the total census population aged 65–69 in the area. Thus the brace labelled 'net migration gain' shows the increase in the cohort's numbers arising from net inward migration.

If net migration losses are reducing the population, the estimates will also reveal this, as Figure 11.7b illustrates. There, the estimated survivors from the initial cohort aged 60–64, exceed the census count of 65–69 year olds in the area five years later, indicating that migration depleted the cohort. Subtracting the estimated survivors from the census population produces the estimate of the net migration loss.

One of the main characteristics of the forward survival approach is that it estimates net migration at the end of the interval, as illustrated in the diagrams. This

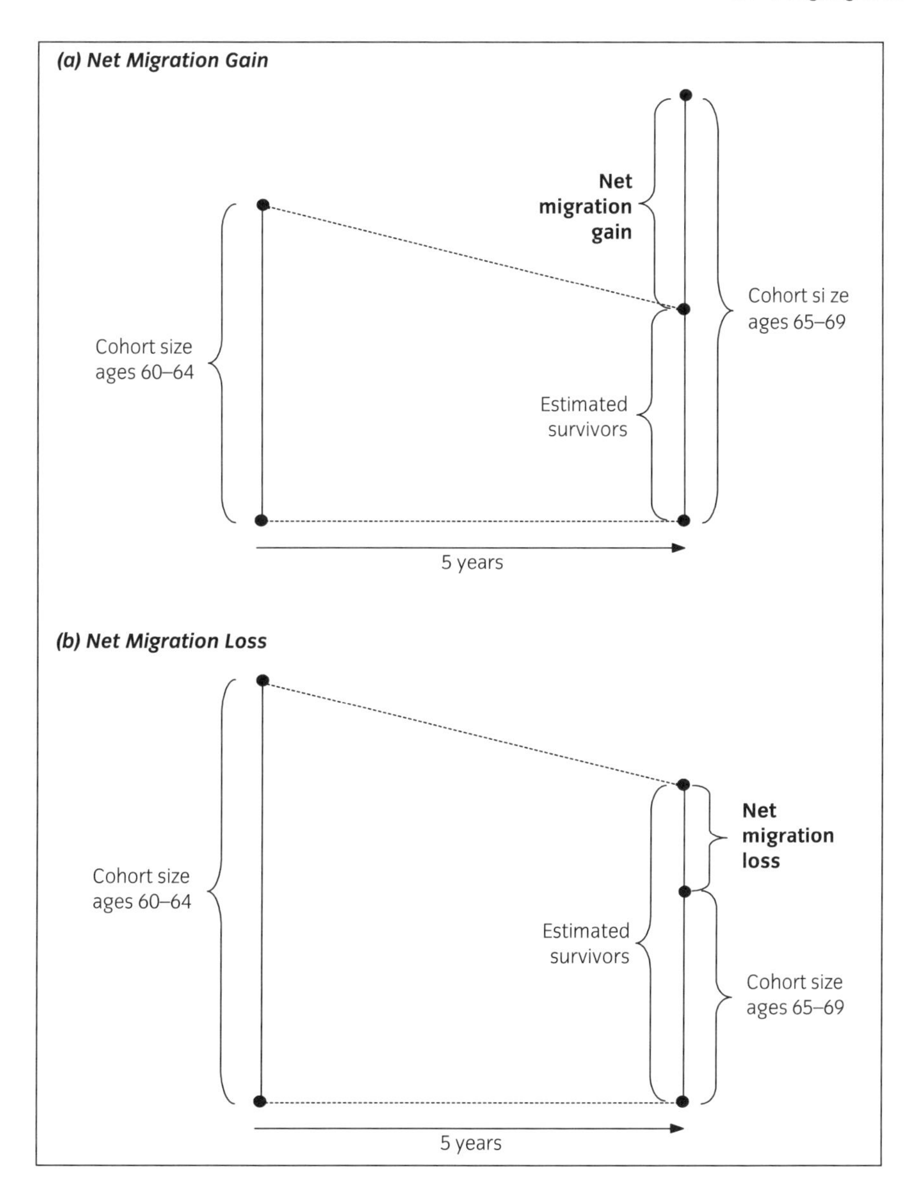

Figure 11.7 Forward survival ratio method of estimating net migration.

reflects that the number of survivors are counted at that point and subtracted from the end-of-period population. Any inward or outward migrants who died during the interval are not taken into account. Forward survival therefore tends to understate the volume of movement. Omitting migrant deaths is of little importance if mortality is low or if there is particular interest in the population at the end of the period.

The formula for forward survival estimates of net migration is (Shryock and Siegel 1973: 630–631):

$$\text{Net } M'_{x+n} = P^n_{x+n} - S \times P^0_x$$

where

Net M'_{x+n} is the estimated net migration for the end-of-period population aged $x + n$, obtained by forward survival
n is the interval in years between the two dates
P^0_x is the initial population aged x
P^n_{x+n} is the end-of-period population aged $x + n$
S is the survival ratio from age x to age $x + n$.

As the formula shows, the numbers in a cohort at the start are multiplied by their survival ratio, then the resulting estimate of survivors is subtracted from the cohort's numbers at the end of the period. The outcome is the net migration estimate.

In calculating these estimates, it is important to keep track of the change in the age of each cohort during the migration interval. The calculations are best understood as referring not to isolated age groups, but to birth cohorts moving through time. Accuracy depends upon setting out the figures to take account of the age shift, such as through showing the cohorts' ages at the start and end of the migration interval. Table 11.5a illustrates the steps in obtaining forward survival estimates of net migration in a five year period for a whole population.

- Columns A and B list the age distribution of the population in 1995 and 2000. Data for age group 60–64 are not needed in 1995, because they would be 65–69 in 2000 whereas the final age group for the population in 2000 is 65 years and over. Similarly, age group 0–4 is omitted in 2000, since the table refers only to the migration of people who were alive in 1995 (Box 11.3 illustrates how to estimate migration in the youngest age group, if needed). The population numbers in columns C and D correspond to the age groups in columns A and B respectively.
- The survival ratios in column E show the proportion surviving from the age group in column A to the age group on the same row in column B. For example, in the first row of the table is the five year survival ratio for the cohort aged 0–4 in 1995 (0.993 33); this was calculated from a life table as ${}_5L_5/{}_5L_0$, denoting survival from age group 0–4 to age group 5–9 (see Section 11.7 for a discussion of survival ratios).
- Estimated survivors in 2000 are obtained by multiplying the cohort's initial numbers by their survival ratio. The answers, in column F, therefore refer to the age groups listed in column B (e.g. for the cohort aged 30–34 in 1995, 1308 × 0.989 62 = 1294 aged 35–39 in 2000).
- Finally, in column G, subtract the number of survivors in 2000 from the census count for 2000: the result is the net migration (e.g. for the cohort aged 5–9 in 2000, 2709 – 2078 = 631).

The figures indicate substantial net movement of families, evident especially in the numbers of children, together with much retirement, or pre-retirement, migration.

Table 11.5 Calculation of net migration to a town by forward survival, 1995–2000 & 1990–2000 (females)

(a) Five year estimates, 1995–2000

Cohort age in 1995 A	Cohort age in 2000 B	Population 1995 C	Population 2000 aged 5+ D	Forward survival ratios (5 year) E	Estimated survivors at end of period (C × E) F	Estimated net migration by age at end of period (D – F) G
0–4	5–9	2092	2709	0.99333	2078	631
5–9	10–14	2250	3137	0.99714	2244	893
10–14	15–19	2235	2559	0.99645	2227	332
15–19	20–24	1952	2429	0.99466	1942	487
20–24	25–29	1700	2135	0.99302	1688	447
25–29	30–34	1473	1780	0.99160	1461	319
30–34	35–39	1308	1692	0.98962	1294	398
35–39	40–44	1463	1894	0.98631	1443	451
40–44	45–49	1546	2151	0.98070	1516	635
45–49	50–54	1546	2111	0.97173	1502	609
50–54	55–59	1538	2169	0.95815	1474	695
55–59	60–64	1518	2299	0.93609	1421	878
60+	65+	5255	5791	0.73243	3849	1942
	Total	25876	32856		24138	8718

(b) Ten year estimates, 1990–2000

Cohort age in 1990 A	Cohort age in 2000 B	Population 1990 C	Population 2000 aged 10+ D	Forward survival ratios (10 year) E	Estimated survivors at end of period (C × E) F	Estimated net migration by age at end of period (D – F) G
0–4	10–14	1885	3137	0.99049	1867	1270
5–9	15–19	2012	2559	0.99360	1999	560
10–14	20–24	2003	2429	0.99113	1985	444
15–19	25–29	1766	2135	0.98771	1744	391
20–24	30–34	1556	1780	0.98468	1532	248
25–29	35–39	1358	1692	0.98131	1333	359
30–34	40–44	1198	1894	0.97607	1169	725
35–39	45–49	1305	2151	0.96727	1262	889
40–44	50–54	1394	2111	0.95297	1328	783
45–49	55–59	1369	2169	0.93106	1275	894
50–54	60–64	1396	2299	0.89691	1252	1047
55–59	65–69	1390	2442	0.84112	1169	1273
60+	70+	4754	3349	0.49200	2339	1010
	Total	23386	30147		20255	9892

Sources: West model life tables (Coale and Demeny 1983) and hypothetical data.

BOX 11.3 **Estimating the net migration of children**

Net migration estimates sometimes refer only to the movement of people alive at the start and end of a period, omitting the migration of children born in the interval. Such figures are comparable with fixed period migration data from censuses. Nevertheless to obtain more complete figures on migration at all ages, procedures are needed to estimate the net migration of children.

Numbers at the 2000 Census	
Children aged 0–4	6100
Children aged 5–9	6000
Females aged 15–44	12100
Females aged 20–49	12000
Net migration 1995–2000	
Females 15–44	2500
Net migration 1999-2000	
Females 15–44	4900
Females 20–49	4800

The above data, which assume a similar volume of migration in each quinquennium, are used to illustrate a method for estimating the net migration of children based on child–woman ratios (CWRs) for each area at the second census. If the migration interval is five years, only the ratio of children 0–4 to women aged 15–44 is needed (CWR_0). Assuming that half the babies were born before their mothers migrated, the estimate of net migration at ages 0–4 is:

$$\text{Net } {}_5M_0' = 0.5 \times CWR_0 \times \text{net migration of females aged 15–44}$$

Using the figures given above, the child–woman ratio in 2000 is:

$$CWR_0 = \frac{6100}{12100} = 0.50413$$

Therefore the net migration of children aged 0–4 in 2000 is:

$$\text{Net } {}_5M_0' = 0.5 \times 0.50413 \times 2500 = 630$$

For a 10-year migration interval (1990–2000), a second ratio of children aged 5–9 to women aged 20–49 (CWR_5) is also required (see United Nations 1970: 34):

$$\text{Net } {}_5M_0' = 0.25 \times CWR_0 \times \text{net migration of females aged 15–44}$$

$$\text{Net } {}_5M_5' = 0.75 \times CWR_5 \times \text{net migration of females aged 20–49}$$

If migration occurred evenly over the 10 years, a quarter of the younger and three-quarters of the older children would have been born before their mothers migrated (ibid. 35). The underlying reasoning is:

continued

BOX 11.3 continued

'The children under 5 years old at the census were born, on the average, 2.5 years earlier; only ¼ of their mothers' migration occurred after that date. The children 5 to 9 years old at the census were born, on the average, 7.5 years earlier; ¾ of their mothers' migration occurred after that date.' (Shryock and Siegel 1973: 632)

Using the figures given above, which specified that there were 6000 children of both sexes aged 5–9 at the second census, the second child–woman ratio is:

$$CWR_5 = \frac{6000}{12000} = 0.50000$$

Therefore, the net migration 1990–2000 of children aged 0–4 (Net ${}_5M_0$) and 5–9 (Net ${}_5M_5$) respectively in 2000 is:

$$\text{Net } {}_5M_0' = 0.25 \times 0.50413 \times 4900 = 618$$

$$\text{Net } {}_5M_5' = 0.75 \times 0.50000 \times 4800 = 1800$$

Alternative approaches employ survival ratios of births and follow the procedures in Table 11.5 (see United Nations 1970: 34). In closed populations, the sum of all the net internal migration estimates for all ages and all areas should be adjusted to produce a zero balance (ibid. 35).

Calculating the estimates for both sexes, and drawing an age pyramid, would clarify such findings.

The principles in calculating net migration are the same when working with data from decennial censuses, as illustrated in Table 11.5b. The age shift there is 10 years, instead of five, and the survival ratios refer to survival for a 10 year period (${}_5L_{x+10}/{}_5L_x$). The results, in column G, refer to persons aged 10 years and over in 2000.

The total net migration for the decade (9892 for ages 10 and over) is relatively low, compared with the five year figure for the second half of the decade (8718 for ages 5 and over). The differences arise from the absence of a figure for age group 5–9 in the 10-year estimates, the greater mortality losses over a longer period in the older ages, together with lower net gains in other ages during the first five years.

An advantage of the forward survival method is its consistency with direct counts of migration in censuses. Censuses count migrants at the end of the migration interval and include only those who are alive at the time of the census – that is, they omit migrant deaths. Also, census questions on place of residence at the start and end of a fixed period necessarily exclude everyone born in the interim, as do forward survival estimates that omit the youngest children.

Another advantage is that the forward survival method is equally applicable at all geographical scales. National estimates necessarily refer to net international migration, however, and raise questions about whether to base the calculations on the resident population or include visitors as well.

For localities, the smaller the population in question, the greater the likelihood that the estimates will produce unpatterned variations between age groups. Small population numbers themselves increase the potential impact of small-scale and random influences, as well as magnifying the effect of (1) temporary absences of residents from the area, (2) errors in the data or (3) inappropriate assumptions about survival, such as assuming that national levels apply locally. At the locality level too, intercensal boundary changes are common. This can necessitate the aggregation of areas to achieve constant boundaries for the start and end of the period.

A further application of forward survival is in deriving components of growth when direct counts of natural increase are unavailable. Subtracting the total net migration from total population change can provide an estimate of natural increase that is reasonably consistent with estimates from vital statistics, because both measure the components of change at the end of the interval. When using forward survival to estimate components of growth, however, the net migration of children born in the interval must be calculated and included in the total net migration (see Box 11.3). This enhances consistency with estimates of net migration from vital statistics. The reason is that vital statistics estimates of net migration include child in-migrants, born elsewhere during the interval; they are not part of the area's natural increase, because their births were not registered there.

Reverse survival

Through measuring change at the end of the period, forward survival offers the advantages of measuring net migration and components of growth in a way reasonably comparable with vital statistics. The alternative approach, reverse survival, does not offer the same advantages, but provides a more complete estimate of the volume of net migration, especially when the migration interval is long or mortality is high. Reverse survival usually produces higher estimates of net migration because the figures include migrants who died. In developed countries, this mainly increases the estimates for older ages. There is a logical problem, however, in stating the end-of-period ages from reverse survival estimates, because some of the migrants would have died: strictly speaking, reverse survival refers to movers' initial ages.

Figure 11.8 illustrates why reverse survival includes migrants who died before the end of the migration interval. The reason is that reverse survival estimates migration at the start, before any of the mortality for the interval has occurred (United Nations 1970: 25). In Figure 11.8, the cohort aged 65–69 in an area is 'survived backwards' to determine how many 60–64 year olds there would have been five years earlier. This produces the 'expected initial population'. If their numbers are greater than the size of the cohort aged 60–64 at the first date, a net migration gain must have augmented their numbers. Conversely, if the expected initial population is less than the size of the cohort at the start, a net migration loss must have depleted their numbers.

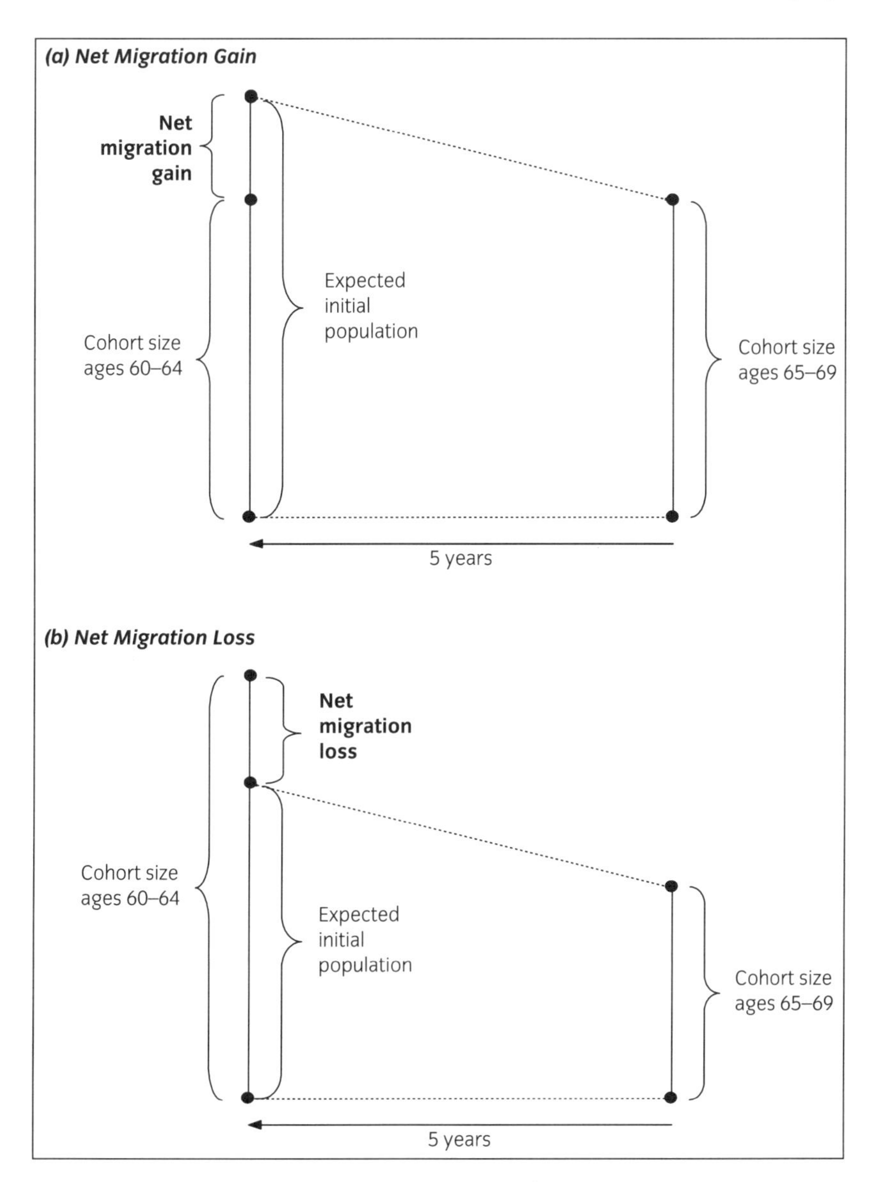

Figure 11.8 Reverse survival ratio method of estimating net migration.

$$\text{Net } M_x'' = \frac{1}{S} \times P_{x+n}^{n} - P_x^0$$

where

Net M_x'' is the estimated net migration for the initial population aged x, obtained by reverse survival

n is the interval in years between the two dates

P_x^0 is the initial population aged x

P^n_{x+n} is the terminal population age $x + n$

$\frac{1}{S}$ is the reciprocal of the survival ratio from age x to age $x + n$.

As evident from the above formula, reverse survival is accomplished through multiplying the end-of-period population by the reciprocal of the survival ratio (i.e. 1 divided by the survival ratio), and subtracting the initial population. For example, using forward survival, if the initial size of a cohort is 500 and $S_x = 0.87543$, the number of survivors will be $500 \times 0.87543 = 437.715$. Conversely, using reverse survival, the expected initial cohort size will be $437.715 \times (1/0.87543) = 500$.

Table 11.6 illustrates the details of reverse survival estimation, using the same data as in the previous examples of forward survival. Compared with the forward survival estimates, all the figures from reverse survival are higher for each cohort, especially at the older ages, because of the inclusion of migrants who died.

Overall, the choice between forward and reverse survival depends on whether the main interest is in:

- the numbers moving during long migration intervals or in populations with high mortality (use reverse survival);
- broad comparability with other fixed-period migration data, or with vital statistics estimates of the components of growth (use forward survival).

Neither approach is more reliable than the other, but forward survival estimates are more easily reconciled with vital statistics estimates, which ought to be the most accurate. Reverse survival is less comparable with vital statistics estimates of net migration, because the vital statistics method counts as deaths inward migrants who died, and hence incorporates them into the natural increase component of population change. Expressing net migration with reference to cohorts' initial ages is also difficult to reconcile with the interest of many studies in the situation at the end of an intercensal period.

The birthplace method

The most versatile procedure for estimating migration is the birthplace method. Scarcity of detailed statistics on birthplaces (apart from states or provinces) within countries is the main disadvantage but, when appropriate, the birthplace method offers the benefits of providing information on life-time migration, net migration, and gross migration, as well as origins and destinations (United Nations 1970: 5ff).

The source data needed to calculate the estimates are:

- statistics on the birthplaces of the population by place of residence;
- statistics on intercensal deaths by birthplace and place of last residence. The deaths may need to be estimated, such as through using life table survival ratios.

With such figures set out in tables, showing birthplace by place of residence (Figure 11.9), the estimates are readily obtained. The birthplace method can also

Table 11.6 Calculation of net migration to a town by reverse survival, 1995–2000 & 1990–2000 (females)

(a) Five year estimates, 1995–2000

Cohort age in 1995	Cohort age in 2000	Population 1995	Population 2000 aged 5+	Reverse survival ratios (5 year)	Estimated survivors at start of period (D × E)	Estimated net migration by age at start of period (F – C)
A	B	C	D	E	F	G
0–4	5–9	2092	2709	1.00671	2727	635
5–9	10–14	2250	3137	1.00287	3146	896
10–14	15–19	2235	2559	1.00356	2568	333
15–19	20–24	1952	2429	1.00537	2442	490
20–24	25–29	1700	2135	1.00703	2150	450
25–29	30–34	1473	1780	1.00847	1795	322
30–34	35–39	1308	1692	1.01049	1710	402
35–39	40–44	1463	1894	1.01388	1920	457
40–44	45–49	1546	2151	1.01968	2193	647
45–49	50–54	1546	2111	1.02909	2172	626
50–54	55–59	1538	2169	1.04368	2264	726
55–59	60–64	1518	2299	1.06827	2456	938
60+	65+	5255	5791	1.36532	7907	2652
Total		25876	32856		35450	9574

(b) Ten year estimates, 1990–2000

Cohort age in 1990	Cohort age in 2000	Population 1990	Population 2000 aged 10+	Reverse survival ratios (10 year)	Estimated survivors at start of period (D × E)	Estimated net migration by age at start of period (F – C)
A	B	C	D	E	F	G
0–4	10–14	1885	3137	1.00960	3167	1282
5–9	15–19	2012	2559	1.00644	2575	563
10–14	20–24	2003	2429	1.00895	2451	448
15–19	25–29	1766	2135	1.01244	2162	396
20–24	30–34	1556	1780	1.01556	1808	252
25–29	35–39	1358	1692	1.01905	1724	366
30–34	40–44	1198	1894	1.02452	1940	742
35–39	45–49	1305	2151	1.03384	2224	919
40–44	50–54	1394	2111	1.04935	2215	821
45–49	55–59	1369	2169	1.07404	2330	961
50–54	60–64	1396	2299	1.11494	2563	1167
55–59	65–69	1390	2442	1.18889	2903	1513
60+	70+	4754	3349	2.03252	6807	2053
Total		23386	30147		34869	11483

Sources: West model life tables and hypothetical data.

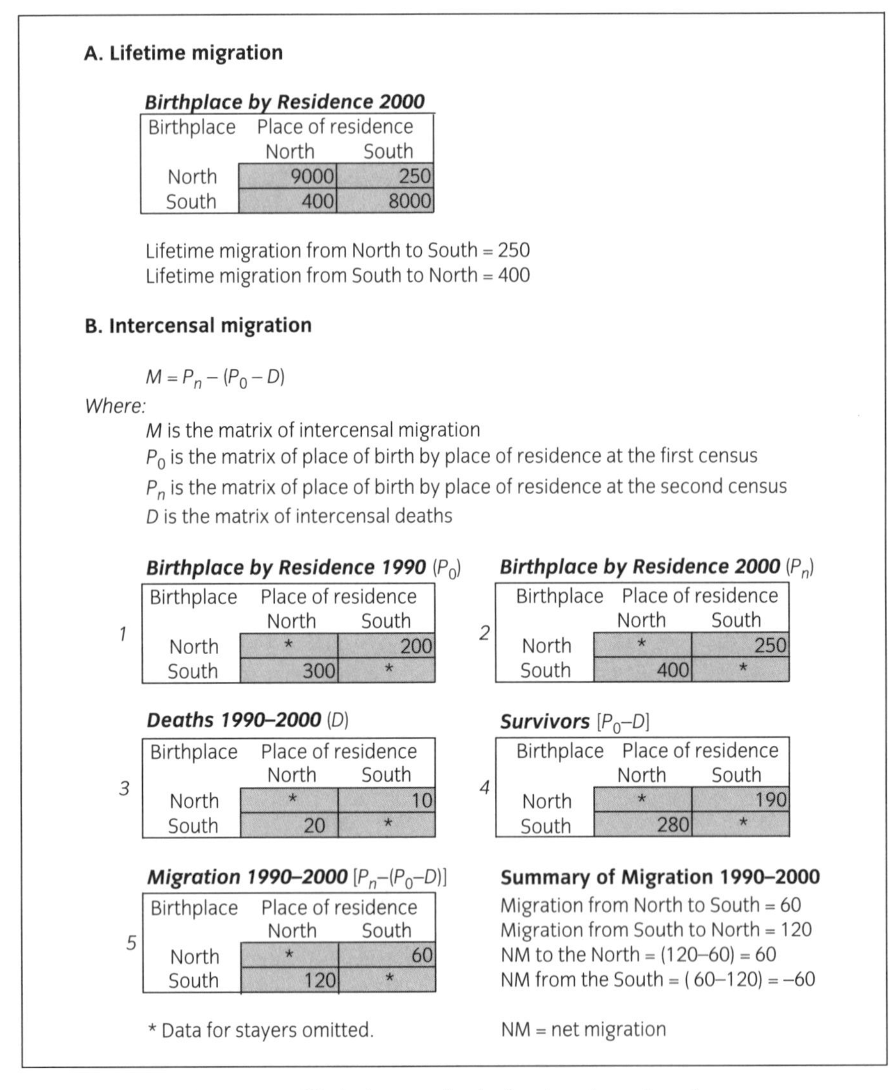

Figure 11.9 Birthplace method of estimating migration.

provide age and sex-specific estimates. For simplicity, the example of the birthplace method in Figure 11.9 uses statistics just for total populations in two regions labelled North and South.

The simplest measure from the birthplace method is lifetime migration. This denotes the number of residents of a place who were born elsewhere (see Table 11.1). In Figure 11.9, the lifetime migration from the North to the South was 250 people, and from the South to the North it was 400. Such figures do not indicate when the people arrived; they are known only to have moved sometime in their lives. Lifetime migration figures similarly provide no information on any other places lived in since birth. These data are useful mainly when there is interest in the long-run outcomes of movement from particular places, rather than in the timing of movement and more immediate origins of migrants.

Birthplace figures also lend themselves to estimating movement during fixed periods, as illustrated in the second example in Figure 11.9, which refers to intercensal migration:

- The first three tables (for P_0, P_n and D) contain the source data, consisting of birthplace and place of residence data for two consecutive censuses and statistics on the number of deaths according to birthplace and place of last residence. For clarity, the numbers of people still living in their region of birth are omitted, since they do not affect the calculation of the total gross and net migration flows.
- The fourth table, 'survivors', illustrates the intermediate step of calculating how many people from the 1990 population were expected to be alive in year 2000.
- The table of 'migration 1990–2000' is the difference between the total population in 2000 and the number of survivors, subtracting corresponding cells in each table.

The result, therefore, is a migration matrix of two-way flows between the regions, from which net migration is easily calculated, as shown at the bottom of Figure 11.9. Extending the calculations to many regions proceeds in the same way and has the potential to supply information on a complex set of spatial linkages. The ability to provide information on both gross and net intercensal migration is the most important advantage of the birthplace method.

Weighing against this, however, are several factors that add to the difficulties of interpreting the migration figures (see United Nations 1970; Shryock and Siegel 1973):

- As in lifetime migration, the birthplace need not be the immediate place of origin during the intercensal period. Someone born in the North may have sojourned in another region or country before moving to the South.
- The intercensal estimates understate the volume of movement; they actually record only the net change in the size of a birthplace group, rather than the total volume of arrivals and departures. For example, if 100 people born in the North move to the South, while another 60 born in the North leave the South during the same period, the estimates will show only 40 Northerners moving South.
- If the census figures refer to de facto populations, rather than the populations usually resident in each region, temporary population movements could affect the findings substantially. This would be so if the South was a popular holiday destination for Northerners at the time of each census.
- The internal migration of foreign-born people is omitted.

11.7 Calculating survival ratios

All of the indirect methods of estimating migration require data on mortality. This section explains the survival ratios used especially in the forward and reverse survival ratio methods, but which have further applications in estimating mortality using the birthplace method. Life table survival ratios are the main approach. They are applicable to both open and closed populations (i.e. open or closed to international migration), and for this reason are widely employed in estimating migration. The alternative approach, using census survival ratios, is applied to closed populations or to open populations where it is possible to subtract international migrants from the base data.

Life table survival ratios

Life table survival ratios (LTSRs) are calculated from the L_x column of a life table for the start or end of the interval. If there were substantial changes in mortality, such as over a lengthy interval, it is advisable to use a life table for the middle year, or to average the survival ratios calculated from life tables for the first and last years of the period (United Nations 1970: 27).

Survival ratios measure the proportion of a birth cohort surviving from one age interval to the next, such as from ages 45–49 in mid-year 1995 to ages 50–54 in mid-year 2000. Since the survival ratios are applied to census statistics, they need to refer to survival from one age interval to the next, rather than from one birthday to the next: census statistics refer to age last birthday, rather than exact age. This explains why L_x, which denotes the average number of survivors in an age interval, is the basis for the survival ratios, rather than l_x, which refers to exact ages – the numbers alive on their birthdays.

Table 11.7 shows that the survival ratios for single years of age are calculated by dividing the L_x value for one age by the L_x for the next younger age, such as L_{51}/L_{50}. For five year age groups, use L_x values from an abridged life table – or add together the single year L_x values in each age interval from a complete (single year) life table (see Table 11.7). An approximation is to calculate the survival ratios using just the L_x value in the middle of each age interval, such as L_{32} instead of ${}_5L_{30}$, and L_{37} instead of ${}_5L_{35}$.

Table 11.7 also gives the formulas for survival ratios for open-ended intervals, such as required when the final age group is 85+, representing the survivors from people aged 80+ five years earlier. The infinity symbol, ∞, denotes an open-ended age interval. For 85+, the survival ratio is the sum of the L_xs from 85 to the oldest age, divided by the sum of the L_xs from 80 to the oldest age. The ratio is equivalent to T_{85}/T_{80}. Finally, the table includes the formula for the survival ratio of births, required if the estimates include the net migration of infants and very young children. For a five year interval, the survival ratio of births is the ratio of the number alive at ages 0–4 (${}_5L_0$, represents the survivors from five years' births), divided by the number of births in a five year period ($5 \times l_0$).

1. **Single year survival ratios**
   - For single year age intervals:

$$S_x = \frac{L_{x+1}}{L_x} \quad \text{e.g.} \quad S_{25} = \frac{L_{26}}{L_{25}}$$

   - For births:

$$S_{births} = \frac{L_0}{l_0}$$

2. **Five year survival ratios**
   - For five year age intervals:

$${}_5S_x = \frac{{}_5L_{x+5}}{{}_5L_x} \quad \text{e.g.} \quad {}_5S_{20} = \frac{{}_5L_{25}}{{}_5L_{20}}$$

   - For births:

$${}_5S_{births} = \frac{{}_5L_0}{5 \times l_0}$$

   - For open-ended age intervals:

$${}_\infty S_x = \frac{{}_\infty L_{x+5}}{{}_\infty L_x} \quad \text{e.g.} \quad {}_\infty S_{80} = \frac{{}_\infty L_{85}}{{}_\infty L_{80}}$$

   and since ${}_\infty L_x = T_x$

$${}_\infty S_x = \frac{T_{x+5}}{T_x} \quad \text{e.g.} \quad {}_\infty S_{80} = \frac{T_{85}}{T_{80}}$$

3. **Approximations for five year survival ratios**

$${}_5S_x = \frac{L_{x+7}}{L_{x+2}} \quad \text{e.g.} \quad {}_5S_{20} = \frac{L_{27}}{L_{22}}$$

   i.e. using the L_x values in the middle of 5 year age groups

4. **Obtaining ${}_5L_x$ values from single year life tables**
   - Add up the L_x values for each age group

$${}_5L_x = L_x + L_{x+1} + L_{x+2} + L_{x+3} + L_{x+4}$$

   e.g.

$${}_5L_0 = L_0 + L_1 + L_2 + L_3 + L_4$$

$${}_5L_{20} = L_{20} + L_{21} + L_{22} + L_{23} + L_{24}$$

$${}_\infty L_{65} = L_{65} + L_{66} + L_{67} + L_{68} + \cdots + L_\infty$$

   - Calculate ${}_5L_x$ from the l_x values (if there are no L_xs in the life table)

$${}_5L_x = 0.5l_x + l_{x+1} + l_{x+2} + l_{x+3} + l_{x+4} + 0.5l_{x+5}$$

   e.g.

$${}_5L_{20} = 0.5l_{20} + l_{21} + l_{22} + l_{23} + l_{24} + 0.5l_{25}$$

   Exceptions:

$${}_5L_0 = 0.3l_0 + 1.1l_1 + 1.1l_2 + l_3 + l_4 + 0.5l_5$$

$${}_\infty L_{65} = 0.5l_{65} + l_{66} + l_{67} + \cdots + l_\infty$$

Census survival ratios

A potential alternative to estimating cohort survival from life table data is to use population statistics by age and sex from two consecutive censuses, such as 1995 and 2000. These produce census survival ratios (CSRs) to measure cohort survival. Since the cohort aged 40–44 in 2000 was aged 35–39 in 1995, the ratio of the two totals (40–44/35–39) is the cohort's census survival ratio. Thus the census survival ratio compares the numbers of males or females in one age group at the start of an interval with the numbers in the next age group at the end:

$$\mathrm{CSR}_x = \frac{P^n_{x+n}}{P^0_x}$$

CSRs may be used in the same way as LTSRs in estimating net migration. For the ratio to be a true indicator of survival, mortality has to be the only cause of change in the cohort's numbers; census survival ratios therefore derive from national population figures, upon which internal migration has no influence. Moreover, the ratios need to be based on population figures that international migration has not affected. While closed populations existed in the past, few contemporary populations are isolated in this way. This calls for adjustments to the base data for the survival ratios.

To eliminate the effects of migration on the census survival ratios, subtract intercensal international departures from the population at the first census and intercensal international arrivals from the population at the second census (Figure 11.10). These adjustments produce an initial population that either remained in or died in the country during the interval, and a terminal population that both stayed in the country and survived over the same period. Although, in principle, such procedures are quite straightforward, in practice they can be problematic unless there are adequate statistics available on the age–sex composition of international arrivals and departures.

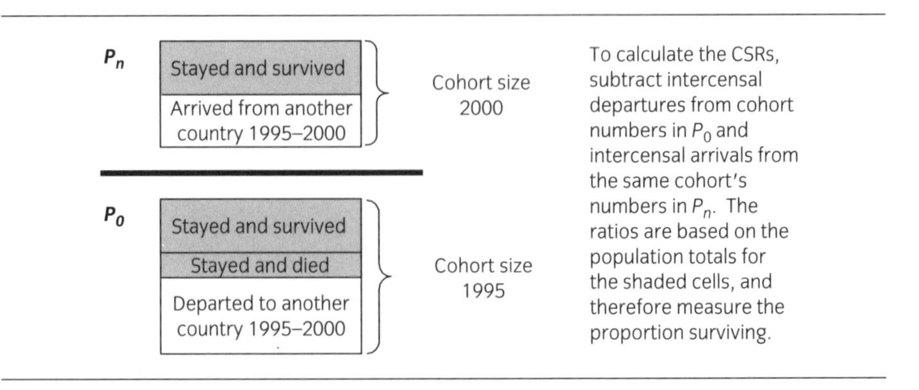

Figure 11.10 Calculating census survival ratios from data for open populations.

Like the LTSR method, the CSR method of estimating net migration assumes that national survival ratios are applicable to sub-national populations, a false assumption if there are marked regional differences in infant mortality, for instance. The CSR method also assumes that census statistics by age are equally accurate for all areas, whereas in developing countries age-misstatement and age heaping have varying impacts on the data from place to place. Survival ratios have further applications in population projections, as discussed in the next chapter.

11.8 Conclusion

Although necessarily considered separately here, migration measures and concepts are integral to the demographic study of population growth, distribution and composition, as well as being of central interest in population geography and applied demography. This chapter has emphasized the importance of distinguishing between different types of population movements and has sought to provide a basis for decisions in relation to defining the population at risk for migration rates and selecting methods to measure or estimate migration. Analyses of migration can have straightforward practical applications in explaining and forecasting population changes, but migration is also a challenging subject area for several reasons: it affects all ages; it is an often repeated event; the motivations and influences underlying migration decisions are difficult to discover; and it is involved in pervasive social and economic changes. Thus beyond the essential demographic tools for investigating migration are many concepts, methods and models developed specifically to address complexities and better reveal the causes and consequences of migration. References listed under 'Further reading' introduce these additional approaches to the study of migration.

Study resources

KEY TERMS

Birthplace method
Census survival ratios
Chain migration
Closed population
CSR method
External migration
Forward survival
Internal migration
International migration
Open population
Partial displacement migration
Residential mobility
Reverse survival
Life table survival ratios
Local migration
LTSR method
Migration
Migration effectiveness
Migration expectancy
Migration interval
Mobility
Movers
Stayers
Survival ratios
Total displacement migration
VS method

FURTHER READING

Biko, Agozino (editor). 2000. *Theoretical and Methodological Issues in Migration Research*. London: Ashgate.

Castles, Stephen and Miller, Mark J. 1998. *The Age of Migration* (second edition). New York: Guilford Press.

Champion, T. and Fielding, T. (editors). 1992. *Migration Processes and Patterns*. London: Belhaven Press.

Clark, W.A.V. 1986. *Human Migration*. Beverly Hills: Sage.

Cohen, Robin (editor). 1996. *Theories of Migration*. Cheltenham, UK: E. Elgar.

Courgeau, Daniel. 1995. 'Migration Theories and Behavioural Models'. *International Journal of Population Geography*, 1 (1): 19–27.

Lewis, G.J. 1982. *Human Migration*. London: Croom Helm.

Massey, Douglas, Arango, Joaquin, Hugo, Graeme, Kouaouci, Ali, Pellegrino, Adela, and Taylor, J. Edward. 1993. 'Theories of International Migration: A Review and Appraisal'. *Population and Development Review*, 19 (3): 431–466.

Plane. D.A. and Rogerson, P.A. 1994. *The Geographical Analysis of Population*. New York: Wiley.

Shaw, R. Paul. 1975. *Migration Theory and Fact; A Review and Bibliography of Current Literature*. Bibliography Series Number Five. Philadelphia: Regional Science Research Institute.

Shryock, Henry S. and Siegel, Jacob S. 1973. *The Methods and Materials of Demography*. Washington: US Bureau of the Census, US Government Printing Office.

Skeldon, R. 1990. *Population Mobility in Developing Countries*. London: Belhaven Press.

Stillwell, John and Congdon, Peter (editors). 1991. *Migration Models: Macro and Micro Approaches*. London: Belhaven Press.

Todaro, M.P. 1976. *Internal Migration in Developing Countries*. Geneva: International Labour Organization.

United Nations. 1970. *Methods of Measuring Internal Migration*. Manuals on Methods of Estimating Population, Manual VI; Population Studies No. 47. New York: United Nations Department of Economic and Social Affairs.

United Nations. 2000. *World Migration Report 2000*. New York: United Nations and International Organization for Migration.

Woods, Robert. 1982. *Theoretical Population Geography*. London: Longman.

Zelinsky, W. 1971. 'The Hypothesis of the Mobility Transition'. *Geographical Review*, 61: 219–259.

INTERNET RESOURCES

Subject	Source and Internet address
Database on migration in ILO member states	**International Labour Organisation** http://www.ilo.org/public/english/protection/migrant/index.htm
Population and migration statistics for the United Kingdom	**National Statistics, HM Government, UK** http://www.statistics.gov.uk/statbase/mainmenu.asp (e.g. select 'Datasets' from the displayed menu)
Information about migration statistics published by the United Nations	**United Nations Statistics Division** http://www.statistics.gov.uk/statbase/mainmenu.asp
Migration statistics for the United States	**US Census Bureau** www.census.gov/population/www/socdemo/migrate.html
Statistics for the European Union, including data on population and migration	**Eurostat** *(Statistical Office of the European Commission)* http://europa.eu.int/comm/eurostat/
Web site of the publisher of *International Migration Review*; includes links to many other organizations concerned with migration	**Center for Migration Studies, New York** http://www.cmsny.org/

Note: the location of information on the Internet is subject to change; to find material that has moved, search from the home page of the organization concerned.

EXERCISES

1 Using the vital statistics method, calculate intercensal net migration to (or from) the following places (hypothetical data):

	Estimated resident population		Intercensal	
	1996 Census	2001 Census	Births	Deaths
Green Lane	22 400	22 100	1872	1018
Kensington	44 150	48 700	4131	1448
Newton	4 150	4 250	359	175
Ponsonby	11 600	11 850	1019	478

2 Identify potential problems in the interpretation of the net migration figures obtained in (1).

3 Employing census survival ratios and the forward survival method, calculate the age-specific net intercensal migration of indigenous females to North Region at ages 5–9 to 65+ in 2001. Assume that the population is closed to external migration; use the National Population figures to calculate the CSRs for indigenous females (hypothetical data).

Ages of indigenous females in the national population and the North region

Age	National population		North region	
	1996 Census	2001 Census	1996 Census	2001 Census
0–4	19 625	17 358	4509	4059
5–9	20 231	18 369	4614	4086
10–14	20 249	19 197	4165	4215
15–19	16 472	17 910	3990	4353
20–24	12 108	14 787	3095	3804
25–29	9 744	11 034	2474	2751
30–34	7 459	8 964	1871	2214
35–39	7 043	7 137	1655	1788
40–44	5 639	6 513	1265	1515
45–49	4 743	5 259	911	1164
50–54	3 479	4 221	718	807
55–59	2 568	3 066	450	594
60–64	1 971	2 211	381	387
65+	2 927	3 321	434	522

4 Under what circumstances can census survival ratios exceed 1?

5 Explain whether census survival ratios are suitable for estimating migration in open populations.

6 When is reverse survival most likely to produce estimates higher than those from forward survival?

7 Sketch a diagram to illustrate the derivation of a forward survival ratio estimate of a net migration loss among men aged 85 years and over in 2000, who moved in the interval 1990–2000.

8 If a community of 30 000 people recorded very low total net migration during the last 20 years, would this indicate that migration was unimportant in maintaining its population?

9 Specify one disadvantage of migration expectancy as a measure of the occurrence of migration.

10 Why is L_x normally used in calculating life table survival ratios rather than l_x?

SPREADSHEET EXERCISE 11: NET MIGRATION ESTIMATES

This exercise provides practice in calculating net migration using the vital statistics method, the census survival ratio method and the life table survival ratio method.

Vital statistics method

Use Excel to answer question 1 in the Exercises (see above).

Census survival ratio method

Use Excel to answer question 3 in the Exercises (see above).

Life table survival ratio method

- Using forward survival, calculate age-specific net migration numbers and rates to complete the table below.
- Draw a bar graph of the net migration estimates.

Net migration estimates for a coastal region

Age in 1996	Age in 2001	${}_5Sx$	Population 1996	Estimated survivors	Population 2001, ages 5 & over	Net migration	Net migration rate/1000[2]
0–4	5–9	0.99737	95456		103452		
5–9	10–14	0.99849	99170		104886		
10–14	15–19	0.99672	98415		101045		
15–19	20–24	0.99207	93957		101544		
20–24	25–29	0.99259	84342		93220		
25–29	30–34	0.99350	83598		92756		
30–34	35–39	0.99301	70191		77853		
35–39	40–44	0.98956	59982		64334		
40–44	45–49	0.98223	53313		55713		
45–49	50–54	0.96979	55775		57715		
50–54	55–59	0.95060	54857		56099		
55–59	60–64	0.92191	46881		48363		
60+	65+	0.72308	128674		103715		
Total[1]							

1. 2001 total omits ages 0–4.
2. Based on the end-of-period population.

Answers:

Age in 2001	Net migration	Net migration rate/1000
5–9	8247	79.7
10–14	5866	55.9
15–19	2953	29.2
20–24	8332	82.1
25–29	9503	101.9
30–34	9701	104.6
35–39	8153	104.7
40–44	4978	77.4
45–49	3347	60.1
50–54	3625	62.8
55–59	3952	70.4
60–64	5143	106.3
65+	10673	102.9
Total	84473	79.6

Section VI

Applied demography

This final section is concerned with techniques employed in planning, policy-making and commercial applications, especially methods of projecting and estimating total populations and methods of projecting and analysing population composition.

Chapter 12 **Population projections and estimates**

Chapter 13 **Population composition**

12 Population projections and estimates

And lastly I took the Map of *London* set out in the year 1658 by *Richard Newcourt*, drawn by a scale of Yards. Now I guessed that in 100 yards square there might be about 54 Families, supposing every house to be 20 foot in the front: for on two sides of the said square there will be 100 yards of housing in each, and in the two other sides 80 each; in all 360 yards: that is 54 Families in each square, of which there are 220 within the Walls, Making in all 11880 Families within the Walls. But forasmuch as there dy within the Walls about 3200 *per Annum*, and in the whole about 13000; it follows, that the housing within the Walls is ¼ part of the whole, and consequently, that there are 47 520 Families in, and about *London*, which agrees well enough with all my former computations: the worst whereof doth sufficiently demonstrate, that there are no Millions of People in *London*, which nevertheless most men do believe, as they do, that there be three Women for one Man, whereas there are fourteen Men for thirteen Women, as else where hath been said.

(Graunt 1662: 68–69)

OUTLINE

12.1 Applications and issues
12.2 Population estimates
12.3 Projection methods
12.4 Calculating cohort component projections
12.5 Elaborating the basic cohort component model
12.6 Further methods for sub-national populations
12.7 Conclusion
Study resources

LEARNING OBJECTIVES

To understand:

- methods of estimation and projection based on growth rates
- the cohort component method of projection and estimation
- the ratio and housing unit methods of projection and estimation

To become aware of:

- the main uses of population projections and estimates
- the diversity of approaches to projection and estimation

COMPUTER APPLICATIONS

Excel module: Cohort Component Projections.xls (Box 12.2)

Spreadsheet exercise 12: Cohort component projections (see *Study resources*)

Interest in practical applications of demography is long established, not least because the *raison d'être* of census and vital statistics collections has been to inform policy-making, planning and administration. Since the 1960s, commercial applications of demography have become a major area of activity, but even here demography's roots extend back much further, notably in the life insurance industry, which has always relied on life tables and related techniques. Many of the methods discussed in this book have applications in planning for the present population as well as in anticipating future needs. Relevance to planning and policy-making constitute major reasons for demography's existence.

Applied demography is today a recognized sub-field of demography, one that emphasizes practical applications of demographic statistics and methods, especially at the state and local levels; it is particularly concerned with present and future developments and spatial variations in population characteristics. Applied demography's future orientation makes population projections a particular interest. Similarly, in its concern for present circumstances there is often emphasis on the numbers and characteristics of (i) client populations for organizations, (ii) consumer markets for businesses, or (iii) workforces for companies or governments. Accordingly, this final section is headed *applied demography* because it focuses on population projections, estimates and population composition – that is, core subject materials of the applied demography sub-field. Nevertheless, most of the content of this book is relevant to applied demography, especially if Jacob Siegel's broad definition, incorporating national and international concerns, is followed:

> A more ample definition of applied demography, then, is that it is the sub-field of demography concerned with the application of the materials and methods of demography to the analysis and solution of the problems of business, private nonprofit organizations, and government, at the local, national, and international levels, with a primary orientation toward particular areas and the present and future. Accordingly, we can characterize applied demography as a decision oriented science concerned with aiding managers, administrators, and government officials in making practical decisions . . . (Siegel 2001: 2)

Population projections, supplying information about prospective developments, represent a vital application of demography in planning and policy-making. While much needed and relied upon, population projections are sometimes misunderstood and their potential under-utilized. This partly reflects the common belief that population projections have but one purpose – to predict the future – whereas projections have varied applications, among which 'prediction' is the most tentative. As the famous physicist Niels Bohr observed: 'it is very difficult to make predictions, especially about the future.'

Comparisons of the age pyramid for China in 2000, and projected to 2050, illustrate outcomes of low fertility and population ageing (Figure 12.1). Yet one of the most important observations to make about the pyramid for 2050 is that it illustrates a future that is hypothetical: a future that may eventuate if actual devel-

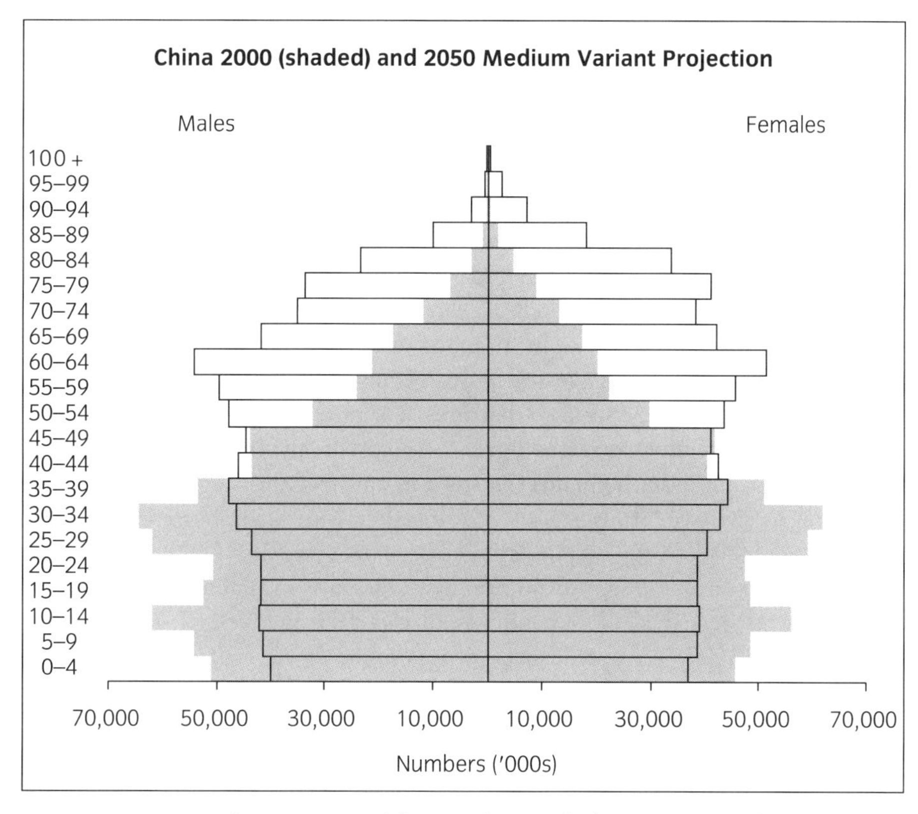

Figure 12.1 Age structure of the population of China in 2000 and 2050.

Source: United Nations Population Division 2002: *World Population Prospects: The 2000 Revision, Volume II: The Sex and Age Distribution of Populations*. p. 309 (on CD).

opments coincide with a battery of assumptions about demographic changes. The projection in Figure 12.1 is one of several for China published by the United Nations in 2002; it differs from earlier United Nations projections, and it differs also from those of other organizations. The availability of many projections for the same date reflects that there are various possible futures for China, depending how events transpire. National and international statistical agencies regularly publish bulletins of population projections, updating them in the light of new information, with the result that there can accumulate a range of assessments of prospective trends.

Methods of preparing population estimates are closely related to methods of projecting populations. Whereas projections mainly look ahead to the future, estimates mainly focus on the present or the recent past. Statistical agencies prepare estimates of the current population, to provide up-to-date information for years in which there is no census. Estimates are based on observed data about population changes (births, deaths and migration), whereas projections refer to dates for which there are no observed data, especially future years. Estimates for historical populations may be made from parish records and other sources. Reverse, or backwards, projection for past years is also possible, generating a series of figures for consecutive years on the basis of a set of assumptions from more recent population changes.

This chapter explains the most frequently used methods for projecting and estimating populations, especially the predominant 'cohort component method', together with their principal applications. Where appropriate, the chapter deals jointly with methods for projections and estimates, as well as with their uses at different geographical scales. This reflects the versatility of some of the procedures and their relevance to varied applications, as summarized in Table 12.1.

Projecting populations by age and sex can entail so many repeated calculations that a computer-based approach becomes essential as the projection period lengthens or the number of areas increases. Accordingly, the chapter includes a spreadsheet exercise that explains the steps in constructing a cohort-component projection model (see 'Study resources'), together with a computer module that calculates and displays projections from entered data (see Box 12.2).

Special-purpose computer projection packages are available for population projections, as described in Appendix A of the electronic version of Bongaarts and Bulatao (2000). A spreadsheet program, however, can produce similar results and output, including age–sex pyramids of projected populations, at the same time serving as a valuable aid for learning the principles involved. Spreadsheet construction also allows the output to be tailored to particular needs, such as through generating, at the same time, projections of sex ratios, dependency ratios, growth rates and other indices. Thus the *Cohort Component Projections module* (Box 12.2) is presented as an adaptable basis for population projections generally.

Table 12.1 Methods for population projections and estimates

Methods	Projections		Estimates	
	National	Sub-national	National	Sub-national
Cohort component	•	•	•	•
Mathematical	•	•	•	•
Model populations	•			
Ratio method		•		•
Housing unit method		•		•
Indicator methods				•
Matrix methods	•	•		
Statistical modelling	•	•	•	•

12.1 Applications and issues

Population projections and estimates make an important contribution to the activities of governments, organizations and businesses. They serve as a basis for allocating funds or determining major expenditure, and they are an aid to understanding the nature of ongoing changes and their implications. This section discusses applications and issues relating to projections and estimates under the headings of 'Planning' and 'Policy-making', which encompass the main considerations. The exploratory and experimental applications of projections for policy-making are important, but are the least recognized.

Planning

Estimates and projections have varied applications in planning, important among which are providing a basis for allocation of resources between areas in relation to population size, and determining the need or scope for investment in particular places. Estimates are also necessary to provide denominators for demographic rates in non-census years, including rates indicative of labour force trends and market behaviour (see Simpson et al. 1997: 265).

Planning requires reliable information on present population numbers and characteristics and the outlook for the future. This reflects that major planning decisions, often entailing considerable financial outlays, may have to be made on the basis of demographic data together with other background information. In these circumstances planners require accurate estimates plus well-considered population forecasts. As Klosterman (1990: 4) noted: 'Population projections and forecasts are among the most important studies prepared by local, state and national planners.'

A forecast involves judgement about the most likely course of events and future trajectory of population changes. 'Forecast', however, implies less certainty than 'prediction'. Like a weather forecast, a population forecast represents what seems probable at the publication date. It is advisable to avoid describing any projection as a 'prediction', because the word conveys certainty about the future, a claim which in most circumstances would be very misleading.

Because of the inherent uncertainty surrounding future trends, it is desirable to present population forecasts in the context of other projections, representing possible alternative futures, together with the highest and lowest plausible outlooks – sometimes called the high and low 'variants'. The projections may include one that extrapolates recent fertility, mortality and migration to serve as an easily understood reference point for comparison with other projections – even though maintenance of the status quo in demographic rates may be improbable. When plotted on line graphs, multiple projections appear as 'projection fans' spanning a range of different trajectories. The existence of the range of uncertainty, denoted by the width of the projection fans, raises several implications for planning:

- The further into the future, the greater the uncertainty. While national population projections may prove fairly reliable for five years ahead, for sub-national populations even five years can allow considerable latitude for variation because of the role of migration in regional and local population changes. Isserman and Fisher (1984: 28) have written that population forecasting requires 'judgement, wisdom, persistence, imagination and luck'. Luck can be particularly important in achieving accurate forecasts at the sub-national level.
- A related point is that the smaller the population, the more difficult forecasting becomes, especially because of the importance of migration as a component of population change. Small populations with high growth rates are especially difficult to project accurately, because high growth implies greater latitude for change to slower growth. Newly settled areas, for instance, tend to have high growth rates, but only temporarily. This is partly because, as numbers grow, the same growth rates require greater increments, and partly because housing construction inevitably reaches completion.
- Where practical, planning strategies should recognize futures other than one preferred forecast, even incorporating contingency arrangements for the high and low variants. Flexibility may be difficult for some major expenditure items, including works on roads, water storage and electricity generation. Yet for others there may be opportunities to keep certain options open, for instance in building construction through erecting temporary structures or designing buildings suitable for future extension or modification for different purposes – such as office blocks convertible to apartment blocks, or schools suited to becoming community facilities.
- Uncertainty in planning arises from the prospective impact of non-demographic influences as well as demographic ones. Population changes are also but one aspect of plans for the future, and other considerations may take precedence. The demand for oil, electricity and water, for example, depends substantially on non-domestic usage in agriculture, commerce and industry, whereas population estimates and projections typically focus on numbers of residents, households and dwellings. Moreover, for social or political reasons, precedence in planning may be given to non-demographic considerations, such as special requirements relating to defence or tourism, or the particular needs of disadvantaged communities.
- Some of the most expensive aspects of public expenditure depend on medium to long-term (10 years plus) projections for regions, that is for the time spans and types of areas that raise major difficulties in forecasting. The smaller the population and the longer the time span, the greater the uncertainty. Installation of power lines, telecommunications lines, water mains and sewerage works may need to anticipate regional demand by many years. Some advise that, in costly developments, it is better to err on the side of over-supply, keeping ahead of demand if funds permit. While this invites criticisms of over-supply or too early development of infrastructure, concerns may be greater if there is an under-supply.

The notion of the range of uncertainty in population projections also raises the question of their reliability – how often do projections coincide with actual experience? From a study of various national population projections over a 15–20 year period since 1950, Keyfitz (1981: 581) found that one third of the observed populations did not fall within the range between the high and low variants. Clearly, projection fans often fail to encompass all probable futures even in the medium term, which underlines the importance of maintaining flexibility where feasible.

Planners using population projections therefore need to be informed about the likelihood that projections will depart from real experience. As Keyfitz (1981: 579) commented:

> Demographers can no more be held responsible for inaccuracy in forecasting population 20 years ahead than geologists, meteorologists, or economists when they fail to announce earthquakes, cold winters, or depressions 20 years ahead.

Despite errors, projections are still more useful than the alternative of assuming that nothing will change. Improving reliability in demographic projections depends fundamentally on improving theories and explanations of population trends. In the absence of adequate theories, reliable forecasting necessarily remains elusive. Also, many practical difficulties arise from the diverse range of factors that influence population – social, economic, political and technological – the effects of which are difficult to incorporate into projections.

Policy-making

As well as assisting planning for ongoing changes, projections can also stimulate innovations in policy making to determine better ways of meeting needs or minimizing problems in the future. For example, projections showing the impending growth in the numbers of the aged in developed countries have served as starting points for rethinking policies for funding pensions and health services for the elderly: the projections have drawn attention to looming rapid growth requiring to be addressed through new strategies for financing aged care and support.

Projections can also serve as warnings of impending developments that would be best avoided through policy interventions to modify existing trends. In less developed countries, projections of population growth have been influential in demonstrating to governments the unsustainablility of continuing high growth rates. In such cases, the usefulness of a projection depends not on its accuracy as a forecast, but on its contribution to shaping an alternative future. Current projections of the long-run effects of below replacement fertility in European countries are similarly portraying undesirable futures for policies to address.

In policy-making, projections also have a valuable analytical or exploratory role. They enable experimentation to demonstrate what would happen to the population if particular changes or interventions occurred. The aim here is not necessarily to show the most likely future, but to aid understanding through illustrating the impact of different courses of events. Questions that may be addressed through analytical projections include: How could zero growth and a stationary age structure be achieved? What would be the consequences of seeking to use migration to offset the effects of low fertility? Is a suggested target population attainable? In what ways would the age structure of the population change if lifestyle improvements and medical interventions raised average life expectancy to 90 years? Analytical projections explore varied outcomes as an aid to improving understanding of population dynamics. The exploratory functions of projections are less well known but, in terms of their potential contribution to policy-making, they are as valuable as their forecasting function.

12.2 **Population estimates**

As the quotation heading this chapter intimates, John Graunt envisaged how population estimates could be obtained through systematic procedures, rather than merely through guessing. His methodology appeared in a few pages of his *Observations* (Graunt 1662: 67–69) and, despite deficiencies, it revealed the potential for an evidence-based approach to population estimation.

Whereas Graunt had no census figures, contemporary methods of estimating populations commonly rely on such sources as a starting point, updated from information on subsequent growth rates or the components of growth. Moreover, since demographic and social surveys are regularly conducted, data from them can often be used in place of national, state, and provincial estimates, especially since they include details of population characteristics. For smaller areas, opportunities to use national survey statistics for estimates are far more limited because sample sizes are insufficient to provide fine detail.

Population estimates fulfil needs for current data on population size, characteristics and distribution. They are used for planning and administration, as well as for calculating current demographic rates – which require up-to-date denominators. *Intercensal estimates* provide figures for years between two existing censuses, such as 1990 and 2000, while *postcensal estimates* provide figures for years since the last census. The former are likely to be more accurate, because there are two reference points together with observed data on births, deaths and migration for all the intervening years. Thus various postcensal population estimates for the same year may differ, more so for small areas. Comparisons of estimates derived in different ways are almost essential, given the inherent deficiencies of particular methods, and the financial implications of decisions that rely on population estimates.

As Table 12.1 illustrates, some methods are equally applicable to estimation and projection, including the use of growth rate formulas to estimate numbers between known figures, or to project numbers for later dates. The cohort component method is ubiquitous in the field of estimates and projections, since it supplies detailed figures by age and sex for both purposes at various geographical scales. Further discussion of population estimates, therefore, is mainly integrated into the ensuing sections on population projections.

12.3 Projection methods

Population projections are empirically based calculations of future, or past, population numbers under specified assumptions about changes in population growth or its components. Unlike population estimates, they are not based on observed data for the intervening period, usually because none exist. Although the cohort component method of projecting populations remains the predominant approach among government agencies in the United States and the United Kingdom, as well as in many other countries, there are additional widely used approaches. This section introduces the principal methods.

Model populations

Previous chapters have already discussed one insightful approach to thinking about demographic futures, namely the construction of stationary and stable population models. Such models are noteworthy from a policy perspective because stable or stationary age structures may represent the desired future of nations, as well as the global population. Stationary age structures also represent supposed conditions at the beginning (pre-transition) and end (post-transition) of the classical demographic transition.

Accordingly, stable and stationary models can serve independently as long-range projections, 70 or more years ahead, for instance to show the implications of maintenance of present birth and death rates (Woods 1982a: 208), or to portray circumstances arising if the population achieved zero growth at present mortality levels. In the mid-1970s, Australia's National Population Inquiry (Borrie 1975) employed this approach to demonstrate what Australia's future age structure would eventually be like assuming constant growth or zero growth, especially since interest groups were campaigning at the time for a zero population growth policy. The differences were dramatic. 'Stable Australia', based on 1971 figures, had a triangular age structure with the potential to double in 60 years (see Figure 9.1), while 'stationary Australia' had a rectangular age structure with a relatively high representation of older people.

As long-range projections, however, stable and stationary models serve exploratory or illustrative purposes, rather than representing the anticipated outcome of events. Extrapolations of the present provide informative benchmarks when considering future prospects, especially in comparison with other projections. The disadvantages of using model populations as projections are that they present information only on age structures, rather than population numbers, and they keep birth and death rates constant – an unlikely situation. Also, they do not refer to a specific date in the future, although it can be assumed that such age structures would be nearly established after about 70 years. For short and medium-term projections and estimates other approaches are needed.

Methods based on growth rates

The so-called 'mathematical methods' provide a means of estimating and projecting total populations either using growth rate formulas (see Chapter 2) or, for more advanced work, through fitting curves representing other mathematical descriptions of population change, such as the logistic curve (see Klosterman 1990). Only the former more common approaches are discussed here. Projections based on growth rates assume constant arithmetic, geometric or exponential growth to estimate populations between dates or to project numbers for a few years ahead. As discussed in Chapter 2, the assumption of constant growth rates is seldom satisfactory; hence projections based on growth rates are either limited to short time spans, or the growth rates are varied from one interval to another. For population estimation within short intervals (e.g. five years), the assumption of constant growth rates is convenient and results will differ little according to whether arithmetic, geometric or exponential growth is assumed.

Assuming arithmetic change, estimates and projections are calculated from the formula:

$$P_x = P_0 + (P_n - P_0) \times \frac{x}{n}$$

where:

P_x = population after x years
n = number of years in the interval between the initial and end-of-period populations
P_0 = initial population
P_n = end-of-period population
x = number of years from P_0 to the year of the estimate.

The same result may be obtained by drawing a graph with a straight line through P_0 and P_n and reading off any intermediate or later value. For longer intervals, a curvilinear trajectory may be more realistic, as represented in geometric and exponential growth curves. In making projections and estimates, the geometric and exponential growth rate formulas are modified to calculate any value of P_n, given P_0, r and n (see Chapter 2, Tables 2.3 and 2.4).

The main disadvantage of the 'mathematical methods' is that they ignore information on annual births, deaths and migration. Thus if such data are available and accurate, more satisfactory intercensal estimates can be derived from data on the components of growth, i.e. $P_n = P_0 + (B - D) + (I - E)$. Estimates of P_n most commonly refer to the resident population.

The cohort component method

Planning for the future would be more straightforward if populations conformed to regular trends, as envisaged in growth curves. However, many changes are sudden and discontinuous, rather than emerging gradually. Changes in the numbers in older age groups, for example, can more closely resemble a roller-coaster ride than a smooth, readily predictable, gradient. This is because changes in population numbers at each age depend on the size of cohorts moving through the age structure. Projection methods that calculate changes in cohorts through time provide a means of anticipating discontinuities – the demographic 'shock waves' arising from the progress of unusually large or otherwise distinctive cohorts. The cohort component method, also known as the cohort survival method, incorporates these effects into projections.

The cohort-component method is the most widely used approach to projecting populations, and is frequently employed also in preparing population estimates by age and sex. It is the preferred method of national and international statistical organizations for projecting national populations, including the United Nations, the World Bank, and the US Census Bureau (Bongaarts and Bulato 2000: 29). For sub-national populations, there is greater diversity in approaches to projection, but the cohort component method remains predominant. There are three main reasons for this:

- It makes adequate use of available statistical information on the components of population change. This means that it has greater analytical value in understanding developments and the relative contributions of growth from different sources. At the same time, the method is flexible: assumptions can be varied according to expectations about future directions of change in specific components.
- It provides projections by age and sex.
- The method is equally applicable at national and sub-national scales, thereby permitting comparability between different series of projections.

Essentially, the cohort component method involves calculating the future size of cohorts, taking into account the effects of fertility, mortality and migration. When working with five-year age groups in five-year steps, each cohort is projected over the first interval, such as from 2000 to 2005, then the projected figures become the base for the next step in the projection, from 2005 to 2010.

As in other demographic calculations, the use of spreadsheets greatly facilitates these operations: often just one or two formulas can be copied across an entire table to calculate the figures for all age groups and as many years as required. The ensuing discussion of the cohort component method, therefore, is based on the spreadsheet tables in Box 12.1, which are also the basis of Spreadsheet Exercise 12; the latter provides an opportunity to build a reasonably detailed projection model.

BOX 12.1 **Worked example of a set of cohort component projections**

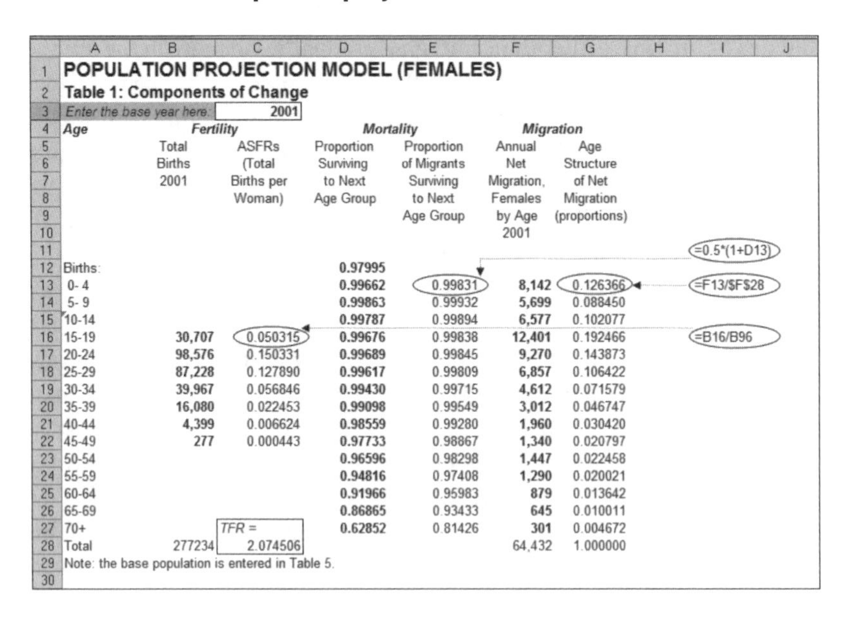

POPULATION PROJECTION MODEL (FEMALES)

Table 1: Components of Change

Enter the base year here: 2001

Age	*Fertility* Total Births 2001	*Fertility* ASFRs (Total Births per Woman)	*Mortality* Proportion Surviving to Next Age Group	*Mortality* Proportion of Migrants Surviving to Next Age Group	*Migration* Annual Net Migration, Females by Age 2001	*Migration* Age Structure of Net Migration (proportions)
Births:			**0.97995**			
0- 4			**0.99662**	0.99831	**8,142**	0.126366
5- 9			**0.99863**	0.99932	**5,699**	0.088450
10-14			**0.99787**	0.99894	**6,577**	0.102077
15-19	**30,707**	0.050315	**0.99676**	0.99838	**12,401**	0.192466
20-24	**98,576**	0.150331	**0.99689**	0.99845	**9,270**	0.143873
25-29	**87,228**	0.127890	**0.99617**	0.99809	**6,857**	0.106422
30-34	**39,967**	0.056846	**0.99430**	0.99715	**4,612**	0.071579
35-39	**16,080**	0.022453	**0.99098**	0.99549	**3,012**	0.046747
40-44	**4,399**	0.006624	**0.98559**	0.99280	**1,960**	0.030420
45-49	**277**	0.000443	**0.97733**	0.98867	**1,340**	0.020797
50-54			**0.96596**	0.98298	**1,447**	0.022458
55-59			**0.94816**	0.97408	**1,290**	0.020021
60-64			**0.91966**	0.95983	**879**	0.013642
65-69			**0.86865**	0.93433	**645**	0.010011
70+		*TFR =*	**0.62852**	0.81426	**301**	0.004672
Total	277234	2.074506			64,432	1.000000

Note: the base population is entered in Table 5.

Table 2: Projected Survivors

Enter the formula for ages 5-9 and copy to all other cells except those for 75+ and the totals.

Year:	2006	2011	2016	2021	2026
Age (end of period)					
5- 9	613,937	664,912	678,926	699,583	722,949
10-14	625,264	644,591	695,496	709,490	730,120
15-19	627,833	645,983	665,268	716,065	730,029
20-24	608,317	651,208	669,299	688,522	739,154
25-29	653,688	654,314	697,072	715,107	734,270
30-34	679,442	686,959	687,583	730,177	748,143
35-39	699,071	701,972	709,446	710,067	752,418
40-44	709,710	710,449	713,323	720,730	721,345
45-49	654,542	710,950	711,678	714,511	721,811
50-54	611,553	647,083	702,211	702,923	705,692
55-59	466,986	595,701	630,021	683,273	683,961
60-64	374,677	448,010	570,052	602,594	653,085
65-69	315,975	349,059	416,501	528,738	558,665
70-74	299,636	277,315	306,054	364,637	462,132
75+	549,194	535,574	512,984	516,849	556,099
Total	8,489,823	8,924,078	9,365,914	9,803,265	10,219,872

=$C$3+5

=B93*$D13

=B106*$D26

=B109*$D27

continued

BOX 12.1 continued

	A	B	C	D	E	F	G	H
52	**Table 3: Projected Total Births**							
53	*Enter the formula for ages 15-19 and copy to all other age groups.*							
54	5 Years to:		2006	2011	2016	2021	2026	
55	Age							=(B96+C96)/2*$C16*5
56	15-19		158,948	166,644	171,352	180,168	188,314	
57	20-24		493,116	509,472	532,391	546,414	572,668	
58	25-29		438,553	441,166	455,037	474,474	486,367	
59	30-34		200,250	201,732	202,889	209,031	217,638	
60	35-39		80,442	80,646	81,229	81,683	84,095	
61	40-44		22,943	23,903	23,963	24,133	24,266	
62	45-49		1,425	1,528	1,591	1,595	1,606	
63	Total		1,395,676	1,425,091	1,468,452	1,517,498	1,574,954	
64								

	A	B	C	D	E	F	G	H
65	**Table 4: Net Migration: Projected Survivors**							
66	*Enter the formula for ages 5-9 and copy to all other age groups.*							
67	5 Years To:		2006	2011	2016	2021	2026	
68	Volume (over 5 years):		**250,000**	**250,000**	**250,000**	**250,000**	**250,000**	
69	Age (end of period)							=(C$68*$G13)*$E13
70	5- 9		31,538	31,538	31,538	31,538	31,538	
71	10-14		22,097	22,097	22,097	22,097	22,097	
72	15-19		25,492	25,492	25,492	25,492	25,492	
73	20-24		48,039	48,039	48,039	48,039	48,039	
74	25-29		35,912	35,912	35,912	35,912	35,912	
75	30-34		26,555	26,555	26,555	26,555	26,555	
76	35-39		17,844	17,844	17,844	17,844	17,844	
77	40-44		11,634	11,634	11,634	11,634	11,634	
78	45-49		7,550	7,550	7,550	7,550	7,550	
79	50-54		5,140	5,140	5,140	5,140	5,140	
80	55-59		5,519	5,519	5,519	5,519	5,519	
81	60-64		4,876	4,876	4,876	4,876	4,876	
82	65-69		3,274	3,274	3,274	3,274	3,274	
83	70-74		2,338	2,338	2,338	2,338	2,338	
84	75+		951	951	951	951	951	
85	Total		248,758	248,758	248,758	248,758	248,758	
86	Note: Arrivals aged 0-4 at the end of the period are calculated as births to migrants.							
87								

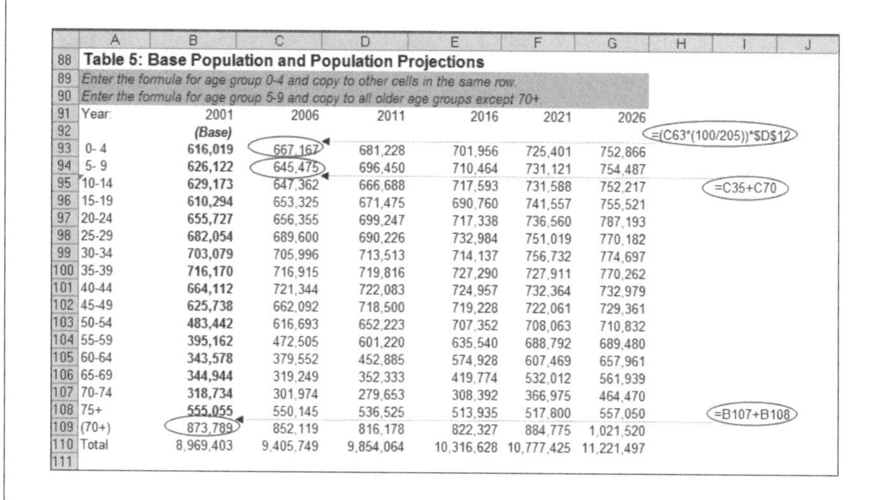

	A	B	C	D	E	F	G	H
88	**Table 5: Base Population and Population Projections**							
89	*Enter the formula for age group 0-4 and copy to other cells in the same row.*							
90	*Enter the formula for age group 5-9 and copy to all older age groups except 70+.*							
91	Year:	2001	2006	2011	2016	2021	2026	
92		***(Base)***						=(C63*(100/205))*$D$12
93	0- 4	**616,019**	667,167	681,228	701,956	725,401	752,866	
94	5- 9	**626,122**	645,475	696,450	710,464	731,121	754,487	
95	10-14	**629,173**	647,362	666,688	717,593	731,588	752,217	=C35+C70
96	15-19	**610,294**	653,325	671,475	690,760	741,557	755,521	
97	20-24	**655,727**	656,355	699,247	717,338	736,560	787,193	
98	25-29	**682,054**	689,600	690,226	732,984	751,019	770,182	
99	30-34	**703,079**	705,996	713,513	714,137	756,732	774,697	
100	35-39	**716,170**	716,915	719,816	727,290	727,911	770,262	
101	40-44	**664,112**	721,344	722,083	724,957	732,364	732,979	
102	45-49	**625,738**	662,092	718,500	719,228	722,061	729,361	
103	50-54	**483,442**	616,693	652,223	707,352	708,063	710,832	
104	55-59	**395,162**	472,505	601,220	635,540	688,792	689,480	
105	60-64	**343,578**	379,552	452,885	574,928	607,469	657,961	
106	65-69	**344,944**	319,249	352,333	419,774	532,012	561,939	
107	70-74	**318,734**	301,974	279,653	308,392	366,975	464,470	
108	75+	**555,055**	550,145	536,525	513,935	517,800	557,050	=B107+B108
109	(70+)	873,789	852,119	816,178	822,327	884,775	1,021,520	
110	Total	8,969,403	9,405,749	9,854,064	10,316,628	10,777,425	11,221,497	
111								

12.4 Calculating cohort component projections

Projections for more than one interval can amount to a sizeable body of statistics, requiring many intermediate calculations. Thus a first consideration in preparing cohort component projections is the structure of the projection model, that is, the arrangement of a series of tables showing the components of change, the intermediate calculations and the projections themselves. The flow chart in Figure 12.2 illustrates the structure of the example in Box 12.1. Although this may be varied if desired, the tables group together the same types of calculations to minimize the likelihood of errors and maximize opportunities to copy spreadsheet formulas.

Arrows in the flow chart show the linkages between the statistics in different tables. Many linkages arise, partly because projected figures are reused as the base

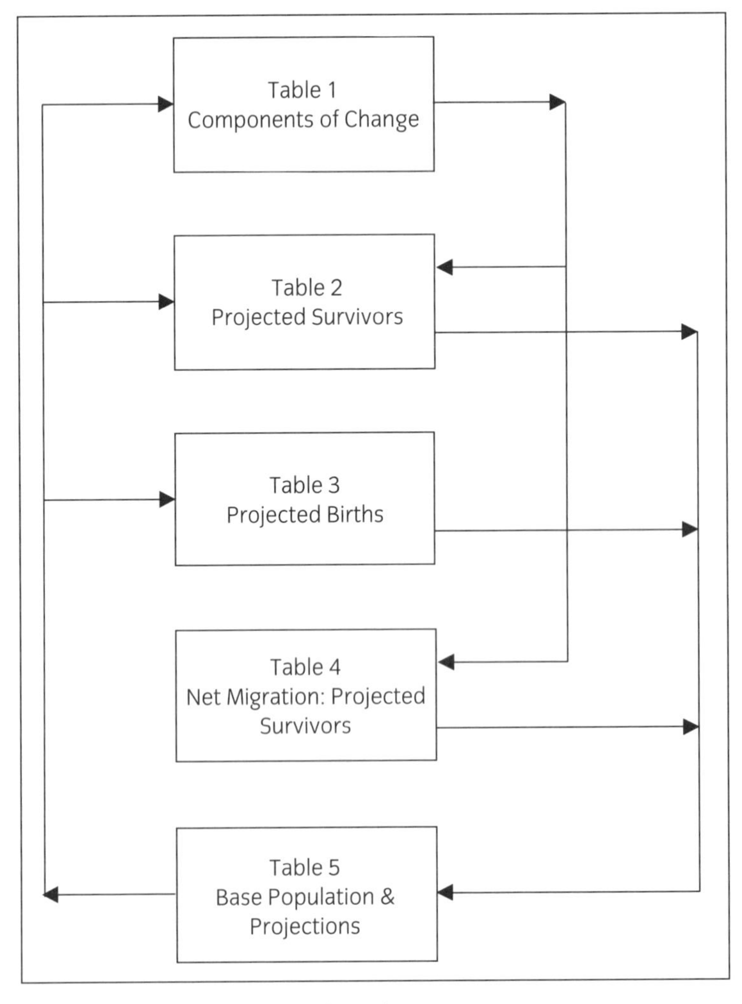

Figure 12.2 Flow chart for cohort component projections.

figures for the next step in the projection: one cycle of calculations permits another. Copying spreadsheet formulas enables these linkages to be handled efficiently.

The following discussion explains the calculation of cohort component projections with reference to each table in turn. The projection is for a hypothetical female population with constant fertility, mortality and migration. Later sections will show how this basic model can be extended to both sexes, as well as incorporating more flexible assumptions about the components of growth. The aim is to explain how to develop cohort component projections suitable for a wide range of applications, at both the national and sub-national levels. In the tables, entered values are shown in bold type, all other calculated values are in normal type: in the spreadsheet, all of the latter cells contain formulas.

Components of change (Table 1)

Projections are based either on the most recent census or on population estimates, if these are considered sufficiently accurate. Projecting forward in five year steps matches the progress of cohorts through five year age groups, and produces a compact and manageable data set. Nevertheless, projections for single years and single years of age may be needed for administrative purposes, to identify other age ranges, including persons eligible to vote, or to anticipate annual changes in age brackets, such as for the school age population or the population of pensionable age.

The first table (Table 1 in Box 12.1) shows the base year (entered in Cell C3) and the components of population change – fertility, mortality and migration – at each age. To save typing column headings in the other tables, i.e. the projection years, these are calculated from the base year in Table 1 (e.g. 2006 = $C$3 + 5). Details of the components of population change are as follows:

- *Fertility* List the total births in the base year alongside the age of the mother and calculate the age-specific fertility rates per woman. The denominators for these rates are given in Table 5, which shows the age structure of the population in the base year (2001) as well as in subsequent years (e.g. the age-specific fertility rate for females aged 15–19 is: 30 707/610 294 = 0.050 315).
- *Mortality* From an appropriate life table, derive five year survival ratios and list them beside each age group. The survival ratios (S_x) are calculated as ${}_5L_{x+5}/{}_5L_x$. For a final open-ended interval, such as from age 70+ to 75+, the survival ratio is equal to T_{x+5}/T_x, while the survival ratio of births is ${}_5L_0/(5 \times l_0)$ (see Chapter 11). Calculating projected survivors in population projections involves procedures identical to those used in 'forward survival estimates' of net migration. Survival ratios for projections can be obtained from official life tables, model life tables (see Chapter 9) or from the *Relational Model Life Tables* module in Chapter 9 (Box 9.2).

  Also needed is a set of survival ratios for migrants. These are estimated from the other survival ratios using the formula $0.5(1 + S_x)$. For age group 0–4, the survival ratio of migrants is 0.5(1 + 0.996 62) = 0.998 31. This assumes that, on average, migrants are present in the destination for half the interval (Pollard et al. 1990: 122). All the migrants are alive at the middle of the interval (i.e. they have a survival ratio of 1 until then), after which they are subject to the mortality of the receiving population. The formula produces the average of the migrant's mortality over the whole interval.

- *Migration* List figures on the net migration of females by age in the base year. In the next column show the proportion of the total net migrants in each age group (e.g. for age group 0–4, 8142/64 432 = 0.126 366).

Projected survivors (Table 2)

The second table shows, for each year in the projection, the number of survivors at each age: age group 0–4 in 2001 yields survivors aged 5–9 in 2006, while those aged 70+ in 2001 yield survivors aged 75+ in 2006. The figures for 2006 in the table are obtained by multiplying the 2001 population (in Table 5) at each age by the corresponding survival ratio in Table 1 (e.g. using age group 0–4 in 2001: 616 019 × 0.996 62 = 613 937 females aged 5–9 in 2006).

Note that the projected survivors in each column are calculated from the populations in Table 5 and the survival ratios in Table 1. Consequently, if the first column of spreadsheet formulas is copied across to other columns of Table 2, as it can be, zeros will be displayed until more cells in Table 5 are completed.

The table is easy to produce in Excel, because the first formula, for ages 5–9 in 2006, is copied to every other age group and column, except for ages 75 and over. The exception occurs because the numbers aged 70+ in 2001 (Table 5) are needed to calculate the survivors aged 75+ in 2006: the two preceding age groups in Table 5 (ages 70–74 and 75+) are not used in calculating Table 2.

Projected total births (Table 3)

Like the previous table, the table showing projected births is also dependent on information in Table 5, namely, the numbers of females in the reproductive age groups. Thus, like Table 2, Table 3 will also contain zero cells until Table 5 is completed. Using Excel, only one formula in Table 3 needs to be typed – for age group 15–19 in 2006 – it can then be copied to all the other age groups.

To calculate the projected total births to women aged 15–19 in the five years to 2006:

- Find the average number of women in age group 15–19 during the interval 2001–2006, using figures for these years in Table 5: (610 294 + 653 325)/2 = 631 809.5.
- Multiply the average by the ASFR for age group 15–19: 631 809.5 × 0.050 315 = 31 789.5. This is an estimate of births based on the average number of women present in the interval 2001–2006.

- Since the above figure represents the average births in one year, multiply the result by 5 to obtain the estimated births in the five year interval 2001–2006: 31 789.5 × 5 = 158 948.

Thus is derived the estimated total births to the 15–19 age group in the five years. The procedure takes account of changes in the number of women of fertile age through time and is the same for every fertile age group (Pollard et al. 1990: 121). In Excel, the above three steps are accomplished using a single formula, which can be copied to the other age groups and other columns. The total births for each interval are then obtained by summing the figures in the columns. The sex ratio at birth is used later (in Table 5) to estimate how many of the total births are female.

Net migration: projected survivors (Table 4)

Whereas Tables 2 and 3 provide data on the projected effects of mortality and fertility on the population, Table 4 provides the net migration data as follows:

- Decide the total net migration for each quinquennium, and type the figure at the top of each column. In the example, an annual net migration of 50 000 females produces a net gain of 250 000 every five years. Although, for clarity, Table 4 shows constant net migration, the figures at the top of each column could be varied.

  Also, in this example, the volume of net migration refers to that of persons aged five years and over at the end of the period; migrant children born in the interval are treated here as part of natural increase. Table 3 includes births both to the existing population and to migrant women.

  Net migration, expressed in absolute numbers, is commonly used as the migration measure in projections. This is because migration can vary greatly through time and is difficult to forecast: there is often insufficient information to justify more detailed assumptions about future inward migration, outward migration or age-specific migration rates. The use of age-specific net migration rates, moreover, can lead to errors because such rates are not based on the population at risk of migrating (Rogers 1990, cited by Plane and Rogerson 1994: 192).
- The total net migration is distributed across the age groups by multiplying the total net migration for the period by the proportion in each age group. Thus for age group 0–4 the calculation is: 250 000 × 0.126 366 = 31 591.5.
- Finally, the estimated migrants in each age are multiplied by the survival ratio of migrants (see Table 1, column E) to obtain the numbers surviving at the end of the interval. For age group 0–4, the working is 31 591.5 × 0.998 31 = 31 538. Since they were five years older at the end of the interval, the result is written beside age 5–9 in Table 4.

In Excel, only the first formula in Table 4 needs to be typed; it can then be copied to all the other age groups. This provides a straightforward approach to projecting net migration; a more detailed analysis could estimate migrant births separately (see Pollard et al. 1990: 122).

An important exception to calculating the survivors of net migration arises when the net migration in an age group is negative, because any deaths among the net outward migrants occur elsewhere. In this case, the survival ratios are not applied and the population loses all of the net migrants. A simplified alternative to treating positive and negative net migration differently is to dispense with using the survival ratios of migrants, instead just adding or subtracting the net migrants for each age group at the end of each interval: this would assume that the net migration figures refer only to survivors at the end of the interval.

Population projections (Table 5)

The preceding operations now make possible the calculation of the age structure of the projected populations. The base population is entered in the first column of data in Table 5. Only three steps remain:

- Derive the projected female population aged 0–4 in 2006 from the total births for the five years to 2006 in Table 3. First multiply the total births by 100/205 to estimate the number of female births; this assumes that the sex ratio at birth is 105 : 100. Next multiply this figure by the survival ratio of female births in Table 1 (i.e. cell D12): 1 395 676 × (100/205) × 0.979 95 = 667 167. The result is the projected number of females aged 0–4 in 2006. The formula can now be copied across to obtain all the other values for age group 0–4, using dollar signs to keep the reference to the survival ratio constant.
- Obtain the projected population aged 5–9 in 2006 by adding together the corresponding cells in Tables 2 and 4. Thus the projected population aged 5–9 in 2006 is: 613 937 (survivors from Table 2) + 31 538 (survivors from net migration in Table 4) = 645 475.
- Copy the formula for age group 5–9 down and across to complete the entries for age groups to 75+. The numbers aged 70+ in Table 5, which are needed for calculating Table 2, are obtained by adding together the figures for ages 70–74 and 75+. The total figures in the last row must exclude the row for 70+ to avoid double counting.

Since, as the flow chart illustrates, there are many linkages between the different tables, the final results are available only when all the required cells contain formulas. In principle, the entire set of projections could be calculated manually, but the process is extremely laborious and inefficient compared with the ease of using a spreadsheet. As many columns as wanted can be copied across the tables to extend the projections into the future. The notes on Spreadsheet Exercise 12 provide further assistance in recreating this example. Other data, including a different base population and different assumptions about fertility, mortality and migration, can then be entered in the spreadsheet to create new projections with relatively little additional work. A useful addition to the basic set of projections is another table presenting summary information on births and deaths in each period, crude birth and death rates and indices such as sex ratios, dependency ratios and growth rates.

12.5 Elaborating the basic cohort component model

Projections for both sexes

Having constructed the projection model for the female population, projections for the male population are readily obtained by copying the whole spreadsheet and modifying the base data and some of the formulas. Even if a projection is wanted only for the total population, it is necessary first to project each sex separately. This is to take adequate account of differences in the numbers of males and females in each birth cohort and differences in their mortality and migration.

Thus, having made a copy of the spreadsheet shown in Box 12.1, replace the data for females with the following age-specific data for males: the base population (Table 5), the survival ratios (Table 1) and net migration (Table 1). The fertility data in Table 1 are not needed this time, as they are already available in the female spreadsheet.

Tables 2 and 4 can remain the same, although it is necessary to ensure that the formulas in them now refer to the statistics for males. The volume of net migration in Table 4 may need to be modified and the age distribution of male migrants may differ, for instance because of the sometimes greater mobility of young single males, or simply because, in family migration, husbands are often older than their wives.

Table 3 remains identical to the corresponding table in the female spreadsheet, since the number of births is always estimated from the female population. Accordingly, set each cell of Table 3 equal to the corresponding cell in Table 3 of the female spreadsheet.

Table 5 also remains the same as before, except that the formula for ages 0–4 must be modified to take account of the number of male births. Replace 100/205 in the formula for ages 0–4 with 105/205: this will give the number of male births from the totals in Table 3.

Varying the projection assumptions

While the basic projection model has constant fertility, mortality and migration, more varied assumptions can be introduced by replacing Table 1 with separate tables for each of the components of change, specifying different rates for each period (see Hinde 1998: 212–215). These are often set by assuming that they vary consistently from one period to another:

- The projected survival ratios may be drawn from a series of life tables wherein life expectancy at birth changes by specified amounts every five years. Alternatively, it might be assumed that the age-specific survival ratios change at constant rates over the projection period, for instance at the same rates they changed during the five years before 2001.
- The volume of net migration may be adjusted in the existing model by altering the figures at the top of each column in Table 4. The future volume of net migration might be set either on the basis of policy goals or with reference to past experience of changes through time. The age structure of net migration would be left constant if there was no evidence of long-term change.
- The fertility assumptions might include a changing pattern of age-specific fertility, based on observed changes in past years. Alternatively, the age pattern of fertility could be left constant while the total fertility rate is varied.

  Age-specific fertility rates can be scaled up or down to produce particular total fertility rates in each period. This entails multiplying each ASFR by the ratio of the required TFR to the observed TFR from the original figures. For instance, in Table 1 of the basic model, the ASFRs produced a TFR of 2.074 506 (cell C28). To produce a set of ASFRs with a TFR of 1.80, multiply each of the original ASFRs by 1.80/2.074 506. The same procedure can be used to scale ASFRs to produce required net or gross reproduction rates.
- Further general strategies for establishing projected demographic rates include: (1) assuming constant rates if there has been little recent change or if the projection period is short; (2) establishing 'target' rates for future years, for instance from extrapolations of past trends; (3) assuming the rates for component areas will converge towards those of a larger or national population (Murdock and Ellis 1991: 229–230).

Demographers usually produce a series of projections from the same base year rather than a single forecast. This reflects the inevitable uncertainty about future developments and the low likelihood that a single 'best guess' will coincide with actual trends for long. Thus the projection assumptions may be varied to produce high, medium and low variants to describe the range of uncertainty in the future. Different combinations of high, medium and low assumptions about fertility, mortality and migration can create many projections, which will probably have to be reduced to a smaller set that better defines the upper and lower limits of anticipated developments.

The versatility of the cohort component method enables it to be applied to projections of many other populations such as urban and rural dwellers, persons belonging to organizations or enterprises as well as diverse sub-groups within society – the labour force, ethnic groups, pensioners and welfare recipients. It is also the principal approach to projecting populations of states, provinces, regions, cities and small areas within cities. This entails some extra considerations and the possibility of alternative approaches, as discussed in the next section.

BOX 12.2 **Module on cohort component projections (Cohort Component Projections.xls)**

This module produces age-specific projections for any base population entered by the user. It also includes a range of national statistics for experimentation. The projections may be printed, or viewed as age–sex pyramids or tables. The projection model is flexible in that the projection assumptions can be varied quite readily, either by entering data or by generating patterns and trends using in-built demographic models. The design of the module is also adaptable, since anyone familiar with Excel could extend or modify the spreadsheets to suit their own purposes.

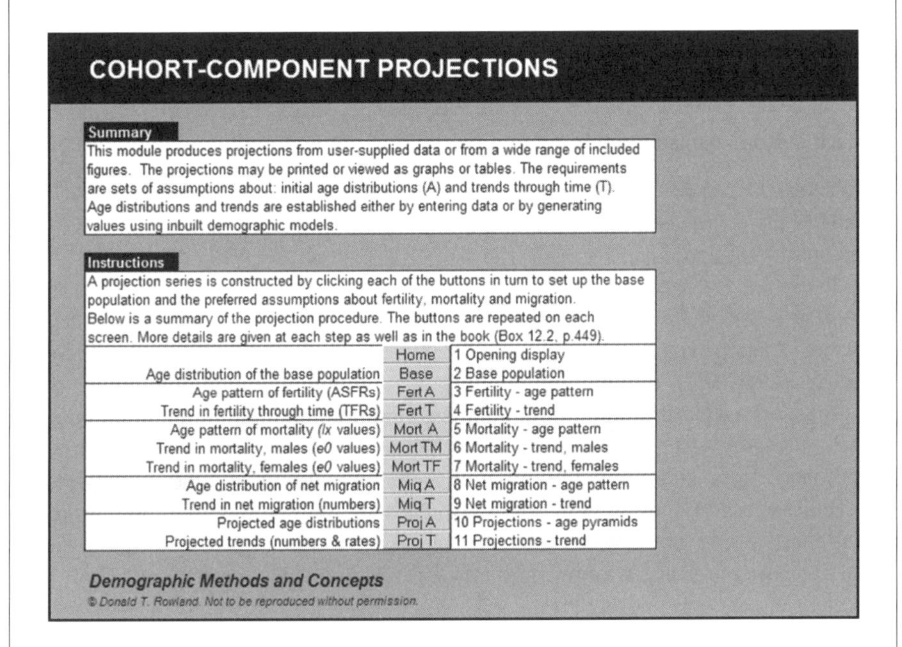

The inputs for the module are initial age distributions (of the base population, births, deaths and net migration) and trends through time (in levels of fertility, mortality and migration). Age distributions and trends are established either by entering data or by generating values. Buttons for changing age distributions and trends are labelled '. . . A' and '. . . T' respectively.

Projections are produced by working through the series of steps indicated by buttons on the opening display. The buttons are repeated throughout the module to enable quick access to different parts of the projection. Each button takes the user to a screen where data are entered or generated. The ensuing discussion explains the steps entailed in using the module.

continued

BOX 12.2 continued

Home (Screen 1)

The opening display summarizes the sequence of steps in projecting a population, as accessed through the series of on-screen buttons.

Base population (Screen 2)

Click button labelled 'Base'

To enter your own data for the projection, select 'User Data' from the top of the menu and type the base year, area name and age data in the *yellow* cells. Data entry must be done carefully, to avoid corrupting or overwriting formulas in other (white) cells. A graph of the data is displayed next to the table. Alternatively, choose a base population from the drop-down menu.

Fertility – Age pattern (Screen 3)

Click button labelled 'Fert A'

To enter your own figures, choose 'User Data' from the menu and type age-specific fertility rates in the yellow cells. These rates are converted to percentages and displayed on the accompanying graph.

To use a model age pattern instead, choose 'Model Data' from the menu and click the 'Change Pattern' button. In the displayed menu, select either the 'high fertility' or 'low fertility' standard pattern, and choose the values of alpha and beta that produce the required age pattern of fertility.

When alpha = 0 and beta = 1, the age pattern is the same as that of the chosen standard pattern. Higher values of alpha shift fertility to older ages, while lower values shift fertility to younger ages. Beta affects the spread or concentration of values. Experimenting with different values, and seeing them displayed in the accompanying graph, will demonstrate the effects of changing alpha and beta. The model age patterns of fertility are generated by a Gompertz relational fertility model (see Newell 1988: 175–178).

Fertility – Trend in total fertility rates (Screen 4)

Click button labelled 'Fert T'

The trend in fertility over the projection period is established either by entering your own data or by using a model to generate the data. To enter your own figures, select 'User Data' from the drop-down menu and, in the yellow cells, type the total fertility rates that apply to each five-year step in the projection period.

Alternatively, select 'Model Data' from the menu and click the button labelled 'Change Trend'. The displayed menu enables the user to select the TFR in years 0, 50 and 100. The spreadsheet plots a curve through these three points to obtain values for the other years. The graph illustrates the trend in the TFRs through time.

continued

BOX 12.2 continued

Mortality – Age pattern (Screen 5)

Click button labelled 'Mort A'

The age pattern of mortality is specified by a set of l_x values for males and females, either specified by the user or selected from the menu.

To type in your own values, select 'User Data' from the top of the drop-down menu and enter l_x values in the yellow cells (e.g. use l_xs from a life table for the base year).

Alternatively, if appropriate, use a set of values from one of the West model life tables listed in the menu. The survival curves for males and females are displayed in the accompanying graph. The spreadsheet calculates other life table functions from the l_xs, using the methods described in Chapter 8. While such methods do not exactly reproduce life expectancy at birth as published in the West model life tables, they provide a close approximation: the maximum error in the examples in the menu is about 0.06 years, or around 22 days of life expectancy.

The male and female l_xs in Screen 5 serve as standard values from which a series of life tables, denoting changes through time, are calculated in Screens 6 (males) and 7 (females).

Mortality – Trend in male life expectancy (Screen 6)

Click button labelled 'Mort TM'

The spreadsheet uses the l_xs in Screen 5 to calculate further life tables for every five-year step in the projections. The spreadsheet accomplishes this by calculating *relational model life tables*, which were discussed in Chapter 9 (Section 9.3) and are the subject of another computer module (see Box 9.2).

Life expectancy at birth for males is shown in the table under the heading 'Selected Data'. To modify these figures, click the 'Change Trend' button and proceed as follows:

- Select the menu box labelled 'Intermediate year for plotting', which is any year between 5 and 95. As in Screen 4, the spreadsheet needs three points through which to plot a curve, this time denoting the trend in male life expectancy through time. In Screen 4, the intermediate year was fixed at year 50, but in this menu more control is provided by enabling the user to choose the intermediate year through which the curve passes. If undecided, choose year 50.
- For Year 0, set the values of alpha and beta. Alpha here denotes the level of mortality, while beta denotes the age pattern of mortality. Setting alpha to 0 and beta to 1 will produce the same life expectancy at birth as in the standard population in Screen 5 – this will normally be the required situation for Year 0. Male life expectancy for Year 0 is displayed in the table under the heading 'Selected Data'.
- For the Intermediate Year, small adjustments to alpha will establish a trend towards an increase or decrease in life expectancy. Beta may also be adjusted if changes in the age pattern of mortality are considered likely. The range of possible values is shown below (see Newell 1988: 157):

continued

BOX 12.2 continued

Values of alpha	
–0.01 to –1.50	Higher life expectancy
0.0	Life expectancy the same as the standard
0.01 to 0.80	Lower life expectancy

Values of beta	
0.99 to 0.60	High infant and child mortality, and low adult mortality, relative to the standard
1.0	The relationship between adult and child mortality remains the same as in the standard life table
1.01 to 1.40	Low infant and child mortality, and high adult mortality, relative to the standard

Often, only minor changes to the values for year zero will be sufficient; extreme changes, especially those producing unlikely combinations of alpha and beta, will generate unrealistic results, as will be evident in the graph showing the survival curves.

- Adjust the alpha and beta values for Year 100 in the same way. The graph and table will show the trend in male life expectancy at birth.

Mortality – Trend in female life expectancy (Screen 7)

Click button labelled 'Mort TF'

The procedure for setting the trend in female life expectancy is the same as described for males. Also, the intermediate year for plotting the curve is the same as males; hence there is no option for altering this on the menu that appears after clicking 'Change Trend'.

Net migration – Age structure at the end of the period (Screen 8)

Click button labelled 'Mig A'

The age pattern of net migration, at the end of each period, is expressed as the percentage of the total net migrants (males plus females) in each age–sex group. To use your own data, select 'User Data' from the menu and type the percentages in the yellow cells.

Just as there are model fertility and mortality schedules, there are also model migration schedules (Rogers and Castro 1981). However, since migration schedules are more complex – requiring between 7 and 11 parameters to define them – they have not been included in the module. Instead, the menu simply provides access to some basic age distributions that might be used for experimentation in place of entering your own data.

continued

BOX 12.2 continued

Net migration – Trend in annual volume (Screen 9)

Click button labelled 'Mig T'

As before, to use your own data, select 'User Data' from the menu and enter the annual volume of net migration in the yellow cells. Alternatively, choose 'Model Data' from the menu and click the 'Change Trend' button. The menu provides a broad range of choices of the annual volume of net migration in years 0, 50 and 100. A curve is fitted through these values and other values are calculated, as displayed in the table and the graph.

The menu values are intended for illustrative purposes only and will not suit all needs. To change the menu values to match your own requirements, replace the figures in cells A844 to A854 of the 'subroutines' worksheet – the new values will then appear in the drop-down menu.

Population projections – Age distributions (Screen 10)

Click button labelled 'Proj A'

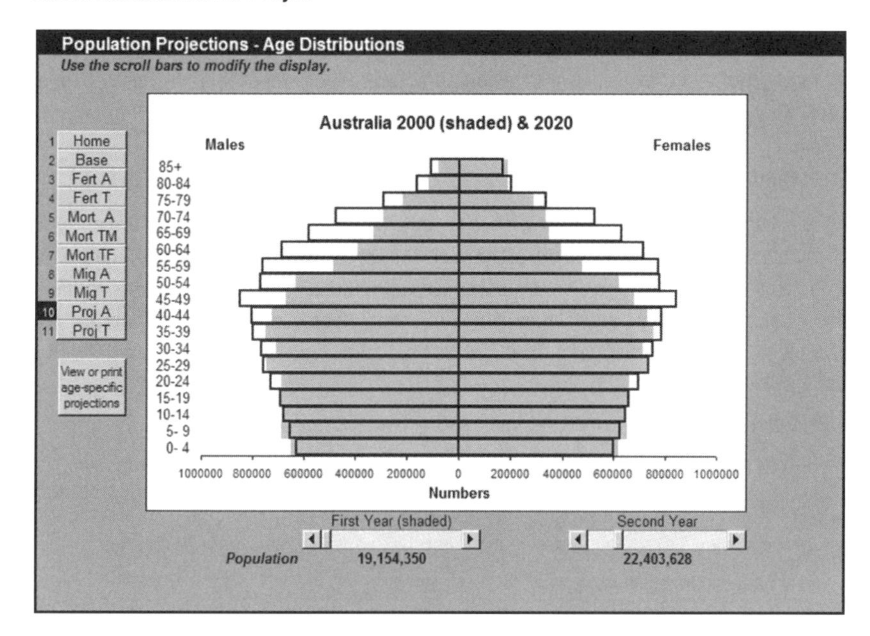

The projections arising from the specified base population and all the assumptions can now be viewed as age–sex pyramids in Screen 10. The display consists of two superimposed graphs, for instance comparing the age structure in the initial year with the age structure at a later date. The display is changed using the scroll bars at the bottom of the display.

To study the complete set of age-specific projections, click the button labelled 'View or print age-specific projections'. The table lists figures for males, females and the total population by age for all of the years. The table, or parts of it, may be printed by using the Excel menu items: File Print. Click a button at the top of the spreadsheet to go back to the previous display.

continued

BOX 12.2 continued

Population projections – Trends through time (Screen 11)

Click button labelled 'Proj T'

The final display is a summary table listing the total population in each year together with details, for the previous five years, of:

- Annual growth rates
- Average annual increase
- Total fertility rates
- Net reproduction rates
- Male and female life expectancy at birth
- Annual net migration
- Annual births, deaths and natural increase
- Crude birth and death rates.

Overall, the table summarizes the projections, the major assumptions and broad outcomes in terms of components of growth and demographic rates. To print the table use the Excel menu items: File Print.

Conclusion

When using a program for population projections, a common problem is that the program does not meet all the specific needs of the user. The spreadsheet approach developed in this module can be modified for different purposes, without calling for specialist knowledge of a programming language. No section of the projection module is hidden or inaccessible, and users with a good knowledge of Excel will be able to adapt it as required. Some potential changes include:

- Extending the age range to higher ages, and/or using other age groups.
- Allowing for changes in the age schedules of fertility and migration through time.
- Calculating the fertility of migrants separately.
- Superimposing a third age structure in Screen 10 to display a sequence of changes.
- Using alternative fertility models.
- Using alternative methods of calculating life tables.
- Using alternative methods of calculating trends through time.
- Including further indices, such as dependency ratios and momentum, in the summary of the projections in Screen 11.
- Enabling the model to generate projections for many areas simultaneously.

12.6 Further methods for sub-national populations

Issues concerning small area projections

Although the cohort component method is used for projections at different geographical levels, it can require supplementation, especially for small areas. Small area populations are subject to commonplace events that nonetheless have dramatic consequences on a local scale, such as the demolition of old houses for new housing developments, or for roads, car parks, offices or shops. The statistically tidy and regular sequence of changes assumed in the cohort component method cannot encompass abrupt and irregular external influences of this type. It is essential to incorporate into projection work knowledge of the local area and prospective developments, even though this may not conform to a standard statistical method.

Consequently, a combined or hybrid approach is often pursued, exploiting the strengths of particular methods for projecting different sections of the population. The 'housing unit method', discussed later in this section, is one approach used in conjunction with others, because it specifically takes into account changes in the nature and size of an area's housing stock. It is therefore useful for projecting the population of districts undergoing urban redevelopment as well as anticipating the future population in 'green field' areas where rural or vacant land is to be converted to residential uses. In practice, the mixing and variation in methods for small area estimates is considerable. Simpson et al. (1997: 272) found, from a survey of small area estimates and procedures used in Britain, that 48 producers of estimates derived their figures in 48 different ways, although many of the differences were matters of detail.

Many agencies and organizations compile population projections, which they use as a basis for their own planning or negotiations for funding. The projections usually refer to the organizations' own areas of interest, and consistency with national projections may not be explored. Contradictory projections arise when local governments or organizations produce their own figures and seek to benefit from optimistic forecasts in order to attract employment and investment, or obtain extra funding from state or federal governments. The inflation inherent in sub-national projections may be intended as a self-fulfilling prophesy, but becomes a source of confusion when seeking to understand the aggregate situation. A United States federal agency once estimated that all the local population forecasts in the country summed to a population tens of millions higher than the highest national projection (Isserman and Fisher 1984: 32). In the United States, the impetus for exaggeration is the many billions of dollars of federal funds allocated to states and localities on the basis of population estimates (Plane and Rogerson 1994: 127).

Aggregation and disaggregation

Broadly speaking, there are two types of approaches to producing consistent national and sub-national projections: aggregating projections for smaller areas or groups to produce an overall projection or, the reverse – disaggregating an overall projection. Applying the aggregation approach, sub-areas are projected individually, then the assumptions may be modified until they produce results consistent with figures projected independently for the total population. In Australia, projections for states are produced separately and subsequently adjusted to match the national figures. A similar procedure is used in Britain (Wood et al. 1999). Projecting the national population independently is advisable as a basis for evaluating all results. Global projections, and projections for broad regions of the globe, however, may be produced by aggregating country projections (Bongaarts and Bulatao 1999: 518). The ratio method and the matrix methods, discussed later, offer other strategies for generating consistent data for interrelated populations and their aggregate total.

The second broad approach to sub-national projections, disaggregation takes a national or total population projection and partitions it into its component sub-populations. This procedure is implemented without detailed assumptions on components of change in the sub-populations, such as through assuming that each maintains its share of the whole population, or its share changes according to a specified trend. Disaggregation may be carried out using the ratio method, which readily produces projected totals for each sub-population.

The ratio method

The ratio method has the advantage of needing only minimal information in order to generate projections for geographic sub-divisions of a country, region or city. The disadvantage is that the projections may not be consistent with the detailed components of population growth. The ratio method assumes either that the proportion of the total population living in each sub-division remains constant through time, or that it varies according to a predefined pattern (Shryock and Siegel 1973: 794; Murdock and Ellis 1991: 213–214). The projected total population is then partitioned using the projected proportions.

Table 12.2 illustrates an application of the ratio method for hypothetical regions:

- Part (a) of the table shows the observed percentages of the total population in each region in 1990 and 2000. From these figures the total change in the percentages is calculated together with the annual average change during the ten years. The North's percentage of the total population declined from 40 to 37 per cent, giving an annual average change of minus 0.3 per cent. The sum of the changes across all regions is 0.
- Part (b) presents the projected percentages in the regions to 2025. These are calculated from the year 2000 data, annual average changes in Section (a) and the

Table 12.2 An example of the ratio method of population projection

(a) Changes in the percentages of the total population in four regions 1990–2000

Region	1990	2000	Change	Annual average change
North	40.0	37.0	−3.0	−0.30
South	30.0	31.5	1.5	0.15
East	20.0	21.0	1.0	0.10
West	10.0	10.5	0.5	0.05
Total	100.0	100.0	0.0	0.00

(b) Projected percentages of the total population in each region

Region	2000	2005	2010	2015	2020	2025
n:	0	5	10	15	20	25
North	37.0	35.5	34.0	32.5	31.0	29.5
South	31.5	32.3	33.0	33.8	34.5	35.3
East	21.0	21.5	22.0	22.5	23.0	23.5
West	10.5	10.8	11.0	11.3	11.5	11.8
Total	100.0	100.0	100.0	100.0	100.0	100.0

(c) Projected populations

Region	2000	2005	2010	2015	2020	2025
n:	0	5	10	15	20	25
North	370 000	391 949	414 458	437 407	460 644	483 979
South	315 000	356 066	402 268	454 231	512 652	578 314
East	210 000	237 377	268 179	302 820	341 768	385 542
West	105 000	118 689	134 089	151 410	170 884	192 771
Total*	1 000 000	1 104 081	1 218 994	1 345 868	1 485 947	1 640 606

* Total population growing at an annual geometric growth rate of 2 per cent.

values of n (the number of years since the base year). Thus the percentage of the population in the East region in 2015 is: $21.0 + (0.10000 \times 15) = 22.5$.

- Part (c) uses these percentages to split the projected total population into figures for each region. The South's population in 2020 is: $1485947 \times (34.5/100) = 512652$.

The ratio method is also applicable to projections by age and sex, but complex adjustments to intermediate calculations are necessary to achieve consistent results (see Shryock and Siegel 1973: 794–5).

The housing unit method

The housing unit method has been one of the more popular alternative approaches to projection and estimation for small areas, alone or in combination with other methods (see Murdock and Ellis 1991: 186–191; Batutis 1993; Smith and Mandell 1993). The housing unit method offers the advantage of requiring data only on dwellings in an area. Through combining information on existing dwellings and projected dwelling commencements and demolitions, the housing unit method produces projections of population totals. Statistics on dwelling numbers may derive from building permits and construction records, electricity connections, field counts of the numbers and types of dwellings, or even from aerial photographs and satellite imagery of settled areas. The population of an area is estimated as:

$$\begin{aligned}\text{Total population} = {} & (\text{number of dwellings} \times \text{proportion occupied} \\ & \times \text{average number of persons per occupied dwelling}) \\ & + \text{numbers in other accommodation}\end{aligned}$$

The proportion of private dwellings that are occupied can be difficult to determine without detailed fieldwork or access to administrative records on electricity supplies – which are usually disconnected when houses become vacant. The average number of persons per occupied dwelling should preferably be calculated with reference to each type of dwelling rather than the total. Thus the bracketed portion of the above equation is repeated for different housing types and the separate results are combined.

Other complicating factors include the age of the dwelling and life cycle variability. The number of people living in single family dwellings – built for parents and their children – inevitably changes over the family life cycle, especially as the children leave home. Single family dwellings that are 10 years old are likely to have peak numbers of occupants, while those that are 20 years old will have declining occupancy. Similarly high levels of mobility and the diversity of contemporary life cycle experience – including staying single, temporary 'living together' arrangements, childlessness and divorce – introduce greater variability into household composition within dwellings of the same type. Hence the housing unit method relies on averages rather than any clear relationship between the type of dwelling and the type of household.

Depending on their purpose, population estimates from the housing unit method may only include usual residents living in their own or rented accommodation. However, if the entire population, including visitors and other non-residents is to be estimated, figures will need to be compiled not only on the occupants of households, but also on the numbers in 'non-private' or 'group' accommodation – such as caravan parks, hotels, motels, hostels, boarding schools, convents, hospitals, military barracks and correctional institutions. The preferred definition of the 'resident population' will determine whether to include in the population estimate such people as students at their term-time addresses, military personnel and patients or inmates in institutions. In the above equation, they comprise most of the 'numbers in other accommodation', or 'group quarters'.

Other indicator methods

The housing unit method utilizes dwellings, and persons per dwelling, as indicators of population size. Other 'indicator methods' utilize electoral rolls, or other directories and lists of residents, to estimate overall numbers. Simpson et al. (1997: 268–9) described the following three approaches to applying indicator methods in estimating small area populations in Britain, where electoral rolls have sometimes been the preferred basis for the procedure:

- *Apportionment* Here estimates are obtained for each sub-area within a district by assuming that each small area's share of the district population is equivalent to each small area's share of the total indicator population, such as people eligible to vote or total dwellings:

$$\begin{aligned}&\text{Estimated small area population}\\&\quad=\text{current district population}\times(\text{current small area indicator}/\\&\quad\ \ \text{current district indicator})\end{aligned}$$

  The method has been popular in Britain because it requires information only for the year of the estimate and boundary changes do not affect it. An estimate of the current population of the whole district is needed in advance.

- *Ratio change* The second procedure updates a past population by multiplying it by the ratio of the current to the past indicator. It requires information only for the small area concerned.

$$\begin{aligned}&\text{Estimated small area population}\\&\quad=\text{small area population at year }0\times(\text{current small area indicator}/\\&\quad\ \ \text{small area indicator at year }0)\end{aligned}$$

- *Additive change* The third procedure also uses information only for the small area in question, adding the change in the indicator variable to the initial numbers in the small area.

$$\begin{aligned}&\text{Estimated small area population}\\&\quad=\text{small area population at year }0+\text{change in small area indicator}\end{aligned}$$

  This method resembles the housing unit method when house completions are the indicator variable. Thus additive population change in private households may be calculated as house completions × persons per dwelling (ibid: 274).

In Britain, Simpson et al. (1997: 272) found that apportionment accounted for 27 per cent of the main approaches used by producers of estimates ($n=48$), while corresponding percentages were 33 for ratio change, 17 for additive change and 8 for the cohort component method, which proved to be the most accurate (ibid: 275). The remainder (15 per cent) consisted of local censuses and other methods. Identification of 'main methods' however, conceals the diversity of approaches used: as noted earlier, hybrid methods were typical.

The indicator method calculations were not usually age-specific, but this breakdown was obtained through partitioning totals with reference to the age distribution at the most recent census (ibid: 271). Supplementary sources for the estimates included statistics on school enrolments, which can provide up-to-date information about younger age groups. A data base of general practitioners' patients, by date of birth, sex and postcode, was noted as a future potential source with comprehensive coverage. Through comparisons between the estimates and the 1991 census, the authors established that, across more than 7000 areas, the estimates of the total populations (spanning 1981–1991) were, on average, 4.3 per cent from the true value. Figures by age were less accurate, especially for persons aged 15–24.

The multiregional projection model

For states, provinces and regions, some consider the 'state of the art' in projection methods to be the multiregional projection model. This uses the methods of matrix algebra to produce cohort component projections simultaneously for a number of sub-populations. Figure 12.3 illustrates a simplified multiregional projection model for a closed population sub-divided into three regions. The population is closed because there is no migration to or from places beyond the three regions. This model consists of:

- A row of cells (called a 'row vector') containing the initial populations, in the year 2000, of each region.
- A transition matrix, with three rows and three columns, showing the probabilities of being in each region in 2000 and in the same or another region in 2010. For example, the probability of being in region x at the start and end is 0.6667, and of being in region x at the start and region y at the end is 0.2000.
- A second row vector, consisting of three cells containing the projected population of each region in 2010.

The product of the vector of initial populations and the transition matrix is the projected population. In matrix algebra, each cell in the initial row vector is multiplied by the cells in the corresponding column of the projection matrix. The sum of these sets of results is the projected population for each region. In Figure 12.3, the projected population in region x is the sum of the product of the population vector in 2000 and the first column of the transition matrix. The projected population of the other two regions is found in the same way, using the second and third

Population in 2000			
x	y	z	
150	120	120	x

Transition Matrix				
Origin	Destination			
	x	y	z	Total
x	0.6667	0.2000	0.1333	1.0000
y	0.0417	0.7500	0.2083	1.0000
z	0.2083	0.0833	0.7038	1.0000

	Population in 2010		
	x	y	z
=	130	130	130

Projected Numbers in Region x			
150	x	0.6667	= 100
120	x	0.0417	= 5
120	x	0.2083	= 25
Population in 2010 =			130

Projected Numbers in Region y			
150	x	0.2000	= 30
120	x	0.7500	= 90
120	x	0.0833	= 10
Population in 2010 =			130

Figure 12.3 Simplified example of a multiregional projection.

columns of the transition matrix respectively. In the example, all three regions have the same population in 2010.

This approach seeks to take account of all interactions between areas and produce regional and total projections that are entirely consistent with one another. It overcomes the problem of studying sub-national populations in isolation, through making full use of the detailed data now available on migration and other components of change in more developed countries.

The multiregional projection model, however, is far from supplanting the cohort component method as the preferred approach of government agencies to preparing national and regional projections. The main reason is the much greater complexity entailed in constructing multiregional models since many extra assumptions are necessary, partly to specify shifts in the detailed components of population change in each area, and partly to bridge gaps in the available data. The procedure becomes more complicated when projections by age and sex are required and the number of regions increases. One of the largest multiregional projection models is for the 50 American states, using five year age groups and five-year time intervals (see Rees 1997).

References at the end of the chapter provide details of the construction of multiregional projection models. The article by Rees (1997) is a lucid introduction, which includes a summary of the main strategies for reducing complexities in such models. As well as identifying the interrelationships between population changes in different areas, the multiregional approach emphasizes the refinement of components of change – that is, examining sources of change in more detail. For methods that do not rely on matrix algebra, the refinement of components of change is itself a suitable direction for improving the data base for projections and deepening understanding of population dynamics.

Modelling using symptomatic variables

A further more specialized approach, for advanced study, seeks to construct a statistical model, especially a regression model, of the relationship between population growth and symptomatic variables, such as during the most recent intercensal period. Results from the model are then employed in projecting or estimating population size. While there is particular appeal in discovering variables that 'predict' population numbers, available data have never adequately specified the underlying causes of demographic change. Thus, as in the housing unit method and other 'indicator methods', the statistical modelling relies particularly on variables that change in association with overall numbers, such as dwellings, automobile registrations, medical insurance clients, school enrolments, births, deaths and tax returns (see Australian Bureau of Statistics 2002: 99; Bogue et al. 1993: V5 p. 20–3; Murdock and Ellis 1991: 184–203; Shryock and Siegel 1973: 802).

12.7 Conclusion

This chapter has discussed projections and estimates of the total population as well as figures disaggregated by age and sex. Chapter 13 discusses further projection techniques where the focus is particular aspects of population composition, such as households and families. Projections for planning and policy-making are an important practical application of demography. The cohort component method is essential knowledge in this area, because it is the predominant method for national and state projections. Moreover, study of the cohort component method promotes an in-depth appreciation of the means of incorporating information on observed changes into the projection process.

For small area projections, there is considerable diversity in the approaches used. This is not necessarily a disadvantage, provided that diversity reflects, firstly, efforts to make the best use of available statistics and local knowledge and, secondly, strategies to improve results for different segments of the population by applying different methods to each (Tayman and Swanson 1996: 527). Utilizing detailed knowledge of past trends and prospective developments prevents the tidiness of standard methodology, but avoids the inflexibility and artificiality of relying on a single method.

Finally, communicating to users the uncertainty inherent within forecasts, projections and estimates, is a necessary part of demographic work; otherwise false expectations will be created, leading to consternation if results prove unfounded.

Forecasts are more reliable for the less distant future, for larger populations, for populations growing slowly and for populations where migration is of least importance as a component of change. Refinement of approaches to specifying migration assumptions is a key topic to address in seeking to reduce uncertainty in sub-national projections and estimates, but one of the most intractable. Thus, as well as forecasting what seems the probable future, providing further information on the range within which numbers seem most likely to fall can enable planning to take account of uncertainty: the future is seldom what it used to be.

Study resources

KEY TERMS

Cohort component method of projection
Housing unit method of projection
Indicator methods of projection
Intercensal estimates
Multiregional projection model
Population estimates
Population forecasts
Population projections
Postcensal estimates
Ratio method of projection

FURTHER READING

Bongaarts, John and Bulatao, Rodolofo A. (editors). 2000. *Beyond Six Billion: Forecasting the World's Population*. Washington: National Academy Press. Appendix A on 'Computer Software Packages for Projecting Population' is available in the on-line version (http://www.nap.edu/catalog/9828.html).

Davis, H.C. 1995. *Demographic Projection Techniques for Regions and Smaller Areas: A Primer*. Vancouver: University of British Columbia Press.

Ghosh, M. and Rao, J.N.K. 1994. 'Small Area Estimation: An Appraisal'. *Statistical Science* 9: 55–93.

Haub, Carl. 1987. 'Understanding Population Projections'. *Population Bulletin*, 42 (4) (December).

Lunn, David J., Simpson, Stephen N., Diamond, Ian, and Middleton, Liz. 1998: 'The Accuracy of Age-Specific Population Estimates for Small Areas in Britain'. *Population Studies*, 52: 327–344.

Lutz, Wolfgang, Vaupel, James W., and Ahlburg, Dennis A. (editors). 1998. 'Frontiers of Population Forecasting'. Supplement to *Population and Development Review*, 24. New York: Population Council.

Murdock, Steve H. and Ellis, David R. 1991. *Applied Demography: An Introduction to Basic Concepts, Methods and Data*. Boulder: Westview Press, Chapter 5.

O'Neill, Brian and Balk, Deborah. 2001. 'World Population Futures'. *Population Bulletin*, 56 (3).

Rees, P. 1994. 'Estimating and Projecting the Populations of Urban Communities'. *Environment and Planning*, A26: 1671–97.

Rees, Philip. 1997. 'Problems and Solutions in Forecasting Geographical Populations'. *Journal of the Australian Population Association*, 14 (2): 145–166.

Smith, Stanley K., Tayman, Jeff, and Swanson, David A. 2001. *State and Local Population Projections: Methodology and Analysis*. The Plenum Series on Demographic Methods and Population Analysis. Norwell, MA: Kluwer Plenum.

Tayman, Jeff and Swanson, David A. 1996. 'On the Utility of Population Forecasts'. *Demography*, 33 (4): 523–528.

INTERNET RESOURCES

Subject	Source and Internet address
International database of population projections	**United Nations Population Division** http://esa.un.org/unpp/ See also: http://www.un.org/popin/wdtrends.htm
Population projections for the United States, together with information about them	**US Census Bureau** http://www.census.gov/population/www/projections/popproj.html
Population projections for Canada	**Statistics Canada** http://www.statcan.ca/english/Pgdb/People/Population/demo23a.htm
Population projections for the United Kingdom and component countries, together with information about the projections	**United Kingdom Government Actuary's Department (GAD)** http://www.gad.gov.uk/population/population.html
Reports and statistics on population projections for the world and for member countries of the European Union, includes a link to software for population projections	**International Institute for Applied Systems Analysis (IIASA)** http://www.iiasa.ac.at/Research/POP/docs/
Population projections for Texas, counties and other sub-areas; includes information on projection methodology	**Texas State Data Center** http://txsdc.tamu.edu/tpepp/txpopprj.php

Note: the location of information on the Internet is subject to change; to find material that has moved, search from the home page of the organization concerned.

EXERCISES

1 Obtain a copy of the latest population projections for your country of residence and write a summary and evaluation of the projection methodology.

2 Make a list of the sources of data needed to implement a cohort component projection for your country of residence.

3 Referring to the sources listed in your answer to question 2, construct a cohort component projection for your country of residence using the computer module described in Box 12.2. Assume that birth rates, death rates and the volume of net migration remain constant over the projection period, and print a copy of the summary table in Screen 11. Use model data, included in the module, wherever statistics are unobtainable, but explain your choices. For example, if information on the age structure of net migration is unavailable, use one of the age structures listed in the menu on Screen 8.

4 Discuss the nature of the information you would draw upon in order to produce a further cohort component projection that assumes a continuation of trends observed either in the most recent intercensal period, or in the period since the last census.

5 Obtain a copy of the latest population projections for the city, town or region in which you live, and write a summary of the projection methodology. In light of your reading on methods of population projection, outline any potential alternatives to, or enhancements of, this methodology.

6 With reference to the tables for the example of cohort component projections (Box 12.1) outline how to produce projections of the total number of deaths in each five-year interval.

SPREADSHEET EXERCISE 12: COHORT COMPONENT PROJECTIONS

This exercise is based on the Tables in Box 12.1. Re-create the five tables for the cohort component projections, but extend the projections to 2031. Type the headings and source data (the latter are shown in bold in Tables 1 and 5) and use formulas in all other cells containing numbers.

Introduction

1. There are many variations of the cohort component model, depending on the availability of data, the purpose of the projections and the selection of assumptions about fertility, mortality and migration.
2. The projection model in this exercise is a simplified example in that it is a one sex model (for females) and it assumes constant rates of fertility and mortality over time, as well as a constant volume of age-specific net migration. As discussed in Section 12.5, the model may be adapted to provide projections for both sexes or projections based on more flexible assumptions about the processes of change.
3. The model is arranged as a series of five interdependent tables (see Figure 12.2) consisting of:
   - Source statistics – components of change
   - Mortality
   - Fertility
   - Migration
   - Projections by age, together with the age structure of the base population.
4. This structure could be varied, but it has the advantage of grouping similar calculations, so that they can be copied from column to column, and there is ample space for extending the projections further into the future by adding more columns.
5. To begin the spreadsheet, type the source data and the row and column headings for all the tables as shown. For consistency in the calculations, the base population is included in Table 5 (which lists the ages of the population at every date) rather than

with the other source statistics in Table 1. Note that age group 70+ is included in Table 5 as the basis for estimating survivors at ages 75+, five years later.

6. Splitting the screen. Because of the size of the spreadsheet, you will be able to have only part of it on the screen at any time. To make it easier to refer to cells at the top of the spreadsheet, while working with the cells at the bottom, split the screen into two horizontal windows. To do this, put the mouse pointer on the small rectangle at the top of the right-hand scroll bar. Press the mouse button and drag the rectangle half-way down the screen. The screen will now be divided in two and you can work with each of the windows independently. To return to a single window, put the mouse pointer on the rectangle again and double click.
7. The notes which follow explain the calculation of each table in the spreadsheet. Work through the instructions in order.

Components of change (Table 1)

8. Calculate the age-specific fertility rates (for total births). Figures from the base population (Table 5) are the denominators. Expressing the figures as rates per woman rather than per 1000 women, saves having to divide the rates by 1000 in later calculations.
9. Calculate the survival ratios for migrants at each age, using $0.5(1 + S_x)$. These will be used later in Table 4.
10. Calculate the age structure of net migration (as proportions). Note that the model will include all births to migrants in the calculations for Table 3 – i.e. in this example they are placed in the births component, rather than the migration component.

Projected survivors (Table 2)

11. Use the five year survival ratios in Table 1 to calculate the number of survivors from the base population in Table 5 (year 2001). The survival ratio for births (the first figure in column D) is not required for Table 2 – it is used to estimate the survivors from total births in Table 3.
12. Copy the first formula across to the other columns, then down to all the other age groups, except 75+ (the formula for which has to refer to the population in the extra row of Table 5 for ages 70+). The projected survivors in 2011, for example, will be based on the projected population in 2006 aged five years younger. After entering all of the formulas, however, Table 2 will not be complete and correct until Table 5 is finished.

Projected survivors from net migration (Table 4)

13. Although Table 3 is next in order, it may be completed later, when needed information on the number of women in the fertile age groups becomes available in Table 5.
14. The model assumes an annual net migration of females of 50 000, or 250 000 over five years. The numbers in each age group are obtained by multiplying 250 000 by the proportion of migrants in the age group (column G of Table 1) and the corresponding survival ratio of migrants (column E of Table 1).

If net migration is negative, the survival ratios are not applied to the figures: the population loses all of the net migrants. This can be handled in Excel by using the function = IF (logical test, value if true, value if false), which will apply the ratio if the net migration is positive, but not if the net migration is negative. The logical test here is whether the net migration in the particular age group is greater than 0.

15. Copy the formulas to all the other columns in Table 4. In this example, the net migration of persons born in each period is included in the estimate of total births.

Projected population at ages 5 and over (Table 5)

16. Add together the figures on projected survivors and projected survivors from net migration, to obtain the total projected population in all age groups, except 0–4 years.
17. The projected figures for females in the reproductive age groups can now be used to calculate projected births.

Projected births (Table 3)

18. Calculate projected births using the method explained in Section 12.4. The annual average figure needs to be multiplied by 5 to produce a statistic for total births in a five-year period. The figure will include births to migrants and non-migrants.

Projected population at ages 0–4 (Table 5)

19. To the total projected births apply the survival ratio of births, and the sex ratio of births (100 female births per 205 total births), to obtain the final statistics for ages 0–4.
20. Add the columns in Table 5, omitting the extra age group 70+, to obtain the total projected population in each year.

Total population in 2031 (Table 5)

21. The projected total female population in 2031 is 11 634 342.

13 Population composition

The *Elements* of true Policy are to understand throughly the *Lands*, and *hands* of any Country.

(Graunt 1662: 15)

OUTLINE

13.1 Projecting population composition

13.2 Labour force projections

13.3 Projections of households and families

13.4 Summary measures of population composition

13.5 Multivariate measures

13.6 Conclusion

Study resources

LEARNING OBJECTIVES

To understand:

- projections of population composition, especially for the labour force, households and families
- key indices of population composition: the index of diversity, the index of segregation, the Gini index and the Lorenz curve

To become aware of:

- the wide range of projections of population composition
- multivariate measures of population composition

COMPUTER APPLICATIONS

Spreadsheet exercise 13: Experiments in data transformation (see Study resources)

This chapter discusses methods for the analysis of population composition that have particular relevance in planning and decision-making. Population composition necessarily encompasses diverse characteristics, including employment, educational attainments, family relationships and ethnicity. While conventional statistical methods widely serve as the basis for the study of population characteristics (see Shryock and Siegel 1973: Chapters 9–12), for some purposes, including projections, further techniques are necessary. The focus of this chapter is special purpose methods and indices that extend the analysis of population composition.

Most of the chapter is concerned with projections of population composition (for the labour force, households and families), and indices that measure concepts employed in understanding implications of population composition (diversity, segregation and inequality). Lifestyles of market segments and the well-being of societies are introduced towards the end of the chapter as illustrations of multivariate approaches which, although beyond the scope of this book, are important subjects for further reading.

13.1 Projecting population composition

Planning for the future of societies calls for many kinds of special-purpose projections to help anticipate needs and budgetary implications arising from changes in the representation of particular groups. These include school children, ethnic minorities, war veterans, nursing home residents and people with health conditions and diseases that have high cost implications, such as spinal injuries, coronary heart disease and dementia. Changes in the market for goods and services can also be anticipated through projections of client populations.

This section is concerned with projections where the focus is the 'supply' of people moving through the age structure, rather than the 'demand' for people with particular characteristics. Instances of the latter include prospective demand for certain occupations or educational qualifications – such as military personnel, doctors, lawyers, social workers and software engineers. Projecting 'demand' cannot proceed largely from current circumstances, but necessarily requires further assumptions about changes in the nature of the economy or the society (Webster 1992).

Sometimes, projections of 'supply' can be obtained readily from existing projections of the total population, by multiplying each age–sex group by the proportion belonging to the population in question. For example, multiplying the projected number of females aged 15–19 by the proportion of females of that age who are still at school will provide a national school enrolment projection for their age and sex, assuming constant enrolment rates (see Shryock and Siegel 1973: 851):

$$\text{Age-specific enrolment rate (females)} = \frac{\text{female enrolments aged } x}{\text{female population aged } x} \times k$$

where $k = 1$ or 100.

Although it is satisfactory to assume constant age-specific rates in some instances, in others it can be misleading because of the distinctive characteristics of particular birth cohorts, or because of age-related changes through time. Thus the proportion of men aged 60–64 who will be war veterans in the year 2020 cannot be expected to be the same as the proportion in 2000, because different birth cohorts have had contrasting experiences of military service. This is an example of *disordered cohort flow*, as discussed in Chapter 4. In such circumstances, cohort-specific rates become necessary.

To take account of age-specific, rather than cohort-specific, changes through time, rates for the initial year might be set from current data and those for the final year from data for the population of a country or region deemed to represent the anticipated future (Shryock and Siegel 1973: 842). Rates for other years can then be calculated from the initial and terminal figures, for instance by assuming linear change between them. Thus if an age-specific rate changed from 6 per cent to 10 per cent during the projection period 2000–2010, the rate for 2005 would be 8 per cent. In practice, however, there are many and varied approaches to devising assumptions about changes in age-specific rates through time. Also, as noted in the previous chapter, projections are normally produced not as a single series, but as a set with at least 'high', 'medium' and 'low' variants. Among the special purpose projections, of widespread interest are those for the labour force and households, both of which have major implications for economic planning and for the nature of the society.

13.2 Labour force projections

The simplest approach to labour force projections, as well as that most consistent with other projections, is to take figures for the total population produced by the cohort component method and calculate from them the projected population in the labour force. This is accomplished by multiplying the projected numbers in each age-sex group by an assumed labour force participation rate (LFPR) for each group. Using data for the base year, the rates themselves may be calculated, separately for males and females, as follows:

$$\mathrm{LFPR}_x = \frac{\mathrm{LF}_x}{P_x} \times k$$

where

LFPR_x = labour force participation rate at age x
LF_x = numbers aged x in the labour force
P_x = the total population aged x
k = 1 or 100.

The labour force participation rate specifies the proportion of the population in the labour force; it may be calculated for the total population as well as for individual age groups. Problems arise, however, in determining the membership of the labour force and, hence, in specifying accurate labour force participation rates. First-time job seekers need to be identified, as do the unemployed actively seeking work. The numbers of unpaid and voluntary workers may also be relevant (see Standing 1982).

One of the most important changes that Western countries have experienced in their labour force participation rates since the 1950s has been the rising participation of women, especially married women, many of whom have dependent children. The graph of labour force participation in the United States illustrates this (Figure 13.1). For men, the lower rates for 1999 compared with those for 1970 particularly reflect the impact of prolonged education at the start of working life,

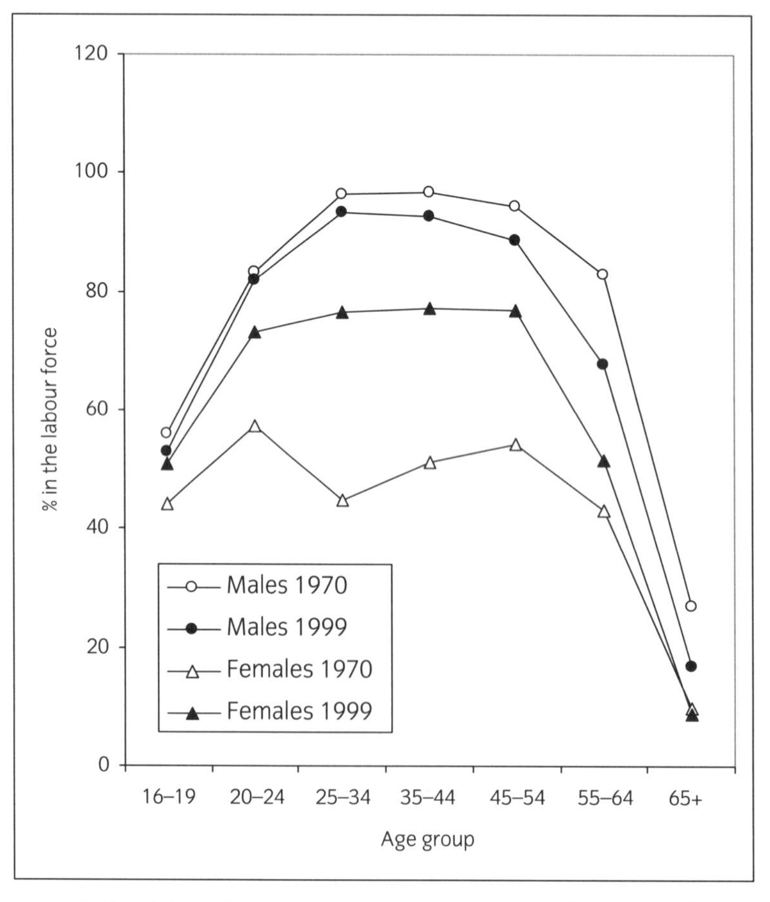

Figure 13.1 Civilian labour force participation rates by age and sex: United States 1970 and 1999.

Data source: US Census Bureau, *Statistical Abstract of the United States: 2000.*

and of early retirement towards its end. Coupled with longer life expectancy, late entries and early retirements imply that many males now spend less than half their lives in the labour force.

Some potential refinements for projections include varying the participation rates over time in light of recent trends, and distinguishing between full-time and part-time workers, as well as the employed and unemployed. Proportions may be used to sub-divide the projected labour force into two or more groups. Labour force projections are also generated from elaborate models that integrate demographic and economic data to provide more detailed analyses of changes in labour supply and demand (see Murdock and Ellis 1991: 217ff).

13.3 Projections of households and families

Projections of households and families similarly call for special consideration, since their numbers by no means depend on population totals alone: additional major influences are the age structure of the population and variations between cohorts in life cycle experience of marriage and childbearing. A significant trend has been the decline in average household size because of developments such as higher proportions remaining single or childless and rising numbers living alone, especially at older ages. In the United States, the average number of persons per household (total population in private dwellings/total households) declined from 3.1 in 1970 to 2.6 in 2000 and, over the same period, the representation of family households fell from 81 per cent to 69 per cent (Fields and Casper 2001: 2–4). Changes are tending to augment growth in the number of households (Figure 13.2), at the same time entailing less efficient use of the housing stock, for instance when lone individuals and couples without children occupy dwellings designed for families of four or more (see Section 10.4 for definitions of dwellings and households).

Separate projections of households and families therefore become necessary, all the more so because household numbers and types (family and non-family households) help to predict both the overall demand for housing as well the demand for particular kinds of dwellings. Additionally, types of households are important indicators of consumer demand for goods and services, since households are the major units for which incomes are earned and spent. Parents with young children tend to allocate relatively more of their income to everyday necessities – food, clothing, housing and transport – while others with no dependents are more likely to have greater discretionary income for luxury goods, dining out, entertainment, holidays and other optional expenditure.

Like projections of the total population, projections of households typically refer to the usual resident population. Hence, existing projections of the total resident population offer a convenient starting point. However, the resident population numbers have to be reduced to those living in private dwellings, omitting persons in non-private dwellings such as hotels, motels, hospitals, nursing homes,

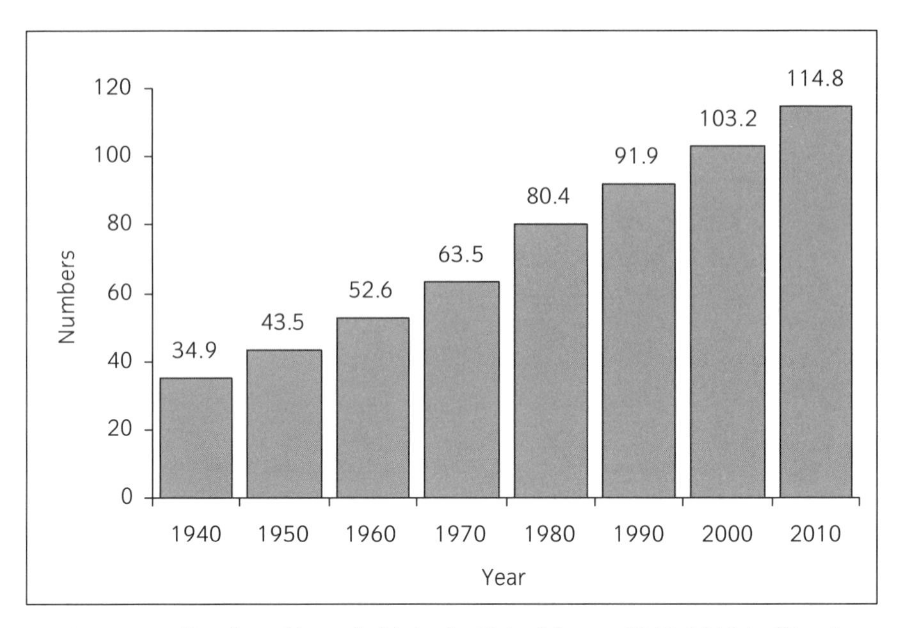

Figure 13.2 Number of households in the United States, 1940–2010 (millions).

Note: figures for 2000 and 2010 projected.

Data source: Day (1996: 6).

military establishments and other institutions. This may be accomplished by multiplying the projected resident population by the projected or current proportion in private dwellings, using age and sex-specific figures.

Projections of households and families, other than those based on complex models of economic and social change, proceed in two main ways. Firstly, like labour force projections they can originate from basic projections by age and sex, through assumptions about the proportions in each type of household or family. Secondly, they may derive from more detailed projections – by age, sex, marital status and other characteristics – to minimize the need for detailed assumptions about the proportions in each type of family or household, or changes in these through time. The best-known method, the *household headship method*, can employ either approach, while the 'living arrangements method' is founded on the former.

Household headship method

In it simplest form, the household headship method calculates the number of households from an existing population projection by age and sex, typically the 'medium variant' of a projection series, together with statistics on the proportion of household heads belonging to each age–sex group. Thus the household headship rates, as the proportions are called, are calculated, separately for each sex, as:

$$\text{Household headship rate} = \frac{\text{household heads aged } x}{\text{population aged } x \text{ living in households}} \times k$$

where $k = 1$ or 100.

Thus, if 30 per cent of women aged 25–29 were household heads at the most recent census, taking 30 per cent of the projected numbers of females in private dwellings in this age group would provide a projection of households headed by females aged 25–29. Combined with similar figures for males and other age groups, these data provide a projection of total households. Average household size for each year can then be obtained by dividing the total projected population in private dwellings by the projected number of households.

If the projection period is short, the headship rates may be kept constant to avoid having to project the rates themselves by age and sex. Change is not ignored, however, since applying the rates to existing cohort component projections at least includes the effects of shifts in the size and age structure of the population.

Table 13.1 illustrates the household headship method for males in selected age groups. Starting with the projected resident population (column B), the projected resident population in households is calculated (column D). Multiplying the latter by the headship rate (column E) gives the projected number of households (column F). Compared with the resident population, the number of households may seem large in the example, but this is because males in their thirties and forties are usually householders. Projections for all age groups in the United States anticipate 2.5 or 2.6 persons per household in 2010 (Day 1996: 21).

Where spouses or partners jointly own or rent their accommodation, it is preferable for statistical purposes that just one of them be identified as the householder. For example, in the United States, 'The householder is the first adult household member listed on the questionnaire. The instructions call for listing the first person (or one of the persons) in whose name the home is owned or rented. Prior to 1980, the husband was always considered the household head (householder) in married-couple households.' (US Census Bureau 2000: 6).

It may be noted that *household head* is now an outmoded term, evoking controversy when two co-residents have claim to this status. In contemporary census

Table 13.1 Household projections by the household headship method (hypothetical data for a male population)

Age group A	Projected resident population 2020 B	Proportion of the resident population in households C	Projected resident population in households 2020 D D = B × C	Household headship rate E	Projected number of households 2020 F F = D × E
30–34	5 287 000	0.992 80	5 248 934	0.855 20	4 488 888
35–39	5 333 000	0.987 59	5 266 817	0.895 00	4 713 802
40–44	5 346 000	0.979 39	5 235 819	0.909 40	4 761 454
45–49	5 381 000	0.977 33	5 259 013	0.917 00	4 822 515

collections, the main purpose of identifying a 'head' is to facilitate the counting of households and to provide the starting point for classifying living arrangements in terms of the relationships of people within households. Accordingly, censuses are abandoning the term 'household head', instead using alternatives such as 'person 1', or 'householder', or 'household reference person' – that is, an individual with reference to whom it is appropriate to code relationships within the household. If a couple happened to list their newborn baby as 'person 1' on their census form, coders would substitute another reference person and classify the relationships as a 'couple with one child', instead of 'household head with parents'.

More detailed projections are available from the headship method when the rates include marital status and are calculated as the ratios of household heads to the total population of the same sex, age and marital status. The new rates are applied to projections similarly by age, sex and marital status (Shryock and Siegel 1973: 848; King et al. 2000). The rates enable the projection of the numbers of households headed by married couples, or by unmarried, separated, divorced or widowed individuals.

Methods for the preparation of base projections that include marital status are described in Shaw (1999), Rowland (1994a), Haskey (1988) and Espenshade (1985). An important reason for producing population projections by marital status is to facilitate household and family projections that take account of changes not only in population size and age structure, but also in marital status distributions. The 'divorce boom', for instance, that began in a number of countries in the 1970s, is transforming the living arrangements (household composition) of many people. Among the young, delayed marriage and consensual unions are similarly having a major impact on trends in household formation.

Nevertheless, inclusion of marital status still provides little information on the projected numbers of different types of family and non-family households, such as couples with children or people living alone. A solution is to employ an even more elaborate form of the headship method utilizing rates by age, sex, marital status and living arrangements, a procedure used in projections of households and families in the United States (Shryock and Siegel 1973: 849; Day 1996).

So far the discussion has assumed the use of constant household headship rates based on a recent census or survey. This can be satisfactory if changes are minimal, or if the projection period is short, or if a static scenario is wanted for comparison with dynamic projection scenarios that recognize changes. Incorporating change is often necessary even though family and household projections typically refer to less protracted projection periods (e.g. 15 years) than projections of the total population (e.g. 50 years), because of the greater uncertainty surrounding more detailed changes in population characteristics. In times of heightened family change, even 10 years can bring major shifts in household formation. The preparation of 'low' (e.g. constant headship rates), 'medium' (e.g. trends consistent with recent changes in headship rates) and 'high' variant projections in itself provides a better appreciation of the uncertainty surrounding future household numbers.

In taking account of change in household headship rates, the basic strategy is to identify trends in them from a number of censuses and surveys, then project future developments, for instance by assuming a linear trend through time. Assumptions about future trends, however, are preferably founded on an understanding of the forces affecting household formation and dissolution, rather than simply on mathematical extrapolation. The outcome of more flexible assumptions is a different set of rates for each year in the projection.

Living arrangements method

An alternative approach to household and family projections, termed here the 'living arrangements method', does not require household headship rates and is capable of generating quite detailed household and family projections in a reasonably straightforward way. *Living arrangements* denote the composition of households, as illustrated by the graph of living arrangements in the United States 1970–2000 (Figure 13.3).

The method is based on proportions of the population in different types of living arrangements. It resembles that described earlier for labour force projections in that projections of families and households are obtained by partitioning projec-

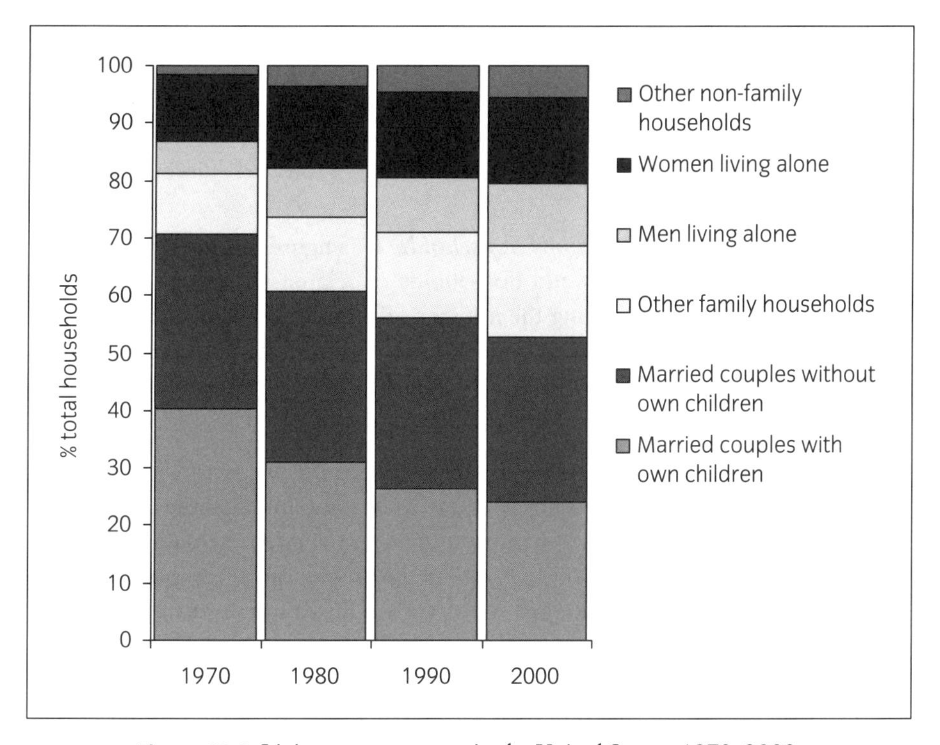

Figure 13.3 Living arrangements in the United States, 1970–2000.

Data source: Fields and Casper (2000: 3).

tions of the total population by age and sex into the required categories, although more steps are involved, as follows (Australian Bureau of Statistics 1999):

1. *Proportions in living arrangements* From census or survey figures (by age and sex), calculate the proportion of the population in each type of living arrangement, such as couples with children, couples without children, single parents and persons living alone.
2. *Projected numbers in living arrangements* Multiply existing projections of the total population (by age and sex) by these proportions, to obtain projections of the numbers in each type of living arrangement – assuming constant proportions in each type. The projections should refer to the resident population in private dwellings, to be consistent with census definitions of households and families as persons living in private dwellings.
3. *Projected numbers of families* From the living arrangement projections, calculate the projected number of families. These consist of couple families (with or without children), lone parent families and other families of related individuals. The number of couple families is half the number of partners in couple families; the number of lone parent families is the sum of the male and female lone parents; while the number of 'other families', consisting of related individuals, may be obtained by dividing the number of persons in such families by the average size of this family type at the last census: the figure for Australia in 1996 was 2.13 (ibid: 112).
4. *Projected number of family households* Multiply the number of families by the ratio of families to households, as shown at the last census – the ratio for Australia in 1996 was 0.98. This allows for households containing more than one family. The resulting figure represents the projected number of family households.
5. *Projected number of non-family households* This figure is obtained as the sum of lone person households and households of related individuals. The latter may be calculated by dividing the number of persons in households of related individuals by the average size of such households at the last census – which was 2.28 for Australia in 1996. Combining the projections for family and non-family households gives the projection for total households.

The procedure can be applied on the assumption that the proportions in each type of living arrangement (see step 1 above) remain constant into the future, or that the proportions vary, in accordance with past changes. Varying the assumptions requires that step 1 be repeated for each year in the projection. Techniques for introducing such variations are described in Australian Bureau of Statistics (1999: 36) and Day (1996: 18–19).

Overall, the 'living arrangements method' accomplishes the same objective as the 'household headship method', without relying on the availability initially of detailed projections by marital status, or marital status and household composition. Trends in living arrangements, however, result from the interaction of many other changes – including marriage. Projecting some of these influences separately provides greater opportunities to consider in detail the processes operating, potentially leading to more reliable results than those obtainable from a more aggregated approach. The shorter the projection period, however, the less the disadvantages become, and the stronger the arguments for minimizing complexities.

Population projections represent a nexus where knowledge of demographic processes and techniques complement each other. The full implementation of household and family projections usually requires the construction of a sizeable spreadsheet or elaborate computer program to perform the calculations. The cohort component projection module in Chapter 12 exemplifies a spreadsheet approach to such tasks (Box 12.2).

13.4 Summary measures of population composition

As well as projecting the future, demographic analysis of population composition is much concerned with current population characteristics. This work can benefit from the use of measures that answer specific questions about the data. An example of such a measure, described in Chapter 3, was the *index of dissimilarity*. This addresses the question of the overall difference between two or more populations in relation to their age distributions, occupational distributions, ethnic composition or other characteristics. The index of dissimilarity also has applications in comparing the same population at different dates. Other indices, based on the same principle are the *index of concentration*, the *index of redistribution* (see Chapter 10), as well as the *index of segregation*, to be discussed shortly.

This section introduces further indices that have a range of applications in analysing population composition. The first is the *index of diversity*, which addresses the question: how homogeneous or heterogeneous is a population in relation to a particular characteristic, such as language or religion. The second is the *index of segregation*, which measures the separation of groups in terms of where they live, that is, differences between their spatial distributions. The final measure is the *Gini index*, a measure of the extent of inequality between the characteristics of populations. The Gini index is widely used in the measurement of income inequality, and also has other applications similar to those of the index of dissimilarity, such as in measuring segregation. An advantage of the Gini index is that it summarizes information plotted as a *Lorenz curve*, itself a valuable graphical device for displaying differences.

Index of diversity

Turning first to the index of diversity within a population, its focus is relative heterogeneity: 'the position of a population along a continuum ranging from homogeneity to heterogeneity' with respect to one or more variables (see Teachman 1980: 341). This section discusses diversity within a population in terms of single variables. Variants of the index may be calculated for diversity between populations, and for two or more variables, as detailed in Lieberson (1969: 852ff).

The index of diversity is employed in the analysis of *nominal data* (Teachman 1980: 342), that is, data divided into discrete categories between which there is no gradation. An example is language, which has discrete categories such as English, French and German. Other examples include occupations, birthplaces, religions, and places of residence. The index is a versatile measure with applications that may extend to 'the structure of imports and exports for a country, the distribution of services between communities, the spatial diversity of industries, and the structure of a gross national product' (Teachman 1980: 341–342). For *continuous data*, such as age and income, conventional statistical measures of variation, such as the standard deviation, are preferred. (Continuous variables are those that can, theoretically, be measured to any degree of precision, such as in terms of time (age), money, height, weight or distance.)

In its standardized form, the index of diversity ranges from 0 (homogeneity – all members of the population are in one category) to 1 (heterogeneity – the population is evenly distributed through all categories). Accordingly, if the variable in question is religion and everyone in the population is Hindu, the index will be 0, whereas it will be 1 if the total population is evenly distributed between Hindu, Sikh and Muslim religions. The greater the number of categories represented, the greater the potential variation; recognizing different castes within the Hindu religion would change the assessment of the population's diversity.

The calculation and interpretation of the index is straightforward (Table 13.2), even though the derivation of the underlying mathematics is complex. The index of diversity within a population (A_w) is calculated by squaring the proportion in each category, summing the squares and subtracting them from 1 (Lieberson 1969; Bohrnstedt and Knoke 1988: 76):

$$A_w = 1 - \left[(x_1)^2 + (x_2)^2 + \ldots + (x_n)^2\right]$$

i.e.:

$$A_w = 1 - \sum x^2$$

The formula measures the probability that randomly paired members of a population will be different on a specified characteristic (Lieberson 1969: 851). The potential maximum value of the index is equal to 1 minus the reciprocal of the number of categories:

$$\text{Maximum } A_w = 1 - \frac{1}{n}$$

where n is the number of categories.

To facilitate comparisons, there is also the *standardized index of diversity*, which has a maximum value of 1 (Lieberson 1969: 860–861; Bohrnstedt and Knoke 1988: 76–77):

$$\text{Standardized index of diversity} = \frac{A_w}{\text{maximum } A_w} = \frac{A_w}{1-\frac{1}{n}} = A_w \frac{n}{n-1}$$

where A_w is the index of diversity, and n is the number of categories.

The standardized index describes the actual level of diversity as a proportion of the maximum level possible with the specified number of categories (Lieberson 1969: 860). As noted above, the standardized index is equal to 1 when the cases are evenly distributed through all categories of the variable, and equal to 0 when all cases are in a single category, as illustrated in Table 13.2.

The table shows four examples of the index of diversity where the variable (language) has three categories (English, French and German). The x values are the pro-

Table 13.2 Examples of the index of diversity (language diversity)

A. High index

Language	x	x^2
English	0.50	0.25
French	0.30	0.09
German	0.20	0.04
Total	1.00	0.38

Index of diversity = **0.62**
Maximum index = 0.67
Standardized index = 0.93

B. Low index

Language	x	x^2
English	0.05	0.0025
French	0.05	0.0025
German	0.90	0.8100
Total	1.00	0.8150

Index of diversity = **0.19**
Maximum index = 0.67
Standardized index = 0.28

C. Maximum index

Language	x	x^2
English	0.33	0.11
French	0.33	0.11
German	0.33	0.11
Total	1.00	0.33

Index of diversity = **0.67**
Maximum index = 0.67
Standardized index = 1.00

D. Minimum index

Language	x	x^2
English	1.00	1.00
French	0.00	0.00
German	0.00	0.00
Total	1.00	1.00

Index of diversity = **0.00**
Maximum index = 0.67
Standardized index = 0.00

portions speaking each language. In panel A, the index of diversity is equal to 1 minus the sum of the squares of the proportions speaking each language, i.e. 1.0 – 0.38 = 0.62, the potential maximum is 1.0 – (1/3) = 0.67, while the standardized index is equal to the index of diversity divided by the maximum index: 0.62/0.67 = 0.93. Panel B of the table shows the same calculations resulting in a low index, reflecting that the majority of the population in this example speak German.

Since the maximum index occurs when there are equal proportions in all categories, the proportions in Panel C are all 0.333 33. The index of diversity is 1.0 – 0.33 = 0.67, the maximum value, for a variable with three categories, is always 1.0 – (1/3) = 0.67, while the standardized index is 0.67/0.67 = 1.0. Panel D illustrates the minimum value of the index, arising where all members of the population speak English.

A disadvantage of the standardized index of diversity is that it can disguise real differences between populations. Imagine two populations with equal proportions in each of their language groups, resulting in a maximum index of 1 for both populations. However, if the first population had six language groups present while the other had three, the standardized indices would reveal nothing about the inherent greater diversity of the first population: a population with more groups is more diverse (see White 1986: 200).

Index of segregation

The index of segregation is another version of the index of dissimilarity (see Sections 3.4 and 10.3). It measures the extent to which an ethnic group differs in its spatial distribution from the rest of the population, or from another group in the population. Typical applications are comparisons of white and non-white populations, or immigrants and the native-born. The greater the dissimilarity between the spatial distributions of the two groups, the greater the ethnic segregation. The most extreme form of segregation is the ghetto, where a high proportion of a group live in a single location, usually to the exclusion of others. Segregation can arise both voluntarily – as a result of preferences to live among people with the same language, religion or way of life – and involuntarily, as a result of deprivation and discrimination.

There are various segregation indices, as reviewed by Taeuber and Taeuber (1965: 195–220) and White (1986), but that based on the index of dissimilarity is one of the more easily comprehended and widely used (Peach 1999). White (1986: 202) describes it as 'the workhorse of segregation measurement'. It is most often employed in the study of urban populations, comparing the residential patterns of groups within cities. Since large statistical units tend to conceal population variations, studies of segregation rely especially on data for small areas, such as street blocks.

When applied to statistics on ethnic groups, the index of dissimilarity has a theoretical range from zero (no segregation) to 100 (total segregation). It represents the percentage of a group that would have to change its residential area in order to achieve zero segregation from the population with which it is being compared (Peach 1999: 323).

Table 13.3 provides a simple illustration of the derivation of the index. The example compares the distribution of whites and 'free colored' across eight wards of Charleston at the start of the American Civil War. The index of segregation is 23.2, indicating that this percentage of the 'free colored' would have needed to live in other areas of Charleston in order for their distribution to match that of the white population. The index is far below many contemporary figures for ethnic minorities in Western cities. Dividing Charleston into just eight areas, however, conceals the full detail of the residential patterns.

Table 13.3 Calculation of the index of segregation for Charleston, South Carolina, 1861

Ward	Population numbers		Percentages		Absolute value of percentage difference
	White	'Free colored'	White	'Free colored'	$\lvert D - E \rvert$
A	B	C	D	E	F
One	2681	121	9.9	3.2	6.7
Two	3102	161	11.5	4.3	7.2
Three	4522	370	16.8	9.8	7.0
Four	5926	815	22.0	21.5	0.4
Five	2739	853	10.2	22.5	12.4
Six	3476	760	12.9	20.1	7.2
Seven	1924	201	7.1	5.3	1.8
Eight	2599	504	9.6	13.3	3.7
Total City	26969	3785	100.0	100.0	46.5
Index of segregation (half the total)					23.2

Source: Taeuber and Taeuber (1965: 46).

The Gini index and the Lorenz curve

Whereas the two previous indices measure diversity and segregation, the Gini index focuses on the concept of inequality. Its best-known application is the measurement of income inequality, but uses extend to subject areas as different as health inequality, segregation (spatial inequality) and population concentration (inequality between population numbers and the area occupied). The index was the invention of Corrado Gini (1884–1965), an Italian statistician and demographer who taught sociology at the University of Rome.

The Gini index (G) is discussed here in the context of the Lorenz curve. The Lorenz curve shows the extent to which a given distribution is uneven compared with an even distribution (Coale and King 1969: 658). The Lorenz curve is plotted on a scatter diagram (an XY or Scatter Chart in Excel) with axes of equal length, from data on the cumulative frequency distributions of two variables. These could be population and land area (measuring population concentration), population and income (measuring income inequality), or the spatial distribution of majority and minority populations (measuring residential segregation).

The square graph has values ranging from 0 to 1 on each axis (or 0 to 100 per cent) (see Figure 13.4). The diagonal line, running at 45 degrees across the graph from the bottom left corner to the top right, denotes the line of zero difference, zero concentration, zero inequality, or zero segregation – depending on the data plotted. For example, if the data referred to population (X) and income (Y), points along the diagonal would denote that 10 per cent of individuals received 10 per cent of income, 20 per cent of individuals received 20 per cent of income and so on (see Creedy et al. 1984). Population is plotted on the horizontal axis, income or land area on the vertical axis. If two populations are being compared, as in segregation indices, the minority population, potentially experiencing segregation, is plotted on the vertical axis (Y) (Plane and Rogerson 1994: 31).

The ensuing discussion uses segregation as an example, comparing the distribution of the white and non-white populations in 10 locations (Table 13.4 and Figure 13.4).

To obtain the Gini index:

1. Calculate the proportions of the white (X) and non-white (Y) populations in each area, (columns B and C) and the ratio of white to non-white (X/Y) (column D).
2. Sort all the rows of data so that the X/Y ratios are in descending order. This will arrange the table appropriately for plotting the cumulative proportions as a Lorenz curve: the curve must extend to the two corners (points 0,0 and 1,1 on the graph) and must be convex to the horizontal axis (Ameil and Cowell 1999: 42).
3. Calculate the cumulative proportions for each population (columns E and F). If the figures at the top of these columns are not zero, insert zero values above the first figures so that the Lorenz curve, when plotted, will start at the origin of the X and Y axes of the graph (Figure 13.4). The extra zeros do not affect later calculations. The shading in columns E and F denotes the data plotted to produce the Lorenz curves.
4. To obtain the cross product $X_i \times Y_{i+1}$ in column G, multiply the first cell of column E (ignoring the zero for the axis origin) by the second cell in column F, then the second by the third and so on. For example, in Table 13.4, $0.04 \times 0.03 = 0.0012$, the first figure in column G of Panel A.
5. To obtain $X_{i+1} \times Y_i$ in column H, multiply the second cell in column E by the first cell in column F (ignoring the zero for the axis origin), then the third by the

Table 13.4 Calculations for the Lorenz curve, the Gini index, and the index of dissimilarity

A. 'Medium' dissimilarity between two distributions (medium segregation)

Location	Proportions		Ratio	Cumulative proportions		Cross products		Absolute differences	
(area)	White	Non-white	X/Y	White	Non-white			Proportions	Cumulative proportions
i			B/C	X_i	Y_i	$X_i \times Y_{i+1}$	$X_{i+1} \times Y_i$	|B – C|	|E – F|
A	B	C	D	E	F	G	H	I	J
(Axis origin):				0.00	0.00				
1	0.04	0.01	4.00	0.04	0.01	0.0012	0.0010	0.0300	0.0300
2	0.06	0.02	3.00	0.10	0.03	0.0110	0.0081	0.0400	0.0700
3	0.17	0.08	2.13	0.27	0.11	0.0459	0.0407	0.0900	0.1600
4	0.10	0.06	1.67	0.37	0.17	0.0962	0.0884	0.0400	0.2000
5	0.15	0.09	1.67	0.52	0.26	0.1612	0.1560	0.0600	0.2600
6	0.08	0.05	1.60	0.60	0.31	0.2460	0.2294	0.0300	0.2900
7	0.14	0.10	1.40	0.74	0.41	0.4070	0.3526	0.0400	0.3300
8	0.12	0.14	0.86	0.86	0.55	0.6450	0.5170	0.0200	0.3100
9	0.08	0.20	0.40	0.94	0.75	0.9400	0.7500	0.1200	0.1900
10	0.06	0.25	0.24	1.00	1.00			0.1900	
Total	1.00	1.00				2.5535	2.1432	0.6600	
Gini index (sum G minus sum H)							**0.4103**		
Index of dissimilarity (half sum I, or maximum value in column J)								**0.3300**	**0.3300**

Table 13.4 *continued*

B. 'High' dissimilarity between two variables (high segregation)

(Axis origin):				0.00	0.00				
1	0.04	0.00	–	0.04	0.00	0.0004	0.0000	0.0400	0.0400
6	0.08	0.01	8.00	0.12	0.01	0.0024	0.0018	0.0700	0.1100
2	0.06	0.01	6.00	0.18	0.02	0.0072	0.0056	0.0500	0.1600
4	0.10	0.02	5.00	0.28	0.04	0.0224	0.0180	0.0800	0.2400
3	0.17	0.04	4.25	0.45	0.08	0.0585	0.0480	0.1300	0.3700
5	0.15	0.05	3.00	0.60	0.13	0.1260	0.0962	0.1000	0.4700
7	0.14	0.08	1.75	0.74	0.21	0.2590	0.1806	0.0600	0.5300
8	0.12	0.14	0.86	0.86	0.35	0.5160	0.3290	0.0200	0.5100
9	0.08	0.25	0.32	0.94	0.60	0.9400	0.6000	0.1700	0.3400
10	0.06	0.40	0.15	1.00	1.00			0.3400	
Total	1.00	1.00				1.9319	1.2792	1.0600	
Gini index (sum G minus sum H)							**0.6527**		
Index of dissimilarity (half sum I, or maximum value in column J)								**0.5300**	**0.5300**

Table 13.4 *continued*

C. No dissimilarity between two variables (zero segregation)

(Axis origin):				0.00	0.00				
1	0.01	0.01	1.00	0.01	0.01	0.0003	0.0003	0.0000	0.0000
2	0.02	0.02	1.00	0.03	0.03	0.0024	0.0024	0.0000	0.0000
3	0.05	0.05	1.00	0.08	0.08	0.0112	0.0112	0.0000	0.0000
4	0.06	0.06	1.00	0.14	0.14	0.0308	0.0308	0.0000	0.0000
5	0.08	0.08	1.00	0.22	0.22	0.0682	0.0682	0.0000	0.0000
6	0.09	0.09	1.00	0.31	0.31	0.1271	0.1271	0.0000	0.0000
7	0.10	0.10	1.00	0.41	0.41	0.2255	0.2255	0.0000	0.0000
8	0.14	0.14	1.00	0.55	0.55	0.4125	0.4125	0.0000	0.0000
9	0.20	0.20	1.00	0.75	0.75	0.7500	0.7500	0.0000	0.0000
10	0.25	0.25	1.00	1.00	1.00			0.0000	
Total	1.00	1.00				1.6280	1.6280	0.0000	
Gini index (sum G minus sum H)							0.0000		
Index of dissimilarity (half sum I, or maximum value in column J)								0.0000	0.0000

A. 'Medium' dissimilarity ('medium' segregation)

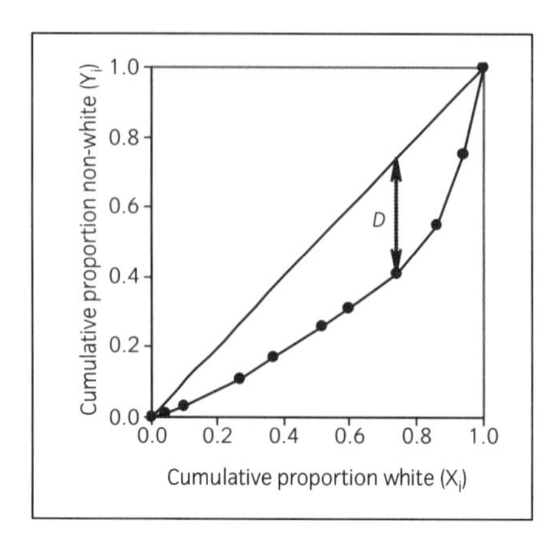

B. 'High' dissimilarity ('high' segregation)

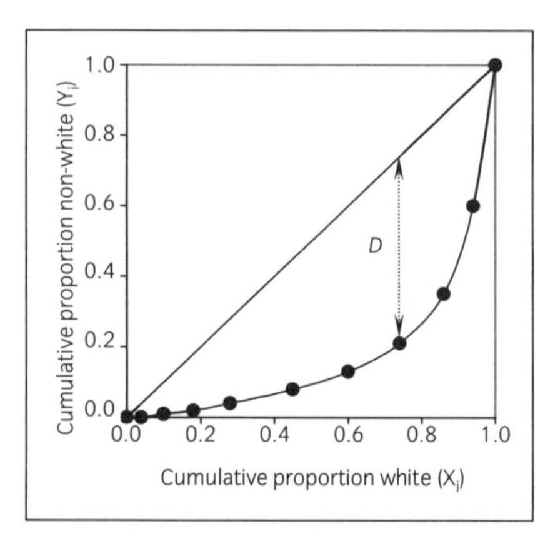

C. No dissimilarity (zero segreagation)

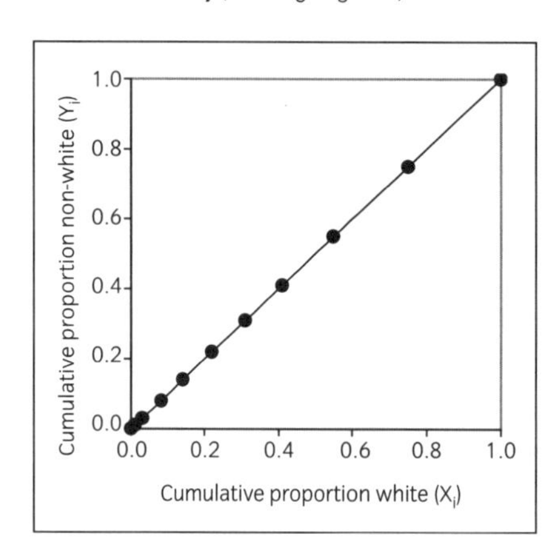

Figure 13.4 Lorenz curves illustrating 'medium', 'high' and zero segregation.

Source: Table 13.4.

Note: D = index of dissimilarity.

second and so on. For example $0.10 \times 0.01 = 0.0010$, the first figure in column H of Panel A.

6. Sum the figures in columns G and H, and subtract the total for column H from the total for column G. The result is the Gini index: 0.4103.

These steps are summarized in the formula for the Gini index (G) (Jones 1967, 278; Shryock and Siegel 1973: 178):

$$G = \left(\sum_{i=1}^{n} X_i Y_{i+1}\right) - \left(\sum_{i=1}^{n} X_{i+1} Y_i\right)$$

where

X_i and Y_i are the cumulative frequency distributions
n is the number of areas or categories.

The Gini index is a summary measure of the deviation in the Lorenz curve from a situation of zero inequality or zero difference. It 'measures the proportion of the total area under the diagonal that lies in the area between the diagonal and the Lorenz curve' (White 1986: 204). Hence the higher the value of the index, the greater the curvature of the line and the greater the mismatch between the two distributions. In the examples of residential segregation of the white and non-white populations in Table 13.4, the higher the Gini index, the greater the bow in the 'segregation curve', as it might be called, and the greater the degree of residential segregation.

The graphs in Figure 13.4 plot the cumulative proportions (highlighted) for the three examples. The diagonal line on each graph is the line representing $X_i = Y_i$. If the corresponding values for each population are the same, plotting them on a scatter diagram produces a straight line.

In the first example in Figure 13.4, the Gini index is 0.41, indicating that about 40 per cent of the area under the diagonal is above the Lorenz curve. In the second example (B), the curvature of the line is greater and the Gini index is just over 0.65. This denotes a fairly high level of segregation – the considerable gap between the diagonal and the Lorenz curve gives a visual impression of the extent of segregation. Finally, the third example (C) illustrates zero segregation: the Gini index is zero and the Lorenz curve coincides with the diagonal because the proportions for the two populations in columns B and C are identical. Each location has the same representation of whites and non-whites.

Taeuber and Taeuber (1965: 203) observed that: 'Most segregation indexes can be regarded as alternative ways of measuring the deviation of the curve from the diagonal.' Thus, for comparison, Table 13.4 and Figure 13.4 also include the index of dissimilarity (*D*) calculated from the same data sets. As in the earlier discussion of this index, the table shows that the index of dissimilarity is equal to half the sum of the absolute differences between the pairs of proportions (column I). It also shows that the index is equal to the highest absolute difference between the pairs of cumulative proportions (the boxed values in column I of Panels A and B); this represents the maximum vertical distance from the diagonal to the curve (Duncan 1957: 30) – as indicated by the double-headed arrows in Figure 13.4.

The mathematical relationship between the index of dissimilarity (*D*) and the Gini index (*G*) is (Duncan 1957: 31):

$$D \leq G \leq 2D - D^2$$

Thus if $D = 0.5$ (50 per cent), then G is at least 0.5 and can be no greater than 0.75. Although the results from the two approaches may differ, Duncan (1957: 31) considered that 'either index serves about as well as the other'.

The Lorenz curve provides a valuable means of visualizing information on inequality, comparing different distributions, and providing a graphical illustration of the magnitude of both the Gini index and the index of dissimilarity. A disadvantage is that different Lorenz curves can produce the same Gini coefficient (see Creedy et al. 1984: 211) and also the same index of dissimilarity: 'The Gini and dissimilarity indexes . . . share the property of most averages – differing distributions can have identical average values' (Taueber and Tauber 1965: 216).

The Gini index is 'the workhorse of income inequality analysis' (White 1986: 203). The US Census Bureau, for instance, regularly uses it to measure how far the United States income distribution is from equality (e.g. Weinberg 1996). Nevertheless, 'the workhorse of segregation measurement' is the index of dissimilarity. This is partly because it is more easily calculated, and partly because its meaning can be expressed very clearly as the percentage of one group that would have to change residences in order to produce an even distribution' (White 1986: 202).

Mathematical debate continues about the relative merits of different measures of inequality (see Coulter 1989: 55ff; White 1986). White (1986: 216), for example, recommended another index, the entropy statistic (*H*), as potentially 'the best general measure of unevenness in population distribution' and one which 'should come into more regular use in demographic analyses of population diversity and segregation' (see Plane and Rogerson 1994: 105–6 and 302; Siegel 2001: 27).

13.5 Multivariate measures

There are many other summary indices of population composition that are of special interest in planning, important among which in current research are some multivariate measures that combine different variables to quantify concepts such as lifestyles, material living standards, quality of life or social and economic well-being. Forerunners of these include early measures of poverty and socio-economic status – the latter mostly employ census information on incomes, occupations and education as the basic indicators. Multivariate measures are computed for cross-national comparisons and studies of social differentiation, but there is also major interest in their commercial applications, which now underpin multimillion dollar industries in the United States and Britain.

While the different types of multivariate measures are subjects for more specialized study, they represent an important frontier in applied demography. Hence this penultimate section introduces them briefly, through a discussion of two major applications: measurement of lifestyles for business decision-making, and measurement of well-being to gauge socio-economic development.

Lifestyles of market segments

The greater abundance of information available on populations, together with the methods and computing technologies of *geographic information systems* (see Section 10.5), have facilitated the development of commercial interest in analyses of population data, from which have also emerged the terms *demographics*, *geode-*

BOX 13.1 **Demographics, psychographics and geodemographics**

Demographics

The noun *demographics* came into use in 1966, referring to 'the whole realm of statistical characteristics of a particular population – especially as used to identify markets'; it denotes a specialized branch of applied demography, employing information on small areas to identify markets for products (Plane and Rogerson 1994: 22 and 299).

Later usage of the term has sometimes made *demographics* a synonym for applied demography, demographic statistics, or even for demography. However, academic writers advocate a definition closer to the original – denoting just a part of applied demography and therefore serving a narrow purpose (Siegel 2001: 5). Thus (Merrick and Tordella 1988: 3) offered the following definition:

Demographics n. (1) *sing*., application of demographic information and methods in business and public administration. (2) *pl*., the demographic information utilized in (1).

Similarly Siegel (2001: 5) rejected the indiscriminate use of the term and defined *demographics* as:

(1) the relatively mechanical production of demographic data on a current basis that are used to serve the marketing needs of business and industry and (2) the resulting data.

Psychographics

The word *psychographics* was coined about the same time as demographics. According to the editors of the magazine *American Demographics*: 'Demographics provide the basic picture of the consumer and psychographic information adds color and depth to that picture. Used together, attribute information can predict what consumers are likely to want and buy in the future.' (quoted in Plane and Rogerson 1994: 300).

Psychographics describes an approach to marketing based on techniques developed in social psychology. From surveys of the tastes and preferences of panels of individuals with known demographic characteristics, inferences are drawn about the location of the most likely consumers of particular products – by identifying geographical areas with households exhibiting the appropriate demographics (ibid).

Geodemographics

This is a newer term employed in business demography, referring to demographic data for spatial units, especially small areas, and the analysis thereof for business applications.

mographics and *psychographics* (Box 13.1). Changes in society have also necessitated such interest because of greater diversity in lifestyles, living arrangements and demographic behaviour: businesses, as well as governments, need to be able to differentiate among sub-populations in order to focus on particular groups (Bogue 1997: 2–3).

Analyses of population characteristics have various commercial applications including targeting advertising and retail product distribution, site analysis and assessing consumer demand for products (Merrick and Tordella 1988: 32–35). In relation to the last of these, one strategy is to seek to find a match between the profile of residents in areas (e.g. from census data) and the profile of a consumer population – as discerned from a survey or data base, such as relating to clothing purchases, dining out, or buying household appliances or sporting equipment. Many businesses and organizations collect customer and client profiles as part of their day-to-day operations. Offering a new service, such as a new type of credit card, is an opportunity to compile detailed financial and other information about the clientele – to inform further marketing strategies.

Underlying this work are the notions of *market segmentation* and *targeting*, which envisage aiming commercial activities at different markets for different products, rather than simply assuming that society consists of a single mass market with common goals and lifestyle preferences. Thus, market segmentation recognizes that there is not a mass market, but a segmented market consisting of groups of consumers with different needs and wants (Merrick and Tordella 1988). Although socio-economic status is a valuable indicator of purchasing power, other indicators are needed of lifestyles and spending habits, to provide a more complete description of the nature of the market. Market research here seeks to categorize populations into behaviourally consistent groups (McNicoll 1992: 406).

> Companies analyze market segments to find niches that will give them a competitive edge. Most segmentation strategies combine demographic and geographic data with information on consumer attitudes and purchasing patterns. Advertising and other marketing efforts can then be targetted to specific groups of likely consumers, saving money and increasing effectiveness at the same time. (Merrick and Tordella 1988: 5)

Statistical analyses may seek to determine, for example, the relationship between fast-food consumption and age, sex, income and occupation – such as through developing a profile of the characteristics of consumers from a survey of fast-food customers. This information is then used to calculate the fast-food purchasing potential of different areas, according to a classification of the areas in terms of their population composition (ibid: 31).

The underlying assumption is that small area populations matching the profile of fast-food consumers are likely to behave similarly to respondents in the survey. The spatial data employed in such classifications needs to refer to the smallest statistical units, in order to identify localities that are fairly homogeneous in terms of their population characteristics.

Classifications of statistical areas, however, can sometimes lead people to draw inferences about the characteristics of individual residents, such as their education, income or lifestyle, from the characteristics of the population of the area in which they live. This is the *ecological fallacy* – making assertions about one unit of analysis, such as an individual, on the basis of another unit of analysis, such as the population of a small area (see Hong 1992; Schwartz 1994). However, a resident of a suburb with a high average income cannot be assumed to be wealthy, nor is a resident of a community with many retired people necessarily retired. Nonetheless, the ecological fallacy does not negate the broad relevance of generalizations about the characteristics of areas (Merrick and Tordella 1988: 30).

Apart from the more traditional types of information sources – censuses and surveys – analyses of population characteristics for commercial applications today are founded on the data bases that companies develop on their customers, be they mobile phone owners, airline passengers, purchasers of computer software, magazine subscribers and consumers of industrial, commercial and domestic products and services. 'Psychographic' information (see Box 13.1) has also become an important supplementary source, adding a new dimension to the analysis of population characteristics. Nevertheless, sound information about the potential market is only a part of the requirement in commercial planning. Other major considerations include the activities of competitors, costs of advertising and distribution, and costs of taxes, land and services – addressing these calls for expertise across a range of disciplines.

Well-being of societies

Turning to multivariate measures employed in cross-national and other general comparisons, there has been mounting dissatisfaction with economic measures as indicators of social progress and quality of life, and growing interest in measures of *well-being*:

> In the last decade the field of social indicators has entered a new era with the development of summary social indicators. The purpose of such indicators is to summarize indicators (objective and/or subjective) from a number of domains into a single index. The motivation for this development is to answer one of the original questions of the social indicators movement, namely, how is a country progressing in terms of social conditions both over time and compared to other countries. The original pioneers of the social indicators movement backed away from this task to concentrate on database development. Now, with the greater availability of social data, a new generation of social indicators researchers has returned to the task of summary index construction. (Sharpe 1999)

One of the better-known examples of well-being measures is the *human development index* (HDI). The limitations of gross domestic product as an indicator of national development led to the publication, in 1990, of the first calculations of the human development index. Through this, the United Nations Development Programme (UNDP) sought to provide a more comprehensive measure of the relative socio-economic progress of nations (Table 13.5).

Table 13.5 Trends in the human development index for selected countries, 1975–1999

Country	1975	1980	1985	1990	1995	1999	HDI Rank
Norway	0.856	0.875	0.887	0.899	0.924	0.939	1
United States	0.861	0.882	0.896	0.912	0.923	0.934	6
Japan	0.851	0.876	0.891	0.907	0.920	0.928	9
United Kingdom	0.839	0.846	0.856	0.876	0.914	0.923	14
Italy	0.827	0.845	0.855	0.878	0.895	0.909	20
Mexico	0.688	0.732	0.750	0.759	0.772	0.790	51
Russian Federation	–	0.809	0.826	0.823	0.778	0.775	55
China	0.522	0.553	0.590	0.624	0.679	0.718	87
Indonesia	0.467	0.529	0.581	0.622	0.662	0.677	102
India	0.406	0.433	0.472	0.510	0.544	0.571	115
Kenya	0.442	0.488	0.511	0.531	0.521	0.514	123
Pakistan	0.343	0.370	0.403	0.441	0.476	0.498	127
Bangladesh	0.332	0.350	0.383	0.414	0.443	0.470	132
Malawi	0.318	0.343	0.356	0.363	0.401	0.397	151
Niger	0.234	0.253	0.244	0.254	0.260	0.274	161

Data source: United Nations Development Programme, *Human Development Report 2001*, (http://www.undp.org/hdr2001/back.pdf).

The index is a composite of three components of human development, namely longevity (life expectancy), knowledge (measured by adult literacy and mean years of schooling) and standard of living (measured by purchasing power) (United Nations Development Program 2002). Lack of data for all countries prevents the inclusion of other aspects, but it is also argued that 'adding more variables could confuse the picture and detract from the main trends' (ibid). The minimum and maximum for each of the three dimensions are 0 and 1. For example 100 per cent literacy receives a score of 1 while, for life expectancy, the minimum is 25 years and the maximum 85 years. The scores for the three components are averaged to obtain the HDI.

The index offers governments and planners an additional basis for evaluating progress through time and determining priorities for policy intervention. It is also calculated for particular social groups or regions to take account of different levels of human development within countries. Table 13.5 shows trends through time in the HDI for a selection of countries. Norway had the highest HDI ranking in 1999 of 0.939, followed closely by Australia, Canada, Sweden, Belgium and the United States. Sierra Leone, with an HDI of 0.258, had the lowest ranking among 162 countries.

13.6 Conclusion

Utilizing demographic information for non-traditional purposes, or in conjunction with other types of data, constitute important changes. Demographic statistics and techniques are now brought to bear in a much broader range of contexts than formerly anticipated. The more conventional applications of demographic methods also have wider importance, because of the information explosion and the growing number of people engaged in research and analytical work on populations.

Chapters 12 and 13 have discussed some of the main concerns of applied demography, namely population projections and estimates and measures of population composition. Each of these has clear applications in planning and decision-making for public administration or private enterprise activity:

- The elaboration of population projections, to refer to the labour force or households or other sections of the population, is essential information in anticipating needs in society, together with prospects for related activities of organizations and enterprises.
- Population composition, in terms of specific characteristics such as age and ethnicity, as well as in terms of broader descriptive concepts – such as diversity, segregation and inequality – is a necessary consideration in decisions about where to build schools or nursing homes, or where to target advertising for new products and how to monitor the effectiveness of policies addressing inequality and discrimination.

Although the above represent major interests within applied demography, the subject matter of this sub-field readily includes many other considerations. This reflects that demography overall is an applied discipline (Keyfitz 1993: 548); analyses of population growth, age structure, fertility, mortality and migration all inform understanding of the nature of society, the processes shaping it and the need for responses or interventions. This was the vision of John Graunt, who saw so much potential in the study of the 'poor despised *Bills of Mortality*' and drew so many conclusions of practical importance from them:

. . . I conceive, That it doth not ill-become a *Peer of the Parliament*, or *Member of his Majestie's Council*, to consider how few starve of the many that beg: That the irreligious *Proposals* of some, to multiply People by *Polygamy*, is withall irrational, and fruitless: That the troublesome seclusions in the *Plague-time* is not a remedy to be purchased at vast inconveniences: That the greatest *Plagues* of the City are equally, and quickly repaired from the Country: That the wasting of *Males* by Wars, and Colonies do not prejudice the due proportion between them and *Females*: That the Opinions of *Plagues* accompanying the Entrance of *Kings* is false, and seditious: That *London*, the *Metropolis* of *England*, is perhaps a Head too big for the Body, and possibly too strong: That this Head grows three times as fast as the Body unto which it belongs, that is, It doubles its People in a third part of the time: That our *Parishes* are now grown madly disproportionable: That our *Temples* are not sutable to our *Religion*: That the *Trade*, and very *City of London* removes

Westward: That the walled City is but a one fifth of the whole Pyle:|| That the old Streets are unfit for the present frequencie of *Coaches*: That the passage of *Ludgate* is a throat too straight for the Body: That the fighting men about *London*, are able to make three as great Armies as can be of use in this *Island*: That the number of Heads is such, as hath certainly much deceived some of our *Senatours* in their appointments of *Pole-money*, &c. (Graunt 1662: 4)

Study resources

KEY TERMS

Gini index
Household headship method of projection
Householder
Human development index
Index of dissimilarity
Index of diversity
Index of segregation
Labour force projections
Living arrangement method of projection
Lorenz curve
Market segmentation

FURTHER READING

Day, Jennifer Cheeseman. 1996. *Projections of the Number of Households and Families in the United States: 1995 to 2010*. US Census Bureau, Current Population Reports, P25-1129. Washington, DC: US Government Printing Office.

Kintner, H., Merrick, Thomas W., Morrison, Peter A., and Voss, Paul R. (editors). 1994. *Demographics: A Casebook for Business and Government*. Boulder, CO: Westview Press.

Merrick, Thomas W. and Tordella, Stephen J. 1988. 'Demographics: People and Markets'. *Population Bulletin*, 43(1) (February) (Population Reference Bureau).

Murdock, Steve H. and Ellis, David R. 1991. *Applied Demography: An Introduction to Basic Concepts, Methods, and Data*. Boulder: Westview Press.

Myers, Dowell (editor). 1990. *Housing Demography: Linking Demographic Structure and Housing Markets*. Madison, WI: University of Wisconsin Press.

Plane, David A. and Rogerson, Peter A. 1994. *The Geographical Analysis of Population: with Applications to Planning and Business*. New York: John Wiley, Chapter 10, 'Demographics'.

Pol, L. and Thomas, R.K. 1997. *Demography for Business Decision Making*. London: Quorum Books.

Rao, K. Vaninadha and Wicks, Jerry W. 1994. *Studies in Applied Demography*. Bowling Green, OH: Population Research Center, Bowling Green State University.

Rives, Norfleet W. and Serow, William J. 1984. *Introduction to Applied Demography: Data Sources and Estimation Techniques*. Beverly Hills: Sage.

Siegel, Jacob S. 2001. *Applied Demography*. San Diego: Academic Press.

Weeks, John R. 2002. *Population: An Introduction to Concepts and Issues* (eighth edition). Belmont, CA: Wadsworth, Chapter 14, 'Demographics'.

INTERNET RESOURCES

Subject	Source and Internet address
Gateway to the US 2000 Census	**US Census Bureau** http://www.census.gov/main/www/cen2000.html
US Census Bureau's *Current Population Reports* and other reports – a wealth of papers on US population and related research methods	**US Census Bureau** http://www.census.gov/prod/www/abs/popula.html#techpap
Geodemographics knowledge base; a directory of material on geodemographics and socio-economic and demographic data sources	**Market Research Society** http://www.geodemographics.org.uk/
Web site of the Claritas Corporation, an example of an international company utilizing demographic information for commercial applications; includes access to samples of data	**Claritas Corporation** http://www.connect.claritas.com/
Resources for teaching population geography	**Department of Social Statistics, University of Southampton** http://www.socstats.soton.ac.uk/popsci/population_links/
United Nations *Human Development Report* and Human Development indicators	**United Nations Development Programme** http://www.undp.org/hdro/index.html http://www.undp.org/hdr2001/back.pdf

Note: the location of information on the Internet is subject to change; to find material that has moved, search from the home page of the organization concerned.

EXERCISES

1 For your country or state of residence, make a list of publications that present projections of aspects of population composition, such as projections of school enrolments, the labour force, welfare recipients and households. Seek to make the list as comprehensive as possible, but exclude projections of the total population by age and sex.

2 Write a summary and evaluation of the projection methodology used in *one* of the publications listed in your answer to the first question. Consult supplementary sources as necessary.

3 From the following statistics on the proportions in 14 religious denominations, calculate the index of diversity, the theoretical maximum value of the index, and the standardized index of diversity:

Religion	Proportion
1	0.062
2	0.004
3	0.014
4	0.049
5	0.009
6	0.000
7	0.002
8	0.020
9	0.778
10	0.003
11	0.049
12	0.005
13	0.003
14	0.002
Total	1.000

Source: Lieberson (1969: 852).

4 From the following data, calculate the index of segregation (dissimilarity) and the Gini index:

Region	Proportion white	Proportion Non-white
A	0.08	0.01
B	0.14	0.02
C	0.17	0.04
D	0.07	0.03
E	0.10	0.05
F	0.15	0.14
G	0.29	0.71
Total	1.00	1.00

5 *For group discussion.* Be prepared to speak for two minutes on the question: 'What can demography do for business?' (Hugo 1991). Base your answer on *one* journal article, book chapter or internet source. Do not attempt to provide a comprehensive statement; rather, aim to present some ideas from a critical perspective as a contribution to a wider-ranging discussion.

SPREADSHEET EXERCISE 13: EXPERIMENTS IN DATA TRANSFORMATION

This exercise illustrates how sets of demographic data may be transformed to make them easier to use or interpret. The two examples to be constructed provide visual demonstrations of transformations and opportunities to experiment with different values. Although it draws on material from previous chapters, the exercise is more complex than most of the others. Knowledge of previous exercises is assumed.

Data transformation entails changing the scale of measurement of a variable, such as from an arithmetic scale to a logarithmic scale, or from proportions to logits (see Armitage and Berry 1987: Chapter 11). The examples enhance understanding of (1) exponential growth (see Table 2.4) and growth projections, and of (2) the relational model life table module presented in Chapter 9. The examples are concerned respectively with transforming, into straight lines, exponential growth curves and survival curves (l_x values). Turning a survival curve into a straight line, for instance, greatly simplifies the calculation of further 'related' life tables, with higher or lower life expectancies.

Example 1: Logarithmic transformation of population growth

Logarithmic transformation replaces the original population totals with their common or natural logarithms; it can be applied only to values greater than zero, since the logarithms of negative numbers do not exist and the logarithm of zero is minus infinity.

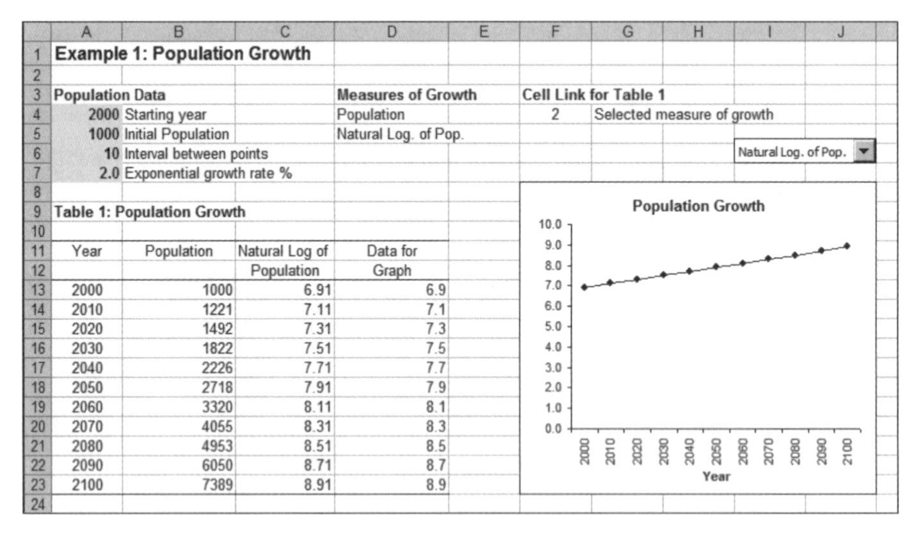

	A	B	C	D	E	F	G	H	I	J
1	Example 1: Population Growth									
2										
3	Population Data			Measures of Growth		Cell Link for Table 1				
4	2000	Starting year		Population		2	Selected measure of growth			
5	1000	Initial Population		Natural Log. of Pop.						
6	10	Interval between points							Natural Log. of Pop.	
7	2.0	Exponential growth rate %								
8										
9	Table 1: Population Growth									
10										
11	Year	Population	Natural Log of	Data for						
12			Population	Graph						
13	2000	1000	6.91	6.9						
14	2010	1221	7.11	7.1						
15	2020	1492	7.31	7.3						
16	2030	1822	7.51	7.5						
17	2040	2226	7.71	7.7						
18	2050	2718	7.91	7.9						
19	2060	3320	8.11	8.1						
20	2070	4055	8.31	8.3						
21	2080	4953	8.51	8.5						
22	2090	6050	8.71	8.7						
23	2100	7389	8.91	8.9						
24										

Figure 13A

To begin the first example, type the information in rows 1 to 7, together with the title and column headings for Table 1. The values in bold, cells A4 to A7, determine the values in Table 1 and may be changed to any other numbers to produce new examples.

To create the values in Table 1:

- *Year*: set the first year (cell A13) equal to A4; calculate the second year (A14) as = A13 + A$6 and copy the formula down.
- *Population*: set the population in B13 equal to A5; calculate the second population as = B13 * (EXP($A$7/100 * $A$6)) and copy the formula down. This generates population projections on the assumption of constant exponential growth, according to the rate per cent specified in A7. In Excel, =EXP() returns *e* raised to the power of the number in the brackets. Excel calculates exponential growth (see Table 2.4) as: $P_0 * \text{EXP}(r * n)$.

- *Natural logarithms of population totals*: set the value in C13 to =LN(B13) and copy the formula down. The =LN() function returns the natural logarithm.
- *Data for the graph*: write the following formula in D13: = INDEX(B13:C13,$F$4) and copy down. The formula uses the index function, employed in other exercises, to choose data from either column B or column C, according to the value in F4 (1 for population, 2 for the natural logarithm of the population).

A population growing exponentially at 2 per cent per annum increases by a constant ratio in each interval, as does a population increasing geometrically. In the spreadsheet exercise, the population totals at the first three points are 1000, 1221 and 1492; the ratio 1221/1000 = 1.221 and 1492/1221 = 1.221. Since the numbers are changing by a constant ratio, when we calculate the natural logarithms of the population totals, the differences between adjacent figures is constant at 0.200, as shown below.

Year	Population	Ratios P_n/P_0	Natural log of population	$\ln P_n - \ln P_0$
2000	1000		6.91	
2010	1221	1.221	7.11	0.200
2020	1492	1.221	7.31	0.200
2030	1822	1.221	7.51	0.200

Since the natural logarithms of the numbers increase by a constant amount in each interval, the points lie on a straight line when plotted on an arithmetic scale. Thus an exponential growth curve becomes a straight line. This is illustrated in Example 1 by plotting a line graph of the data in columns A and D of Table 1. To change the data displayed in the graph, create a drop-down menu box ('combo-box') in which the 'input range' is D4:D5 and the 'cell link' is F4 (see Spreadsheet Exercises 6 and 10).

The spreadsheet now permits the display of population projections based on any source data in cells A4 to A7. The projections can be presented in the graph either as exponential curves or as straight lines: the population increases by a constant ratio, while the natural log of the population increases by a constant amount.

Example 2: Logit transformation of proportions surviving

When working with proportions, logarithmic transformation is not appropriate and the logit transformation may be used instead (Armitage and Berry 1987: 367).

In the second part of this exercise, a scatter diagram is produced comparing the l_x values of two life tables. When the l_xs are plotted the points lie on a curve, but when the logits of the l_xs are plotted, the points lie on a straight line. This does not mean that a scatter plot of the logits of l_xs from any two life tables will produce a perfect linear relationship, but the match is sufficiently close to permit modelling on the basis of this assumption.

To start the second example, click on the tab for Sheet 2, at the bottom of the screen, and type the information in rows 1 to 13, together with the row labels and the values shown in bold in Table 2. In the scatter diagram for this example, the l_xs for the standard population will be plotted along the X axis, those for the second population along the Y axis.

	A	B	C	D	E	F	G	H	I	J	K	L	M
1	**Example 2: Proportions Surviving**												
2													
3	**Measures**		**Mortality Levels**		**Cell Links for Table 2**								
4	lx		e0 = 40 years		1	Life table for x axis							
5	lx/100000		e0 = 60 years		2	Life table for y axis							
6	Logits		e0 = 80 years		3	Selected measure							
7													
8	**Table 2: Proportions Surviving**												
9													
10	Life Tables (lx values)				Measures From Selected Life Tables						Data for Graph		
11	from relational models based					X axis			Y axis		X axis	Y axis	
12	on Brass's 'general' standard				lx	lx/100000	Logits	lx	lx/100000	Logits			
13	e0:	40	60	80	40	40	40	60	60	60	40	60	
14	0	**100000**	**100000**	**100000**	100000	1.00000		100000	1.00000				
15	1	**82256**	**94448**	**99285**	82256	0.82256	-0.76689	94448	0.94448	-1.41695	-0.77	-1.42	
16	5	**73169**	**90914**	**98791**	73169	0.73169	-0.50161	90914	0.90914	-1.15159	-0.50	-1.15	
17	10	**71088**	**90022**	**98661**	71088	0.71088	-0.44983	90022	0.90022	-1.09984	-0.45	-1.10	
18	15	**69557**	**89343**	**98560**	69557	0.69557	-0.41315	89343	0.89343	-1.06313	-0.41	-1.06	
19	20	**67040**	**88184**	**98386**	67040	0.67040	-0.35500	88184	0.88184	-1.00499	-0.35	-1.00	
20	25	**63778**	**86597**	**98140**	63778	0.63778	-0.28287	86597	0.86597	-0.93289	-0.28	-0.93	
21	30	**60588**	**84942**	**97875**	60588	0.60588	-0.21501	84942	0.84942	-0.86503	-0.22	-0.87	
22	35	**57428**	**83192**	**97586**	57428	0.57428	-0.14967	83192	0.83192	-0.79965	-0.15	-0.80	
23	40	**54069**	**81201**	**97243**	54069	0.54069	-0.08156	81201	0.81201	-0.73156	-0.08	-0.73	
24	45	**50371**	**78832**	**96816**	50371	0.50371	-0.00742	78832	0.78832	-0.65741	-0.01	-0.66	
25	50	**46068**	**75812**	**96240**	46068	0.46068	0.07880	75812	0.75812	-0.57120	0.08	-0.57	
26	55	**40942**	**71781**	**95407**	40942	0.40942	0.18318	71781	0.71781	-0.46681	0.18	-0.47	
27	60	**34977**	**66372**	**94158**	34977	0.34977	0.31003	66372	0.66372	-0.33996	0.31	-0.34	
28	65	**27905**	**58682**	**92062**	27905	0.27905	0.47459	58682	0.58682	-0.17542	0.47	-0.18	
29	70	**20364**	**48409**	**88456**	20364	0.20364	0.68185	48409	0.48409	0.03183	0.68	0.03	
30	75	**12763**	**34930**	**81425**	12763	0.12763	0.96104	34930	0.34930	0.31106	0.96	0.31	
31	80	**6377**	**19994**	**67114**	6377	0.06377	1.34329	19994	0.19994	0.69333	1.34	0.69	
32	85	**2271**	**7857**	**41049**	2271	0.02271	1.88099	7857	0.07857	1.23097	1.88	1.23	
33	90	**484**	**1752**	**12710**	484	0.00484	2.66299	1752	0.01752	2.01337	2.66	2.01	
34	95	**49**	**180**	**1451**	49	0.00049	3.81031	180	0.00180	3.15908	3.81	3.16	
35													

Figure 13B

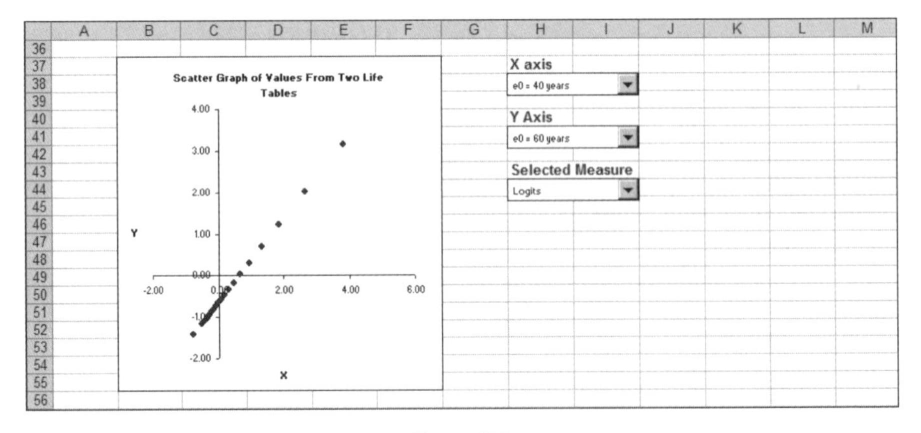

Figure 13C

The next step is to enter formulas in the middle section of the table to show which two of the three sets of life table values have been selected for comparison, together with the proportions and logits calculated from them:

- In column E, use the index function to choose the X values from columns B to D, according to the number in F4.
- In column F, calculate proportions, by dividing the l_xs in column E by 100 000.
- In column G, calculate the logits of the proportions in column F using the formula (Newell 1988: 153):

$$\text{logit}(p_x) = 0.5\ln\left(\frac{1-p_x}{p_x}\right)$$

where ln denotes the natural logarithm, x is an exact age, and p_x is a proportion >0 and <1.

Thus if $p = 0.900\,00$, logit $= -1.098\,61$. In Excel, the formula is written: logit = 0.5 * LN((1 – p_x)/p_x). There is no logit for the proportion surviving at age zero, since the figure is 1.

- Repeat the above three steps in columns H, I and J, this time using the value in F5 for the index function that chooses the Y values from columns B to D.
- In columns K and L, use the index function again to choose the X values from columns E to G and the Y values from columns H to J, according to the value in F6.

These two columns of data (omitting headings and row labels) are now plotted in a scatter diagram using Excel's Chart Wizard and the chart type labelled 'XY (Scatter)', a description of which is accessible from the Excel Help menu.

Three menu boxes are then created to the right of the chart, to permit selection of the data plotted on the X axis (cell link to F4) and the Y axis (cell link to F5), together with the required measure (cell link to F6). The menus permit the display of different pairs of life table values as l_xs, proportions or logits of proportions.

Although the exercise is now complete, it may be noted that the second example illustrates the basis for calculating relational model life tables, which assume a linear relationship between the logits of the l_xs in a 'standard' life table and those in a second life table, such as one with higher life expectancy. The second life table is calculated from the logits of the standard using the equation for a straight line: $y = \alpha + \beta x$. To generate a new life table from the standard there are only two parameters to adjust: the line's *intercept* on the Y axis (α or alpha, which affects life expectancy), and the *slope* of the line (β or beta, which affects the relationship between childhood and adult mortality). Further information about relational model life tables is presented in Chapter 9 (see Section 9.3 and Box 9.2).

BODMAS

1. $10 + (7+3)/2 \times 5 = 35$

$$7+3 = \frac{10}{2} = 5 \times 5 = 25 + 10 = 35$$

2. $4 + 2^5 \times 6 + 9 = 205$

$$2^5 \times 6 = 192 + 4 + 9 = 205$$

↑
(32)

1 2 4 8 16 32
2 2 2 2 2

Appendix A Basic maths

Order of operations

When a formula contains two or more mathematical operations, they are performed in the following order:

- brackets and fractions
- powers and roots
- division and multiplication, left to right
- addition and subtraction, left to right.

Where there are a number of operations of equal priority, work from left to right. Thus $10 + (7 + 3)/2 \times 5 = 35$ and $4 + 2^5 \times 6 + 9 = 205$.

Rounding

Numbers are usually rounded:

- to make them easier to read (e.g. in text and tables, figures are often rounded to one decimal place);
- to avoid a false impression of precision (e.g. real demographic rates are seldom measured so accurately that more than two decimal places are justified).

To round decimals:

0.50 and above, round up, e.g. 6.55 becomes 6.6, or 7.

0.49 and less, round down, e.g. 6.444 becomes 6.44 or 6.4, or 6.

Thus, rounding to two decimal places, 7.5151 = 7.52 and 3.4999 = 3.50; rounding to whole numbers, 6.4999 = 6 and 4.5000 = 5.

Exceptions to rounding

- In intermediate calculations, retain at least five decimal places;
- where readers may need to use the data for further calculations, retain up to five decimal places.

Note: although Excel spreadsheets normally display rounded numbers, the program stores and calculates with 15 significant digits of precision.

Powers of numbers

A number raised to a power is multiplied by itself the number of times indicated by the power, e.g. $8 \times 8 = 8^2$, $7 \times 7 \times 7 = 7^3$.

To multiply two powers of the same number, add the powers, e.g. $8^2 \times 8^3 = 8^5$.

To divide two powers of the same number, subtract the powers, e.g. $8^5/8^3 = 8^2$.

Thus when working with powers of numbers substitute addition and subtraction for multiplication and division.

Powers in Excel: e.g. 2^4 is written 2∧4 (i.e. using the caret symbol at shift-6); similarly:

$$\sqrt{2} \text{ or } 2^{1/2} \text{ is written : } = 2\wedge(1/2)$$

Logarithms

Logarithms take advantage of the rule for multiplying and dividing powers of numbers. Logarithms substitute addition and subtraction for multiplication and division to facilitate those processes.

Operation	Operation with logs
multiplication	Addition: to multiply numbers add their logs
division	Subtraction: to divide numbers subtract their logs
powers	Multiplication: to obtain powers of numbers multiply the log by the power
roots	Division: to obtain roots of numbers divide the log of the number by the power

Common logarithms represent the power to which 10 must be raised to give the number, e.g. log(1000) = 3, since $1000 = 10^3$; also log(180) = 2.255 27 since $180 = 10^{2.255\,27}$.

Antilogarithms tell us what number is represented by the value of 10 to a particular power; e.g. antilog(3) = 1000 and antilog(2.255 27) = 180.

In Excel:

- log = LOG(number)
- antilog = 10∧(number).

For example, to raise a number to a power, multiply the log of the number by the power; to calculate 5^3: log(5) × 3 = 0.698 97 × 3 = 2.096 91; antilog(2.096 91) = 125. Similarly, to find the root of a number, divide the log of the number by the power: to calculate the cube root of 343: log(343)/3 = 2.535 29/3 = 0.845 10; antilog(0.845 10) = 7.

Some operators

< less than

> greater than

Σ Greek capital sigma stands for the summation process, the adding together of all the cases to the right of the Σ sign. Thus Σx means add up the values of x, from the first to the last.

$|x|$ vertical rules indicate the absolute value of x, that is the number without its sign; e.g. $|10 - 14| = 4$. Absolute values are used, for example in calculating the index of dissimilarity. In Excel use = ABS(x).

Appendix B Using the Excel modules

Technical requirements

The modules are intended for use with Excel 5.0 through Excel 2002.

A PC with a Pentium processor, or equivalent, is recommended for running the modules, especially the *Cohort Component Projections* module and others with animated graphics. While less powerful computers will handle many of the tasks, slow performance or screen flickering may occur. Making a graph smaller can reduce or eliminate screen flickering for animated graphics (click on the graph and use the sizing handles on the frame of the graph).

Installing and running the modules

The modules are Excel files which may be run directly from the CD, although no changes will be saved, since the CD is Read Only. To enable changes to be saved, first copy the files to the computer's hard drive.

The files are opened in the usual ways: either by double clicking on the file name in My Computer; or by clicking File Open (from the standard toolbar in Excel) and locating the file in the directory.

Problems confronting new users

For first-time users of Excel, the following are potential problems, some of which may be due to unfamiliarity with using a mouse:

The screen display disappears, partially or completely

The probable cause is inadvertently scrolling across or down; press the Home key on the keyboard to put the cursor in column A, and press the Page Up key on the keyboard to see the top of the spreadsheet again.

Objects on the screen, such as charts or buttons, move out of their original locations

This is caused by accidentally dragging objects with the mouse; to correct this, either click Edit then Undo (or Undo Move Object or Undo Paste), or close and reopen the spreadsheet to restore the original appearance of the display.

Text or formulas disappear or become corrupted – e.g. #REF!, #NUM! or #VALUE! may appear

This is due to inadvertently deleting or overwriting information in cells other than the yellow shaded cells where user information may be entered; to correct this, either click Edit then Undo (or Undo Drag and Drop or Undo Paste), or close and reopen the spreadsheet to restore the original values and formulas.

The display will not fit on the screen, or the display is too small

If the image on the screen is too big or too small, go to **View** and choose **Zoom** to increase or decrease the magnification of the display. For further help, refer to the next section.

Optimizing the display

The following actions can increase the size of the display and enhance the readability of information:

1. Run the *Excel window* full-screen: click the 'maximise' button (the middle button) at the top right of the Excel window.
2. Run the *spreadsheet window* full-screen: click the 'maximise button' (if displayed) at the top right of the spreadsheet window.
3. Run the application full-screen (the toolbars will no longer be displayed): click View then Full Screen.
4. Increase, or decrease, the magnification of the display: click View then Zoom.
5. If the display is still too small, adjust the screen resolution: see advice on 'screen resolution' in the Excel Help menu.
6. Switch off the row and column headers if present, together with the sheet tabs, on the spreadsheet: click Tools Options View and uncheck the boxes for 'Row & column headers' and 'Sheet tabs'. This has little effect on the size of the display, but reduces the amount of unnecessary information.

On most of the CD files, the row and column headers are not displayed.

Worksheet protection

To protect displays from accidental changes, some worksheets have 'Protection' enabled. The main exceptions are worksheets where user data may be entered – they cannot be protected. If you wish to switch off protection, for instance to modify a worksheet, choose Tools Protection Unprotect Sheet.

'Chart tips'

When using the mouse to point to information on charts, series names and values will appear if the following options are set: from the standard toolbar Tools Options Chart Show Names Show Values.

Scroll bars

The scroll bar controls do not normally cause problems, but if they do, for example if a single click moves the control several steps instead of one, the following are possible causes: insufficient memory; other files or programs open; other files or programs running full-screen.

Screen Resolution

The preferred screen resolution, for both PCs and Macs, is 1024 × 768.

Using Apple Mac Computers

The modules will run on Macs, but text boxes and buttons may not be displayed correctly. Also ticks and crosses in the Abridged Life Tables module may not be displayed if the required font is not available. Because on restrictions on the length of file names, modules will have to be renamed when saved.

Further Information

Further information about the Excel modules is provided in the file Readme.rtf on the CD.

Appendix C Introduction to Excel

There are often several ways of performing the same procedure in Excel. This summary provides a description of the most common operations and the easiest ways to do them on a PC. Clicking the right mouse button on a cell will also display a menu with many common options.

Detailed information about Excel is accessed by clicking Help on the menu bar, or pressing F1 on the keyboard. Internet support for Excel users is available at http://www.microsoft.com/office/excel/support/.

1. Preliminaries

1.1 Loading Excel

Choose: Start Programs Microsoft Office Excel.

1.2 Moving the cursor

Use the arrow keys on the keyboard or move the mouse pointer and click the mouse (left button) in the required cell.

1.3 Getting out of trouble

If you make a mistake while typing or something goes wrong, press the ESC key

1.4 Getting help or running a tutorial

Click the Help menu at the right-hand end of the menu bar at the top of the screen.

1.5 Displaying toolbars

The Standard and formatting toolbars are needed much of the time. If these are not displayed click: View – Toolbars – Standard, then View – Toolbars – Formatting. Repeat these steps to remove a toolbar from the screen.

1.6 Cell identification

- Columns are labelled A, B, C, . . .
- Rows are labelled 1, 2, 3, . . .
- The cell ID (column and row identifiers) and the contents of the cell in which the cursor is positioned are displayed in the Formula Bar immediately above the column labels.

2. Entering labels, values and formulas

2.1 Entering text

- Type table titles, row and column headings, sources and notes much as if using a word-processing program.
- Use the backspace key to delete mistakes.

- If the text begins with a number, it may be necessary to type a single quote (') before the number. If the quote is omitted, Excel may convert the entry into a date (e.g. the labels for age groups 5–9 and 10–14).
- Use mouse or arrow keys, or press Enter, as entries are completed.

2.2 Positioning text within cells (centred, or left or right-justified)

- Either: use the alignment buttons on the Formatting Toolbar (align left, centre, align right or justify)
- Or: in the Menu bar click Format Cells Alignment and choose the required horizontal alignment from the menu. The Alignment menu also gives access to other formatting options.
- To centre text across a number of cells, highlight the cells and click the Merge and Centre button on the formatting toolbar.

2.3 Positioning text in many cells simultaneously

Click the mouse on the first cell and drag the mouse (left button down) to highlight other cells; then proceed as in 2.2.

2.4 Entering values

- Type the number as in a word-processor.
- Use the backspace key to delete mistakes.
- Use arrow keys to go to the next cell, or press Enter.

2.5 Changing the number of decimal places displayed in value cells

Either: select the cell or cells, and click the 'Increase decimal' or 'Decrease decimal' buttons on the Formatting toolbar.

- Or: select the cell or cells; click Format on the menu bar, select Cells then the Number tab.

2.6 Entering formulas

- All formulas begin with an equals sign; e.g. to add the contents of cell C3 and H9, type = C3 + H9
- Formulas contain operators, brackets, cell numbers and function symbols.
- Operators include: +, –, * (multiply), / (divide), ^ (powers, roots)
- Functions include: = SUM(A1:A10) [add up numbers in cells A1 to A10]; =LOG(*x*) [common logarithm of *x*]
- To enter formulas, type normally and press Enter
- For maximum accuracy, and to save having to type cell IDs, click the required cell, then type the next term in the formula – the selected cell ID will be retained in the formula. The formula is displayed in the Formula Bar above the column headers as you type. E.g. to type the formula = R2 – D2, type = then click on cell R2, type the minus sign, click on cell D2 and press Enter.
- Similarly, to define a range in a formula (e.g. D5 . . . D27) without typing cell IDs, click the first cell (e.g. D5) and drag the mouse to highlight all the required cells, then type the next term in the formula. E.g. to add up the numbers in cells D5 to D27, type =SUM(, click on D5 and drag to D27, then close the bracket and press Enter. The

formula will read =SUM(D5:D27). Shortcut: highlight the cells to be added then click the summation button (Sigma sign) on the Standard toolbar.

2.7 Copying text and values

- Use the mouse to highlight the cells to be copied.
- Click the Copy button on the Standard toolbar.
- Click the mouse on the cell you wish to copy to and press Enter, or click the Paste button on the Standard toolbar. (Use of the Paste button enables the same information to be pasted into a number of different cells, without having to copy it again.)

Copying formulas: relative references (cell IDs change relative to the row ID)

- E.g. =SUM(A1:A5) in cell A6 becomes =SUM(B1:B5) when copied to cell B6.
- That is, the cell references change relative to the position of the cell containing the formula.
- For relative copying, use the procedure described above in 2.7.

Copying formulas: absolute references (cell IDs remain constant when copied)

- For example, when calculating percentages, the total population (e.g. in cell A35) might be needed throughout. To prevent the cell ID from being changed, put dollar signs in the cell ID to make it an absolute (unchanging) reference (e.g. $A$35). The cell references will remain constant irrespective of the position of the cell containing the formula.
- Before copying the formula, double click on the cell to edit the formula – the cursor will appear in the cell; the cursor may be repositioned using the left or right arrow keys or by clicking the mouse. Now insert $ signs in front of any term which is to remain constant [shortcut: press F4 to insert $ signs]. Finally, use the same copying procedure described above in 2.7.

3. Editing

3.1 Inserting rows

- Click row ID(s) at the point where rows are to be inserted; e.g. highlighting four rows will be a preliminary to inserting four rows.
- Click Insert in the Menu Bar and select Rows.

3.2 Editing cells without retyping

- Double click on the required cell – the cursor will appear in the cell.
- Use the left or right arrow keys (or mouse) to move the cursor to the point where changes are to be made and type corrections (use the Backspace key to delete characters). When the editing is completed, press Enter.

3.3 Blanking out cells

- Select cells with the mouse.
- Click Edit, on the menu bar, and select Clear All (or press the Delete key).

3.4 Deleting rows

- Select row numbers (on left of the screen) with the mouse.
- Click Edit on the Menu bar and click Delete.

3.5 Changing column widths

- Click column ID(s) along the top of the screen; then click Edit, on the menu bar, and select Column Width.
- Alternatively, move the mouse pointer on to a boundary between columns (the pointer will change to a cross with two arrow heads), hold down the left mouse button and 'drag' the boundary to the required width.

4. Printing a spreadsheet

- Click File in menu, select Print and press Enter
- To print part of a spreadsheet: highlight the cells you wish to print; from the File menu, choose Print Area – Set Print Area, then proceed as above.

5. Some commonly needed Excel functions

- Summation (adds the selected cells): = SUM( )

  Type = SUM, open brackets then use the mouse to highlight the cells to be added, close brackets and press Enter.

  Alternatively, use the AutoSum button, which has a sigma (Σ) sign on it, and modify the selection of cells as necessary.
- Absolute value (returns the value without the sign): = ABS ( )
- Common logarithms (see Appendix A): = LOG ( )

  Note since a logarithm is the power to which 10 must be raised to give the number (e.g. log(100) = 2, since $10^2 = 100$), the antilogarithm in Excel is expressed as 10 to the power of the log. Thus the antilog of 2 is written = 10^(2). The caret symbol (^) is located on the keyboard at Shift 6.
- Averages (for individual cases, not groups): = AVG( )
- To see the full set of functions available in Excel, click on the Insert Function button (labelled *fx*) on the standard tool bar.

Appendix D Answers to exercises

Chapter 1

1 Because of the pervasive influence of migration at the sub-national level. International and internal migration not only affect overall growth rates, but also have a great impact on local and regional birth rates, for instance through the inward migration of young couples. Migration is commonly the dominant factor in population change at the sub-national level – in street blocks, suburbs, towns and regions (see Chapter 11).

2 Because the terms on each side of the equation balance: population growth is equal to the sum of the components of growth. In practice, however, population statistics on observed growth, natural increase and net migration never match up exactly. For example, natural increase and net migration may sum to a total greater than the growth apparent between censuses. This difference is known as the 'error of closure'; it affected the pre-1990 statistics in Figure 1.4, as described in the footnotes to the source table in the *Statistical Abstract of the United States*.

3 See Table 1.2.

4 To facilitate comparisons and provide summary measures.

5 The mid-period population approximates the average numbers exposed to the risk of experiencing the demographic event.

6 Because the United States had an older population. Other things being equal, the older the population (i.e. the higher the percentage of the total in older ages), the higher the crude death rate.

7 Crude rates are based on the total population without regard for the true population at risk of experiencing the event. Crude divorce rate: divorces in a calendar year/total mid-year population × 1000. Crude marriage rate: marriages in a calendar year/total mid-year population × 1000. The main disadvantage is that the two rates take no account of the population at risk of marriage or divorce; they are based on the total mid-year population, including children.

Chapter 2

1 Average annual increase.

2 In an arithmetic progression, each number differs from the succeeding one by a constant *amount*, as in 100, 200, 300. In a geometric progression, each number differs from the succeeding one by a constant *ratio*. For example, the numbers double in each interval in the series: 1, 2, 4, 8, 16 and the constant ratio is 1:2.

3 Base the answer on Figure 2.6. In this example, the numbers to be replaced are greater than the gross growth; the difference between them is the net decline.

4 World population change, geometric growth:

Year	Period	Mid-year population	Percentage change	Average annual increase	Geometric growth rate %	Doubling time	Year to reach:
1960		3037					*4 billion:*
1970	1960–1970	3696	21.70	65.90	1.98	35.30	1974.85
1980	1970–1980	4432	19.91	73.60	1.83	38.17	*5 billion:*
1990	1980–1990	5321	20.06	88.90	1.84	37.92	1987.10
2000	1990–2000	6067	14.02	74.60	1.32	52.83	*7 billion:*
							2011.40

To calculate the year in which the population reached 4 billion, find the decade in which the population passed 4 billion (i.e. 1970–80) and use the growth rate for that decade. Since the data refer to the mid-year population (i.e. 1970.5), add the value of n (the interval between two populations) to 1970.5. Similarly, for the year the population reached 5 billion, use the growth rate for 1980–90 and 1980.5 as the base date. For the projected population of 7 billion, use the most recent growth rate (i.e. 1990–2000); the base date is 2000.5.

5 World population change, exponential growth:

Year	Period	Mid-year population	Exponential growth rate %
1960		3037	
1970	1960–1970	3696	1.96
1980	1970–1980	4432	1.82
1990	1980–1990	5321	1.83
2000	1990–2000	6067	1.31

6 Doubling times in years:

Growth rate %	From the geometric growth rate	From the approximation: $70/r\%$
1	69.7	70.0
2	35.0	35.0
3	23.4	23.3
4	17.7	17.5
5	14.2	14.0
6	11.9	11.7

7 Projections at 3 per cent per annum:

Year	Arithmetic growth	Geometric growth	Exponential growth
2005	1 000 000	1 000 000	1 000 000
2006	1 030 000	1 030 000	1 030 455
2007	1 060 000	1 060 900	1 061 837
2008	1 090 000	1 092 727	1 094 174
2009	1 120 000	1 125 509	1 127 497
2010	1 150 000	1 159 274	1 161 834

Chapter 3

1 Because the graph will not necessarily reveal the relative numbers of males and females at each age. For example, in long-lived populations, the higher representation of older women is likely to be concealed.

2 See Section 3.4 pp. 88–9.

3 Lower birth rates and population ageing are the main processes.

4 Small population numbers increase the potential for extreme variations, due to chance or institutional influences – such as boarding schools, hostels, nunneries, monasteries, homes for the aged and military bases.

5 Summary figures:

		Japan	Hong Kong
Sex ratios			
	30–34	83.4	118.8
	60–64	92.6	50.0
	Total	96.2	105.8
Median ages			
	Total	22.3	23.7
	15–64	32.2	30.0
Dependency ratios			
	Child	59.5	45.2
	Aged	8.3	3.8
	Total	67.8	49.1
Ageing index		13.9	8.5

6 Index of dissimilarity = 14.5.

Chapter 4

1 (a) 6.50, (b) 0.74, (c) 6.48.

2 (a) 9.96, (b) 1.35, (c) 9.93.

Chapter 6

1

Age	ASDRs			Sex ratio of ASDRs
	Males	Females	Total	
Under 1	13.4	10.2	11.8	131
1–4	0.5	0.4	0.5	127
5–9	0.2	0.2	0.2	95
10–14	0.3	0.2	0.3	143
15–19	1.3	0.5	0.9	285
20–24	1.7	0.5	1.1	309
25–29	1.9	0.7	1.3	284
30–34	2.2	0.8	1.5	269
35–39	2.8	1.1	2.0	259
40–44	3.5	1.7	2.6	205
45–49	4.8	2.7	3.7	180
50–54	7.6	4.3	5.9	178
55–59	12.1	6.7	9.3	180
60–64	19.0	10.8	14.6	177
65–69	28.2	16.0	21.5	177
70–74	43.4	24.6	32.6	176
75–79	65.8	38.4	49.2	171
80–84	101.0	63.5	76.5	159
85+	177.2	140.0	150.3	127
CDR	9.2	8.1	8.6	

5 The death would be classified by the external cause, that is, drowning. Classification of the death as a suicide would require further evidence, such as a suicide note.

Chapter 7

2

Age	ASFRs per 1000	Female ASFRs /1000 women	Expected survivors of female births/1000
15–19	20.13	9.63	9.55
20–24	64.58	31.41	31.09
25–29	116.64	56.75	56.06
30–34	106.09	51.67	50.93
35–39	43.80	21.11	20.74
40–44	7.61	3.67	3.59
45–49	0.20	0.10	0.09
Total	359.04	174.35	172.07
Times 5	1795.19	871.73	860.36
Rate per woman:	1.80	0.87	0.86
	= TFR	= GRR	= NRR

4 The level of childbearing at which women of reproductive age have sufficient daughters to replace, exactly, their own numbers in the population.

5 No, the replacement level depends on the level of mortality.

6 Because NRR includes the effects of mortality.

7 Since they are calculated from the sum of age-specific rates, this is equivalent to applying direct standardization in which the standard population has one woman in each age group (see Newell 1988: 67; Shryock and Siegel 1973: 525).

Chapter 8

1 (a) 22.02 years.

(b) 36.988 per cent.

(c) 98 563.

(d) 75 338.

(e) 0.003 04.

(f) 937 (out of an original 100 000 live births).

(g) 2400.

(h) 908.

(i) 0.

(j) 9.741 29 at age 62.

2 (a) q_0 Probability of dying between exact ages 0 and 1.

(b) q_x Probability of dying between exact ages x and $x + 1$.

(c) p_{21} Probability of surviving from exact age 21 to exact age 22.

(d) p_x Probability of surviving from exact age x to exact age $x + 1$.

(e) M_x Age-specific death rate of persons aged x last birthday (from observed data).

(f) e_0 Life expectancy at age 0.

(g) e_{41} Life expectancy at age 41.

(h) e_{84} Life expectancy at age 84.

(i) e_x Life expectancy at age x.

(j) m_x Age-specific death rate of persons aged x last birthday (from life table data).

3

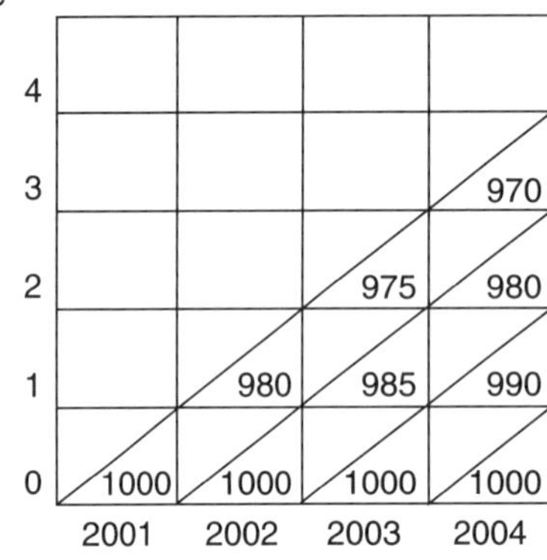

4

Age	Pop.	Deaths	M_x	q_x	p_x	l_x	d_x	L_x	T_x	e_x
0	117 724	1769	0.015 03	0.014 91	0.985 09	100 000	1491	98 956	6 956 234	69.56
1	121 143	143	0.001 18	0.001 18	0.998 82	98 509	116	98 439	6 857 279	69.61
2	125 743	101	0.000 80	0.000 80	0.999 20	98 392	79	98 353	6 758 840	68.69
3	129 258	88	0.000 68	0.000 68	0.999 32	98 313	67	98 280	6 660 487	67.75
4	137 559	69	0.000 50	0.000 50	0.999 50	98 246	49	98 209	6 562 207	66.79

5

(a) No. the form of the age profile of a stationary population depends on the mortality rates: high mortality at young ages produces a triangular profile, since relatively few survive to middle and older ages. Conversely, low mortality at all but the oldest ages produces a more rectangular profile.

(b) Because survival is selective: those who live to age 60 are healthier, or better adapted to survive, than those who die at younger ages. Life expectancy at birth is therefore reduced by the presence of some who will die prematurely.

(c) The sum equals T_0.

(d) The sum equals l_0.

(e) The crude birth rate of the life table population.

(f) Because l_x refers to an exact age, not to an interval between two exact ages.

(g) Because T_x refers to the total population aged exactly x or more.

Chapter 9

1 (a) Yes. It is a special case of a stable population, because it has the same characteristics, except that its growth rate is zero and, therefore, the numbers in each age group remain constant.

(b) The increase, or decrease, in population size that would occur if fertility changed immediately to replacement level. Whereas the intrinsic rate of natural increase indicates the growth rate implicit in current fertility and mortality, ignoring the age structure, population momentum shows the growth potential implied by the age structure alone (Pressat 1985: 150).

(c) Reduce the birth rate. The higher the level of natural increase, the higher the momentum. More particularly, reduce the birth rate to below replacement level, which leads to negative momentum and population decline. China's 'One Child Policy' is an example of a strategy with the potential to reduce fertility below replacement level.

(d) Population ageing entails an increase in the percentage of the population in older ages, often taken as 65 years and over.

(e) Model life tables are sets of hypothetical life tables spanning a wide range of life expectancies as well as major patterns of age-specific mortality. They have varied applications, including providing mortality assumptions for demographic estimates and projections and enabling the estimation of new life tables from limited data.

(f) The median depends on the actual ages of people in the group. If everyone in age group 20–24 was aged 23 or 24, the median would be greater than the mid-point (22.5 years).

(g) The intrinsic growth rate, or the intrinsic rate of natural increase, is the constant growth rate (r) of a stable population. The rate is 'intrinsic' to, or inherent within, the population concerned because it is based solely on its age-specific fertility and mortality in the year of observation.

(h) This can occur when a population has a birth rate below replacement level, but its age structure still has positive momentum.

(i) Only part of the growth is due to momentum arising from the movement of larger cohorts to higher ages (assuming replacement level fertility); usually, more of the growth is due to self-reinforcing processes, whereby larger generations of parents beget even larger generations of children (assuming above-replacement fertility).

2 Mean length of a generation = 28.94, $r = -0.00712$.

3 Gross reproduction rate: 1.04

Net reproduction rate: 1.02

Mean length of a generation: 27.38

Intrinsic rate of natural increase: 0.08%

Age distribution of the stable population (percentages):

Age	0–4	5–9	10–14	15–19	20–24	25–29	30–34	35–39	40–44	45–49	50–54
Males	3.4	3.4	3.4	3.3	3.3	3.3	3.3	3.2	3.2	3.1	3.0
Females	3.2	3.2	3.2	3.2	3.2	3.2	3.1	3.1	3.1	3.1	3.0
Age	**55–59**	**60–64**	**65–69**	**70–74**	**75–79**	**80–84**	**85–89**	**90–94**	**95–99**	**100+**	**Total**
Males	2.9	2.7	2.5	2.1	1.6	1.0	0.5	0.2	0.0	0.0	49.3
Females	2.9	2.8	2.7	2.5	2.1	1.6	0.9	0.4	0.1	0.0	50.7

4 99 789 480

5

% < 30	Momentum
50	1.35
40	1.08
30	0.81

Note: $(T_0 - T_{30})/T_0 = 0.37$

6 Across all four life expectancies, the higher the GRR the lower the percentages aged 65 years and over. When fertility is very high (GRR = 4), variations in life expectancy have little effect on population ageing. At lower levels of fertility, very high life expectancy (80 years) augments population ageing. This effect is most pronounced when fertility is below replacement level (e.g. GRR = 0.5). Thus long-lived populations with very low fertility have the highest levels of ageing.

	% Total population aged 65 years and over			
Female e_0	GRR			
	0.5	1.0	2.0	4.0
20	23	12	5	2
40	27	14	5	2
60	30	15	5	2
80	38	19	7	2

Chapter 10

1 (a) LQ > 1: a relatively high percentage, compared with that for the base population.

(b) LQ = 1: the percentage is identical to that for the base population.

(c) LQ < 1: a relatively low percentage, compared with that for the base population.

2 No, the results would be affected by contrasts between the numbers and characteristics of areas in each country; for instance, the larger the number of provinces or states the greater the apparent redistribution.

3 Either aggregate the areas into larger units with fixed boundaries through time, or aggregate small area data to coincide with the areas required.

4

Region	Density 2000 (square miles)	Density 2000 (square kilometres)
Western	98.7	38.1
Eastern	100.2	38.7
Northern	52.6	20.3
Middle	37.6	14.5
Southern	48.4	18.7
Total	68.3	26.4

5 (a) It is affected by the number of areas – more detailed data are likely to show a higher level of redistribution.

(b) Whereas a value of 1 for a location quotient is a benchmark figure distinguishing between relatively high and relatively low values, there is no such benchmark for the index of redistribution.

(c) Contrasting pairs of percentage distributions can produce similar indices of redistribution (see Box 10.1).

6 Index of redistribution = 6.5; index of concentration = 47.7.

Chapter 11

1

	Change 1996–2001	Natural increase	Net migration
Green Lane	–300	854	–1154
Kensington	4550	2683	1867
Newton	100	184	–84
Ponsonby	250	541	–291

2 No information is provided on origins or destinations, gross migration, age–sex composition, or the volume of net internal migration compared with that of net external migration. Also, the figures refer only to the resident population – the effects of temporary migrations are excluded.

3

Age	Net migration
5–9	−134
10–14	−163
15–19	669
20–24	222
25–29	−69
30–34	−62
35–39	−2
40–44	−15
45–49	−16
50–54	−4
55–59	−39
60–64	0
65+	−31
Total	355

4 If the younger age group was under-enumerated at the earlier census, such as due to an undercounting of infants and young children. An advantage of the CSR method is that the ratios, although inaccurately measuring survival, may correct for under-enumeration in the census (see United Nations 1970: 29).

5 Yes, provided that closure is achieved by eliminating the effects of external migration during the intercensal period (see Section 11.7, p. 420).

6 When the migration interval is long or mortality is high.

7 Refer to Figure 11.7.

8 This is unlikely, given the widespread occurrence of migration. In a developing country, circular mobility may be vital to sustaining a community economically and enabling families to continue to live in a particular location. In Western societies, populations may experience net migration gains in certain age groups and, at the same time, losses in others. A rural community, for example, may lose young people leaving school and older people retiring from work, while gaining young families returning home or attracted by employment, cheaper housing or environmental characteristics. Although the sum of all the age-specific figures may produce a total net migration around zero, the balance between inward and outward migration can be essential to population replacement and maintaining a community's economic and demographic viability.

9 Migration expectancy is a synthetic cohort measure that does not take account of changes through time in migration probabilities; it also assumes no more than one move per person in the time period.

10 Because L_x approximates the average size of the cohort in the age interval and because the survival ratios are applied to census data on age last birthday (i.e. the numbers in an age interval, rather than at exact ages).

Chapter 12

6 Add together the following separately obtained components:

- Male deaths from the initial population (total initial male population minus total male survivors).
- Female deaths from the initial population (total initial female population minus total female survivors).
- Deaths of male children born in the interval (total male births minus total male births × survival ratio of male births).
- Deaths of female children born in the interval (total female births minus total female births × survival ratio of female births).
- Deaths of male net migration gains (total male net inward migration minus total male survivors in the net migrant population).
- Deaths of female net migration gains (total female net inward migration minus total female survivors in the net migrant population).

Note: if there is an overall net migration loss, the deaths of migrants are not counted because the net migrants died elsewhere. In Box 12.2, the projection methodology assumes that the net migration losses are subtracted (from survivors) at the end of the interval and, hence, that all net migration losses occurred at the end of the interval – as in the forward survival ratio method of estimating net migration (see Section 11.6). A more complex projection methodology could allow for net migration losses among both survivors and non-survivors from the initial population, but this would have to be weighed against the benefits of additional complexity.

Chapter 13

3

Religion	Proportion	x^2
1	0.062	0.003 84
2	0.004	0.000 02
3	0.014	0.000 20
4	0.049	0.002 40
5	0.009	0.000 08
6	0.000	0.000 00
7	0.002	0.000 00
8	0.020	0.000 40
9	0.778	0.605 28
10	0.003	0.000 01
11	0.049	0.002 40
12	0.005	0.000 03
13	0.003	0.000 01
14	0.002	0.000 00
Total	1.000	0.614 67

Index of diversity = 0.39
Maximum index = 0.93
Standardized index = 0.41

4

Region	Proportion White	Proportion Non-white	Ratio White/ Non-white	Cumulative proportion White (X)	Cumulative proportion Non-white (Y)	Cross Products $X_i \times Y_{i+1}$	Cross Products $X_{i+1} \times Y_i$	Absolute differences between proportions
A	0.08	0.01	8.00	0.08	0.01	0.0024	0.0022	0.07
B	0.14	0.02	7.00	0.22	0.03	0.0154	0.0117	0.12
C	0.17	0.04	4.25	0.39	0.07	0.0390	0.0322	0.13
D	0.07	0.03	2.33	0.46	0.10	0.0690	0.0560	0.04
E	0.10	0.05	2.00	0.56	0.15	0.1624	0.1065	0.05
F	0.15	0.14	1.07	0.71	0.29	0.7100	0.2900	0.01
G	0.29	0.71	0.41	1.00	1.00			0.42
Total	1.00	1.00	0.00			0.9982	0.4986	0.84
Gini index =							**0.4996**	
Segregation index (index of dissimilarity) =								**0.42**

Bibliography

Abler, R., Adams, J.S., and Gould, P. 1972. *Spatial Organization*. New Jersey: Prentice-Hall.

Ameil, Y. and Cowell, F.A. 1999. *Thinking about Inequality*. Cambridge: Cambridge University Press.

Anon. 1984. 'The European Parliament on the Need for Promoting Population Growth', (Documents). *Population and Development Review*, 10 (3): 569–570.

Anon. 1998. 'Metropolitan Areas in the 1990s'. *American Demographics* (May, on the Internet).

Ariès, Philipe. 1980. 'Two Successive Motivations for the Declining Birth Rate in the West'. *Population and Development Review*, 6 (4): 645–650.

Armitage, P. and Berry, G. 1987. *Statistical Methods in Medical Research*. Oxford: Blackwell Scientific Publications.

Australian Bureau of Statistics (annual). *Causes of Death*. Canberra: Australian Bureau of Statistics.

Australian Bureau of Statistics (annual). *Deaths*. Canberra: Australian Bureau of Statistics.

Australian Bureau of Statistics. 1997. *Developing an Approach to the Analysis and Presentation of Multiple Causes of Death Data*. Canberra: Australian Bureau of Statistics.

Australian Bureau of Statistics. 1999. *Household and Family Projections, Australia 1996–2021*. ABS Catalogue No. 3236.0. Canberra: Australian Bureau of Statistics.

Australian Bureau of Statistics 2000. *Causes of Death, Australia*. Canberra: Australian Bureau of Statistics.

Australian Bureau of Statistics 2002. *Regional Population Growth*. Canberra: Australian Bureau of Statistics.

Bailey, T.C. and Gattrell, A.C. 1995. *Interactive Spatial Data Analysis*. Harlow: Longman.

Baker, Therese L. 1988. *Doing Social Research*. New York: McGraw-Hill.

Baldassare, Mark. 1979. *Residential Crowding in Urban America*. Berkeley: University of California Press.

Barclay, George W. 1966. *Techniques of Population Analysis*. New York: John Wiley.

Batutis, Michael J. 1993 'Estimates preparation: methods and procedures'. In Bogue et al. 1993, V5, pp. 20-14 to 20-19.

Becker, Howard S. 1986. *Writing for Social Scientists: How to Start and Finish Your Thesis, Book or Article*. Chicago: University of Chicago Press.

Bell, Martin. 1996. 'How Often Do Australians Move? Alternative Measures of Population Mobility'. *Journal of the Australian Population Association*, 13 (2): 101–124.

Bell, Martin and Hugo, Graeme 2000. *Internal Migration in Australia 1991–1996: Overview and the Overseas-Born*. Canberra: Department of Immigration and Multicultural Affairs.

Berry, Brian J.L. 1974. *The Human Consequences of Urbanisation*. London: Macmillan.

Betts, K. and Seitz, A. 1994. *Writing Essays and Research Reports*. Melbourne: Nelson.

Bhrolchain, Maire Ni. 1992. 'Period Paramount? A Critique of the Cohort Approach to Fertility'. *Population and Development Review*, 18 (4): 599–629.

Biko, Agozino (editor). 2000. *Theoretical and Methodological Issues in Migration Research*. London: Ashgate.

Bogue, Donald J. 1997. *Defining a New Demography: Curriculum Needs for the 1990's and Beyond*. Chicago: Social Development Center.

Bogue, Donald J. et al. 1993. *Readings in Population Research Methodology* (eight volumes). Chicago: Social Development Center.

Bohrnstedt, George W. and Knoke, David. 1988. *Statistics for Social Data Analysis*. Itasca, IL: Peacock Publishers.

Bongaarts, John. 1975. 'Why High Birth Rates are so Low'. *Population and Development Review*, 1: 289–296.

Bongaarts, John and Bulato, Rodolfo A. 1999: 'Completing the Demographic Transition'. *Population and Development Review*, 25 (3), 515–529.

Bongaarts, John and Bulatao, Rodolofo A. (editors). 2000. *Beyond Six Billion: Forecasting the World's Population*. Washington: National Academy Press. Full text available on-line (http://www.nap.edu/catalog/9828.html).

Bongaarts, John and Potter, Robert G. 1983. *Fertility, Biology and Behaviour: An Analysis of Proximate Determinants*. New York: Academic Press.

Boorstin, Daniel J. 1984. *The Discoverers*. London: Dent.

Borrie, W.D. 1973. 'The Place of Demography in the Development of the Social Sciences', in *International Population Conference, Liège*. Liège: International Union for the Scientific Study of Population, pp. 73–93.

Borrie, W.D. 1977. 'Some social aspects of ageing'. *Proceedings of the Thirteenth Annual Conference of the Australian Association of Gerontology*. Melbourne: Australian Association of Gerontology, pp. 12–20.

Borrie, W.D. (chairman). 1975. *Population and Australia: A Demographic Analysis and Projection*. First Report of the National Population Inquiry. Canberra: Australian Government Publishing Service.

Bouma, Gary D. 2000. *The Research Process*. Melbourne: Oxford University Press.

Bourgeois-Pichat, J. 1989. 'From the 20th to the 21st Century: Europe and its Population after the Year 2000.' *Population*, English Selection No. 1, 44: 57–90.

Bouvier, Leon F. and De Vita, Carol J. 1991. 'The Baby Boom: Entering Midlife'. *Population Bulletin*, 46 (3).

Brass, W. 1971. 'On the scale of mortality', in Brass, W. (editor), *Biological Aspects of Demography*. London: Taylor and Francis.

Brass, William. 1996. 'Demographic Data Analysis in Less Developed Countries, 1946–1996'. *Population Studies*, 50 (3): 451–467.

Broe, G.A. and Creasey, Helen. 1995. 'Brain Ageing and Neurodegenerative Diseases: a Major Public Health Issue for the Twenty-First Century'. *Perspectives in Human Biology*, 1: 53–58.

Bryman, Alan. 2001. *Social Research Methods*. Oxford: Oxford University Press.

Caldwell, John C. 1996. 'Demography and Social Science'. *Population Studies*, 50: 305–333.

Carmichael, Gordon A. 1995. 'Consensual Partnering in the More Developed Countries'. *Journal of the Australian Population Association*, 12 (1): 51–86.

Carr-Saunders, A. M. 1936. *World Population: Past Growth and Present Trends*. Oxford: Clarendon Press.

Castles, Ian 1986. 'Social Statistics and Social Change'. Paper presented at the Third National Conference of the Australian Population Association, Adelaide.

Castles, Ian 1998. 'Malthus as Scientist'. Paper presented at the National Academies Forum on Malthus and His Legacy: 200 Years of the Population Debate. Canberra.

Castles, Stephen and Miller, Mark J. 1998. *The Age of Migration* (second edition). New York: Guilford Press.

Champion, T. and Fielding, T. (editors). 1992. *Migration Processes and Patterns*. London: Belhaven Press.

Chesnais, Jean-Claude. 1992. *The Demographic Transition: Stages, Patterns and Economic Implications: A Longitudinal Study of Sixty-Seven Countries Covering the Period 1720–1984*. Oxford: Clarendon Press.

Christie, David, Gordon, Ian, and Heller, Richard. 1997. *Epidemiology: An Introductory Text for Medical and other Health Science Students*. Sydney: University of New South Wales Press.

Clark, W.A.V. 1986. *Human Migration*. Beverly Hills: Sage.

Coale, A. J. and Trussell, T. J. 1974. 'Model Fertility schedules: Variations in the age structure of childbearing in human populations'. *Population Index*, 40: 185–258.

Coale, Ansley J. and Guo, Guang, 1989. 'Revised regional model life tables at very low levels of mortality'. *Population Index* 55: 613–643.

Coale, Ansley J. 1973. 'The Demographic Transition', in *International Population Conference, Liège*. Liège: International Union for the Scientific Study of Population, pp. 53–72.

Coale, Ansley J. and Demeny, Paul. 1983. *Regional Model Life Tables and Stable Populations* (second edition). New York: Academic Press.

Coale, Ansley and Guo, Guang. 1990. 'New Regional Model Life Tables at High Expectation of Life'. *Population Index*, 56: 26–41.

Coale, Ansley and Trussell, James. 1996. 'The Development and Use of Demographic Models'. *Population Studies*, 50 (3): 469–484.

Coale, Ansley J. and Watkins, Susan Cotts (editors). 1986. *The Decline of Fertility in Europe: The Revised Proceedings of a Conference on the Princeton European Fertility Project*. Princeton, NJ: Princeton University Press.

Coale, John P. and King, Cuchlaine A.M. 1969. *Quantitative Geography: Techniques and Theories in Geography*. London: John Wiley.

Cohen, Robin (editor). 1996. *Theories of Migration*. Cheltenham, UK: E. Elgar.

Coleman, D.A. 1998. 'Reproduction and Survival in an Unknown World'. *NIDI Hofstee Lecture Series*, 15. The Hague: Netherlands Interdisciplinary Demographic Institute.

Coleman, David. 1999. 'Demographic Data for Europe: A Review of Sources'. *Population Trends* (Winter): 42–52.

Coulter, Philip B. 1989. *Measuring Inequality: A Methodological Handbook*. Boulder: Westview Press.

Courgeau, Daniel. 1995. 'Migration Theories and Behavioural Models'. *International Journal of Population Geography*, 1 (1): 19–27.

Cox, Peter, R. 1970. *Demography*. Cambridge: Cambridge University Press.

Creedy, John et al. 1984. *Economics: An Integrated Approach*, Englewood Cliffs, NJ: Prentice Hall.

Crimmins, Eileen M. 1993. 'Demography: the Past 30 Years, the Present, and the Future'. *Demography*, 30: 579–591.

Daugherty, Helen G. and Kammeyer, Kenneth C.W. 1995. *An Introduction to Population.* New York: Guilford Press.

Davis, H.C. 1995. *Demographic Projection Techniques for Regions and Smaller Areas: A Primer.* Vancouver: University of British Columbia Press.

Davis, K., Bernstam, M. S., Ricardo-Campbell, R. (editors) 1986. 'Below-replacement fertility in industrial societies: causes, consequences, policies'. *Supplement to Population and Development Review*, Volume 12.

Day, Jennifer Cheeseman. 1996. *Projections of the Number of Households and Families in the United States: 1995 to 2010*, US Census Bureau, Current Population Reports, P25–1129. Washington, DC: US Government Printing Office.

Defoe, Daniel 1772. *A journal of the plague year: being observations or memorials of the most remarkable occurrences, as well publicke as private, which happened in London during the last great visitation in 1665.* (Oxford English novels, 1969). London: Oxford University Press.

Demko, G. J. Rose, H. M. and Schnell G. A. (editors) 1970. *Population geography: a reader.* New York: McGraw-Hill.

Dent, Borden D. 1985. *Cartography: Thematic Map Design.* Dubuque, IA: William C. Brown.

Dorling, D. 1995. *A New Social Atlas of Britain.* London: John Wiley.

Dorling, D. 1996. 'Area Cartograms: Their Use and Creation.' *Concepts and Techniques in Modern Geography*, 59: 1–68.

Downs, Anthony. 1994. *New Visions for Metropolitan America.* Washington, DC and Cambridge, MA: The Brookings Institution and Lincoln Institute of Land Policy.

Duncan, Otis Dudley. 1957. 'The Measurement of Population Distribution'. *Population Studies*, 11 (1): 27–45.

Easterlin, Richard A. 1996. 'Economic and Social Implications of Demographic Patterns', in Robert H. Binstock and Linda K. George (editors), *Handbook of Aging and the Social Sciences* (fourth edition). San Diego: Academic Press, pp. 84–93.

Economic and Social Commission for Asia and the Pacific (ESCAP). 2001. *Population Headliners*, September–October: 3.

Ehrlich, Paul. 1970. *The Population Bomb: Population Control or Race to Oblivion?* New York: Ballantine Books.

English, Brian A. and King, Raymond J. 1983. *Families in Australia.* Sydney: Family Research Unit, University of New South Wales.

Espenshade, T.J. 1985. 'Multistate Projections of Population by Age and Marital Status'. *International Population Conference*, Florence. Liège: International Union for the Scientific Study of Population.

Fan Jingjing. 1995. 'China Population Structure by Age and Sex'. *Data User Service Series*, No. 6. Beijing: China Publishing House.

Fields, Jason and Casper, Lynne M. 2000. *America's Families and Living Arrangements.* US Census Bureau, Current Population Reports, P20-537. Washington, DC: US Government Printing Office.

Friedlander, D. 1969. 'Demographic Responses and Population Change'. *Demography*, 6: 359–381.

Ghosh, M. and Rao, J.N.K. 1994. 'Small Area Estimation: An Appraisal'. *Statistical Science*, 9: 55–93.

Gibaldi, Joseph. 1998. *MLA Style Manual and Guide to Scholarly Publishing*. New York: Modern Language Association of America.

Giddens, Anthony 1997. *Sociology*. (Third edition). Cambridge: Polity Press.

Glass, D.V. 1950. 'Graunt's Life Table'. *Journal of the Institute of Actuaries*, 26: 60–64.

Glenn, Norval D. 1977. *Cohort Analysis, Quantitative Applications in the Social Sciences*, Sage University Papers. Beverly Hills: Sage.

Goode, W.J. 1963. *World Revolution and Family Patterns*. New York: The Free Press.

Gottmann, Jean. 1961. *Megalopolis: The Urbanized North-Eastern Seaboard of the United States*. New York: Twentieth Century Fund.

Gould, W.T.S. and Prothero, R.M. 1975: 'Space and Time in African Population Mobility', in Leszek A., Kosinski, and R. Mansell Prothero (editors), *People on the Move: Studies in Internal Migration*. London: Methuen, Chapter 2, pp. 39–49.

Graunt, John. 1662. *Natural and Political Observations Made Upon the Bills of Mortality*, edited by Walter F. Willcox, 1939. Baltimore: Johns Hopkins University Press.

Grebenik, E. 1989. 'Demography, Democracy and Demonology'. *Population and Development Review*, 15 (1): 1–22.

Hajnal, J. 1965. 'European Marriage Patterns in Perspective', in D.V. Glass and D.E.C. Eversley (editors), *Population in History: Essays in Historical Demography*. London: Edward Arnold.

Hajnal, J. 1974. 'European Marriage Patterns in Perspective', in D.V. Glass and D.E.C. Eversley (editors), *Population in History: Essays in Historical Demography*. London: Edward Arnold.

Hakim, Catherine. 1987. *Research Design: Strategies and Choices in the Design of Social Research*. London: Allen and Unwin.

Halli, Shiva S. and Rao, K. Vaninadha. 1992. *Advanced Techniques of Population Analysis*. New York: Plenum Press.

Hannagan, T.J. 1982. *Mastering Statistics*. Sutton, Surrey: Macmillan.

Harper, Andrew C., Holman, C.D.J., and Dawes, Vivienne P. 1994. *The Health of Populations*. Melbourne: Churchill Livingstone.

Hartley, Shirley F. 1982. *Comparing Populations*. Belmont, CA: Wadsworth Publishing Company.

Haskey, J. 1988. 'Mid-1985 Based Population Projections by Marital Status'. *Population Trends*, 52: 30–32.

Haub, Carl 1987: 'Understanding Population Projections'. *Population Bulletin*, 42 (4) (December).

Haupt, Arthur. 1986. 'Halley's Other Comet'. *Population Today*, 14 (1): 3 and 10–11.

Hauser P. M. and Duncan O. D. (editors) 1959. *The study of population: an inventory and appraisal*. Chicago : University Of Chicago Press.

Hauser, Philip M. 1976. 'Aging and World-Wide Population Change'. In Robert H. Binstock and Ethel Shanas (editors), *Handbook of Aging and the Social Sciences*. New York: Van Nostrand Reinhold, pp. 59–86.

Hennekens, Charles H. and Buring, Julie E. 1987. *Epidemiology in Medicine*. Boston: Little, Brown and Company.

Henry, Louis. 1976. *Population: Analysis and Models*. London: Edward Arnold.

Hessler, Richard M. 1992. *Social Research Methods*. St Paul: West Publishing Company.

Hill, K. and Zlotnik, H. 1982. 'Indirect Estimation of Fertility and Mortality', in J.A. Ross (editor), *International Encyclopaedia of Population.* New York: Free Press, pp. 324–34.

Hinde, Andrew. 1998. *Demographic Methods.* London: Arnold.

Hobcraft, J., Menken, Jane, and Preston, Samuel. 1982. 'Age, Period and Cohort Effects in Demography: A Review'. *Population Index*, 48: 4–43.

Hollingsworth, T.H. 1969. *Historical Demography.* London: The Sources of History Limited in association with Hodder and Stoughton.

Hong, Lawrence K. 1992. 'Simple Procedures for Laboratory Demonstrations of Ecological Fallacy and Psychological Reductionism'. *Teaching Sociology*, 20: 292–297.

Hugo, Graeme J. 1982. 'Circular Migration in Indonesia'. *Population and Development Review*, 8 (1): 59–83.

Hugo, Graeme 1986. *Australia's Changing Population: Trends and Implications.* Melbourne: Oxford University Press.

Hugo, Graeme J. 1991. 'What Population Studies Can Do for Business'. *Journal of the Australian Population Association*, 8 (1): 1–22.

Isserman, Andrew M. and Fisher, Peter S. 1984. 'Population Forecasting and Local Economic Planning: The Limits on Community Control Over Uncertainty'. *Population Research and Policy Review*, 3: 27–50.

Jackson, John A. 1986: *Migration.* Aspects of Modern Sociology. London: Longman.

Jain, S.K. 1979. *Basic Mathematics for Demographers.* Demography Teaching Notes 1. Canberra: Development Studies Centre, Australian National University.

Jones, F. Lancaster. 1967. 'A Note on "Measures of Urbanization," with a Further Proposal'. *Social Forces*, 46 (2): 275–279.

Kertzer, David I. 1983. 'Generation as a Sociological Problem'. *Annual Review of Sociology*, 9: 125–149.

Keyfitz, Nathan. 1971. 'On the Momentum of Population Growth'. *Demography*, 8: 71–80.

Keyfitz, Nathan. 1972. 'Population Theory and Doctrine: A Historical Survey', in William Petersen (editor), *Readings in Population.* New York: Macmillan, pp. 41–49.

Keyfitz, Nathan. 1981. 'The Limits of Population Forecasting'. *Population and Development Review*, 7 (4): 579–593.

Keyfitz, Nathan. 1993. 'Thirty Years of Demography and *Demography*'. *Demography*, 30 (4): 533–550.

Keyfitz, Nathan and Flieger, Wilhelm. 1971. *Population: Facts and Methods of Demography.* San Francisco: W.H. Freeman.

Kim, Young J. and Schoen, Robert. 1997. 'Population Momentum Expresses Population Aging'. *Demography*, 34 (3): 421–427.

King, Dave, Hayden, Janet, Jackson, Roger, Holmans, Alan, and Anderson, Dorothy. 2000. 'Population of Households in England to 2021'. *Population Trends*, 99 (Spring): 13–19.

Kinsella, Kevin and Velkoff, Victoria A. 2001. *An Aging World: 2001*, US Census Bureau, Series P95/01-1. Washington, DC: US Government Printing Office.

Kintner, H., Merrick, Thomas W., Morrison, Peter A., and Voss, Paul R. (editors). 1994. *Demographics: A Casebook for Business and Government.* Boulder, CO: Westview Press.

Kirk, Dudley. 1972. 'Population (Part 1: The Field of Demography)', in David S. Sills (editor), *International Encyclopedia of the Social Sciences.* New York: Macmillan and the Free Press, Volume 12, pp. 342–348.

Kirk, Dudley. 1996. 'The Demographic Transition'. *Population Studies*, 50: 361–387.

Kleinbaum, David G., Kuper, Lawrence L., and Morgenstern, Hal. 1982. *Epidemiologic Research: Principles and Quantitative Methods*. New York: Van Nostrand Reinhold.

Klosterman, Richard E. 1990. *Community Analysis and Planning Techniques*. Savage, MD: Rowman and Littlefield.

Knodel, John. 1999. 'Deconstructing Population Momentum'. *Population Today*, 27 (3) (March): 1, 2, 7.

Kosinski, L.A. and Prothero, R. Mansell (editors). 1975. *People on the Move: Studies in Internal Migration*. London: Methuen.

Krebs, Charles J. 1978. *Ecology: The Experimental Analysis of Distribution and Abundance*. New York: Harper and Rowe.

Laroche, Benoit. 1993. 'The Future of Population Census', in *International Population Conference, Montreal*, Volume 3. Liège: International Union for the Scientific Study of Population, pp. 151–155.

Laslett, Peter (editor). 1973. *The Earliest Classics: John Graunt and Gregory King*. Pioneers of Demography Series. Farnborough, Hants: Gregg.

Laurillard, Diana. 1993. *Rethinking University Teaching: a Framework for the Effective Use of Educational Technology*. London: Routledge.

Lawton, Richard. 1986. 'Population', in John Langton and R.J. Morris (editors), *Atlas of Industrializing Britain 1780–1914*. London: Methuen.

Lee, E.S. 1966: 'A Theory of Migration'. *Demography*, 1: 47–57.

Lewins, F. 1987. *Writing a Thesis*. Canberra: ANU Press.

Lewis, G.J. 1982. *Human Migration*. London: Croom Helm.

Lieberson, Stanley. 1969. 'Measuring Population Diversity'. *American Sociological Review*, 34: 850–862.

Lilienfeld, David E. and Stolley, Paul D. 1994. *Foundations of Epidemiology* (third edition). New York: Oxford University Press.

Lin, Nan. 1976. *Foundations of Social Research*. New York: McGraw-Hill.

Long, Larry H. 1970. 'On Measuring Geographic Mobility'. *Journal of the American Statistical Association*, 65: 1195–1204.

Long, Larry H. 1988. *Migration and Residential Mobility in the United States*. New York: Russell Sage Foundation.

Lopez, Alan D. 1983. 'The Sex Mortality Differential in Developed Countries', in Alan D. Lopez and Lado Rukicka (editors), *Sex Differentials in Mortality: Trends, Determinants and Consequences*. Canberra: Department of Demography, Australian National University, pp. 53–120.

Lucas, David. 1994. 'World Population Growth and Theories', in Lucas and Meyer (1994), Chapter 2, pp. 13–28.

Lucas, David and Meyer, Paul (editors). 1994. *Beginning Population Studies*. Canberra: National Centre for Development Studies, Australian National University.

Lucas, David, McMurray, Christine, and Streatfield, Kim. 1989. *Looking at the Population Literature*. Demography Teaching Notes 6. Canberra: National Centre for Development Studies, Australian National University.

Lunn, David J., Simpson, Stephen N., Diamond, Ian, and Middleton, Liz. 1998: 'The Accuracy of Age-Specific Population Estimates for Small Areas in Britain'. *Population Studies*, 52: 327–344.

Lutz, Wolfgang, Vaupel, James W., and Ahlburg, Dennis A. (editors). 1998. 'Frontiers of Population Forecasting'. Supplement to *Population and Development Review*, 24. New York: Population Council.

Magri, Solange. 1994. 'Urban Density Definitions'. *Urban Futures* 4 (2 and 3): 49–53.

Malthus, T.R. 1798. *An Essay on the Principle of Population*, edited by Geoffrey Gilbert, 1993, World's Classics Series. Oxford: Oxford University Press.

Malthus, T.R. 1830. 'A Summary View of the Principle of Population', in George J. Demko, Harold M. Rose, and George A. Schnell. 1970. *Population Geography: A Reader*. New York: McGraw-Hill, pp. 44–79.

Massey, Douglas, Arango, Joaquin, Hugo, Graeme, Kouaouci, Ali, Pellegrino, Adela, and Taylor, J. Edward. 1993. 'Theories of International Migration: A Review and Appraisal'. *Population and Development Review*, 19 (3): 431–466.

McDonald, Peter. 1995. *Families in Australia*. Melbourne: Australian Institute of Family Studies.

McDonald, Peter and Kippen, Rebecca. 2001. 'Labor Supply Prospects in 16 Developed Countries, 2000–2050'. *Population and Development Review*, 27 (1): 1–32.

McFalls, Joseph A. 1998. 'Population: a Lively Introduction' (third edition). *Population Bulletin*, 53 (3).

McNeill, Patrick. 1985. *Research Methods*. London: Tavistock Publications.

McNicoll, Geoffrey. 1984. 'Consequences of Rapid Population Growth: Overview and Assessment'. *Population and Development Review*, 10 (2): 177–240.

McNicoll, Geoffrey. 1992. 'The Agenda of Population Studies: a Commentary and Complaint'. *Population and Development Review*, 18 (3): 399–420.

McVey, Wayne W. and Kalbach, Warren E. 1995. *Canadian Population*. Ontario: Nelson Canada.

Meadows, Donella H., Meadows, Dennis L., Randers, Jorgen, and Behrens, William W. 1974. *The Limits to Growth: A Report for the Club of Rome's Project on the Predicament of Mankind*. London: Pan Books.

Meadows, Donella H., Meadows, Dennis L., and Randers, Jorgen. 1992. *Beyond the Limits: Global Collapse or a Sustainable Future?* London: Earthscan Publications Limited.

Merrick, Thomas W. and Tordella, Stephen J. 1988: 'Demographics: People and Markets'. *Population Bulletin*, 43 (1) (February).

Minchiello, Victor. 1990. *In-depth Interviewing: Researching People*. Melbourne: Longman Cheshire.

Mitchell B.R. 1975. *European Historical Statistics 1750–1970*. London: Macmillan.

Morrill, Richard L. 1974. *The Spatial Organization of Society*. North Scituate, MA: Duxbury Press.

Moser, C. and Kalton, G. 1975. *Survey Methods in Social Investigation*. London: Heinemann.

Murdock, Steve H. and Ellis, David R. 1991. *Applied Demography: An Introduction to Basic Concepts, Methods, and Data*. Boulder: Westview Press.

Murray, Christopher J.L. and Chen, Lincoln C. 1992. 'Understanding Morbidity Change'. *Population and Development Review*, 18: 481–503.

Myers, Dowell (editor). 1990. *Housing Demography: Linking Demographic Structure and Housing Markets*. Madison, WI: University of Wisconsin Press.

Myers, George C., Torrey, Barbara Boyle, and Kinsella, Kevin. 1989. 'The Paradox of the Oldest Old in the United States: An International Comparison', in Richard M. Suzman, David P. Willis, and Kenneth G. Manton (editors), *The Oldest Old*. New York: Oxford University Press.

Nam, Charles B. 1994. *Understanding Population Change*. Itasca, IL: F.E. Peacock.

Namboodiri, Krishnan. 1991. *Demographic Analysis: a Stochastic Approach*. San Diego: Academic Press.

National Centre for Health Statistics. 2002. *National Vital Statistics Report*, 50 (5).

National Institute on Aging. 1998. *Aging in the Americas into the XXI century* (wall chart). National Institute on Aging, PAHO/WHO and US Bureau of Census (place of publication not specified).

Neugarten, Bernice L. and Hagestad, Gunhild O. 1976. 'Age and the Life Course', in Robert H. Binstock and Ethel Shanas (editors), *Handbook of Aging and the Social Sciences*. New York: Van Nostrand Reinhold Company, pp. 35–55.

Neutze, Max. 1977. *Urban Development in Australia: a Descriptive Analysis*. Sydney: Allen and Unwin.

Newell, C. 1988. *Methods and Models in Demography*. London: Frances Pinter.

Newman, David M. 2000. *Sociology: Exploring the Architecture of Everyday Life*. Thousand Oaks (California): Pine Forge Press.

O'Neill, Brian and Balk, Deborah. 2001. 'World Population Futures'. *Population Bulletin*, 56 (3).

Olshansky, S. J. and Ault, A. B. 1986. 'The fourth stage of the epidemiologic transition: the age of delayed degenerative disease'. *Milbank Memorial Fund Quarterly*, 64 (3): 355–391.

Omran, A.R. 1971. 'The Epidemiologic Transition: a Theory of the Epidemiology of Population Change'. *Milbank Memorial Fund Quarterly*, 49: 509–538.

Omran, A.R. 1981. 'The Epidemiologic Transition', in J.A. Ross (editor), *International Encyclopaedia of Population*. New York: The Free Press, pp. 172–175.

Palmore, James A. and Gardner, Robert W. 1983. *Measuring Mortality, Fertility and Natural Increase: a Self-Teaching Guide to Elementary Measures*. Honolulu: East-West Population Institute, East-West Center.

Peach, Ceri. 1999. 'London and New York: Contrasts in British and American Models of Segregation'. *International Journal of Population Geography*, 5: 319–351.

Plane, D.A. and Rogerson, P.A. 1994. *The Geographical Analysis of Population*. New York: Wiley.

Pol, L. and Thomas, R.K. 1997. *Demography for Business Decision Making*. London: Quorum Books.

Pol, Louis G. and Thomas, Richard K. 2000. *The Demography of Health and Health Care*. New York: Plenum Press.

Pollard, A.H., Yusuf, Farhat, and Pollard, G.N. 1990. *Demographic Techniques* (third edition). Sydney: Pergamon Press.

Population Reference Bureau. 1999. *World Population Data Sheet*. Washington: Population Reference Bureau.

Population Reference Bureau. 2000. *World Population Data Sheet*. Washington: Population Reference Bureau.

Population Division of the Department of Economic and Social Affairs of the United Nations Secretariat. 2001a. *World Population Prospects: The 2000 Revision*, Vol. I, *Comprehensive Tables*. New York: United Nations.

Population Division of the Department of Economic and Social Affairs of the United Nations Secretariat. 2001b. *The Sex and Age Distribution of Populations*, United Nations Publications, ST/ESA/Ser.A/199. New York: United Nations.

Population Reference Bureau. 2001a. *World Population Data Sheet*. Washington: Population Reference Bureau.

Population Reference Bureau. 2001b. *United States Population Data Sheet*. Washington: Population Bureau.

Poston, D.L. and Rogers, R.G. 1985. 'Towards a Reformulation of the Neonatal Mortality Rate'. *Social Biology*, 32: 1–12.

Pressat, Roland. 1972. *Demographic Analysis: Methods, Results, Applications*. Chicago: Aldine Publishing Company.

Pressat, Roland. 1973. *Population*. Harmondsworth, Middlesex: Penguin Books.

Pressat, Roland. 1985. *The Dictionary of Demography*, edited by Christopher Wilson. Oxford: Blackwell.

Preston, Samuel H. 1986. 'The Relation between Actual and Intrinsic Growth Rates'. *Population Studies*, 40 (3): 343–351.

Preston, Samuel H. 1993. 'The Contours of Demography: Estimates and Projections'. *Demography*, 30 (4): 593–606.

Preston, Samuel H., Heuveline, Patrick, and Guillot, Michel. 2001: *Demography: Measuring and Modeling Population Processes*. Oxford: Blackwell.

Price, Charles A. 1963. *Southern Europeans in Australia*. Melbourne: Oxford University Press.

Price, Charles A. 1976. 'Some Measures of Population Growth and Replacement', in Charles A. Price and Jean I. Martin (editors), *Australian Immigration: a Bibliography and Digest*, Number 3, Part 1. Canberra: Department of Demography, Australian National University.

Rajbanshi, B.S. and Sharma, K.R. 1980. 'Age and Sex Composition', in United Nations, *Population of Nepal*, ESCAP Country Monograph Series, No. 6. Bangkok: Economic and Social Commission for Asia and the Pacific, pp. 22–30.

Rao, K. Vaninadha and Wicks, Jerry W. 1994. *Studies in Applied Demography*. Bowling Green, OH: Population Research Center, Bowling Green State University.

Ravenstein, E.G. 1885. 'The Laws of Migration'. *Journal of the Statistical Society*, 48: 167–219.

Rees, P. 1994. 'Estimating and Projecting the Populations of Urban Communities'. *Environment and Planning*, A26: 1671–1697.

Rees, Philip. 1997. 'Problems and Solutions in Forecasting Geographical Populations'. *Journal of the Australian Population Association*, 14 (2): 145–166.

Richardson, A.J., Brunton, P.J., and Roddis, S.M. 1998. 'The Calculation of Perceived Residential Density'. *Road and Transport Research* 7 (2): 3–15.

Richmond, Anthony H. 1969. 'Sociology of Migration in Industrial and Post-Industrial Societies', in J.A. Jackson (editor), *Migration*, Sociological Studies 2. Cambridge: Cambridge University Press.

Ridley, Matt. 1994. 'Why Presidents Have More Sons'. *New Scientist*, 3 December: 28–31.

Riley, Nancy E. 1997. 'Gender, Power and Population Change'. *Population Bulletin*, 52 (1) (May).

Rives, Norfleet W. and Serow, William J. 1984. *Introduction to Applied Demography: Data Sources and Estimation Techniques*. Beverly Hills: Sage.

Rockett, Ian R.H. 1999. 'Population and Health: An Introduction to Epidemiology' (second edition). *Population Bulletin*, 54 (4).

Rogers, A. 1990. 'Requiem for the Net Migrant'. *Geographical Analysis*, 22: 283–300.

Rogers, A. and Castro, L.J. 1981. *Model Migration Schedules*. Laxenberg, Austria: International Institute for Applied Systems Analysis (IIASA).

Rogers, R.G. and Hackenberg, R. 1987. 'Extending Epidemiologic Transition Theory: a New Stage'. *Social Biology*, 34 (3–4): 234–243.

Roseman, C.C. 1971. 'Migration as a Spatial and Temporal Process'. *Annals of the Association of American Geographers*, 61: 589–598.

Rosenfield, A. and Maine, D. 1985. 'Maternal Mortality: a Neglected Tragedy. Where is the M in MCH?' *Lancet*, 13 July, 2 (8446): 83–85.

Ross, John A. (editor). 1982. *International Encyclopedia of Population*. New York: Free Press.

Rowland, D.T. 1979. *Internal Migration in Australia*. Census Monograph Series. Canberra: Australian Bureau of Statistics.

Rowland, D.T. 1980. 'Migration between Australian Colonies in the 1880s', in *1888 Bulletin*, No. 5, Bicentennial History Project, pp. 160–169.

Rowland, D.T. 1994a. 'Family Change and Marital Status in Later Life'. *Australian Journal on Ageing*, 13: 168–171.

Rowland, D.T. 1994b. 'Population Policies and Ageing in Asia: a Cohort Perspective', in United Nations, *The Ageing of Asian Populations*. New York: United Nations Department of Economic and Social Information and Policy Analysis, ST/ESA/SER.R/125, pp. 15–32.

Rowland, D.T. 1996. 'Population Momentum as a Measure of Ageing'. *European Journal of Population*, 12: 41–61.

Rowland, D.T. 2002. 'Childlessness', in *International Encyclopedia of Marriage and Family Relationships*. New York: Macmillan.

Ryder, N.B. 1964. 'The Process of Demographic Translation'. *Demography*, 1: 74–82.

Ryder, N.B. 1965. 'The Cohort as a Concept in the Study of Social Change'. *American Sociological Review*, 30: 842–861.

Ryder, N.B. 1972. 'Cohort Analysis', in D.E. Sills (editor), *International Encyclopaedia of the Social Sciences*. New York: Macmillan and the Free Press, pp. 546–550.

Ryder, N.B. 1986. 'Observations on the History of Cohort Fertility in the United States'. *Population and Development Review*, 12 (4): 617–643.

Schachter, Jason. 2001a. 'Geographical Mobility'. *Current Population Reports* P20-538 (http://www.census.gov/population/www/socdemo/migrate/p20-538.html).

Schachter, Jason. 2001b. 'Why People Move: Exploring the March 2000 Current Population Survey'. *Current Population Reports* P23-204 (http://www.census.gov/prod/2001pubs/p23-204.pdf).

Schwartz, Sharon. 1994. 'The Fallacy of the Ecological Fallacy: the Potential Misuse of a Concept and the Consequences'. *American Journal of Public Health*, 84 (5): 819–824.

Sharpe Andrew. 1999. *A Survey of Indicators of Economic and Social Well-being*. Ottawa, Ontario: Centre for the Study of Living Standards (http://www.csls.ca/pdf/paper3a.pdf).

Shaw, Chris. 1999. '1996-Based Population Projections by Legal Marital Status for England and Wales'. *Population Trends*, 95 (Spring): 23–32.

Shaw, R. Paul. 1975. *Migration Theory and Fact; A Review and Bibliography of Current Literature*. Bibliography Series Number Five. Philadelphia: Regional Science Research Institute.

Shryock, Henry S. and Siegel, Jacob S. 1973. *The Methods and Materials of Demography* (two volumes). Washington: US Bureau of the Census, US Government Printing Office.

Siegel, Jacob S. 2001. *Applied Demography*. San Diego: Academic Press.

Simpson, Stephen, Middleton, Liz, Diamond, Ian, and Lunn, David. 1997. 'Small-Area Population Estimates: a Review of Methods used in Britain in the 1990s'. *International Journal of Population Geography*, 3: 265–280.

Skeldon, R. 1990. *Population Mobility in Developing Countries*. London: Belhaven Press.

Slome, Cecil, Brogan, Donna R., Eyres, Sandra J., and Lednar, Wayne. 1986. *Basic Epidemiologic Methods and Biostatistics: a Workbook*. Boston: Jones and Bartlett.

Smith, David M. 1977. *Patterns in Human Geography*. Harmondsworth: Penguin Books.

Smith, David P. 1992. *Formal Demography*. New York: Plenum Press.

Smith, Stanley K. and Mandell, Marylou 1993. 'A comparison of population estimation methods: housing unit versus component II, ratio correlation, and administrative records'. In Bogue et al. 1993, V5, pp. 20–20 to 20–27.

Smith, Stanley K., Tayman, Jeff, and Swanson, David A. 2001. *State and Local Population Projections: Methodology and Analysis*. Plenum Series on Demographic Methods and Population Analysis. Norwell, MA: Kluwer Plenum.

Spiegelman, Mortimer. 1973. *Introduction to Demography*. Cambridge, MA: Harvard University Press.

Standing, Guy. 1982. 'Labor Force', in John A. Ross (editor), *International Encyclopedia of Population*. New York: The Free Press, Volume 1, pp. 391–398.

Stillwell, John and Congdon, Peter (editors). 1991. *Migration Models: Macro and Micro Approaches*. London: Belhaven Press.

Stockwell, Edward G. and Groat, H. Theodore. 1984. *World Population: An Introduction to Demography*. New York: Franklin Watts.

Sullivan, Thomas J. 1992. *Applied Sociology: Research and Critical Thinking*. New York: Macmillan.

Tabah, Leon. 1980. 'World Population Trends, a Stocktaking'. *Population and Development Review*, 6 (3): 355–389.

Taeuber, K.E. 1966. 'Cohort Migration'. *Demography*, 3: 416–422.

Taeuber, K.E. and Taeuber, A.F. 1965. *Negroes in Cities: Residential Segregation and Neighborhood Change*. Chicago University Population Research and Training Center Monographs. Chicago: Aldine.

Tayman, Jeff and Swanson, David A. 1996. 'On the Utility of Population Forecasts'. *Demography*, 33 (4): 523–528.

Teachman, Jay D. 1980. 'Analysis of Population Diversity: Measures of Qualitative Variation'. *Sociological Methods and Research*, 8 (3): 341–362.

Teachman, Jay D., Paasch, Kathleen and Carver, Karen Price 1993. 'Thirty Years of Demography', *Demography*, 30 (4): 523–532.

Teitelbaum, M.S. 1975. 'Relevance of Demographic Transition Theory for Developing Countries'. *Science*, 188: 420–425.

Thomas, W.I. and Znaniecki, F. 1958. *The Polish Peasant in Europe and America*. New York: Dover Publications.

Tien, H. Yuan et al. 1992. 'China's Demographic Dilemmas'. *Population Bulletin*, 47 (1) (June).

Timms, D. 1965. 'Quantitative Techniques in Urban Social Geography', in R.J. Chorley and P. Haggett (editors), *Frontiers in Geographical Teaching*. London: Methuen, pp. 239–265.

Todaro, M.P. 1976. *Internal Migration in Developing Countries*. Geneva: International Labour Organization.

Trlin, Andrew. 1994. 'Extending Epidemiologic Transition Theory', in John Spicer, Andrew Trlin, and Jo Ann Walton (editors), *Social Dimensions of Health and Disease: New Zealand Perspectives*. Palmerston North: Dunmore Press Ltd.

Troy, Patrick. 1995. *Australian Cities: Issues, Strategies and Policies for Urban Australia in the 1990s*. Cambridge: Cambridge University Press.

Tufte, Edward R. 1983. *The Visual Display of Quantitative Information*. Cheshire, CT: Graphics Press.

Tufte, Edward R. 1990. *Envisioning Information*. Cheshire, CT: Graphics Press.

Tufte, Edward R. 1997. *Visual Explanations: Images and Quantities, Evidence and Narrative*. Cheshire, CT: Graphics Press.

Uhlenberg, Peter R. 1979. 'A Study of Cohort Life Cycles: Cohorts Of Native Born Massachusetts Women, 1830–1920', reprinted in Donald W. Hastings and Linda G. Berry (editors), *Cohort Analysis: A Collection of Interdisciplinary Readings*. Oxford, OH: Scripps Foundation for Research in Population Problems.

United Nations (annual). *Demographic Yearbook*. New York: United Nations.

United Nations (biennial). *World Urbanization Prospects*. New York: United Nations Department of Economic and Social Affairs.

United Nations. 1970. *Methods of Measuring Internal Migration*. Manuals on Methods of Estimating Population, Manual VI; Population Studies No. 47. New York: United Nations Department of Economic and Social Affairs.

United Nations. 1982a. *Model Life Tables for Developing Countries*. New York: United Nations Department of International Social and Economic Affairs.

United Nations. 1982b. *Stable Populations Corresponding to the New Model Life Tables for Developing Countries*. New York: United Nations Department of International Social and Economic Affairs.

United Nations. 1983. *Manual X: Indirect Techniques for Demographic Estimation*. New York: Department of International Economic and Social Affairs.

United Nations 1993. *World Population Prospects: the 1992 Revision*. New York: United Nations, Department of Economic and Social Information and Policy Analysis.

United Nations. 2000a. 'Replacement Migration: Is it a Solution to Declining and Ageing Populations?' *Population Newsletter*, No. 69. New York: Population Division, Department of Economic and Social Affairs, United Nations Secretariat.

United Nations. 2000b. *Demographic Yearbook: Historical Supplement 1948–1997*, New York: United Nations (CD).

United Nations. 2000c. *World Migration Report 2000*. New York: United Nations and International Organization for Migration.

United Nations Development Program. 2002. 'Analytical Tools for Human Development' (http://www.undp.org/hdro/statistics/anatools.htm).

United Nations, Population Division of the Department of Economic and Social Affairs of the United Nations Secretariat. 2001a. *World Population Prospects: The 2000 Revision*, Vol. I, *Comprehensive Tables*. New York: United Nations.

United Nations, Population Division of the Department of Economic and Social Affairs of the United Nations Secretariat. 2001b. *The Sex and Age Distribution of Populations*, United Nations Publications, ST/ESA/Ser.A/199. New York: United Nations.

United States Census Bureau. 1976. *The Statistical History of the United States: From Colonial Times to the Present*. New York: Basic Books.

United States Census Bureau. 2000. *Statistical Abstract of the United States: 2000*. Washington: US Census Bureau.

United States Census Bureau. 2001. 'Geographical Mobility March 1999 to March 2000, Detailed Tables'. *Current Population Reports*, P20-538. Washington: US Census Bureau.

van de Kaa, Dirk J. 1987. 'Europe's Second Demographic Transition'. *Population Bulletin*, 42 (1).

van de Walle, Etienne and Knodel, John. 1970. 'Teaching Population Dynamics with a Simulation Exercise'. *Demography*, 7 (4): 433–448.

Vandeschrick, Christophe. 2001. 'The Lexis Diagram, a Misnomer'. *Demographic Research*, 4 (3) (http://www.demographic-research.org/).

Wade, Winnie (editor). 1994. *Flexible Learning in Higher Education*. London: Kogan Page.

Ware, Helen. 1981. *Women, Demography and Development*. Canberra: Development Studies Centre, The Australian National University.

Waring, Joan M. 1975. 'Social Replenishment and Social Change: The Problem of Disordered Cohort Flow'. *American Behavioural Scientist*, 19: 237–256.

Watkins, Susan. 1987. 'The Fertility Transition: Europe and the Third World Compared'. *Sociological Forum*, 2 (4): 645–673.

Webster, E.M. 1992. 'Labour Market Forecasting in Australia'. *Journal of the Australian Population Association*, 9 (2): 185–205.

Weeks, John R. 1994. *Population: An Introduction to Concepts and Issues* (fifth edition). Belmont, California: Wadsworth.

Weeks, John R. 2002. *Population: An Introduction to Concepts and Issues* (eighth edition). Belmont, California: Wadsworth.

Weller, Robert H. and Bouvier, Leon F. 1981. *Population: demography and policy*. New York: St. Martin's Press.Weeks, John R. 2002. *Population: An Introduction to Concepts and Issues* (eighth edition). Belmont, CA: Wadsworth.

Weinberg, Daniel H. 1996. 'A Brief Look at Postwar U.S. Income Inequality'. *Current Population Reports* US, P60-191. Washington: US Census Bureau.

Weinstein, Jay and Pillai, Vijayan K. 2001. *Demography: The Science of Population*. Boston: Allyn and Bacon.

White, Michael J. 1986. 'Segregation and Diversity Measures in Population Distribution'. *Population Index*, 52 (2): 198–221.

Wilber, George L. 1963. 'Migration Expectancy in the United States'. *Journal of the American Statistical Association*, 58: 444–453.

Willcox, Walter F. (editor). 1939. *Natural and Political Observations Made Upon the Bills of Mortality*, by John Graunt, 1662. Baltimore: Johns Hopkins University Press.

Wilmoth, John R. and Horiuchi, Shiro. 1999. 'Rectangularization Revisited: Variability of Age at Death within Human Populations'. *Demography*, 36 (4): 475–495.

Windschuttle, Keith and Windschuttle, Elizabeth. 1988. *Writing, Researching, Communicating: Communication Skills for the Information Age*. Sydney: McGraw-Hill.

Winter, J.M. 1977. 'Britain's "Lost Generation" of the First World War'. *Population Studies*, 31: 449–466.

Winter, J.M. 1986. *The Great War and the British People*. Cambridge, MA: Harvard University Press.

Wood, Jenny, Horsfield, Giles, and Vickers, Lucy. 1999. 'The New Subnational Population Projections Model: Methodology and Projection Scenarios'. *Population Trends* (Winter): 21–28.

Woods, Robert. 1982a. *Population Analysis in Geography*. London: Longman.

Woods, Robert. 1982b. *Theoretical Population Geography*. London: Longman.

World Bank. 1994. *World Population Projections*. Baltimore: Johns Hopkins University Press.

Wunsch, G.J. and Termote, M.G. 1978. *Introduction to Demographic Analysis: Principles and Methods*. New York: Plenum.

Yaukey, David and Anderton, Douglas L. 2001. *Demography: the Study of Human Population* (second edition). New York: Waveland Press.

Zelinsky, W. 1971. 'The Hypothesis of the Mobility Transition'. *Geographical Review*, 61: 219–259.

Zopf, P. 1984. *Population: An Introduction to Social Demography*. Palo Alto, CA: Mayfield Publishing Co.

Index

A

absolute change 59
Africa 77, 200, 353, 362–3
age 77, 79
age group mid-point 309
age heaping 79
age structure 76–115, 97–101, 107–8, 129, 431
aged population 70
ageing in place 101
ageing index 86, 90
ageing 100–1, 102, 105–6, 332–4
age-specific death rate 194, 195, 198, 275–6, 288
age-specific divorce rate 233
age-specific enrolment rate 470
age-specific fertility rate 231, 236–8
age-specific marriage rate 233, 254
age-specific migration rate 399–400
age-specific rate 126
AIDS, see HIV/AIDS
Anaxagoras 87
answers to exercises 513–24
applied demography 348, 430, 490, 495
area graphs 83–5
Ariès, Philippe 224–5
arithmetic growth 46–50, 60, 438
Asia 380–1
Australia 22, 27, 63, 80–1, 303, 316–19, 328–30, 355, 374, 437
Australian Bureau of Statistics 478
average annual increase 60

B

baby boom 222, 223, 233
baby boom cohorts 141
Bangladesh 228
behavioural approach to migration 390–1
below-replacement fertility 89, 224, 234, 246, 435
Bills of Mortality 14
birth cohorts, see cohorts
birth rates 17–18, 22–3, 242
birthplace method (migration estimates) 403, 414, 415–17
births 230, 233
Black Death 14, 194
Bongaarts, John 226–229, 242
Borrie, W. D. 180
boundary changes 60
Brass, William 319–20, 336
Brazil 80–1, 316–19, 328–30
Britain, see United Kingdom

C

Caldwell, J. C. 155
California 68–9
Canada 22, 27, 80–1, 316–19, 328–30, 356–7
Canberra 63
caretaker ratio 86, 90–1
cartogram 375–6
causes of death 181–3
cause-specific death rates 203
census atlases 368
census survival ratios 420–1
census tracts 350
censuses 25–6, 60, 79, 363, 364
chain migration 389–90
Charleston 483
childlessness 223
child-woman ratio 231, 235–6, 410, 411
China 77–8, 80–1, 87, 95–6, 316–19, 328–30, 431
choropleth maps 370–2, 373
circular mobility 387, 388
closed population 385, 387
Coale, Ansley J. 21, 313, 325, 327, 348
cohort 53, 120, 135, 136
cohort analysis 120–1, 135–42, 229–30, 439, 471
cohort component projections 141, 439–48, 449–54, 455, 462, 466–8
cohort life tables 269–70

Coleman, D. A. 156, 225
community, definition 392–3
completed fertility 250
components of growth 29–30, 412, 442–4
conference posters 166–7
consensual unions 251, 253
convergence 156
crude birth rate 33–4, 100, 231, 234–5
crude death rate 33–4, 100, 193–4, 195
crude divorce rate 232
crude marriage rate 232, 253
cumulative frequencies 92–3

D

data bases, commercial 493
de facto marriages, see consensual unions
de facto population 25, 405
de jure population 25, 405
death rates 17–18, 22–3
deductive research 153–6
dementia 192
Demeny, Paul 313, 325, 327
demographic ageing, see population ageing
Demographic and Health Surveys 336
demographic balancing equation 29–30
demographic rates 31–3, 123–4, 200
demographic statistics, sources 24–9
demographic transition 14, 16–24, 54, 89–90, 98–101, 301, 391
Demographic Yearbook 28
demographics 491
demography 2, 3, 14, 16
Denmark 228
density 354–5, 373
dependency ratios 85–6, 88–91, 100
descriptive statistics 382
desktop demography 5–6
direct standardization 125–30
disordered cohort flow 141–2, 471
distribution, see population distribution
divorce rates 254–5
dot maps 369–70
doubling time 62, 64, 65
dwellings 363–4

E

ecological fallacy 493
economic dependency ratio 86, 89–90
Egypt 66, 80–1, 316–19, 328–30
Ehrlich, Paul 57–8
endogenous causes of death 201
entropy statistic 490
epidemiologic transition 185–92, 193, 206
estimates, see population estimates
ethnic groups 60
Europe 24, 26, 73–5, 77
European marriage pattern 222
Eurostat 28
Excel (see also spreadsheet exercises) 4, 5–6, 8, 301, 432, 507–8, 507–8, 509–12
Excel modules:
  Age Structure Data Base 80–1
  Age Structure Simulations 102–7, 302
  Cohort Component Projections 449–54
  Computer Assisted Learning Exercise on Abridged Life Tables 287
  Demographic Transition 19–20
  Epidemiologic Transition 187–91
  Growth Rates 67–8
  Measures of Population Distribution 357–60
  Migration in the United States 395–6
  Population Clocks 55–7
  Population Dynamics 22–3
  Population Momentum 328–30
  Proximate Determinants of Fertility 226–9
  Pyramid Builder 161–2
  Relational Model Life Tables 320–1
  Standardization of Rates 124–5
  West Model Life Tables and Stable Populations 316–19
exercises 42, 71–3, 110–11, 144–5, 170–2, 214–15, 257–8, 295–6, 339–41, 378–9, 423–5, 465–6, 498–9
exogenous causes of death 201
exponential growth 46, 50–3, 64–8, 438

F

familism 225
family 221–5, 364, 473–9
fertility 220–61
fertility compared with mortality 252
fertility differentials 225
Finland 228
fixed period migration 387, 398
flexible delivery 6–7
flow maps 368–9, 372, 374
forward survival (migration estimates) 406–12, 414
France 32, 80–1, 121–2, 316–19, 328–30
Friedlander, Dov 391
further reading 39–41, 71, 109, 143, 169, 213, 256, 294, 338–9, 377, 422–3, 464–5, 497

G

gender 88
general divorce rate 232, 254
general fertility rate 231, 235
general marriage rate 232, 253–4
generation 138
generational life tables 269
geodemographics 368, 491
geographic information systems (GIS) 368
geographical areas, see spatial units
geometric growth 46, 50–1, 61–4, 438
Germany 80–1, 316–19, 331, 328–30
Gini index 483–90
Gould, W. T. S. 388–9
Graunt, John 3, 14–15, 37, 38, 77, 155, 168, 266–7, 436, 495
Graunt, John: quotations 13, 38, 45, 76, 119, 150, 179, 220, 265, 300, 347, 384, 429, 469, 495–6
Greece 112
gross migration 387
gross reproduction rate 243–6, 325–6
growth 16–24, 45–75
growth rates 62–3, 65

H

Hajnal, J. 222, 348
Halley, Edmund 267
Harrison, Peter 365
health 206–12
health statistics 183–4
HIV/AIDS 120
Ho Chi Minh City 120
Hong Kong 110–11
household head, definition 475–6
household headship rate 474–5
household projections 473–9
householder, definition 475, 476
households 363–4, 473
housing 363–7
housing density 365–7
housing unit method of projection 455, 458–9
hubristic stage 192
Hugo, Graeme 388
human development index 493–4
Hungary 228
Hutterite fertility 241
hypothesis, definition 155–6

I

incidence of disease 208–10
incidence rate 208
index of concentration 362–3
index of dissimilarity 95–7, 357, 360, 361, 482, 485, 486–90
index of diversity 479–82
index of redistribution 353, 361–2
index of segregation 482–3
India 57–8, 80–1, 89, 316–19, 328–30, 334–5
indicator methods of projection 459–60, 462
indirect standardization 130–4
Indonesia 80–1, 228, 316–19, 328–30, 387–8
inductive research 153–6
inequality 483
infant mortality rate 196, 198–201, 276
information literacy 156
intercensal estimates 436
internal migration 386, 387, 392, 394
International Classification of Diseases 182
international migration 386, 387, 391–2
internet resources 41–2, 71, 110, 144, 170, 214, 257, 295, 339, 378, 423, 465, 498
interpreting statistics 35–7, 98, 129–30, 150–68
intrinsic birth rate 312
intrinsic death rate 312
intrinsic rate of natural increase 302, 304–7, 342
Italy 80–1, 89, 100–1, 112, 237, 238, 239, 241, 245, 331, 316–19, 328–30

J

Jamaica 228
Japan 18, 21, 35, 80–1, 110–11, 316–19, 328–30

K

Kenya 228
key terms 39, 71, 109, 143, 169, 213, 256, 294, 338, 377, 422, 464, 497
Keyfitz, Nathan 435
Kirk, Dudley 154
Kuwait 121–2, 125–33

L

labour force 60, 472
labour force participation rate 471–2
labour force projections 471–3
Lee, Everett 390

Less Developed Regions 80–1, 89, 173–5, 316–19, 328–30
Lexis diagram 136–7, 140, 200, 269
life expectancy 100, 270, 282, 283, 291–2, 494
life table survival ratios 418–19
life tables 265–99;
abridged 285–92
applications 266, 335
applications in NRR 247–9
birth rate 284
combined for both sexes 292
complete 270–84
death rate 284
definitions of functions 274
formatting conventions 279
formulas 284, 286
function d_x 279–80, 290
function e_x 283, 291–2
function l_x 268, 278–9, 289
function L_x 269, 277, 280–1, 290–1, 292, 418–19
function m_x 283
function p_x 276, 289
function q_x 276, 277, 289
function T_x 281–2, 291
M_x 275–6, 288
origins 266–7
paradox of the life table 283
radix 278, 290
spreadsheets 274–5, 285, 287
types of functions 279
lifestyles 490–3
lifetime migration 387
Limits to Growth 51–2
living arrangements 477–9
local migration 387, 392, 394
location quotients 355–61
logarithmic transformation 500–1
logarithms 506
logistic curve 46, 50, 54–7
logit transformation 501–3
logits 322–4
London 347, 370, 384, 429, 495–6
longitudinal analysis 135
Lorenz curve 484, 488–9
Lotka, Alfred 302–3

M

Malthus, Thomas Robert 49–51, 64, 151, 304
maps, see population maps
marital status 205–6, 207
marital status projections 476
market segmentation 490–3
marriage 251, 253
maternal mortality rate 204–5
mathematical methods of projection 438
mathematical operators 506
maths 505–6
mean age 91
mean age at childbearing 312
mean length of a generation 306, 312
median age 79, 91, 92–4, 255
megalopolis 351
Merrick, Thomas W. 492
metropolitan areas 349, 352
Mexico 58–60, 80–1, 146–7, 316–19, 328–30
mid-year population 31, 198, 199
migration (see also net migration) 22–3, 87, 101, 304–5, 384–426
migration effectiveness 385, 400–1
migration estimates 403–17
migration exchanges 385, 401
migration expectancy 401–3
migration interval 387
migration matrix 400
migration of children 398, 410–11, 412, 445
migration rates 387, 393, 394, 397–401
migration statistics 391–4
migration stream 387
migration, risk of 397–8
mobility 386–9
mobility transition 391
modal age 91
model life tables 248, 293, 312–25
model stable populations 325–7
momentum, see population momentum
More Developed Regions 80–1, 89, 173–5, 316–19, 328–30
mortality 179–219, 444, 445–6
mortality differentials 194, 204–6, 207
mortality statistics 180–4
multiregional projection model 460–2
murder victims 93–4

N

Napoleon 368–9
natural fertility 236
natural increase 18, 22–3, 29–30
neonatal mortality 201–2
neonatal mortality rate 196, 201
Nepal 87
net growth 68–9
net migration 22–3, 29–30, 35, 387, 403–17, 445–6
net reproduction rate 246–50, 306
New York 58, 355
New Zealand 27, 80–1, 316–19

nodal regions 351–2
nominal data 480

O

occupancy rates 364–5
Omran, Abdul 185–6, 191
open population 385–6, 387
order of operations 93, 505
origin-destination matrix 400
outlining 164

P

Pakistan 65–6, 80–1, 316–19
paradox of the life table 283
partial displacement migration 389
Pearl, Raymond 54
percentage change 59–60
percentages 32–3
perinatal mortality rate 197, 202
period analysis 120, 137, 229–38
period life tables 267–9
pie graphs 83–4
planning 433–5, 455
Poland 228
policy making 435–6
population composition 469–503
population composition, projection methods 470–9
population decline 24, 46
population distribution 347–83
population estimates 335–7, 338–9, 431–2, 433–5, 436–7
population maps 367–76
population momentum 304, 327–32
population projections 334–5, 429–68
population projections, methods 437–62, 470–9
population pyramids 77, 80–3, 111–15, 145–9, 151, 161–3
Population Reference Bureau 29, 35–6
population surveys 27
population turnover 69–70
postcensal estimates 436
post-neonatal mortality rate 196–7, 201
powers of numbers 505
prediction 430
Pressat, Roland 194
Preston, Samuel 154–5
prevalence of disease 208–10
prevalence rate 208
Price, Charles A. 69, 389–90
primary data 152
probabilities 32–3
probability of dying 276, 277, 289
probability of surviving 276, 278, 289
projections, see population projections
proportional symbols 372–5
proportions 32–3
proportions surviving 278, 289
Prothero, R. M. 388–9
proximate determinants of fertility 226–9, 241
pyschographics 491, 493

Q

Quebec 357

R

radix of a life table 278, 290
rate of natural increase 34–5, 234
rate of net migration 35
rate ratios 121
rates, see demographic rates
ratio method of projection 456–8
ratios 32–3
Ravenstein, E. G. 390
real cohorts 140, 230, 250
rectangularization of the survival curve 186
Rees, Philip 461
regions 351–2
relational model life tables 319–25
relative risk 122, 211
replacement 69–70, 385
replacement-level fertility 225, 241, 243, 245
research design 157
resident population 25, 405, 473
residential mobility 387
return migration 387
reverse survival (migration estimates) 412–14, 415
risk 31, 210–11, 230, 397–8
risk factors 210–11
room occupancy rate 365
Roseman, C. C. 389
rounding 505
Ryder, Norman 120, 138

S

school enrolment projections 470–1
second demographic transition 223–5

secondary data 151–2
segregation 482, 484, 485–9
self-reinforcing growth 327, 330
settlement hierarchy 350
sex 88
sex differentials in mortality 204–5
sex ratio 32, 77, 85, 87–8
sex ratio of age-specific death rates 122, 204
Simpson, Stephen 459–60
small area projections 455–62
Snow, John 180, 367
social indicators 493
South Africa 80–1, 316–19, 328–30
space-time typology 388–9
spatial units 349–52
spreadsheet exercises:
 1 Basic Demographic Rates 43–4
 2 Population Growth 73–5
 3 Population Pyramids 111–15
 4 Superimposed and Compound Population Pyramids 145–9
 5 Line Graphs and Pie Charts 172–5
 6 A Graphical Database 216–19
 7 Measures of Fertility 258–61
 8 Abridged Life Tables 296–9
 9 Stable Population Models 341–3
 10 A Descriptive Statistics Module 379–83
 11 Net Migration Estimates 425–6
 12 Cohort Component Projections 466–8
 13 Experiments in Data Transformation 499–503
Sri Lanka 43–4
stable population 53, 300–12, 316–19, 333–4, 341–3, 437–8
standardization 121, 123–34, 142, 194, 234–5
standardized index of diversity 480–2
standardized mortality ratio 133–4
stationary population 268–9, 301, 302, 307, 334–5, 437–8
stillbirth rate 197, 203
stillbirths 202
suburban populations 101
surveys, see population surveys
survival ratio methods (migration estimates) 405–14
survival ratios 411, 418–21
Sweden 89
synthetic cohorts 140–2, 230, 238–50, 268

T

Tabah, Leon 222
Texas 58
theory, definition 155
time, concepts of 135–6
Tordella, Stephen J. 492
total displacement migration 389
total fecundity 227
total fertility rate 227, 239–43
total marital fertility rate 241
Tufte, Edward 367, 368

Uhlenberg, Peter 139
uniform regions 351–2
United Kingdom 27, 52–3, 80–1, 87, 121–2, 125–33, 162, 228, 316–19, 328–30, 371, 375
United Nations 28
United States of America:
 age structure 80–1, 95–6, 146–9, 162
 components of growth 30
 death rates 124
 fertility 228, 250–1
 growth 22, 52–4, 61–4, 304
 households 473, 475
 housing density 366–7
 housing occupancy rates 365, 366
 human development index 494
 infant mortality 199
 labour force participation rates 472–3
 life cycles of cohorts 139
 life expectancy 87
 living arrangements 477
 median age 94
 metropolitan regions 351
 migration 394–6, 399
 migration expectancy 401–3
 murder victims 93–4
 population ageing 102, 105–6
 population cartogram 376
 population density 354, 355, 366, 373
 population momentum 328–30
 population projections 455
 sex ratio of age-specific death rates 204, 205
 spatial units 350
 stable populations 316–19
 Statistical Abstract of the United States 27
 urban areas 352
 vital statistics 43–4, 216–18
urban populations 350–1, 352
urban regions 351–2

van de Kaa, Dirk 223
vital statistics 26–7

vital statistics method (migration estimates) 404–5, 414
world population 63, 66, 68, 80–1, 83, 84–5, 173–5, 316–19, 328–30, 332

W

Waring, Joan 141
Washington 373
well-being, measures 493
Wells, H. G. 351
Wilber, George 401, 403
Winter, J. M. 140
World Bank 28
World Fertility Survey 336
World Health Organization

Y

Yugoslavia 228

Z

Zelinsky, Wilbur 391
zero population growth 70